|     |     |     |     |     | Groups |     |     | VIIIA |
|-----|-----|-----|-----|-----|--------|-----|-----|-------|
|     |     |     | IIIA | IVA | VA | VIA | VIIA | He 2 |
|     |     |     | B 5 | C 6 | N 7 | O 8 | F 9 | Ne 10 |
|     | IB  | IIB | Al 13 | Si 14 | P 15 | S 16 | Cl 17 | Ar 18 |
| Ni  | Cu ++ 29 | Zn 30 | Ga 31 | Ge 32 | As 33 | Se 34 | Br 35 | Kr 36 |
| Pd  | Ag + 47 | Cd 48 | In 49 | Sn 50 | Sb 51 | Te 52 | I 53 | Xe 54 |
| Pt ++ | Au + 79 | Hg ++ 80 | Tl + 81 | Pb ++ 82 | Bi 83 | Po 84 | At 85 | Rn 86 |

# Introductory chemistry

Edmund J. Leddy

Don Roach

MIAMI-DADE JUNIOR COLLEGE

# Introductory chemistry

QD
31.2
L4

RINEHART PRESS  SAN FRANCISCO

*Introductory chemistry* by Edmund J. Leddy and Don Roach

**Copyright © 1972 by Rinehart Press**
*A division of Holt, Rinehart and Winston, Inc.*

A preliminary edition of this text was published by
Holt, Rinehart and Winston, copyright © 1970

All rights reserved

Library of Congress Catalog Card Number: 76-165029

**ISBN: 0-03-078295-3**

PRINTED IN THE UNITED STATES OF AMERICA

2  3  4  5     071     1  2  3  4  5  6  7  8  9

To chris, rita, charissa, jennifer, david and paul

# Preface

Many students who enter college today have had little or no background in chemistry. In order to introduce these students to chemistry, many colleges are now offering one-semester (or two-quarter) introductory chemistry courses. This text has been designed for such a course. It can also be used satisfactorily in a terminal course, especially where the objective is to give the student a feel for scientific models. Since the text is designed as an introduction to chemistry, there is no prerequisite beyond basic algebra.

The mental concepts which scientists use to "explain" chemical behavior are called models. One of the objectives in writing this text is to present the experimental observations upon which these models are based. Our experience has shown that this approach is both desirable and successful with beginning students.

Throughout the book unity terms are used to solve problems. We believe that the consistent application of this method gives the student insight into problem solving. The mole concept and balancing chemical equations are introduced early in the text (Chapter 3). This gives the student an early introduction to problem solving and the use of chemical equations.

The basis of much of the book is electronic configuration, electronic structure, and how electronic structure determines the properties of elements and compounds. The discussion of electronic orbitals (s, p, d, f) is not a difficult topic and yet it does provide a systematic way to explain reactions of the more common elements. Some topics are more easily understood if they are studied in sequence. For example, electronic configurations in atoms leads smoothly into the periodic table and then into bonding (Chapters 3–6). Basic nomenclature (Chapter 7) logically follows bonding.

Few people (including chemists) can fully understand and appreciate a new concept the first time they are exposed to it. In writing this text, we have strived to make new concepts more understandable and enjoyable by placing appropriate solved examples within the body of the text. These examples are worked out step by step so that the student understands what has been accomplished in the solution. Several short questions and/or problems follow many sections so the student may test his understanding of the material in the section. These problems range from the relatively simple to the more complex; thus both student and instructor are provided with some measure of whether the student understands the material. These problems provide a form of self-testing, which, after successful solution, will encourage the student and help build his self-confidence. Answers to selected odd-numbered end-of-chapter exercises are located in the back of the text so the student can easily check his answers.

At the end of each chapter there is a list of behaviorial objectives. These behaviorial objectives tell the student what he should be able to do after he has thoroughly studied the chapter. Thus, after reading the objectives, the student knows what is expected of him. For this reason, it might be preferable to read the behaviorial objectives before reading the chapter.

In writing the text, we have included considerably more material than is usually covered in a beginning course. This gives the instructor freedom to choose the material to be covered. We feel certain chapters are necessary for the nucleus of a beginning course. These are Chapters 1–10, 12, 13, and 14. After covering these chapters, students should be able to understand any of the concepts in the later chapters. Thus, the instructor can choose those which he feels best meet the needs of his particular class.

This edition of the text is a revision of the preliminary edition which was class tested by many of our colleagues. We would like to acknowledge the help and advice particularly of Bob

Drobner, Doug McLean, Ronda Waldinger, Mike Guttman, and Rasma Derums at Miami-Dade, Frank Reeves, Don Schumacher, Bill Mooney, Robert Carola, Peggy Park, Gene Smith, Paul Schmitt, and Felix Cooper at Holt, Rinehart and Winston, and our typists Arlene Johnson (who has since changed her major to chemistry), Cavel Kyser, and Mary Ellen Newcomer.

*edmund j. leddy*
*don roach*

# Contents

## 1

### Introduction  1

*What is chemistry?    Matter    Particles    Changes in matter (energy)
Models and science    Systems    Chemistry and the scientific world
Observations on a chemical system    The sublimation of iodine, a model
Compounds and elements    Mixtures    States of matter    Phases
The electrical nature of matter    Forms of energy    A model for chemical
changes*

## 2

### Measurement  28

*Measurement    The metric system    Uncertainty and significant figures
Length    Volume    Mass    The use of unity terms in calculations
Density    Significant figures in calculations*

## 3

### Atoms and the mole  47

*Distribution of the elements    Names and symbols of the elements
Particulate theory    Atoms    Formulas    Subatomic particles*

The nuclear atom   The modern atomic model   Isotopes   Atomic mass units   Atomic weight   Avogadro's number   The mole   Molecular weight and moles   Writing balanced equations   Percent composition   Empirical formulas

# 4

## Electronic structure   76

Discrete packets—quanta   Energies of electrons in atoms   Main energy level   Energy sublevels—second quantum number   Orbitals   Electron spin   Valence electrons   Electron-dot structures

# 5

## Periodicity   98

Historical development   Structural basis of the periodic chart   Periods   Groups   Electron-dot structures for atoms   Electronic configuration and the periodic chart   The periodic law or periodic model   Ionization energy—a periodic property   Stable configurations—the noble gases   A classification for the elements   Prediction of properties

# 6

## Bonding   114

Chemical bonding   Electronegativity   Covalent bonding   Formulas and names of covalent compounds   Ionic bonding   Formulas and names of binary ionic compounds   The polar covalent bond   Intramolecular and intermolecular forces   Moles of atoms, molecules, and ions

# 7

## Oxidation numbers and nomenclature   142

Oxidation numbers   Naming binary compounds   Naming ternary compounds

# 8

## States of matter   160

Kinetic molecular theory   Pressure   Temperature   Specific heat   Solids   Liquids   Gases

## 9

### Water and its elements  183

*Physical properties of water     The water molecule, a model     Chemical properties of water     Chemical properties and structure—a model     Decomposition of water     Mole ratios     Hydrogen     Oxygen*

## 10

### Chemical changes and equations  208

*Interpretation of a balanced equation     Collision theory     Reactions—a rearrangement of atoms     Reactions—classification by mechanism*

## 11

### Selected elements and compounds  228

*Group IA elements     Group VIIA elements     Reactions of metals with acids     Practice in determining products of a reaction*

## 12

### The gas laws  255

*Gas densities     Pressure, temperature, and volume changes     Gas law calculations     General gas laws     Partial pressures     Relative rates of diffusion     Real and ideal gases*

## 13

### Calculations involving chemical equations  285

*Chemical unity terms—a review     The mole relationship in a chemical equation     Calculations involving chemical equations*

## 14

### Solutions  305

*Examples of solutions     Types of solutions     Factors affecting solubility     Factors affecting the rate of dissoultion     Concentrations of solutions     Properties of solutions*

## 15

### Theory of ionization and ionic reactions  326

*Conductivity of electrolytic solutions    Freezing points and boiling points of electrolytes    Solubility    Conductivity and the solvent    Theory of ionization    Mixing solutions—reactions    Dissolving a precipitate*

## 16

### Acid-base reactions  345

*Aqueous solutions of acids and bases    Proton transfer reactions—the Brønsted-Lowry model    Electron donor reactions—the Lewis model    Preparation of Arrhenius acids and bases    Normality of acids and bases    The pH concept    Titrations    Salts—Brønsted acids and bases*

## 17

### Chemical kinetics and chemical equilibrium  369

*Reaction rates    The concept of dynamic equilibrium    Shifting the equilibrium    Common equilibrium situations    The equilibrium constant*

## 18

### Oxidation-reduction  388

*A review of oxidation-reduction reactions    Half-cells    The hydrogen electrode    Oxidation potentials    Balancing redox equations    Stoichiometry of redox reactions    Redox reactions as a source of chemical energy*

## 19

### Chemistry of periodic groups  410

*Group VIIIA—the noble gases    Group VIIA—the halogens    Group VIA—the oxygen family    Group VA—the nitrogen family    Group IVA—the carbon family    Group IIIA—the boron family    Group IIA—the alkaline earth family    Group B elements—the transition elements*

## 20

### Organic chemistry  437

*Chemical bonds in organic chemistry   Saturated and unsaturated hydrocarbons   Alkanes   Alkenes   The effects of structure on properties—alcohols and acids   Polymers   Organic compounds and their functional groups*

### Appendix I
### A review of mathematics  460

*Significant figures   Scientific notation*

### Appendix II
### Answers to selected odd-numbered questions  465

### Index  471

# Introductory chemistry

# 1

## 1.1 What is chemistry?

**Introduction**  A noted chemist once said that chemistry is what chemists do. Have you ever wondered why a candle doesn't burn until you put a match to it, why ice floats on water, or why your body needs certain foods? These questions are the concern of chemists.

What do chemists do? A poet might describe a candle as a symbol of light or inspiration; the same candle might be a source of light or warmth to a man lost in a cave; whereas a chemist might note the color and height of the flame and the height of the candle. These factors might help the chemist to determine how long the candle would burn. One activity of a chemist is the making of *observations*.

How would a chemist accurately determine the height of the flame or the height of the candle? He would need a ruler to measure them. This is an example of another activity of the chemist, *measurement*. Observations and measurements must be recorded in writing to communicate them to other persons and enable them to be recalled at later dates. This *communication* of observed and measured facts is another activity of the chemist.

In addition to watching a candle burn and recording his observations, a chemist may want to find out how the candle behaves in a different environment. What happens to a burning

■ EXERCISE  Light a candle and watch it burn. Make a list of your observations. Compare this list with that of several classmates.

candle when one blows on it gently, harder? If the candle goes out, why does it go out? By blowing on the candle, one changes its environment. The careful control of the environment during a process such as the burning of a candle is one way in which a chemist experiments. ■

## 1.2  Matter

How do items such as a piece of silver, a copper wire, a cement slab, or a balloon full of air differ from concepts such as hope, beauty, God, love, death, and peace? The first list contains items that make up our material world and are composed of matter. A chemist notes and records observations and measurements of matter. *Matter* is the "stuff" of which the world and every material thing in it is made.

What do these items have in common: a gold watch, a gold ring, a gold wire, a gold nugget, a gold bracelet? Each of these items contains the substance known as gold. *A substance* is a particular kind of matter which has a specific set of characteristics or *properties*. Some substances that you may be familiar with are gold, silver, water, sugar, salt, baking soda, and alcohol. No two different substances have yet been found that have *exactly* the same set of properties. ■ A box of salt contains other substances in addition to table salt; it is therefore impure. If a sample contains salt and no detectable amounts of other substances, the sample is said to be pure. If detectable amounts of other substances are also present, then the sample is impure. ■

■ EXERCISE  Name the substance with these properties: A solid, nonbrittle substance which has a metallic luster, conducts electricity well, and does not corrode readily.

■ EXERCISE  Read the labels on the following household items to determine if any of them are pure substances: salt, sugar, baking soda, baking powder, washing soda, liquid bleach, white vinegar, and clear household ammonia.

## 1.3  Particles

Consider a sample of water in a glass. On close examination there appear to be no boundaries separating the water at the top of the glass from water at the bottom. Therefore the water can be described as being *continuous*. Consider samples of these pure substances: a diamond, a grain of salt, a bottle of oxygen, and a cup of alcohol. These samples all appear to be continuous.

Although bulk samples of substances in nature appear to be continuous, there is evidence to support the existence of very small, discrete particles within these samples of matter. How might one explain why leaves move in the wind, or why a dye spreads through water? One might try to explain these observations by assuming that matter consists of tiny particles, so small that they cannot be seen. When the wind moves leaves one might imagine air particles striking the leaf, causing it to move. A dye placed in a beaker of water spreads because moving water particles hit dye particles, causing them to move slowly throughout the water.

Compare the readily observable properties of two pure substances, iron and oxygen. Iron is a solid, shiny metal. Oxygen is an invisible, odorless gas. Chemists attribute the difference in properties of iron and oxygen to differences in their particles. In Chapters 3 and 4 some of the differences in particles of iron and oxygen will be discussed.

The composition of a substance also determines the properties of that substance. Compare samples of carbon monoxide and carbon dioxide. Carbon monoxide is produced in small amounts in car engines and by poorly ventilated fires. Carbon dioxide is also produced in car engines and by the burning of many substances. Both carbon monoxide and carbon dioxide are colorless, odorless gases which contain carbon and oxygen. However, small amounts of carbon monoxide cause death whereas small amounts of carbon dioxide can be tolerated. Why are some of the properties of these two gases different? Chemists believe that a particle of carbon monoxide contains half as much oxygen as a particle of carbon dioxide (Fig. 1.1). The amount and type of simpler substances that make up a particle of a compound substance is called the *composition of that substance*.

one particle of carbon monoxide
1 carbon particle
1 oxygen particle

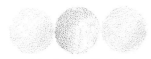

one particle of carbon dioxide
1 carbon particle
2 oxygen particles

FIGURE 1.1

*A particle of poisonous gas, carbon monoxide, contains half as much oxygen as a particle of carbon dioxide.*

Consituent
↓
whole or given particles which make one.

Carbon is a common substance. Most lead pencils contain carbon in the form of graphite, a soft, dark, solid substance. Diamond is another form of carbon. Diamonds are clear, hard solids. Why does the same substance, carbon, have two different, solid forms? The chemist believes that the different forms of carbon are due to different arrangements of the carbon particles (Fig. 1.2). The arrangement of the constituent particles which make up a given sample of matter is known as the *structure of that sample*.

Why is it that when a man becomes engaged to a woman he gives her a diamond ring instead of a graphite ring? Two properties of a well-cut diamond are its ability to reflect light and its hardness. One property makes it beautiful and the other makes its beauty last. These two properties determine the use of diamonds. The use of any substance is determined by its properties. Graphite, which is a dark, slippery substance, is used in lead pencils and as a lubricant.

The work of many chemists involves the development of sub-

1.3 Particles

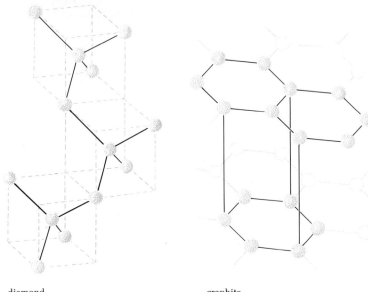

FIGURE 1.2
*The arrangement of carbon particles in diamond and in graphite are different. The spheres represent carbon atoms; the lines connecting the carbon atoms represent forces that hold atoms together.*

diamond                    graphite

stances with new properties. <u>The chemist may alter the composition and structure of substances in order to produce new substances with desired properties</u>. Soaps do not work well in hard water and they produce too many suds for washing machines. New substances, synthetic detergents, were produced by chemists. Detergents work well in hard water, have a low sudsing action, and can be decomposed by organisms in rivers and streams. ■

■ EXERCISE  *Water and hydrogen peroxide are two substances that contain hydrogen and oxygen. List several properties and uses of water. What are the properties and uses of hydrogen peroxide?*

## 1.4  Changes in matter (energy)

Is there a difference in the appearance of a flash bulb before and after it has been used? What does a candle look like before and after it has been burned? Compare a fresh piece of iron with a rusted sample of iron. Each of the alterations demonstrated in Fig. 1.3 involves a chemical change. How can you recognize a chemical change?

A new flash bulb contains tiny pieces of a metallic-looking substance. The used flash bulb contains a powdery, gray substance. There has been a change in properties during this chemical change. In a *chemical change* one or more substances produces one or more new substances. Ony way to recognize a chemical change is by the appearance of a substance with new properties.

■ EXERCISE  *How many changes in properties are involved when a fresh piece of iron changes into rust?*

In addition to a change in properties, what else do you observe when a flash bulb is set off? The bulb produces a bright light and becomes hot. Both heat and light are forms of energy. Chemical changes either produce or use energy. What about rusting iron? Does it produce energy? The event occurs so slowly that it is difficult to observe the production of energy, but energy *is* produced. ■

A battery is constructed in such a way that electricity is produced when a chemical change occurs inside. Electricity, like heat and light, is a form of energy.

What is the purpose of a flash bulb? Why do we refrigerate milk? The light from a flash bulb is used to help produce a chemical change on the photographic film in a camera. The removal of heat from milk causes the slowing down of chemical changes that make milk sour. These examples show that energy such as heat or light can cause chemical change. For this reason the chemist also studies the energy effects that accompany chemical changes.

■ EXERCISE  *Energy affects and causes chemical changes in the human body. Describe what you might observe when a person is exposed to excessive amounts of sunlight.*

■ EXERCISE  *How would coffee and sleeping pills individually affect the rate of chemical change in the body?*

An open jar of mayonnaise left in a hot room quickly turns rancid. The same mayonnaise will keep for weeks in a refrigerator. A solution of hydrogen peroxide keeps for a long time in a dark room but decomposes rapidly in sunlight. Such chemical changes occur at different *rates*. Heat and light affect the rate at which changes occur. By knowing what factors influence changes, the chemist is better able to control these changes. ■

## 1.5 Models and science

What are some of the properties of a piece of nickel? It is a solid that feels heavy and has a metallic luster. Nickel is attracted by a magnet. It dissolves in a solution of the substance, hydrochloric acid, liberating a colorless gas and turning the solution green (Fig. 1.4). Matter that we can observe directly, such as a bar of nickel, is part of the *macroscopic* world. Many of the samples of substances that a chemist studies are macroscopic in size.

Substances viewed on a *microscopic* level look different than when viewed on a macroscopic level. Using special techniques, it is now possible to magnify substances a millionfold.

The chemist attempts to explain events in the macroscopic and microscopic world differently. The explanations given by the chemist are based on *assumptions* made about the invisible, submicroscopic world. The *submicroscopic* world is too small to be seen through the most powerful microscope.

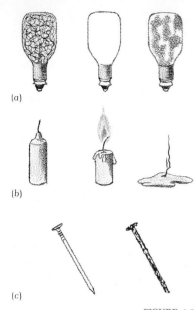

**FIGURE 1.3**
*Three samples of matter before and after they have undergone a chemical change:* (a) *flashbulb,* (b) *candle,* (c) *iron nail. New substances can be recognized by a new set of properties.*

■ **EXERCISE** *Give an operational definition for each of the following: salt, sugar, a solid, a liquid, and a gas.*

How does a chemist explain events which occur on a macroscopic level? A possible reason given in Section 1.3 for the spreading of a dye throughout the water in which it is placed was that both the dye and the water consist of particles. In this case, "particle" is a concept developed by the chemist to help explain a phenomenon. A concept, a mental picture, a graph, or a mathematical equation which helps us to explain or understand observations is called a *model*. The idea of a particle is a model. A model was used in Fig. 1.2 to show why samples of carbon in graphite and diamond behave differently.

Models are subject to change. An architect's model, for instance, may undergo change during the construction of the building. The models used by the chemist undergo modification and change as the chemist discovers new and different properties of substances. Rarely does he discard a model entirely; the chemist usually modifies the model so that it fits the new facts.

If an explanation or model is adequate, the scientist should be able to *predict* the behavior of matter under a different set of conditions. If the model successfully predicts the behavior of a substance, the model becomes more acceptable. If the model fails in a number of predictions, the scientist will try to modify it so that the observed facts are explained.

Consider again the models shown for graphite and diamond in Fig. 1.2. According to the model, graphite and diamond are different because their particles are arranged differently; the particles in graphite are farther apart than particles in diamond. Using the information in this model and the fact that graphite is inexpensive and diamonds are expensive, chemists reasoned that it might be possible to make diamonds by compressing graphite and forcing the particles closer together. Today, millions of carats of man-made diamonds are produced each year.

Define a diamond. Define graphite. A diamond is a clear, glasslike mineral of exceptional hardness. Graphite is a soft, dark-colored solid. Each of these definitions is based on the properties of the substance. Definitions based on the properties of substances, or on what one observes to occur are called *operational definitions*. A possible operational definition of nickel is: a solid substance with a metallic luster. It feels heavy and is easily magnetized. It dissolves in hydrochloric acid, turning the solution green and releasing a colorless gas. ■

Are there other definitions besides operational definitions? Look at Fig. 1.2 again. A diamond is that substance composed of particles of carbon. At equal distance from each particle are four more particles. The angle formed between any three

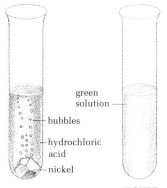

**FIGURE 1.4**
A sample of nickel will dissolve in a liquid known as hydrochloric acid. As it dissolves, it produces bubbles of gas and turns the solution green.

■ EXERCISE (a) A chemist selected a man as the system for study over a period of time. Is this system open or closed? (b) In a bowl, mix some vinegar with some baking soda or washing soda. Is this system open? (c) What evidence do you have for your answers?

nearby carbon atoms is 109°. Graphite can be defined as a substance composed of particles of carbon. At equal distances from each particle of carbon are three more carbon particles. The angle formed between three nearby carbon atoms is 120° and each atom lies in the same plane. Both these definitions are based on models. Definitions based on models are said to be *theoretical definitions*. Later in this chapter we will develop operational and theoretical definitions for solids, liquids, and gases.

## 1.6 Systems

The part of the material world that the chemist selects for study is called a *system*. A system in which the amount of matter remains constant is said to be a *closed system*. A system where the amount of matter changes with time is called an *open system*. Systems are chosen by chemists for convenience. A test tube containing sugar would be a convenient system for observing the macroscopic properties of sugar. The human body might be a convenient system for the observation of some of the life processes which take place within it. A chemist might even consider the entire world as a system if, for example, he were interested in the chemistry of the formation of the earth. Once the chemist selects a system for study, he calls everything else in the environment the *surroundings*. ■

## 1.7 Chemistry and the scientific world

What relationship does chemistry have to the rest of the scientific world and the world in general? Consider for a moment a system consisting of a large lake formed by a dam with a hydroelectric plant. The system includes a nearby community and surrounding territory. Who would be interested in studying this system?

A biologist might be interested in the plants and animals that live in, near, or on the lake. He is concerned not only with the living organisms of the region but also with the chemical processes which take place in these organisms. In order to understand the life processes, a biologist must be aware of the substances and chemical changes that occur inside and outside of organisms.

A geologist would be interested in what land formations are present before the dam is constructed and on how the dam will affect the environment. He is concerned with the chemical and physical processes which led to the formation of the surrounding rocks and minerals. Before the dam was built, a geologist was probably consulted about the chances of a landslide occurring in the future. This prediction would require a knowledge and understanding of the chemical substances and rock formations in the surrounding hills.

A physicist would be concerned with the effect of a high-voltage current on material in the power lines from the hydroelectric plant. A knowledge of the properties of the materials and their chemical nature would be essential in order to understand the effect of electricity on the lines.

People in the areas of biology, giology, physics, agriculture, engineering, and medicine use chemistry as a tool to understand their areas of study. Like the chemist, they choose a system on which to make observations and measurements. To understand these observations they formulate models. Just as the chemist does, they test the model by trying to make predictions about future phenomena. The ability to observe carefully and to use models is also necessary in such diverse fields as sociology, psychology, philosophy, history, economics, and education. ■

■ EXERCISE  Two well-known theories or models in economics are capitalism and communism. How can economics be "understood" in terms of each of these models?

## 1.8 Observations on a chemical system

Note what happens and try to explain the observations in the following chemical system using the *particulate* or particle model of nature.

A closed bottle of hydrochloric acid was placed next to a closed bottle of aqueous ammonia (aqueous ammonia is often called ammonium hydroxide) as shown in Fig. 1.5. Hydrochloric acid is made by bubbling the colorless, poisonous gas hydrogen chloride into water. The gas dissolves readily, forming a clear, colorless solution called hydrochloric acid. Aqueous ammonia is made by bubbling the colorless gas ammonia into water. The gas dissolves readily, forming a clear, colorless solution called aqueous ammonia.

When the top to the ammonia bottle was removed nothing could be seen, but in a few seconds an odor characteristic of ammonia was noticed. The top was replaced. Next, the top to the hydrochloric acid bottle was removed. In a few seconds a sharp odor was noticed. The top was replaced.

Finally, the tops to both bottles were removed at the same time. In a few seconds white fumes formed close to the bottle of hydrochloric acid. If the top to either bottle was replaced, the white fumes disappeared.

What model would explain the observations? Perhaps when the tops of the bottles are removed, particles of gaseous ammonia and gaseous hydrogen chloride can escape from the solutions. These particles may then leave the bottle and start to spread about the room. Some ammonia particles could eventually collide with some hydrogen chloride particles. Their interaction could produce the white fumes.

Further observation of the white fumes indicates that they have a unique set of properties entirely different from either ammonia or hydrogen chloride gas. The white fumes actually consist of small solid particles. This solid does not have the odor of ammonia or of hydrogen chloride and its water solution behaves differently from either of the original substances. Evidently a new substance has been formed. Chemists call the substance which makes up the white fumes ammonium chloride. Ammonium chloride particles are formed when ammonia particles combine with hydrogen chloride particles. This process is illustrated in Fig. 1.6.

Why do the white fumes form closer to the bottle of hydrochloric acid? Does this suggest anything about the relative speed of ammonia particles and hydrogen chloride particles? After developing our model further, we will look at that question in Chapter 12, when we study gases.

How does the chemist indicate what happens when open bottles of aqueous ammonia and hydrochloric acid are placed next to each other? Particles of ammonia left the aqueous ammonia to form ammonia gas. He can indicate this change using a word equation:

$$\text{ammonia(aqueous)} \longrightarrow \text{ammonia(gas)} \tag{1.1}$$

The arrow ($\rightarrow$) in the equation means "yields" or "produces." The term (aqueous) means that initially the ammonia was dissolved in water. The term (gas) describes the final state of the ammonia. Substances written to the left of the arrow are the initial substances and are called *reactants*. Ammonia(aqueous) is a reactant. The arrow points to the final substances, which are called *products*. Ammonia(gas) is a product. The following are word equations that describe the other obser-

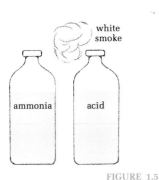

FIGURE 1.5
(a) *The top to the ammonia bottle is removed and then replaced. An odor, characteristic of ammonia is noticed.*
(b) *The top to the hydrochloric acid bottle is removed and then replaced. A sharp odor is noticed.* (c) *The tops to both bottles are removed. White fumes appear close to the hydrochloric acid bottle.*

vations. Particles of hydrogen chloride left an aqueous solution of hydrogen chloride:

hydrogen chloride(aqueous) $\longrightarrow$
$$\text{hydrogen chloride(gas)} \quad (1.2)$$

When the ammonia particles collided with the hydrogen chloride particles they combined to form solid ammonium chloride:

ammonia(gas) + hydrogen chloride(gas) $\longrightarrow$
$$\text{ammonium chloride(solid)} \quad (1.3)$$

The plus (+) sign indicates that the two substances, ammonia and hydrogen chloride, had been brought into contact. ■

■ EXERCISE *Carefully list your observations made on an ice cube melting at room temperature in a tall glass. Explain these observations with the particle model of matter using the ideas of particles, motion, heat, and gravity.*

## 1.9 The sublimation of iodine, a model

A student wished to study some of the properties of the substance iodine. Some iodine was placed on a watch glass as shown in Fig. 1.7. The iodine was in the form of small, dark purple crystals with a characteristic luster. Since iodine reacts with the skin and is poisonous, the student used a spatula to place a small amount of the crystals into two test tubes. To one test tube he added some water; to the other he added carbon tetrachloride. He stoppered both test tubes and shook them for a few seconds. The water turned brown and some solid iodine remained on the bottom of the tube. The carbon tetrachloride turned purple and no more solid iodine could be seen in the test tube.

Next, some iodine crystals were placed in a flask and covered with a watch glass containing cold water (Fig. 1.8). The flask was then heated gently over the low flame of a Bunsen burner. As the flask was heated, a purple gas began to diffuse throughout the entire flask. Crystals of a dark purple solid began to form on the underside of the cold watch glass. At the same time iodine crystals in the flask were growing smaller. On further heating, the iodine crystals at the bottom of the flask disappeared.

What was the purple gas in the flask? What was the dark purple solid that formed on the bottom of the watch glass? Assume that some particles of iodine left the bottom of the flask and re-formed as solid iodine on the watch glass. How can the solid on the watch glass be tested to see if it is iodine? The solid was dark purple and shiny. A sample of the substance on the

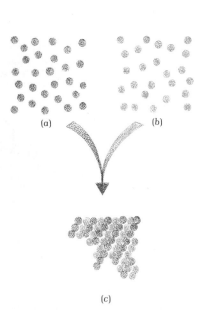

FIGURE 1.6
(a) Particles of ammonia. (b) Particles of hydrogen chloride. (c) Ammonium chloride, a new substance, forms when ammonia particles combine with hydrogen chloride particles.

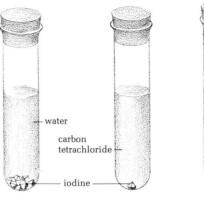
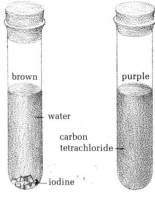

(a) iodine (purple)  (b) before shaking  (c) after shaking

**FIGURE 1.7**
(a) Crystals of dark purple, solid iodine. (b) Before shaking, the crystals lie at bottom of test tubes containing water and carbon tetrachloride. (c) After shaking, the iodine gives the water a brown color and the carbon tetrachloride a purple color.

watch glass was mixed with water; the water turned brown. A second sample of the substance was mixed with carbon tetrachloride; the carbon tetrachloride turned purple. A third sample was placed in a flask, and gently heated. Purple gas was formed. Since the substance on the bottom of the watch glass had the same properties as iodine, we can assume that it was iodine.

Upon heating, solid iodine forms a purple gas. When this gas strikes the cool watch glass, the solid iodine re-forms. The word equation for this change is:

$$\text{iodine(s)} \xrightleftharpoons[\text{removal of heat}]{\text{addition of heat}} \text{iodine(g)} \tag{1.4}$$

The double arrow indicates that while some of the reactants are changing to products, the reverse is also occurring—products are changing back to reactants. The symbol (s) means solid and the symbol (g) means gas. Other symbols commonly used when writing equations are (l) for liquid, (v) for vapor, and (aq) for aqueous solution.

When a substance such as iodine goes directly from a solid to a gas, the substance is said to *sublime*. Dry ice, snow below the freezing point, and moth balls are all examples of substances that sublime.

Why doesn't the iodine sublime noticeably before it is heated? Assume that solid iodine is composed of tiny, unique particles. When heated, some of the iodine particles could break away from the solid. Heating causes iodine particles to move rapidly, so they break away from the solid and form a purple gas (Fig. 1.9). Upon continued heating, all of the iodine

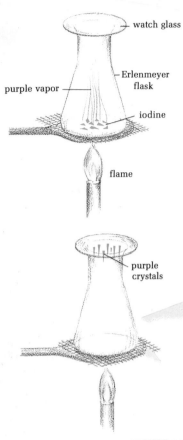

FIGURE 1.8

*Iodine is in an Erlenmeyer flask covered with a watch glass. (a) As the flask is heated, a purple vapor forms. (b) Dark purple crystals form under the watch glass and the iodine crystals at the bottom disappear.*

particles break away from the solid and we observe on a macroscopic level that the solid disappears. When particles of gaseous iodine strike the cool underside of the watch glass they lose some heat, causing their rapid motion to slow down. The slower moving particles re-form the solid iodine.

When a particle is in motion, one form of energy is being used. If particles of gaseous iodine are moving about more rapidly than particles of solid iodine, where does this extra energy come from? The *energy was added* to the system *by heating* with a Bunsen burner. When gaseous particles hit the cool watch glass they gave up some of their energy of motion to particles in the watch glass. *Energy was removed* from the system *by cooling. Heating increases motion; cooling decreases motion.*

*Some properties of substances are due to the macroscopic size of their sample. A single particle of iodine is not a solid. A solid contains many particles close together.*

Are particles in solid and gaseous iodine in motion? To the naked eye, solid iodine appears to be motionless because it has a definite size and shape. We conclude that *particles of iodine in the solid state are either motionless or so limited in their motion that we cannot visibly observe it.* Gaseous iodine moves throughout and completely fills its container. From this observation we conclude that particles of gaseous iodine are in motion and free to move about their container.

Why didn't the particles of solid iodine fly apart by themselves? We suggest that there are attractive forces between iodine particles. These attractive forces between particles of matter are analogous to, but not the same as, the force of gravity. In a sample of solid iodine, these attractive forces confine the particles to the rigid structure of the solid. Since the particles in gaseous iodine are free to move around, the force of attraction between the particles in the gas must be less than the force of attraction between iodine particles in the solid.

Are the particles closer together in solid or gaseous iodine? Since gaseous iodine moved throughout and completely filled the flask in our experiment, it occupies more space than solid iodine. If gaseous iodine occupies more space than the solid sample from which it was formed, the particles must be farther apart in the gas than they are in the solid. The larger distance between iodine particles in the gas is the reason why the attractive forces in the gas are less than they are in the solid. *Attractive forces between particles decrease as the distance between particles increases.*

(a)

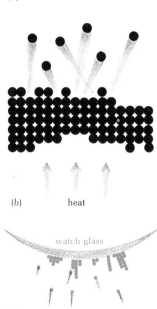

(b) heat

(c)

**FIGURE 1.9**
(a) *Solid iodine consists of many particles closely packed together.* (b) *Heating causes these particles to move about.* (c) *When some iodine particles strike the underside of the cold watch glass they adhere to it and re-form the solid.*

Summarizing:

1. Iodine is composed of small invisible particles.
2. Iodine particles are relatively close together in the solid phase.
3. Iodine particles are widely separated and free to move about in the gaseous phase.
4. The forces of attraction between iodine particles in the gaseous phase are small.

## 1.10 Compounds and elements

What clear, colorless, odorless liquid readily dissolves such substances as sugar, salt, ammonia gas, and hydrogen chloride gas? The substance water has these characteristics.

Can water itself be broken down or *decomposed* into simpler substances? Consider the apparatus shown in Fig. 1.10(a). Two wires from a battery were placed into a large beaker of water. A small amount of sulfuric acid was added to the water in order to increase its ability to conduct electricity. After a short period of time, bubbles of a colorless gas could be seen rising from each of the ends of the wire. The formation of new substances with different properties indicates that a chemical change is occurring in the water. Note that the gases do not appear to be soluble in water.

Larger amounts of gas were collected by placing inverted test tubes of water over the ends of the wire. Samples of a colorless, odorless gas were collected in each test tube as shown in Fig. 1.10(b). Notice that the gas in one test tube occupies twice the *volume* of the other gas. Are the gases identical? Each was tested by means of a burning splint plunged into each test tube as shown in Fig. 1.11. The splint burned brighter in the test

**FIGURE 1.10**
(a) *A beaker containing a solution of sulfuric acid in water. Wires from a battery have been placed in the beaker.* (b) *The bubbles of gas are collected in test tubes by the displacement of water.*

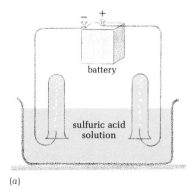

(a)

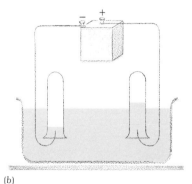

(b)

tube containing the lesser amount of gas. The burning splint caused a small pop or bang when it was introduced into the other test tube. The two gases differ in their properties.

Other experiments indicate that many substances burn well in the gas that caused the splint to burn brighter. This gas is the substance *oxygen*. The second test tube contains a gas that will itself burn, but does not cause other substances to burn. This substance is *hydrogen*. The pure substance, water, can be decomposed by electrical energy into two other substances, oxygen and hydrogen. The word equation for this change is:

$$\text{water(l)} \xrightarrow{\text{electrical energy}} \text{hydrogen(g)} + \text{oxygen(g)} \quad (1.5)$$

Substances, like water, that can be decomposed into simpler substances are called *compounds*. Water is a compound that contains oxygen and hydrogen. This is an *operational definition* of a compound. It is defined by what it does, or by the way it reacts under specific conditions. Other examples of substances that we have discussed that can be decomposed under the proper conditions are: salt, sugar, carbon dioxide, and carbon tetrachloride.

Can oxygen or hydrogen gas be decomposed into other substances? Repeated attempts to decompose hydrogen and oxygen by using heat, light, electricity, or other ordinary means of applying energy have been unsuccessful. Hydrogen and oxygen are elements. *Elements* are operationally defined as substances that cannot be decomposed into simpler substances by ordinary means. (We will not classify nuclear energy as ordinary means.) Some common elements are: nitrogen, carbon, aluminum, magnesium, silicon, iodine, chlorine, sodium, iron, nickel, and copper. So far, chemists and physicists have found 105 such substances. All the other substances in the universe are believed to be composed of these elements. In Chapter 3 we will study combinations of elements to form compounds.

What is a theoretical definition of a compound? An element? Since a sample of hydrogen gas is assumed to consist of hydrogen particles and a sample of oxygen gas of oxygen particles, a sample of water decomposes into hydrogen particles and oxygen particles.

$$\text{water particles} \xrightarrow{\text{electrical energy}}$$
$$\text{hydrogen particles} + \text{oxygen particles} \quad (1.6)$$

It seems reasonable, therefore, to assume that a particle of water is made up of particles of hydrogen and particles of oxygen. Since hydrogen and oxygen do not decompose into simpler

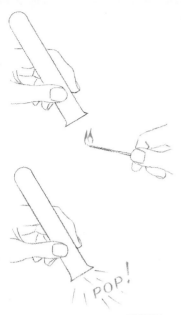

**FIGURE 1.11**
*Test tubes containing gases formed by passing electricity through water. A splint burns brighter in one gas and causes the other to explode.*

| TABLE 1.1 | Mixture | Components |
|---|---|---|
| *Two common mixtures and their major components.* | Air | Nitrogen gas |
| | | Oxygen gas |
| | | Carbon dioxide gas |
| | | Water vapor |
| | | Argon gas |
| | Sea water | Water |
| | | Sodium chloride dissolved in water |
| | | Magnesium chloride dissolved in water |
| | | Potassium chloride dissolved in water |
| | | Sodium bromide dissolved in water |

substances, a particle of hydrogen or a particle of oxygen does not contain other particles.

The theoretical definition or model of a compound states that a particle of a compound contains two or more different kinds of particles. An element is made up of only one kind of particle. The models of compounds and elements will be developed further in Chapter 3.

## 1.11 Mixtures

In nature, some elements are found uncombined with other elements. Some examples are gold, diamond, silver, copper, neon, and sulfur.

Some examples of naturally occurring compounds are halite (salt, that is, sodium chloride), calcite (calcium carbonate), galena (lead(II) sulfide), and quartz (silicon(IV) oxide). Most compounds occur in nature as mixtures. Some examples of naturally occurring mixtures are sea water, air, granite, petroleum, natural gas, and all forms of matter in living organisms, including yourself. The components of two mixtures are shown in Table 1.1.

A number of mixtures were prepared by mixing some substances together in test tubes. The first test tube contained solid salt crystals and water; the second contained sand and water; the third, solid iodine crystals and water; the fourth, solid iodine crystals and liquid carbon tetrachloride; the fifth, water and liquid carbon tetrachloride; and the sixth, solid sugar crystals and water. These six chemical systems are shown in Fig. 1.12.

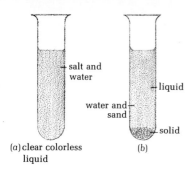
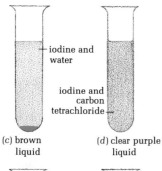
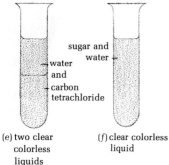

**FIGURE 1.12**
*Six chemical systems.*

■ **EXERCISE** *Classify the following systems as homogeneous or heterogeneous: powdered sugar and sand, a*

What similarities exist among the contents of the first, fourth, and sixth test tubes? In the first, a clear, colorless, continuous liquid is observed. In the fourth, a clear, purple, continuous liquid is observed. In the sixth, a clear, colorless, continuous liquid is observed. Each of these mixtures is continuous. Systems such as these, which contain two or more substances and which appear uniform throughout, are called *homogeneous* mixtures.

How can the apparent disappearance of the macroscopic properties of solid salt, iodine, and sugar be explained? Consider the sugar and water system. Since it is no longer possible to distinguish visibly between the water and the sugar, a reasonable assumption is that the two substances have mixed on the submicroscopic level. Because sugar particles are intimately mixed with water particles on the submicroscopic particulate level, neither the solid sugar nor the sugar particles are visible even with the most powerful microscope. Homogeneous mixtures of the particles of two or more substances are called *solutions*. Sugar in water, salt in water, and iodine in carbon tetrachloride are examples of solutions.

What is the difference between a solution such as salt and water and a compound such as pure water? When water is decomposed it always produces hydrogen gas and oxygen gas. Not only that, but it *always* produces two volumes of hydrogen gas for each one volume of oxygen gas. *Elements occur in compounds in definite ratios.*

What is the ratio of salt to water in a solution of salt and water? Samples of three different salt-water solutions were analyzed in a laboratory and found to contain the following percentages of salt: 5.13 percent, 8.27 percent, and 15.3 percent. These results show us that the *ratio of components of a mixture may vary.*

What similarities exist among the contents of the second, third, and fifth test tubes? After the second tube has been shaken, the sand falls to the bottom. After the third tube has been shaken, the iodine falls to the bottom. The water turns brown, indicating that some iodine has mixed with the water. In the fifth, there appear to be two different liquid layers. Evidently water and carbon tetrachloride do not mix well. Since we can visually detect different parts of these mixtures, these mixtures are not continuous. Mixtures of two or more substances that have distinctly observable boundaries between two of the components are called *heterogeneous* mixtures.

What are some common homogeneous and heterogeneous mixtures? Vinegar, gasoline, and the air are examples of homo-

*closed bottle of beer, an open bottle of beer, the air over Miami during a hurricane, the air over Los Angeles on a smoggy day, vinegar and water, oil and water, an egg, and a crystal glass.*

geneous mixtures. Your blood stream, a smoke-filled room, and a granite rock are examples of heterogeneous mixtures. In order to recognize a heterogeneous mixture it may be necessary, as in the case of smoke in air, to examine the mixture under a hand lens or microscope. ∎

## 1.12 States of matter

A solid substance, such as a grain of sand or an ice cube, has a rigid form. The *size* of the solid and its *shape* do not depend on its container. In explaining the sublimation of iodine a model was developed to account for the rigid nature of solid iodine. Particles of iodine in the solid were assumed to be held together by attractive forces. These attractive forces keep the particles in the solid bound together (Fig. 1.13(a)).

What do you observe when an ice cube melts in a glass? A clear liquid forms, flows down off the ice, and begins to fill the bottom of the glass. Water, like other liquids, *flows* and *takes the shape* of its container. The volume or size of a sample of liquid water does not depend on its container.

How can one explain the observation that liquids flow and take the shape of their container? One could assume that when ice is heated, the particles in the solid gain sufficient energy to increase their motion to partially overcome their attractive forces. As the solid is heated, the individual particles gain energy of motion. In the liquid state, the particles possess enough mobility to be able to move past one another freely (Fig. 1.13(b)). Why doesn't a small sample of liquid completely spread out and fill its container? Evidently there are still attractive forces between the particles in the liquid that keep the particles from filling the whole container.

What happens if we continue heating the water formed from the melting ice? After a time the water begins to boil and eventually all of the liquid disappears. Steam, like purple iodine gas

**FIGURE 1.13**

*(a) A model of particles in a solid. The particles are close together, and do not move past one another readily. (b) A model of particles in a liquid. The particles are close together but they can move about past one another easily. (c) A model of particles in a gas. The particles are far apart, move rapidly, and fill their container.*

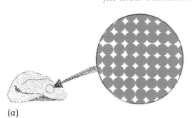
(a)

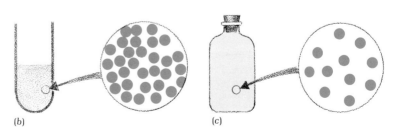
(b)   (c)

■ **EXERCISE** Classify each of the following as predominantly solid, liquid, or gas: a cement slab, wet cement, air, beer, foam on a beer, a car's exhaust, and sand being poured from a bottle.

and other gases, takes the *shape* and *volume* of its container (Fig. 1.13(c)). In steam, as in iodine gas, the particles are moving rapidly in random directions. As with iodine vapor, the particles of steam are widely separated from one another. ■

## 1.13 Phases

Consider the chemical system of ice cubes floating in water. In this system one can see that the ice is physically distinct from the water. One could remove the ice by using a pair of forceps or his hand, or by carefully pouring the water off. A physically distinct part of a system which has definite boundaries and which is physically separable from other parts of the system operationally defines a *phase*. A mixture of ice and water contains two phases, a solid ice phase and a liquid water phase. The word "phase" may be used instead of the word "state" when one is referring to the three common forms of matter—solid, liquid, gas.

A chemical system might contain several solids, liquids, and gases. In many cases the phases can be separated from each other rather simply. Consider the case of a test tube of sand in water. If the sand grains are large and have sunk to the bottom the water can be poured off. This process is called *decantation*. If the sand grains are smaller it may take a while for the solid to settle out. This process may be speeded up by using a *centrifuge*. This instrument spins the test tube and its contents around as shown in Fig. 1.14. This action causes the solid to settle to the bottom rapidly. The liquid may then be poured out.

slow-settling solid

test tubes spin at high speed

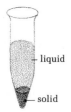
liquid
solid

FIGURE 1.14
A centrifuge can be used to separate a slow-settling solid from a liquid.

■ **EXERCISE** Name two different methods for separating a mixture of sand and iodine. Sand does not dissolve in carbon tetrachloride.

**Two solid phases**

If ground sugar crystals and fine white sand were mixed, the system would consist of two solid phases. Even though the two phases appear to be similar, they could be separated by means of tweezers and use of a magnifying glass. Is there an easier way to separate them? ■

**Saturated aqueous solutions**

What happens when a few crystals of salt are added to a glass half full of water? When a few more crystals are added to the solution, they too will eventually dissolve with stirring. If this process is repeated several times, a point will finally be reached when no more salt will dissolve. If more salt is added after this point, the excess salt will fall to the bottom and remain as a solid phase. When as much solid as possible has dissolved in a liquid at a given temperature, the solution is said to be a saturated solution.

**Gases**

Consider the saturated solution described in the previous section. How many phases are there in this chemical system? There is a solid phase, a liquid phase, and also a gaseous phase above the solution since the glass was only about half filled with liquid. This gaseous phase consists of air and water vapor. ■

■ EXERCISE  *Name two common liquids that do not mix with water. Determine which layer is on top when the two liquids are added to water.*

## 1.14 The electrical nature of matter

When the electrical generator shown in Fig. 1.15(a) is turned, sparking occurs at the two electrodes indicating electrical activity. What is the nature of this electrical activity? A lightweight, metallic-coated sphere was suspended from each electrode by a thin wire (Fig. 11.5(b)). When the generator was turned, the two spheres slowly began to move toward one another as if they were being attracted by an invisible force (Fig. 1.15(c)). Even though the generator was stopped, the two spheres remained close to one another. When a metal rod was placed between the two spheres, sparking occurred and the spheres returned to their original positions.

FIGURE 1.15
(a) Sparking occurs between the electrodes of an electrical generator. (b) and (c) Spheres on the two different electrodes attract one another.

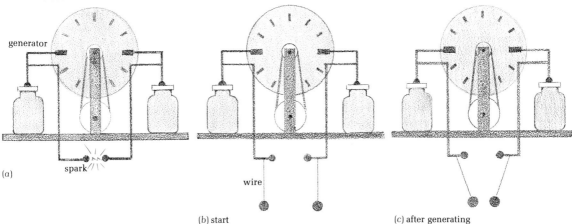

start

after generating

FIGURE 1.16
Spheres on the same electrode repulse each other.

When the two spheres were suspended from the same electrode and the generator was turned on, the two spheres moved apart (Fig. 1.16). When the two spheres were suspended from the other electrode and the generator was turned on, again the spheres moved apart. Since the spheres move away from one another, it is concluded that there are repulsive forces between them.

Can electricity cause chemical substances to move also? Consider an experiment in which a yellow solution is poured into the apparatus shown in Fig. 1.17. The yellow solution contains the solid substance iron (III) chloride dissolved in water. The solution can be seen in the glass tube. On either side of the solution is a gelatin-like substance which prevents liquids from flowing about easily. Next to the gel are electrodes connected to a battery. After an hour, it is noticed that the yellow color has moved through one of the gels and toward the negative electrode. Again, it is concluded that electricity causes motion.

How can the motion of the spheres and of the yellow solution by electricity be explained? A model has been developed that assumes that all matter consists of two types of charged particles. One charge is called positive and one charge is called negative. The positively charged particles are *protons*. The negatively charged particles are *electrons*. The metal spheres placed on the electrodes are assumed to possess many billions of protons and an equal number of electrons. When the number of positive and negative charges is equal, the material is said to be neutral. (Fig. 1.18(a)).

If negative particles are removed from one sphere, there is an excess of positively charged particles (Fig. 1.18(b)). The sphere is said to be positively charged. If negative particles are added to one sphere, there is an excess of negatively charged particles (Fig. 1.18(c)). The sphere is said to be negatively charged.

When a sphere is placed on each electrode of the generator, electrons are removed from one sphere and added to the other sphere. It has been observed that the two spheres attract each other (Fig. 1.19(a)). Therefore, it may be concluded that *oppositely charged particles attract one another*.

When two spheres are placed on the same positive electrode, electrons are removed from each sphere. Both spheres become positively charged and it is observed that they repulse each other. When two spheres are placed on the same negative electrode, electrons are added to each sphere. Both spheres become negatively charged and it is observed that they repulse each other (Fig. 1.19(b) and 1.19 (c)). It is concluded, then, that *like-charged particles repulse one another*.

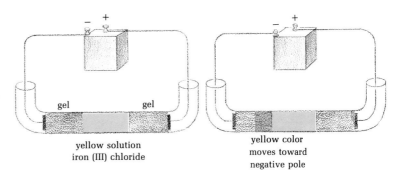

**FIGURE 1.17**
An apparatus demonstrating the motion of a substance caused by electricity.

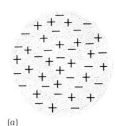

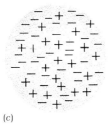

**FIGURE 1.18**
(a) A neutral sphere contains equal numbers of positive (+) and negative (−) charges. (b) When electrons are removed from a neutral sphere, there is an excess of positive charges. (c) When electrons are added to a neutral sphere, there is an excess of negative charges.

When oppositely charged spheres are connected by a metal rod, sparking occurs and the spheres return to their original position. Electrons flow from the negatively charged sphere through the metal rod to the positively charged sphere. Both spheres become neutral. It is concluded that there are no forces between neutral samples of matter.

Why does the yellow color in a solution of iron (III) chloride move toward the negatively charged electrode? Chemists believe that solutions of many substances contain charged particles called *ions*. The yellow color in the solution is believed to be due to a *positive* ion. When the battery is connected to the electrodes, the positive ions move toward the negatively charged electrode. The negative ions which are colorless in this case move toward the *positively charged* electrode. Chemists believe that all chemical interactions can be explained in terms of two kinds of charged particles.

## 1.15 Forms of energy

**Kinetic energy**

What happens when a moving baseball strikes a glass window? The attractive forces holding the glass particles together are overcome and the glass shatters. The pushing apart of attracting particles is a form of *work*. The ability of an object or system to do work is referred to as *energy*. The baseball possesses energy because it is moving. Energy of motion is called *kinetic energy*.

What was used to overcome the attraction of particles in solid iodine, causing iodine gas to form? Heat from a Bunsen burner flame was used. The flame consists of many rapidly moving particles that possess kinetic energy. This rapid motion was transferred to the glass flask and then to the solid iodine. Rapidly moving particles of iodine then left the solid and formed a gas. *Heat* is kinetic energy on the submicroscopic level. The kinetic energy of a particle is given by the formula:

$$K.E. = \tfrac{1}{2}mv^2 \qquad (1.7)$$

where K.E. is kinetic energy, $m$ is the mass of the particle, and $v$ is the velocity of the particle.

### Potential energy

Does the boulder shown in Fig. 1.20 possess energy? The boulder is located behind a mound on top of a high hill. Although it does not possess kinetic energy, it could be pushed over the mound and then down the hill. The boulder has a potential for kinetic energy. Any object or system, that because of relative position could produce kinetic or some other form of energy, is said to possess potential energy.

Consider a flash bulb and its contents. Examination of the bulb shows that it is a sealed glass bulb containing shiny, fine, metallic threads of magnesium. Not visible to the eye is the gas, oxygen. When wires from a battery are touched to it, a bright light and heat are emitted. A white ash known as magnesium oxide is left in the bulb. Since both heat and light are forms of energy, we conclude that chemicals in the flash bulb have a potential energy which is released when they react. Like the boulder on a hill, the chemicals need a push before they react. The push, in the form of electrical energy from the battery, starts a change in which magnesium and oxygen form magnesium oxide.

$$\text{magnesium(s)} + \text{oxygen(g)} \xrightarrow{\text{electrical energy}} \text{magnesium oxide(s)} + \text{energy} \qquad (1.8)$$

A potential energy diagram for this reaction is shown in Fig. 1.21. Note the similarity between it and the potential energy diagram for the boulder. Once given a push, many chemical systems will continue to undergo change if the products have a lower potential energy. The relative potential energy of reactants and products is one factor that indicates whether a chemical change will take place.

### Light

Light is a form of energy that can travel through a vacuum. In addition to visible light there are also ultraviolet and infrared light rays. Even X-rays and radio waves are forms of light rays. Can light cause a chemical change? Sunburn is one con-

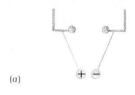
(a)

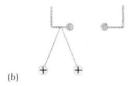
(b)

(c)

**FIGURE 1.19**
(a) *Objects with opposite charges attract one another.* (b) *and* (c) *Objects with the same charge repel one another.*

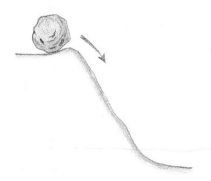

**FIGURE 1.20**
*A boulder on a hill possesses potential energy. Given a push or activated, the boulder can move down the hill.*

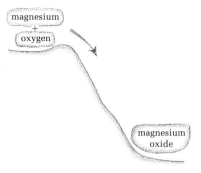

FIGURE 1.21

The reactants, magnesium and oxygen, possess potential energy. Given a push or activated, they release this energy and form a substance with lower potential energy, magnesium oxide.

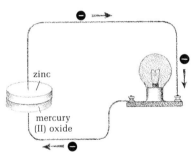

FIGURE 1.22

A chemical change in a mercury cell causes electrons to leave the zinc metal, flow through a wire, and re-enter the mercury(II) oxide section of the cell.

sequence of the human chemical system undergoing a change because of ultraviolet rays. Light reflected from a person causes a chemical change on a piece of photograpic film. Growing plants demonstrate the effect of light on their chemical systems: they grow.

**Electrical energy**

Earlier it was shown that electrical energy can cause water to undergo chemical change. Can chemical systems be used to supply electrical energy? Consider the mercury cell, which is often used to power hearing aids, transistor radios, and electric watches. A fresh battery contains granules of shiny, metallic zinc powder in one section and mercury(II) oxide in another section. In a used cell the zinc metal has turned into a white powder known as zinc oxide. In the section containing mercury(II) oxide, shiny metallic drops are found which resemble liquid mercury. These observations indicate that as the battery generated electricity the zinc and mercury(II) oxide were transformed into zinc oxide and mercury.

$$\text{zinc(s)} + \text{mercury(II) oxide(s)} \longrightarrow$$
$$\text{zinc oxide(s)} + \text{mercury(l)} + \text{electrical energy} \quad (1.9)$$

What does the electrical current that flows in a wire consist of? In Section 1.14 it was noted that electrical interactions can be explained by assuming that matter consists of protons and electrons. Electricity in a wire is believed to be due to the flow of electrons from one end of the wire to the other as shown in Fig. 1.22. In a mercury cell, electrons originate in the zinc metal. They flow out one end of the cell through a wire and back into the mercury(II) oxide section of the cell. This flow of electrons is caused by the chemical change in the cell.

## 1.16 A model for chemical changes

When electrical energy is passed through water, it decomposes into the elements hydrogen and oxygen. The simplest known hydrogen particles are called hydrogen atoms. The simplest known particles of any element are known as *atoms*. Actually, in hydrogen and oxygen gas the atoms combine in pairs to form hydrogen molecules and oxygen molecules as shown in Fig. 1.23. A particle consisting of two or more atoms is called a *mol-*

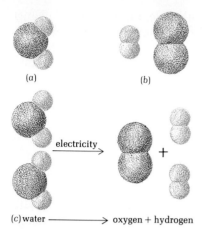

FIGURE 1.23
(a) A water molecule. (b) Hydrogen and oxygen molecules. (c) Water can be changed into hydrogen and oxygen.

■ EXERCISE  Are the melting of ice and the boiling of water chemical changes? Give reasons for your answer.

ecule. The simplest particle of water contains one oxygen atom and two hydrogen atoms and is called a water molecule.

During decomposition, water molecules are broken apart with the application of electrical energy.

electrical energy + water(l) molecules ⟶
    hydrogen(g) molecules + oxygen(g) molecules   (1.10)

Can a model be used to define a chemical change? In a chemical change atoms or molecules either combine, separate, or are rearranged.

Is the sublimation of iodine a chemical change? Iodine, like hydrogen and oxygen, consists of molecules. Each iodine molecule contains two iodine atoms. When solid iodine is heated, the molecules separate from one another, but this change does not result in a change in the composition of iodine. Changes which involve no change in composition are called physical changes by most chemists. Physical changes involve changes in size, shape, appearance, or state of matter. Some scientists would say that this is a physical change because there are iodine molecules in both the solid and vapor phases. Other scientists, however, do not agree. They say that iodine molecules are held together in the solid by attractive forces. The overcoming of these forces when iodine gas is formed constitutes a chemical change. ■

## Summary

Chemistry is what chemists do. Chemists make observations on matter and the changes that matter undergoes. They experiment with chemical systems to find out what factors affect the systems; they measure and record their observations and they communicate their observations to other chemists. They attempt to explain the behavior of chemical systems by developing and using models.

A substance is a particular type of matter. Substances are recognized by their particular properties. Properties of substances depend upon the kind of particles, the ratio of particles, and the arrangement of particles in a sample of a substance. The possible uses of a substance depend upon its properties.

A chemical reaction can be recognized when a substance or substances with one set of properties changes into a substance or substances with a new set of properties. During a chemical reaction, atoms are either combining, separating, or rearranging themselves. Often chemical reactions are accompanied by the absorption or evolution of energy in the form of heat, light, or electricity. Many chemical reactions occur at a noticeable rate only when energy is added.

An element is a substance that cannot be decomposed by ordinary means. The simplest type of particle in an element is an atom. So far all the atoms in an element have been assumed to be identical.

A compound is a substance that can be decomposed by ordinary means. Compounds consist of two or more different types of atoms bound together.

A mixture contains two or more substances. A continuous mixture is said to be homogeneous and is called a solution. A discontinuous mixture is said to be heterogeneous. The term heterogeneous is also applied to samples of pure substances existing in two or more phases. An ice and water system is an example. The behavior of solid, liquid, or gaseous phases of matter can be explained by the relative motion and attractive forces between the particles in each phase.

The electrical behavior of matter can be explained by assuming the existence of positively charged particles called protons and of negatively charged particles called electrons. Like charges repel each other and unlike charges attract each other. All chemical interactions can be explained using the electrical model of matter. In order to separate attracting bodies in a system, work must be done on the system. The energy needed to separate attracting bodies in chemistry is often supplied by heat, light, or electricity.

## Objectives

By the end of this chapter you should be able to do the following:

1. List five activities of a scientist. (Question 20)

2. Describe a sample of matter by listing its important properties. (Question 3)

3. Describe a chemical change by listing observations made before, during, and after the change. (Question 4)

4. Describe a phase change for a pure substance by listing changes undergone by the particles. (Question 5)

5. Identify chemical systems as being either open or closed. (Question 13)

6. Cite evidence for the existence of submicroscopic particles of matter. (Question 6)

7. List the general properties of solids, liquids, and gases. Using the particulate model of matter, explain the properties of each one. (Question 7)

8. Given the reactants and products, write the word equation for a chemical change. (Question 8)

9. State the operational and theoretical definitions of a compound and of an element. (Question 1)

10. Operationally define and give examples of two types of mixtures. (Question 1)

11. Use a model to state when there is attraction and when there is repulsion between electrically charged bodies. (Question 10)

12. Give a theoretical definition of heat. (Question 12)

## Questions

1. Explain in your own words the meaning of the following terms. Where possible give both operational and theoretical definitions: chemist; observations; measurement; substance; property; state; matter; solid; liquid; gas; structure; composition; macroscopic; microscopic; submicroscopic; operational definition; theoretical definition; system; open system; closed system; products; reactants; sublimation; compound; element; homogeneous; solution; mixture; phase; proton; electron; kinetic energy; potential energy; heat; work; heterogeneous; phase change; chemical change.

2. List ten observations or measurements that have at some time been made on you. What model was used to explain the results of these measurements?

3. Obtain a piece of lead from a mechanical pencil. Examine the lead and observe its behavior under different environments. For example, does it write on greasy paper or does it react with vinegar? List ten observations made on the lead.

4. Obtain a small piece of steel wool. List some of its properties. If possible, weight it. Roll it into a small ball and heat it in a flame for several minutes. Do any of of its properties change? Does its weight change?

5. Watch two ice cubes melt in two different environments. Explain the changes that occur using the particulate model of matter. What role does heat play on the particles?

6. Place a lump of sugar in a glass of warm water without stirring. Observe it. After a while, taste it. Cite evidence for the existence of particles of sugar and particles of water.

7. Compare and contrast the general properties of solids, liquids, and gases. Use a model to explain the properties of these states.

8. When vinegar is mixed with baking soda, a colorless, odorless gas is produced, the taste of vinegar disappears, and the baking soda is no longer useful. Vinegar is a solution of acetic acid in water. Baking soda is mainly sodium hydrogen carbonate. The gas produced is carbon dioxide. Other products are water and sodium acetate. Write the word equation for the reaction.

9. Use a model to state what makes an object positively charged, negatively charged, or neutral.

10. Using an electrical model of matter, state when objects will attract each other, repel each other, and have no effect on each other.

11. According to the particulate model of matter, in what way do particles of matter in cold water differ from particles of matter in hot water?

12. What effect does heat have on submicroscopic particles?

13. Consider your own body; is it an open or closed chemical system?

14. Use the particulate model to explain why a gas occupies a relatively larger volume than does a liquid.

15. Use the particulate model to explain why a solid is more rigid than either a liquid or a gas.

16. "Brake fluid" is used in hydraulic braking systems. Use the particulate model to explain why a liquid is preferable to a solid or a gas for this purpose.

17. Develop and conduct an experiment which will demonstrate whether tap water is a mixture or a pure substance.

18. In which container in the following chemical systems is there (1) one phase, (2) two phases, (3) three phases, or (4) four phases? (a) *a full glass of water;* (b) *a filled and sealed test tube of oxygen;* (c) *a balloon full of air;* (d) *a filled and sealed test tube of sand, water, and carbon tetrachloride;* (e) *a half-filled and sealed test tube of sand, water, and carbon tetrachloride;* (f) *a full glass of fresh beer;* (g) *a full glass of fresh soda and ice cubes.*

19. Ask a geologist or other knowledgeable person what local minerals or rocks have been or could be mined. What are the chemical names of these substances or mixtures?

20. Obtain an account of one of the space programs. Make a list of each of the following: observations; measurements; experiments; predictions; new models.

21. Which, if any, of the following are not essential for human life? Carbon; nitrogen; oxygen; iron; sulfur; hydrogen; iodine; chlorine; sodium; potassium; calcium; copper; gold; phosphorus; cobalt.

# Measurement

## 2.1 Introduction

"Qualitative" and "quantitative" are two terms which are important to chemists. These two terms can be defined operationally. When a person is said to be tall, his height is expressed *qualitatively*. If it is stated that he is 6 feet and 8 inches tall, his height is expressed *quantitatively*. Chemists make measurements to describe quantitatively the properties of matter and the changes which matter undergoes.

Chemists measure length, volume, mass, density, temperature, pressure, and energy. They also measure the length of time required for a change to occur.

If you hold a block of aluminum and a block of lead of the same size, the lead will feel heavier. Why does the lead appear to be heavier than the aluminum? How much heavier? Measurements are required to determine this. First, determine the mass of each block; then measure the size of each block, that is, their volumes or the space they occupy. Chemists use a derived property of matter, *density*, to compare the quantity of matter contained in the same volume of two different substances. A derived property is a property which is not measured directly but determined from other measurements.

Measurements are of value in many cases where comparisons are desired. If a certain fuel is being considered for a particular use and the user wants to know *how fast* it burns, *how much* heat is released, or *how hot* the flame gets, so that he can compare it with other fuels before he makes a decision, he would have to use measurements for such comparisons.

Advertising takes advantage of our reliance on comparative measurements when they cite information on the length of cigarettes, the wheelbases of cars, the dimensions of girls, and the length of skirts.

Many properties of substances are derived from measurements. Such measurements will help answer such questions as: What is the size and mass of the fundamental units of matter? What volume does a given sample of a gas occupy at room temperature and atmospheric pressure? What is the volume occupied by one gram of water in the gaseous state? in the liquid state? in the solid state? Why does a lemon taste sour? How long does it take for a piece of steel wool to rust?

## 2.2  Measurement

How would you measure the length of a block of wood? This can be done by placing a ruler beside the block and comparing the block with the markings on the ruler, as shown in Fig. 2.1. The length of the block is compared with a standard length, a ruler in this case.

*Measurement* is the act of determining the extent, dimensions, quantity, degree, or capacity of a sample of matter. When a measurement is made, an unknown quantity is compared with a reference standard. Reference standards from two systems of measurement, the *English system* and the *metric system*, are in common use today. The original standard for measurement in the English system was the length of an average man's foot. Since the length of feet varies, a standard foot was initiated for more accurate measurement. The standard of measurement in the metric system was originally the length of a platinum-iridium bar stored at Sèvres, France. The standard meter is now defined in terms of the wavelength of a particular orange-red light.

What is the length of the block in Fig. 2.1? Using the top ruler which is calibrated in inches and tenths of inches, note that the block appears to be 1.89 inches. The units are inches and the magnitude or size of 1.89 indicates the number of inches. *Every measurement, when it is recorded, must include both* magnitude and units.

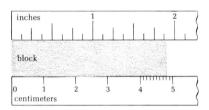

**FIGURE 2.1**

*A block measured in inches and in centimeters. The distance between the two ends of the block is the same, whatever the system of measurement used. 1.89 inches are equivalent to 4.80 centimeters.*

## 2.3 The metric system

What volume does the block in Fig. 2.1 occupy? What is its surface area? What is the mass of the block? The answer to each of these questions requires a measurement. What system of measurement would be used in making these measurements? Americans generally use the English system which uses units of inches, feet, yards, miles, quarts, gallons, and pounds. Scientists almost universally use the metric system of measurement with units of meters, liters, and grams.

The metric system is a decimal system of measurement in which all of the units of one class are related to one another by powers of ten.

10 millimeters = 1.0 centimeter (cm)
10 centimeters = 1.0 decimeter
10 decimeters  = 1.0 meter (m)

Note the similarity. The common prefixes of the metric system are included along with their meanings in Table 2.1.

TABLE 2.1

Common prefixes in the metric system and their meaning.

| | | | |
|---|---|---|---|
| micro | $= \frac{1}{1,000,000}$ | $= 0.000001$ | $= 1 \times 10^{-6}$ |
| milli | $= \frac{1}{1000}$ | $= 0.001$ | $= 1 \times 10^{-3}$ |
| centi | $= \frac{1}{100}$ | $= 0.01$ | $= 1 \times 10^{-2}$ |
| deci | $= \frac{1}{10}$ | $= 0.1$ | $= 1 \times 10^{-1}$ |
| kilo | $= \frac{1000}{1}$ | $= 1000$ | $= 1 \times 10^{3}$ |

## 2.4 Uncertainty and significant figures

If one student measured the block in Fig. 2.1, he might obtain a value of 1.89 inches, but other students from the same class could measure this same block and several of them would obtain values different from the measurement made by the first student. The measurements made by one group of eight students on such a block are recorded in Table 2.2.

Why were slightly different values obtained for the length of the block? Which value is correct? All of the values reported by the students actually are acceptable. When compared to the

**TABLE 2.2**

*A set of measurements recorded by a group of eight chemistry students for the length of the block shown in Fig. 2.1.*

| Student | Measurement recorded (inches) |
|---|---|
| 1 | 1.87 |
| 2 | 1.89 |
| 3 | 1.88 |
| 4 | 1.89 |
| 5 | 1.89 |
| 6 | 1.88 |
| 7 | 1.87 |
| 8 | 1.89 |

ruler in Fig. 2.1, the length of the block is seen to be greater than 1.8 inches but less than 1.9 inches. The values reported by the students were 1.87, 1.88, and 1.89 inches. The first two figures of each of these values are the same and show that each student knew the value was greater than 1.8 but less than 1.9. The third figure could not be determined precisely by reference to the ruler and had to be estimated. There is said to be some *uncertainty* in the third figure.

Whenever measurements are recorded, only one figure may contain uncertainty. In the measurement reported by the students, two of the figures, 1 and 8, were known with certainty. The last figure was uncertain. A chemist interprets a reported value of 1.88 inches for the length of the block to mean that 1 and 8 are known with certainty. The only uncertainty in 1.88 inches is in the last figure, 8. We say that the measurements 1.87, 1.88, and 1.89 have three *significant figures* even though the last figure in each case contains some uncertainty. Scientists have agreed to report only one figure with any uncertainty; the rest of the figures should be known with certainty.

The accuracy of a measurement depends on the technique or skill of the person making the measurement and on the particular instrument being used. No measurement is ever exact. Every reported measurement contains some uncertainty but there should be uncertainty in only one figure.

## 2.5 Length

*Length* is defined as the distance between two points. As was noted earlier, the length of an object is determined by comparison with a standard length.

Length measurements may vary from 0.00000001 ($1 \times 10^{-8}$) inch for the diameter of atoms, the fundamental particles of the elements, to 93,000,000 ($9.3 \times 10^7$) miles between the sun and the earth. To measure very large or very small distances one must use instruments which are more sophisticated than a ruler.

Beginning chemistry students will generally record their length measurements in centimeters, which can be converted to the other common units of length in the metric system—the meter, the millimeter, and the *Angstrom* (See Table 2.3).

How do units of length in the metric system compare with units of length in the English system? How could such a comparison be made experimentally? Refer to Fig. 2.1 and note that

**TABLE 2.3**

*Equivalencies for units of length.*

| | |
|---|---|
| 1 meter = 10,000,000,000 Angstroms | = $1.00 \times 10^{10}$ Angstroms |
| 1 meter = 1,000,000 micrometers | = $1.00 \times 10^{6}$ micrometers |
| 1 meter = 1000 millimeters | = $1.00 \times 10^{3}$ millimeters |
| 1 meter = 100 centimeters | = $1.00 \times 10^{2}$ centimeters |
| 1 meter = 0.001 kilometer | = $1.00 \times 10^{-3}$ kilometer |
| 1 meter = 39.4 inches | = $3.94 \times 10^{1}$ inches |
| 1 inch = 2.54 centimeters | = $2.54 \times 10^{0}$ centimeters |

the two "rulers" both measure the distance between the same two points, that is, the distance between the two ends of the block.

When the student measured the distance between the two ends of the block using the English system, he obtained a value of 1.89 inches. Using the metric system, he obtained a value of 4.80 centimeters. Since the distance between the two ends of the block is the same, whatever the system of measurement used to measure it, 1.89 inches must be equivalent to 4.80 centimeters. Expressed in equation form, this relationship is:

$$1.89 \text{ in.} = 4.80 \text{ cm} \tag{2.1}$$

The "=" expresses a "physical equivalency." This equation can be simplified by dividing both sides by 1.89.

$$\frac{1.89 \text{ in.}}{1.89} = \frac{4.80 \text{ cm}}{1.89} \tag{2.2}$$

$$1.00 \text{ in.} = 2.54 \text{ cm} \tag{2.3}$$

According to Eq. 2.3, 1.00 inch is equivalent to 2.54 centimeters. This is true because 1.00 inch represents the same distance as does 2.54 centimeters.

What happens if both sides of Eq. 2.3 are divided by 1.00 in.?

$$\frac{2.54 \text{ cm}}{1.00 \text{ in.}} = \frac{1.00 \text{ in.}}{1.00 \text{ in.}} \tag{2.4}$$

$$\frac{2.54 \text{ cm}}{1.00 \text{ in.}} = 1 \tag{2.5}$$

Equation 2.5 is called a unity term. On the left side of the equation, the numerator 2.54 cm is physically equal to the denominator, 1.00 in.; that is, both quantities represent the same distance. When this is the case, the expression may be considered as unity (1) just as $5/5 = 1$ or $a/a = 1$. A fraction whose numerator and denominator are equivalent is said to be a *unity term*. The unity term, 2.54 cm/1.00 in. = 1, expresses the physical relationship that 2.54 centimeters and 1.00 inch represent the same length. Unity terms will be used in making later calculations.

A list of metric length units and their corresponding English equivalents is included in Table 2.3. Note that a centimeter is equal to 100,000,000 ($10^8$) Angstroms, a unit used to indicate the sizes of atoms.

### Reading and recording length measurements

How accurately can you read the centimeter scale in Fig. 2.1? The second digit after the decimal is in doubt. The length of the block is 4.80 cm. A meterstick like the one used in this measurement is read to three significant figures. If more than three significant figures are desired in a length measurement, a more refined instrument must be used. ■

■ EXERCISE  1. Use the centimeter scale in Fig. 2.1 to measure: (a) the length and width of a dollar in centimeters, and (b) the diameters of several coins. 2. Compare your answers with those of a classmate. 3. Calculate the area of a dollar bill in square centimeters ($A = l \times w$, where $l$ = length and $w$ = width). 4. Calculate the area of the face of a quarter ($A = 3.14 \times r^2$, where $r$ = radius).

## 2.6 Volume

The *volume* of an object is defined as the space which the object occupies. How could you measure the volume occupied by a sample of liquid? One instrument, similar to a measuring cup and used to measure liquid volumes, is called a *graduated cylinder* (Fig. 2.2). The volume of a liquid such as water is determined by reading the bottom of the curved surface or *meniscus* as is shown in Fig. 2.2.

The volumes of regular solids like cubes, cylinders, or spheres can be calculated using the appropriate mathematical formula in each case, provided one knows the necessary dimensions.

How would you measure the volume of a handful of pebbles? Pebbles are too irregular to have their dimensions measured. What happens to the water level in the graduated cylinder in Fig. 2.3 when a pebble is dropped into the water? The water level rises because the pebbles occupy space. By measuring the water level in a graduated cylinder before adding the pebbles and then measuring the water level again after the pebbles have *displaced* the water, the volume of the pebbles may be determined. What is the volume of the pebbles in Fig. 2.3?

The *liter* is the standard volume unit in the metric system. The liter and the milliliter are used in most scientific volume measurements. One liter is equal to 1000 milliliters (ml). Table 2.4 lists some volume equivalencies.

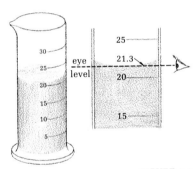

FIGURE 2.2

A graduated cylinder used for measuring the volume of liquids. The liquid occupies 21.3 ml.

## TABLE 2.4

Equivalencies for units of volume.

1 liter = 1000 cubic centimeters = $1.00 \times 10^3$ cubic centimeters
1 liter = 1000 milliliters = $1.00 \times 10^3$ milliliters
1 liter = 1.06 quarts
1 liter = 61.0 cubic inches = $6.10 \times 10^1$ cubic inches

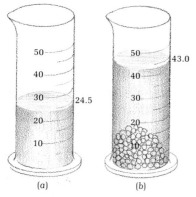

(a)  (b)

FIGURE 2.3

Measuring the volume of pebbles using the displacement of water. The volume occupied by the water before the addition of the pebbles was 24.5 ml. The volume of the water plus the pebbles was 43.0 ml.

A cube which is 1.0 cm on a side has a volume of 1.0 cubic centimeter (cm³).

$$V = 1.0 \text{ cm} \times 1.0 \text{ cm} \times 1.0 \text{ cm} = 1.0 \text{ cm}^3 \qquad (2.6)$$

The milliliter is equal to the cubic centimeter, and consequently ml and cm³ are used interchangeably.

How big is a liter? We know that 1.000 liter is equal to 1000 ml, but how does this compare to volume units in the English system? As seen in Table 2.4, a liter is approximately the size of a quart.

The accuracy possible with a graduated cylinder depends on the graduations and on the size of the cylinder. The graduated cylinder shown in Fig. 2.3 can be read to within 0.1 ml. The volume of the water in (a) is 24.5 ml and the volume of the water plus the pebbles in (b) is 43.0 ml. The volume of the pebbles is the difference between these two volumes, or 18.5 ml. There are three significant figures in each of these numbers.

When a liquid is poured from a graduated cylinder, some of the liquid remains behind. *A graduated cylinder is calibrated to hold the measured volume, not to deliver the measured volume to another container.*

In addition to the graduated cylinder, instruments such as the pipet, the buret, and the volumetric flask are used to measure the volume of liquids. These instruments are shown in Fig. 2.4.

How could one experimentally arrive at a unity term involving liters and quarts? To do this requires measuring the same volume of water using both the English system and the metric system. If one found that 0.542 quart was physically equivalent to 511 milliliters, then:

0.542 qt = 511 ml

$$0.542 \text{ qt} = (511 \text{ ml}) \frac{1.000 \text{ l}}{1000 \text{ ml}} = 0.511 \text{ l}$$

$$\frac{0.542 \text{ qt}}{0.511 \text{ l}} = 1 = \frac{1.06}{1.00 \text{ l}}$$

34  2: Measurement

If this compares quarts to liters, what would be the unity expression comparing liters to quarts?

## 2.7 Mass

Mass is the quantity of matter in an object. Each particle of matter has an attraction for other particles of matter. The force of attraction depends on how much matter is present.

One laboratory instrument used to measure mass is an equal arm *balance*, like the one shown in Fig. 2.5. The object to be weighed is placed on one pan. Standardized masses are added to the other pan. When the beam is just horizontal, the object is balanced.

More commonly found in a beginning chemistry laboratory is a *triple beam balance*, like the one in Fig. 2.6. The object to be weighed is placed in the pan. The appropriate weights are then moved across one or more of the three beams until the object is balanced. The beams are graduated; one beam may be read in increments of 10 grams, one in increments of 1 gram, and still another in increments of 0.01 gram. What is the reading on the balance in Fig. 2.6? To four significant digits, the weight is 56.45 grams.

The *gram* (g) is the unit of mass scientists most commonly use. A list of mass equivalencies is found in Table 2.5. Note that the English pound (1.00) is equivalent to 454 grams. Read the labels of several commercial products around your home and note the weight designations which are given in both the English system and the metric system, a practice dictated by international trade.

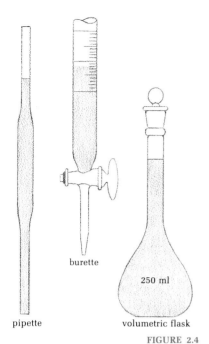

FIGURE 2.4

*The pipette, burette, and volumetric flask are some of the instruments which the chemist uses to make volume measurements. The instrument selected to measure a particular volume depends on the size of the volume and the accuracy desired.*

TABLE 2.5

*Equivalencies for units of mass.*

| | | |
|---|---|---|
| 1 gram | = 1,000,000 micrograms | = $1 \times 10^6$ micrograms |
| 1 gram | = 1000 milligrams | = $1 \times 10^3$ milligrams |
| 1 gram | = .001 kilogram | = $1 \times 10^{-3}$ kilogram |
| 1 pound | = 454 grams | = $4.54 \times 10^2$ grams |

The capacity of common laboratory balances varies from thousands of grams to a small fraction of a gram. Common laboratory balances weigh to about 0.1 or 0.01 g. If a quarter weighs 5.56 grams, how many significant figures are in the measurement? (Answer: three)

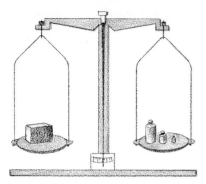

FIGURE 2.5
An equal-arm balance. An object is "weighed" by placing it in one pan; standardized masses are then placed in the opposite pan until the beam is balanced.

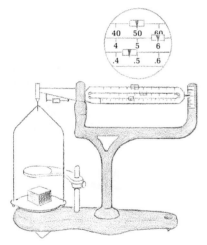

FIGURE 2.6
A triple-beam balance in use.

## 2.8 The use of unity terms in calculations

How many dozens are 48 eggs? The logic used in solving this problem is similar to that which is applied in solving chemical problems. To determine the number of dozens in 48 eggs one recognizes the fact that there are 12 eggs in one dozen and can write:

$$12 \text{ eggs} = 1 \text{ dozen} \tag{2.7}$$

Dividing both sides of this equation by 1 dozen gives:

$$\frac{12 \text{ eggs}}{1 \text{ dozen}} = \frac{1 \text{ dozen}}{1 \text{ dozen}} = 1$$

$$\frac{12 \text{ eggs}}{1 \text{ dozen}} = 1 \tag{2.8}$$

Both sides of Eq. 2.7 could also be divided by 12 eggs to give Eq. 2.9.

$$\frac{12 \text{ eggs}}{12 \text{ eggs}} = \frac{1 \text{ dozen}}{12 \text{ eggs}} = 1$$

$$\frac{1 \text{ dozen}}{12 \text{ eggs}} = 1 \tag{2.9}$$

Equations 2.8 and 2.9 are both unity terms. On one side of each equation the numerator and denominator are physically equivalent, since a dozen eggs is physically equivalent to 12 eggs; they are both symbols for the same group of eggs. Therefore, on the opposite side of the equation, the factor must be unity. Equation 2.9 is the unity term which we will use in our calculation.

When we multiply a physical entity by unity, we do not change the physical entity; nor does multiplying a physical entity by a unity term change the physical entity. We want to know how many dozens 48 eggs are or how many dozens are equivalent to 48 eggs. Multiplying the physical entity 48 eggs by the unity term 1 dozen/12 eggs does not physically change the 48 eggs, but gives us the number of dozens equivalent to 48 eggs.

$$48 \text{ eggs} \times \frac{1 \text{ dozen}}{12 \text{ eggs}} = 48 \times \frac{1}{12} \times \frac{\text{eggs}}{\text{eggs}} \times \text{dozen} = 4 \text{ dozen}$$

The unity term 1 dozen/12 eggs converts the number of eggs to the number of dozens of eggs.

Notice that the unit "eggs" appears in the numerator of one

term and the denominator of the other term. Units are subject to the mathematical operations of division, multiplication, and raising to powers, just as numbers are. The unit "eggs" divided by the unit "eggs" is equal to unity.

$$\frac{\text{eggs}}{\text{eggs}} = 1$$

By proper use of units you can check the answer to a chemical problem. You should always take advantage of this fact, not only to check your answer, but also to help provide some insight into the solution of the problem. If the wrong unity term had been used, the units would be absurd,

$$48 \text{ eggs} \times \frac{12 \text{ eggs}}{1 \text{ dozen}} = 576 \frac{\text{eggs}^2}{\text{dozen}}$$

because eggs²/dozen has no physical significance, whereas both the units eggs and dozen have physical significance. Since eggs²/dozen has no physical significance, we conclude that the answer is incorrect and that the problem was solved incorrectly. If a problem is analyzed and set up correctly, the proper units will result.

EXAMPLE 1   What unity term should be used to solve any given problem? How many pounds does a 80,800-gram boy weigh? The original units, the units desired, and a relationship between the given and wanted units must be known. For example, to determine how many pounds a 80,800-gram boy weighs, one would analyze the problem as:

*Given:* 80,800 *grams*

*Find:* The equivalent number of *pounds*

*Relationship:* From Table 2.4, 454 g = 1.00 lb

*Solution:* To convert grams to pounds we write:

$$80{,}800 \text{ g} \times \boxed{\text{Unity Term}} = ? \text{ pounds}$$

The unity term which is necessary to make this conversion must have grams in the denominator and pounds in the numerator.

$$80.800 \text{ g} \times \boxed{? \frac{\text{lb}}{\text{g}}} = ? \text{ lb}$$

Since 454 g = 1.00 lb, write:

$$80{,}800 \text{ g} \times \boxed{\frac{1.00 \text{ lb}}{454 \text{ g}}} = 178 \text{ lb}$$

The unity term 1.00 lb/454 g converts 80,000 g to 178 lb; thus the weight of the 80,800-gram boy expressed in pounds is 178 lb.

The solution to the same problem can be demonstrated using scientific notation.

$$8.08 \times 10^4 \text{ g} \times \boxed{\frac{1.00 \text{ lb}}{4.54 \times 10^2 \text{ g}}} = 1.78 \times 10^2 \text{ lb}$$

Refer to the Appendix for proper use of scientific notation.

Just as there are various physical equivalencies, such as 12 eggs = 1 doz or 454 g = 1.00 lb, there are numerous chemical equivalencies. Throughout this book unity terms will be employed as factors in the solution of problems. The problems at the end of each chapter should be solved using the *factor-unit method*.

The following examples help illustrate the use of unity terms.

EXAMPLE 2  The shortest dash in the Olympic Games is 100 meters. How many yards is this length?

Given: Distance = 100 meters

Find: The equivalent number of yards

Relationships: Table 2.3 does not give a relationship for yards and meters. However, a relationship for meters and inches is listed, and with this, one can do the conversion. Inches can then be converted to yards provided the number of inches equivalent to one yard is known.

(1) From Table 2.3: 1.00 meter = 39.4 inches

(2) From recall: 1.00 yard = 36.0 inches

Solution: Relationship (1) converts meters to inches. Relationship (2) converts inches to yards.

$$100 \text{ meters} \times \boxed{\frac{39.4 \text{ in.}}{1.00 \text{ meter}}} \times \boxed{\frac{1.00 \text{ yd}}{36.0 \text{ in.}}} = 109 \text{ yd} = 1.09 \times 10^2 \text{ yd}$$

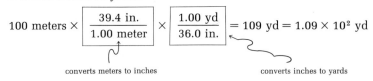

converts meters to inches           converts inches to yards

EXAMPLE 3  The human body normally excretes about 2.0 grams of potassium per day via the urine. Express this weight of potassium in (a) milligrams (mg), (b) kilograms (kg), (c) pounds.

(a) Given: Mass of potassium = 2.0 g

Find: The equivalent number of milligrams

Relationship: 1.0 g = 1.0 × 10³ mg

$$\text{Solution: } 2.0 \text{ g} \times \boxed{\frac{1.0 \times 10^3 \text{ mg}}{1.0 \text{ g}}} = 2.0 \times 10^3 \text{ mg}$$

(b) *Given:* Mass of potassium = 2.0 g

*Find:* The equivalent number of kilograms

*Relationship:* 1.0 g = 1.0 × 10⁻³ kg

*Solution:* $2.0 \text{ g} \times \boxed{\dfrac{1.0 \times 10^{-3} \text{ kg}}{1.0 \text{ g}}} = 2.0 \times 10^{-3} \text{ kg}$

(c) *Given:* Mass of potassium = 2.0 g

*Find:* The equivalent number of pounds

*Relationship:* 454 g = 1.00 lb

*Solution:* $2.0 \text{ g} \times \boxed{\dfrac{1.00 \text{ lb}}{454 \text{ g}}} = 4.4 \times 10^{-3} \text{ lb}$

**EXAMPLE 4**  The average body output of water through the skin and lungs is about 900 ml per day. How much is this, expressed in (a) liters, (b) cubic centimeters (cm³), (c) quarts (qt), (d) cubic inches (in.³)?

(a) *Given:* Volume of water = 900 ml

*Find:* The equivalent number of liters

*Relationship:* 1000 ml = 1.00 liter

*Solution:* $900 \text{ ml} \times \boxed{\dfrac{1.00 \text{ liter}}{1.00 \times 10^3 \text{ ml}}} = 0.900 \text{ l} = 9.00 \times 10^{-1} \text{ liter}$

(b) *Given:* Volume of water = 900 ml

*Find:* The equivalent number of cm³

*Relationship:* 1.00 liter = 1.00 × 10³ ml and 1.00 liter = 1.00 × 10³ cm³. Therefore, 1.00 × 10³ ml = 1.00 × 10³ cm³ or 1.00 ml = 1.00 cm³.

*Solution:* $9.00 \times 10^2 \text{ ml} \times \boxed{\dfrac{1.00 \text{ cm}^3}{1.00 \text{ ml}}} = 9.00 \times 10^2 \text{ cm}^3 = 900 \text{ cm}^3$

(c) *Given:* Volume of water = 900 ml

*Find:* The equivalent number of quarts

*Relationships:* 1.00 × 10³ ml = 1.00 liter and 1.06 qt = 1.00 liter

*Solution:* $900 \text{ ml} \times \boxed{\dfrac{1.00 \text{ liter}}{1.00 \times 10^3 \text{ ml}}} \times \boxed{\dfrac{1.06 \text{ qt}}{1.00 \text{ liter}}}$
$= 0.954 \text{ qt} = 9.54 \times 10^{-1} \text{ qt}$

(d) *Given:* Volume of water = 900 ml

*Find:* The equivalent number of cubic inches

*Relationship:* 1000 ml = 1.000 liter, 1.00 liter = 61.0 in.³

*Solution:* $900 \text{ ml} \times \boxed{\dfrac{1.000 \text{ liter}}{1000 \text{ ml}}} \times \boxed{\dfrac{61.0 \text{ in.}^3}{1.00 \text{ liter}}} = 54.9 \text{ in.}^3$

## 2.9 Density

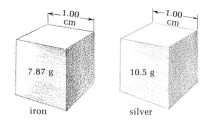

FIGURE 2.7
The mass of 1.00 cubic centimeter of iron and of 1.00 cubic centimeter of silver.

Which is heavier, a 1.00 cm³ block of iron or a 1.00 cm³ block of silver? Blocks of silver and iron, each of which has a volume of 1.00 cm³, are shown in Fig. 2.7. The mass of the iron block is indicated to be 7.87 g, the mass of the silver block is 10.5 g. Both of these blocks have the same volume, 1.00 cm³, but since the silver block has a greater mass, it must contain more matter than does the iron block. The ratio of the mass of a sample of a substance to the volume of that sample is called the *density* of the substance. Density is the quantity of matter, mass, contained in a unit volume. Equation 2.10 gives density $d$ as mass per unit volume.

$$\text{density} = \frac{\text{mass}}{\text{volume}} \quad \text{or} \quad d = \frac{m}{v} = m/v \qquad (2.10)$$

Mass is usually expressed in grams, while volume is expressed in cubic centimeters, milliliters, or liters. The units of density, then, are g/cm³, g/ml, or g/liter.

The mass of a sample of substance does not vary with the temperature. The volume of a sample of substance, however, will vary with temperature, since the particles of matter move farther apart as the temperature is increased. Since volume varies with the temperature, density also is subject to a variance. The density of water at three different temperatures has been experimentally determined to be:

(at 10°C) $d_{H_2O}$ = 0.9997 g/ml
(at 25°C) $d_{H_2O}$ = 0.9971 g/ml
(at 100°C) $d_{H_2O}$ = 0.9584 g/ml

Since density varies with temperature, the temperature at which a density value is determined should be reported along with the value.

The densities of gases are highly dependent on both the temperature and pressure. For this reason, both the temperature and the pressure *must* be reported along with *all* density values for gases.

From the data in Fig. 2.7, the density of iron is 7.87 g/cm³ and the density of silver is 10.5 g/cm³. The density of a substance is actually a physical unity. For silver, 10.5 g is the mass of the same sample which has a volume of 1.00 cm³; therefore, these two measurements refer to the same sample, so they represent a physical unity.

$$\frac{10.5 \text{ g silver}}{1.00 \text{ cm}^3 \text{ silver}} = \text{unity}$$

**TABLE 2.6**

*The densities of some common substances. These substances are classified as E, element; C, compound; or M, mixture.*

| Substance | Classification | Density(g/ml) | Physical state |
|---|---|---|---|
| Lithium | E | 0.53 | s |
| Sodium | E | 0.97 | s |
| Magnesium | E | 1.74 | s |
| Sulfur | E | 2.07 | s |
| Aluminum | E | 2.70 | s |
| Iron | E | 7.87 | s |
| Silver | E | 10.5 | s |
| Lead | E | 11.3 | s |
| Mercury | E | 13.5 | l |
| Tungsten | E | 19.3 | s |
| Osmium | E | 22.5 | s |
| Sand | M | 2.6 | s |
| Milk | M | 1.03 | l |
| Blood | M | 1.05 | l |
| Gasoline | M | 0.67 | l |
| Ethyl alcohol | C | 0.79 | l |
| Ice | C | 0.917 | s |
| Balsa wood | M | 0.11 | s |
| Ebony | M | 1.3 | s |
| Hydrogen | E | 0.090 | g |
| Nitrogen | E | 1.25 | g |
| Oxygen | E | 1.43 | g |
| Carbon dioxide | C | 1.96 | g |
| Xenon | E | 5.89 | g |

A density value does not reveal the actual mass or volume of a given sample of a substance. It only gives the relationship of the mass to the volume of a sample. The following examples illustrate some typical density problems.

EXAMPLE 5   A sample of metal from Table 2.6 weighs 37.4 g and occupies 1.94 ml. What is the density of the metal? What metal is it?

Given: Sample of metal, $m = 37.4$ g, $V = 1.94$ ml

Find: d

Relationship: density of substance $= \dfrac{\text{mass of sample}}{\text{volume of sample}}$  or  $d = \dfrac{m}{V}$

Solution: $d = \dfrac{37.4 \text{ g}}{1.94 \text{ ml}} = 19.3 \dfrac{\text{g}}{\text{ml}}$

Since tungsten has a density of 19.3 g/ml, (Table 2.6), the metal is assumed to be tungsten.

EXAMPLE 6   A graduated cylinder was filled up to the 16.7 ml mark with a liquid which had a density of 1.23 g/ml. What is the mass of the liquid?

Given: V of liquid sample = 16.7 ml
        d of liquid = 1.23 g/ml

Find: m of liquid sample

Relationship: $d = \dfrac{1.23 \text{ g}}{1.00 \text{ ml}}$ (a unity term)

Solution: 16.7 ml × $\boxed{?}$ = ? g

A unity term is needed which contains g in the numerator and ml in the denominator. The unity term 1.23 g/1.00 ml satisfies the requirements.

16.7 ml × $\boxed{\dfrac{1.23 \text{ g}}{1.00 \text{ ml}}}$ = 20.5 g

The unity term 1.23 g/1.00 ml converts the volume occupied by the sample to the mass of the sample. The mass of 16.7 ml of a liquid with a density of 1.23 g/ml is 20.5 g.

EXAMPLE 7   An irregular piece of lead weighs 94.8 g. What is the volume of this piece of lead? (*Hint:* obtain needed data from Table 2.6.)

Given: m of lead sample = 94.8 g
        d of lead = 11.3 g/ml (from Table 2.6)

Find: V of lead sample.

Relationship: $d =$ 11.3 g/1.00 ml (a unity term)

Solution: 94.8 g × $\boxed{?}$ = ? ml

A unity term is needed which contains ml in the numerator and g in the denominator. The unity term 1.00 ml/11.3 g satisfies these requirements.

94.8 g × $\boxed{\dfrac{1.00 \text{ ml}}{11.3 \text{ g}}}$ = 8.47 ml

The volume of 94.8 g of lead is 8.47 ml.

## 2.10   Significant figures in calculations

In science it is important to know how accurate a measurement is. The block in Fig. 2.1 is 1.89 inches long. It would be incorrect to say that it is 1.890 inches long, since the 9 is in doubt. The addition of a zero after the 9 is meaningless. The number 1.89 has three significant figures, 1, 8, and 9. The 9 is considered to be significant even though it is in doubt. Significant figures are important to scientists, since the result of a calculation can never be more significant than the measurements used in the calculation. The following example will demonstrate the proper use of significant figures.

EXAMPLE 8   The density of water at 25°C has been determined to five significant figures. The value is $d_{H_2O} = 0.99707$ g/ml. A quantity of water was measured at 25°C in a 25 ml-graduated cylinder. The volume reported was 17.4 ml since the graduated cylinder could be read to within 0.1 ml. What is the mass of the sample to the correct number of significant figures?

Given: 17.4 ml of $H_2O$

Find: m of sample of $H_2O$

Relationship: $d = \dfrac{m}{V} = \dfrac{0.99707 \text{ g}}{1.0000 \text{ ml}}$

Solution: $m = 17.4 \text{ ml} \times \boxed{?} = ? \text{ g}$

$m = 17.4 \text{ ml} \times \boxed{\dfrac{0.99707 \text{ g}}{1.0000 \text{ ml}}} = 17.3 \text{ g}$

Since one of the measurements used in this calculation has only three significant figures, the answer calculated using this measurement can have no more than three significant figures. The solution to a problem can contain no more significant figures than the *least* significant measurement used in obtaining the solution.

## Summary

The comparison of a property of unknown magnitude with a reference standard is the process which we call measurement. Chemists measure such properties as length, volume, mass, density, temperature, pressure, energy, and the time required for reactants to be changed into products. Every measurement is expressed in units. The chemist uses the units of the metric system for most of his measurements. The metric system is a decimal system where each unit of a class is related to every other unit of the class by powers of ten. The most important prefixes in the metric system for the student are *centi*, *milli*, and *kilo*.

Every measurement contains some uncertainty; that is, no measurement can ever be exact. The chance of obtaining an accurate measurement depends on the technique or skill of the person making the measurement and on the particular instrument being used. The uncertainty of a measurement should be reflected in only the last figure of the number recorded.

The unity terms which can be derived from measurements are useful in calculations. A unity term is a term in which the numerator is physically equivalent to the denominator. One such unity term is 2.54 cm/1.00 in. = 1, where 1 is unity. Multiplication of a physical entity by a unity term is just the same as multiplying the entity by unity.

The proper use of units is essential to the correct solution of problems. Units are subject to mathematical operations just as numbers are. The units which are obtained from the solution to a problem should match what is asked for in the problem. The student should use this fact to advantage since this helps him check his answer. An insight into the solution is also provided by the units asked for in the problem.

The proper use of significant figures should always be observed. Whenever measurements are used in calculations, the answer should contain only as many significant figures as the measurement which is known to the *least number of significant figures*.

In the next chapter, the sizes and masses of the submicroscopic particles of matter will be introduced. Throughout the entire text, the *factor-unit* method of calculation will be used in the solution of problems.

## Objectives

*By the end of this chapter, you should be able to do the following:*

1. Give examples of measurements a chemist might make. (Questions 14–16)

2. Use a ruler to make a length measurement. (Question 14)

3. Use a graduated cylinder to make a volume measurement. (Question 16)

4. Use a balance to make a mass measurement. (Question 15)

5. Convert one type of unit to another, using a unit factor. (Question 2)

6. Calculate the density of an object, given mass and volume data. (Questions 3, 4, 9, 16)

7. Calculate the mass of an object, given density and volume data. (Questions 6, 7, 11, 13)

8. Calculate the volume of an object, given density and mass data. (Questions 5, 8, 12, 17)

9. Demonstrate the correct use of significant figures. To test your knowledge and ability, complete the Questions that follow.

## Questions

1. Define and explain each of the following terms without referring to the text. After completing the list, check your definitions and explanations with those in the chapter.

*measurement; standard; significant figure; milli-; centi-; kilo-; length; centimeter; millimeter; kilometer; volume; graduated cylinder; liters; milliliters; cubic centimeters; mass; balance; triple beam balance; gram; milligram; kilogram; unity term; factor-unit method; density.*

2. Make the conversions indicated. Set up the solution using unity terms as factors.

Example 1: 4.83 in. = ? cm
$$4.83 \text{ in.} \times 2.54 \text{ cm/in.} = 12.0 \text{ cm}$$

Example 2: 73.6 cm = ? yd
$$73.6 \text{ cm} \times \frac{1.00 \text{ in.}}{2.54 \text{ cm}} \times \frac{1.00 \text{ yd}}{36.0 \text{ in.}} = 0.804 \text{ yd}$$

(a) 8.01 in. = ? cm
(b) 79.4 cm = ? in.
(c) 24.6 cm = ? m
(d) 0.943 cm = ? mm
(e) 0.00431 m = ? mm
(f) 1.50 ml = ? liter
(g) 4.00 qt = ? liter
(h) 160 lb = ? g
(i) 94 lb = ? kg
(j) 1.090 g = ? kg
(k) 0.153 g = ? mg
(l) 592 mg = ? lb
(m) 4.8 kg = ? g
(n) 10.1 cm$^3$ = ? liter
(o) 152 cm$^3$ = ? ml
(p) 2.02 liter = ? ml
(q) 2.02 liter = ? cm$^3$
(r) 15.2 g/liter = ? g/ml
(s) 76.2 g/liter = ? g/cm$^3$
(t) 1.89 g/ml = ? g/liter
(u) 24,000 in.$^2$ = ? ft$^2$
(v) 10,000 cm$^3$ = ? m$^3$
(w) 1.00 g/ml = ? lb/ft$^3$

**Note:** For Questions 3–12, refer to Table 2.6 for densities of substances.

3. Calculate the density of water in g/ml if 0.546 liter weigh 546 g.

4. What is the density of a coin that weighs 1.5 g and occupies 0.212 ml?

5. What is the volume of a student who weighs 160 lb and has a density of 0.95 g/ml?

6. What is the weight of 1.0 liter of gasoline?

7. If you donate 240 ml of blood, how much weight do you lose?

8. A student lost 4000 g. Assuming the loss was mostly water, what volume did the student lose?

9. The density of sand was determined by a student using the displacement of water to measure the volume of the sand. The water level in a graduated cylinder rose from 8.90 ml to 13.6 ml. The sand weighed 12.2 g. What was the density of the sand?

10. A quart of milk costs 35 cents. Calculate the cost per gram. *Hint:* How many ml in a quart? How many grams in a ml?

11. A student collected a flask full of dry nitrogen gas. What is the weight of the nitrogen if the flask has a volume of 500 ml?

12. What is the volume of a balloon if the gas inside weighs 3.01 g and has a density of 1.30 g/liter?

13. A student claimed to be able to carry a liter of mercury in each hand. What do you think about the prospects of performing this experiment? The density of mercury is 13.5 g/ml. Would it be easier for the student to carry two liters of water instead of two liters of mercury?

14. Measure the length of a given object—the length of this textbook for instance—using both the English system and the metric system. Use your data to derive a unity term relating feet and centimeters.

15. Examine the containers of several products around your home to determine the mass of the contents in both grams and pounds. Use your data to derive a relationship between ounces and grams. If a balance is available, determine the mass of several objects in grams. Use your derived unity term to convert their masses to ounces.

16. A student determined the mass of a small metal bar to be 27.6 g. He measured the volume of the bar by the displacement of water as was done for the pebbles. The volume he found was 15.9 ml. What is the density of this metal? If the metal is known to be either magnesium or aluminum which is it? If a balance is available, determine the density of both magnesium and aluminum and compare your value with the value determined by this student.

17. Use your weight expressed in grams (454 g = 1.00 lb) to determine the volume you occupy. Assume that the density of the human body is approximately 1.0 g/ml.

18. The term parts per million (ppm) represents the number of grams of one substance per million grams of a mixture. A certain tuna fish contains 0.50 ppm of mercury. How many grams of mercury are contained in the fish if it weighs 50,000 g? How many micrograms of mercury is this?

# Atoms and the mole

## 3.1 Introduction

Observations from experiments such as those discussed in Chapter 1 indicate that matter is particulate. What is the nature of these particles of matter? What is their structure? How are particles of gold similiar to and different from particles of iron? These are some of the questions considered in this chapter.

## 3.2 Distribution of the elements

In Chapter 1 we operationally defined an element as a substance which cannot be decomposed into simpler substances by ordinary means.

Ninety elements have been found to occur in nature. Another fifteen elements have been man-made. All material substances contain one or more of these 105 elements individually, as free elements, or combined in some way.

Are the elements distributed equally in nature? Examine Table 3.1. Although oxygen makes up 49.2 percent of the mass of the earth's crust, most of the oxygen is in compounds that make up the minerals and rocks on the earth's surface.

What do the elements iron, aluminum, copper, silver, gold, nickel, mercury, and zinc have in common? How can you recognize a metal? Freshly cut metal surfaces have a character-

■ EXERCISE  Using a chemistry reference text such as the Handbook of Chemistry and Physics, determine whether the following elements are metals or nonmetals: calcium, helium, fluorine, sodium, potassium, neon, manganese, and argon.

istic metallic luster. Copper and gold have a yellow color, but all the other common metals have a silvery color. Metals are good conductors of heat and electricity. Most metals can be hammered into shapes (*malleable*) and drawn into wires (*ductile*). At room temperature all metals except mercury are solids.

Some nonmetals like hydrogen, nitrogen, oxygen, and chlorine are gases at room temperature. Others like carbon, sulfur, iodine, and phosphorus are solids. One nonmetal, bromine, is a liquid. Nonmetals generally lack a metallic luster. Solid nonmetals are brittle, and most are poor conductors of heat and electricity. One exception, carbon in the form of graphite, is a good electrical conductor and is often used in batteries to improve electrical conduction. Approximately 25 percent of the 105 known elements are nonmetals. Seventy-five percent are metals. ■

TABLE 3.1

The distribution of the elements in the earth's crust.

| Element | Percentage of crust |
| --- | --- |
| Oxygen | 49.2 |
| Silicon | 25.7 |
| Aluminum | 7.5 |
| Iron | 4.7 |
| Calcium | 3.4 |
| Sodium | 2.6 |
| Total | 93.1 |
| Other elements | 6.9 |
|  | 100.0 |

## 3.3  Names and symbols of the elements

How have names been assigned to the elements? Some elements are named in honor of a country or area: americium, californium, francium, germanium, and polonium. Some are named after famous scientists: curium, einsteinium, fermium, and mendelevium. Some are named for their properties: oxygen (from the Greek, "acid former"), so named by Lavoisier because he thought all acids contained oxygen; hydrogen (from the Greek, "water former"), because when it burns in air, water is the product.

For convenience, a chemist uses a *symbol* for an element. Symbols for some of the elements already mentioned are: hydrogen, H; carbon, C; oxygen, O; iron, Fe; and lead, Pb. What is the relationship between the names of elements and their symbols? For some of the elements, the symbol is the first one or two letters of the name, for example, Ca for calcium, and O for oxygen. Some symbols, such as Pb for lead, are derived from other, more common, names for the element. For instance, lead, or plumbum, was used long ago for plumbing. The symbol for an element should suggest that element and its properties to you.

A complete listing of the 105 known elements and their symbols is inside the back cover. The elements which are most important in this course are included in Table 3.2.

How does the use of symbols help the chemist? Consider the

**TABLE 3.2**

*A list of important elements.*

| Element | Symbol | Element | Symbol |
|---|---|---|---|
| Aluminum | Al | Lead | Pb |
| Argon | Ar | Lithium | Li |
| Arsenic | As | Magnesium | Mg |
| Beryllium | Be | Manganese | Mn |
| Barium | Ba | Mercury | Hg |
| Boron | B | Neon | Ne |
| Bromine | Br | Nickel | Ni |
| Cadmium | Cd | Nitrogen | N |
| Calcium | Ca | Oxygen | O |
| Carbon | C | Phosphorus | P |
| Chlorine | Cl | Platinum | Pt |
| Chromium | Cr | Potassium | K |
| Cobalt | Co | Radium | Ra |
| Copper | Cu | Silicon | Si |
| Fluorine | F | Silver | Ag |
| Gold | Au | Sodium | Na |
| Helium | He | Sulfur | S |
| Hydrogen | H | Tin | Sn |
| Iodine | I | Uranium | U |
| Iron | Fe | Zinc | Zn |

chemical change that takes place when powdered zinc and powdered sulfur are heated together. Zinc is a metal. It does not appear to change when it is added to water or to the colorless, clear, liquid carbon disulfide. When mixed with hydrochloric acid it dissolves, producing an odorless, colorless gas and a clear, colorless solution. As shown in Fig. 3.1(a), a burning splint held over a test tube as the bubbles of hydrogen are forming produces a "pop."

Sulfur is a yellow solid. When it is added to a solution of hydrochloric acid it slowly sinks to the bottom of the container. No apparent change occurs. Sulfur also sinks in water without reacting. As shown in Fig. 3.1(b), small amounts of sulfur dissolve in the clear, colorless liquid carbon disulfide. Caution should be used in handling this liquid because it is poisonous and very flammable.

What happens when powdered zinc and powdered sulfur are mixed? At first nothing appears to happen. When the heterogeneous mixture is heated, however, it burns violently. A white solid remains. This solid does not dissolve in water or carbon disulfide. When added to hydrochloric acid it readily dissolves forming a clear, colorless solution. A foul smelling, rotten egg odor permeates the area (Fig. 3.1(c)). These are characteristic properties of zinc sulfide. Evidently, when heated, zinc and sulfur combine to form zinc sulfide. This may be expressed by a word equation.

$$\text{zinc(s)} + \text{sulfur(s)} \xrightarrow{\text{heating}} \text{zinc sulfide(s)} + \text{energy} \quad (3.1)$$

Instead of using a word equation to represent a reaction, the chemist substitutes symbols for the substances involved—Zn is written for zinc and S for sulfur. Since zinc sulfide is believed to contain atoms of zinc and sulfur combined in a one-to-one ratio, it is symbolized by the formula ZnS. A *formula* represents a compound. A formula includes not only the symbols of the elements that make up that compound, but also the number of atoms of each element in the compound. Rewriting Eq. 3.1, using symbols and formulas, gives:

$$\text{Zn(s)} + \text{S(s)} \xrightarrow{\text{heating}} \text{ZnS(s)} + \text{energy} \quad (3.2)$$

■ EXERCISE  For each of the following symbols and formulas, (1) write the name of the element for each symbol and (2) write the name of the elements appearing in each formula. CO and $CO_2$, Po and $P_2O_5$, Cs and $CS_2$, NO and NO.

That zinc atoms combine with sulfur atoms, releasing energy and forming a new substance, is all indicated by Eq. 3.2. Symbols, formulas, and equations communicate ideas in concise form. They are part of the language of the chemist. ■

## 3.4  Particulate theory

In Chapter 1 we developed a particulate model to explain some of our observations. At first, we started with a hypothesis or "educated guess" about one set of experimental observations, and then developed a model to explain observations from several experiments. Since it does explain a number of observations, we have more faith in this particulate model. A model which correlates and explains the available experimental observations is often called a *theory*.

How can we extend the particulate theory? How are atoms of one element similar to those of another and how do they differ from one another?

## 3.5  Atoms

The idea that matter consists of particles was proposed at least as far back as 400 B.C. by Greeks such as Democritus. He proposed, among other things, that particles of water must be

| TABLE 3.3 | Sample | Grams of carbon | Grams of oxygen |
|---|---|---|---|
| The weight ratios of carbon to oxygen in samples of gases that contain only carbon and oxygen. | 1 | 1.00 | 2.66 |
| | 2 | 1.00 | 1.33 |
| | 3 | 1.00 | 1.33 |
| | 4 | 1.00 | 2.66 |
| | 5 | 1.00 | 2.66 |

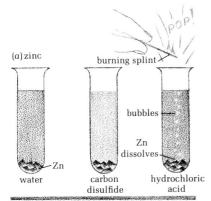

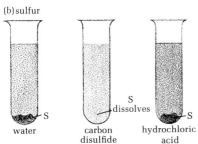

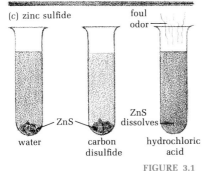

FIGURE 3.1

The behavior of (a) zinc, (b) sulfur, and (c) zinc sulfide in water, carbon disulfide, and hydrochloric acid.

round and smooth because they slip over one another so easily. Particles of iron were thought to be rough and jagged, supposedly explaining why they stuck to one another.

In 1803, John Dalton proposed the first comprehensive particulate model. Let us examine some hypothetical data similar to that which Dalton tried to explain with his model.

Samples of gases that contain only carbon and oxygen were collected from various sites and analyzed. The results are shown in Table 3.3. The weight of oxygen per one gram of carbon found in each sample is listed. Note that in samples 1, 4, and 5, 2.66 grams of oxygen were found for every one gram of carbon. These samples contain the substance known as carbon dioxide. In samples 2 and 3, 1.33 grams of oxygen were found for each gram of carbon. These samples contain the substance carbon monoxide. One compound contains twice as much oxygen as the other. Dalton proposed the following model.

1. Each element consists of tiny, indestructible particles called *atoms*.
2. All atoms of an element are identical in mass, size, and behavior.
3. The differences in properties of elements are a result of the differences in their atoms.
4. In a compound, atoms of two or more different elements are combined.
5. A chemical reaction involves the union, disunion, or rearrangement of atoms.

According to Dalton both carbon monoxide and carbon dioxide consist of atoms of carbon and oxygen bound together. Dalton assumed that one compound, the one with less oxygen, consisted of *one atom of oxygen* combined with *one atom of carbon*. He assumed that the compound with twice as much oxygen contained *two atoms of oxygen* for each *one atom of carbon*. In writing the formulas, the line or lines between the symbols represent the forces which hold the atom together; these are called bonds.

Carbon monoxide:    C=O        CO
Carbon dioxide:     O=C=O      $CO_2$

The ratio of carbon and oxygen atoms that Dalton assumed for carbon monoxide and carbon dioxide are still accepted as correct today. The ratio that he proposed for water was one atom of hydrogen for one atom of oxygen, HO. Logical arguments put forth by Avogadro in 1812, however, and subsequently accepted, indicated that water contains two atoms of hydrogen for one atom of oxygen, $H_2O$.

Dalton's model has been modified somewhat. With the discovery of radioactivity at the start of the twentieth century it became apparent that atoms were not indestructible. Some atoms decompose by themselves and are said to be radioactive. Other atoms can be made to decompose by bombarding them with high-energy particles. These high-energy changes are out of the realm of ordinary chemical change. During an ordinary chemical change we can still consider, as Dalton did, that atoms are indestructible particles.

Scientists in the early twentieth century also found that all atoms of an element are not exactly alike. As we shall see a little later, they may vary in mass.

## 3.6 Formulas

In the preceding section we introduced the use of formulas to represent compounds. The formula for carbon monoxide, which has one atom of carbon combined with one atom of oxygen, is CO. The formula for carbon dioxide, which has one atom of carbon combined with two atoms of oxygen, is $CO_2$. The small subscript "$_2$" written after the symbol for oxygen means two oxygen atoms.

Listed in Table 3.4 are the names, diagrams, and formulas of three common compounds. Look at the formula for ammonia, $NH_3$. It tells us (1) what elements are combined together: nitrogen and hydrogen; and (2) what the atom ratio is: three atoms of hydrogen for one nitrogen atom.

In ethyl alcohol $CH_3CH_2OH$ there are two carbon atoms, six hydrogen atoms, and one oxygen atom. In calcium hydroxide $Ca(OH)_2$ which is used in mortar for building, the subscript "$_2$" applies to each symbol within the brackets ( ). There is one calcium atom to two oxygen atoms to two hydrogen atoms.

The symbol for hydrogen is H. Chemists believe that in a sample of hydrogen gas, hydrogen atoms are bound together into pairs as shown in Fig. 3.2. A pair of hydrogen atoms is called a *hydrogen molecule* and the molecule has the formula

FIGURE 3.2
*In a sample of hydrogen gas, atoms are combined into pairs.*

**TABLE 3.4**

*The names, diagrams, and formulas of three common compounds.*

| Name | Diagram | Formula |
|---|---|---|
| Ammonia | H—N—H<br>       &#124;<br>       H | $NH_3$ |
| Nitric acid | H—O—N—O<br>          &#124;<br>         O | $HNO_3$ |
| Ethyl alcohol |     H  H<br>    &#124;  &#124;<br>H—C—C—OH<br>    &#124;  &#124;<br>    H  H | $CH_3CH_2OH$ |

$H_2$. Hydrogen is thus said to be *diatomic* or to consist of diatomic molecules. The prefix "di" is derived from the Greek and means two.

Other elements consist of diatomic molecules. The most important are: nitrogen, $N_2$; oxygen, $O_2$; fluorine, $F_2$; chlorine, $Cl_2$; bromine, $Br_2$; and iodine, $I_2$. These elements are necessarily diatomic only as pure elements or in mixtures. For example, chlorine atoms exist in sodium chloride in a one-to-one ratio with sodium atoms. The formula of sodium chloride is NaCl. ■

■ **EXERCISE** *What elements are present in the following compounds? What are the atom ratios?* $NaCl$, $MgCl_2$, $Mg(OH)_2$, $Al(OH)_3$, $H_2SO_4$, $Fe_3(PO_4)_2$.

## 3.7 Subatomic particles

Dalton's atomic model included the idea that atoms are indestructible and cannot be broken down into smaller units. This idea was consistent with all observed facts until the discovery of the *subatomic particles* of an atom. Many subatomic particles—mesons, neutrinos, positrons, antiprotons, electrons, protons, and neutrons—have been "discovered." Only three particles are necessary to explain most chemical reactions: the electron, the proton, and the neutron.

In Chapter 1 we noted that the concept of electrons and protons had been proposed in order to explain electrical interactions in chemistry. Electrons and protons are subatomic particles that possess a characteristic electrical charge as one of their properties. An electron, symbolized as $e^-$, possesses a 1— charge; a proton, symbolized as $p^+$, possesses a 1+ charge. The unit of charge is referred to simply as an electronic charge. A neutron, symbolized by n, possesses no net charge; it is a neutral particle.

| TABLE 3.5 | Symbol | Particle | Charge | Mass (grams) | Mass (amu) |
|---|---|---|---|---|---|
| The symbols, assigned charges, and masses of the three most important fundamental particles. | $e^-$ | electron | 1− | $9.107 \times 10^{-28}$ | 0.0005486 |
| | $p^+$ | proton | 1+ | $1.672 \times 10^{-24}$ | $1.007 \approx 1.0$ |
| | n | neutron | 0 | $1.675 \times 10^{-24}$ | $1.009 \approx 1.0$ |

Protons, neutrons, and electrons are so small that it is inconvenient to express their masses in grams. When expressing the masses of subatomic particles, atoms, or molecules, chemists use *atomic mass units* (amu). The standard for this scale is discussed in Section 3.11. The mass of a proton has been found to be 1.0073 amu. The mass of a neutron is 1.0087 amu. Since the masses of both the proton and the neutron are close to unity, we say that their masses are approximately 1 amu.

Experiments performed by J. J. Thomson in 1897 and Millikan in 1909 indicate that the mass of an electron is much less than that of the proton or neutron. At the same time, the charges on the electron and proton are equal, but of opposite character. The mass of an electron is 0.000549 amu or about 1/1840 as much as the mass of a proton or neutron. The masses, charges, and symbols of all three fundamental particles are shown in Table 3.5.

## 3.8 The nuclear atom

What observations have been made which might tell us what an atom looks like? In the absence of a battery, generator, or other source of electricity, most samples of elements were found to be electrically neutral. This led scientists to conclude that the atoms that make up these neutral samples are themselves neutral. A neutral atom contains equal numbers of electrons and protons.

At the beginning of this century, many scientists were studying the effect of fast moving, charged particles on matter. In 1912, E. Rutherford studied the behavior of alpha particles when they struck a gold foil. The results of his work led to a nuclear model for the atom.

Figure 3.3(a) shows Rutherford's experiment. A radioactive material was used as a source of high-energy alpha particles. Alpha particles contain two neutrons and two protons; they possess a 2+ charge. A slit in a lead block allowed a narrow beam of alpha particles to strike a piece of gold foil. Behind the gold foil was a zinc sulfide screen. One property of zinc sulfide

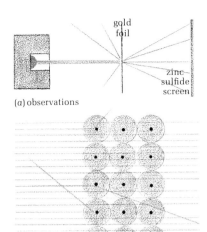

**FIGURE 3.3**
Rutherford's experiment with a gold foil. (a) Rapidly moving, positively charged alpha particles were shot at gold foil. Most alpha particles were not deflected at all, a few particles were deflected slightly, and fewer particles were deflected back. (b) Rutherford's model of an atom to explain the above observations.

is that it emits a ray of light when struck by an alpha particle. The zinc sulfide screen could be moved, which helped to determine if any alpha particles were being deflected.

Some alpha particles were deflected as shown in Fig. 3.3(a). More surprising to Rutherford, some alpha particles were deflected back, that is they were reflected. He compared this to shooting a 15-inch shell at a piece of tissue paper and having it come back and hit you.

What type of charge would repel the positive charge of an alpha particle? Rutherford proposed that the positive alpha particles had been deflected by other positive particles in the atom. In order to deflect a high-energy alpha particle, the positive particle in the atom must have a very high mass. This positive part of the atom he called the nucleus. According to Rutherford's calculations *over 99 percent of the mass of an atom must reside in the nucleus.*

Why did most of the alpha particles pass through the gold foil without being deflected? Rutherford concluded that there must be nothing in their path to cause deflection; *the atoms consist largely of empty space.* According to Rutherford's calculations, the diameter of the nucleus is 100,000 times smaller than the diameter of the whole atom. The rest of the space of the atom contains the negatively charged electrons (Fig. 3.3(b)).

## 3.9 The modern atomic model

The essentials of Rutherford's nuclear atom are still accepted today. The model of the atom is summarized as:

1. An atom consists of two sections, a nucleus and a region of space outside the nucleus.
2. The nucleus contains protons and neutrons.
3. The region of space outside the nucleus contains electrons.
4. The volume of the nucleus is minute compared to the volume of the atom.
5. Most of the mass of the atom resides in the minute nucleus.

The number of positive charges or protons in the nucleus is called the *atomic number*. Atoms of each of the 105 elements have a different atomic number. The number of protons in the nucleus ranges from 1 for hydrogen to 105 for the latest "manmade" element, hahnium. (See the table inside the back cover of this book.)

When the symbol for an element is written, its atomic number is sometimes included, and written at the lower

| TABLE 3.6 | Element | Symbol | Atomic number | Number of protons | Number of electrons |
|---|---|---|---|---|---|
| Atomic numbers of six elements. | Hydrogen | $_1$H | 1 | 1 | 1 |
| | Helium | $_2$He | 2 | 2 | 2 |
| | Lithium | $_3$Li | 3 | 3 | 3 |
| | Iron | $_{26}$Fe | 26 | 26 | 26 |
| | Gold | $_{79}$Au | 79 | 79 | 79 |
| | Lawrencium | $_{103}$Lw | 103 | 103 | 103 |

■ EXERCISE  How many protons and how many electrons are in a neutral atom of each of the following elements? $_7$N, $_{12}$Mg, $_{53}$I.

left hand corner of the symbol. Thus hydrogen, with an atomic number of 1, can be written as $_1$H. Lawrencium, atomic number 103, can be written as $_{103}$Lr. Some other examples are listed in Table 3.6. ■

## 3.10 Isotopes

Are all atoms in an element identical, as proposed by Dalton? All neutral atoms of an element must have the same number of protons and electrons. An atom must contain one positive charge in the nucleus to be a hydrogen atom. If it contains any other number of positive charges, then it is some element other than hydrogen. Accurate mass determinations using an instrument called a mass spectrometer have shown that there are hydrogen atoms with three different masses. If the number of protons cannot vary, what causes the variance in mass? A variable number of neutrons in the nucleus would account for the different masses (Fig. 3.4). The atoms of an element which differ in mass due to different numbers of neutrons in the nucleus are called *isotopes*.

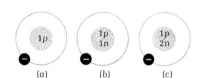

FIGURE 3.4
(a) Hydrogen-1; (b) hydrogen-2; (c) hydrogen-3.

The three isotopes of hydrogen shown in Fig. 3.4 are hydrogen-1, hydrogen-2, and hydrogen-3. Often the hydrogen-2 and hydrogen-3 isotopes are called deuterium and tritium, respectively. The sum of the number of protons and neutrons in an atom is called the *mass number*. Hydrogen-1 has one proton and no neutrons; its mass number is one. Hydrogen-2 has one proton and one neutron; its mass number is two. Hydrogen-3 has one proton and two neutrons; its mass number is three.

$$\text{mass number} = \text{protons} + \text{neutrons} \tag{3.3}$$

To represent an isotope of an element, the symbol of the element is written with the atomic number on the lower left side and the mass number on the upper left side. For example, the symbol for hydrogen-3 is:

mass number ⟶ $^{3}_{1}H$ ⟵ atomic number

The "1" indicates one *proton* in the nucleus. The "3" indicates three *particles* in the nucleus.

Many elements have two or more isotopes. Some of these isotopes are unstable; they disintegrate to yield other elements. These changes must involve changes in the nucleus. Isotopes which decompose to yield other elements are known as *radioactive isotopes*.

EXAMPLE 1   One isotope of uranium is $^{235}_{92}U$. How many protons, electrons, and neutrons are there in this isotope?

For $^{235}_{92}U$:

Atomic number $= p^{+} = 92$

Mass number $= p^{+} + n = 235$

$n = 235 - p^{+} = 235 - 92 = 143$

In a neutral atom, $p^{+} = e^{-} = 92$

■ EXERCISE   Give the number of electrons, protons, and neutrons for each of the following isotopes: $^{12}_{6}C$, $^{14}_{6}C$, $^{16}_{8}O$, $^{17}_{8}O$, $^{35}_{17}Cl$, $^{37}_{17}Cl$.

## 3.11   Atomic mass units

The approximate mass in amu of a particular isotope can be determined by adding the number of protons and neutrons in that isotope.

What is the reference standard when the unit of mass is the atomic mass unit? Scientists have agreed that the international reference standard for atomic mass units will be the carbon-12 isotope, $^{12}_{6}C$. This isotope contains six protons and six neutrons. It has been assigned a mass of exactly 12 atomic mass units. The masses of all other isotopes are related to carbon-12 as the reference standard. This means that a hydrogen atom is about 1/12 the mass of a carbon atom.

## 3.12   Atomic weight

A sample of naturally occurring hydrogen consists of the hydrogen-1 and hydrogen-2 isotopes. Table 3.7 shows that about 99.984 percent of naturally occurring hydrogen is hydrogen-1. The other 0.016 percent of the element is hydrogen-2.

**TABLE 3.7**

*The average weight of the two naturally occurring chlorine isotopes and two naturally occurring hydrogen isotopes.*

| chlorine-35 | 75.4% | |
| chlorine-37 | 24.6% | Average value is 35.453 amu |
| hydrogen-1 | 99.984% | |
| hydrogen-2 | 0.016% | Average value is 1.00797 amu |

The atomic weight of hydrogen is the average weight of both isotopes, taking into account the fact that there is 99.984 percent of one isotope and only 0.016 percent of the other isotope.

How is the atomic weight of chlorine listed in Table 3.7 obtained? Chlorine-35 makes up 75.4 percent of the naturally occuring element and chlorine-37 makes up the remaining 24.6 percent. The atomic weight of chlorine is the average mass of chlorine-35 and chlorine-37, taking into account the fact that there is 75.4 percent of one isotope and 24.6 percent of the other isotope. In this case:

(75.4% of chlorine-35) + (24.6% of chlorine-37)
= (0.754 × 35.0 amu) + (0.246 × 37.0 amu) = 35.5 amu

The answer, 35.5 amu, agrees with the value listed in Table 3.7 to three significant figures. ■

■ EXERCISE  *Calculate the atomic weight of boron. Naturally occurring samples of this element consist of two isotopes, 80.4 percent boron-11 and 19.6 percent boron-10. Compare your answer to the atomic weight of boron listed inside the back cover of this book.*

## 3.13 Avogadro's number

We have already seen that hydrogen has an atomic weight which is about 1/12 of the atomic weight of carbon. The atomic weight of naturally occurring hydrogen is 1.0, that of naturally occurring carbon is 12.0. If we take the same number of carbon atoms and hydrogen atoms, then the carbon atoms will have a mass which is twelve times the mass of the hydrogen atoms. For example, what is the mass of 1000 carbon atoms? What is the mass of 1000 hydrogen atoms? The mass of 1000 carbon atoms is 12,000 amu. The mass of 1000 hydrogen atoms is 1000 amu. The ratio of carbon mass to hydrogen mass is 12 to 1.

This suggests that there should be some number of carbon atoms, such that if we take this number, we will have *12.0 grams of carbon.* If we take exactly the same number of hydrogen atoms, we should have *1.0 gram of hydrogen.* The number of atoms in 12.0 grams of carbon and in 1.0 gram of hydrogen has been determined and is known as Avogadro's number. This number, which was named in honor of Count

Amedeo Avogadro, is $6.02 \times 10^{23}$. The atomic weight of hydrogen, expressed in grams, contains $6.02 \times 10^{23}$ hydrogen atoms. The atomic weight of carbon, expressed in grams, contains $6.02 \times 10^{23}$ carbon atoms.

1.0 gram hydrogen = $6.02 \times 10^{23}$ hydrogen atoms
12.0 grams carbon = $6.02 \times 10^{23}$ carbon atoms

The atomic weight of any element expressed in grams contains $6.02 \times 10^{23}$ atoms of that element.

## 3.14 The mole

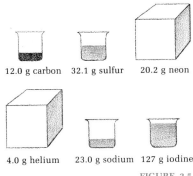

FIGURE 3.5
*What do each of these samples have in common?*

Examine the samples of elements shown in Fig. 3.5: a 12.0 gram sample of carbon, a 4.0 gram sample of helium, a 23.0 gram sample of sodium, a 32.1 gram sample of sulfur, a 20.2 gram sample of neon, and a 127.0 gram sample of iodine. What do each of these samples have in common? Notice that the mass of each sample is the atomic weight of the element expressed in grams. Therefore each sample contains $6.02 \times 10^{23}$ atoms of that particular element. This number of particles, $6.02 \times 10^{23}$, is defined as a *mole*. Each sample in Fig. 3.5 contains one mole of atoms of that element (Table 3.8).

A mole of atoms is Avogadro's number of atoms, or $6.02 \times 10^{23}$ atoms. We can also speak of a mole of electrons, a mole of molecules, a mole of dollars, or even a mole of people. In order to get an idea of the size of a mole, let us calculate the number of moles of people who live on the earth's surface. The current world population is approximately 3 billion or $3 \times 10^{9}$ people. If 1 mole of people = $6.02 \times 10^{23}$ people, then:

$$3 \times 10^{9} \text{ people} \times \frac{1 \text{ mole people}}{6 \times 10^{23} \text{ people}} = 5 \times 10^{-15} \text{ mole}$$

This is a rather tiny fraction of a mole—0.000000000000005 mole. ■

■ **EXERCISE** *One drop of sea water contains fifty billion gold atoms. Calculate the number of moles of gold in one drop of sea water.*

What is the purpose of the mole concept? It is not possible in the ordinary laboratory to work with only five or six atoms. The mole concept allows us to calculate the number of particles present in macroscopic amounts of matter. For example, the atomic weight of silver is 108. We know that if we weigh 108 grams of silver we will have $6.02 \times 10^{23}$ silver atoms. If we weigh any other quantity of silver, we can still determine the number of atoms in that particular quantity using the mole concept. Examine the following problem.

TABLE 3.8

Each sample in Fig. 3.5 contains 1.00 mole of atoms.

| Element | Mass | Atoms | Moles of atoms |
|---|---|---|---|
| Carbon | 12.0 g | $6.02 \times 10^{23}$ | 1.00 |
| Helium | 4.0 g | $6.02 \times 10^{23}$ | 1.00 |
| Sodium | 23.0 g | $6.02 \times 10^{23}$ | 1.00 |
| Sulfur | 32.1 g | $6.02 \times 10^{23}$ | 1.00 |
| Neon | 20.2 g | $6.02 \times 10^{23}$ | 1.00 |
| Iodine | 127.0 g | $6.02 \times 10^{23}$ | 1.00 |

EXAMPLE 2  How many atoms are there in 5.10 grams of silver?

*Given:* Weight Ag = 5.10 g

*Find:* The number Ag atoms

*Relationship:* 1.00 mole of silver = 108 g silver = $6.02 \times 10^{23}$ silver atoms

*Solution:* 5.10 g Ag × Unity Term = ? atoms Ag

We know that 108 g Ag = $6.02 \times 10^{23}$ atoms Ag; therefore putting atoms Ag in the numerator and grams Ag in the denominator gives us the unity term: $\dfrac{6.02 \times 10^{23} \text{ atoms Ag}}{108 \text{ g Ag}} = 1$

$$5.10 \text{ g Ag} \times \frac{6.02 \times 10^{23} \text{ atoms Ag}}{108 \text{ g Ag}} = 2.85 \times 10^{22} \text{ atoms Ag}$$

■ EXERCISE  *Which of the following samples contains a greater number of atoms? How many atoms are in each sample? 5.71 g He, 7.71 g S; 9.71 g Fe.*

## 3.15 Molecular weight and moles

When a stream of burning hydrogen gas is plunged into a bottle of pale green chlorine gas, the hydrogen continues to burn. As the flame burns, the pale green color of chlorine disappears. A new gas forms which is soluble in water and which forms white fumes when mixed with ammonia gas. This new gas is hydrogen chloride.

hydrogen(g) + chlorine(g) ⟶
        hydrogen chloride(g) + energy   (3.4)

Samples of hydrogen, chlorine, and hydrogen chloride consist of diatomic molecules: $H_2$, $Cl_2$, and $HCl$. How many particles are in a mole of such molecules? A mole of hydrogen molecules is $6.02 \times 10^{23}$ hydrogen molecules. A mole of chlorine molecules is $6.02 \times 10^{23}$ chlorine molecules. A mole of hydrogen chloride is $6.02 \times 10^{23}$ hydrogen chloride molecules. Note that there are two hydrogen atoms per $H_2$ molecule, two

**TABLE 3.9**

The relationship between a mole of hydrogen atoms and a mole of hydrogen molecules.

| Substance | Formula | Weight of a particle | Particles in a mole | Weight of a mole | Atoms in a mole |
|---|---|---|---|---|---|
| Hydrogen atoms | H | 1.0008 amu | $6.02 \times 10^{23}$ atoms | 1.008 g | $6.02 \times 10^{23}$ |
| Hydrogen molecules | $H_2$ | 2.0016 amu | $6.02 \times 10^{23}$ molecules | 2.016 g | $12.04 \times 10^{23}$ |

chlorine atoms per $Cl_2$ molecule, and one hydrogen atom and one chlorine atom per HCl molecule.

What is the weight of a mole of hydrogen molecules? Each hydrogen molecule contains two hydrogen atoms with a mass of 1.008 amu.

$$2 \times \text{hydrogen mass} = 2 \times 1.008 \text{ amu} = 2.016 \text{ amu}$$

The mass of a hydrogen molecule is 2.016 amu. The mass of any molecule is called the *molecular weight* of that molecule. Just as the weight of a mole of atoms of an element is the atomic weight of that element in grams, the weight of *a mole of molecules of a substance is the molecular weight of the substance in grams*. A mole of hydrogen molecules weighs 2.016 grams (Table 3.9).

What is the weight of a mole of hydrogen chloride? The molecular weight of hydrogen chloride can be calculated by adding the weight of each atom.

hydrogen atom mass = 1.0 amu
chlorine atom mass = 35.5 amu
---
molecular weight of HCl = 36.5 amu
1 mole HCl = 36.5 g

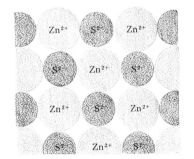

FIGURE 3.6
A piece of zinc sulfide consists of an equal number of positive zinc ions and negative sulfide ions.

Some compounds do not consist of actual molecules. For example, a sample of the compound zinc sulfide (ZnS) is believed to consist of many billions of positive zinc ions combined with an equal number of negative sulfide ions (Fig. 3.6). Recall that ions are atoms bearing a charge. There is no discrete ZnS molecule. In discussing substances that contain ions it is common to talk of a formula unit. A *formula unit* is the simplest ratio of ions in the compound. For zinc sulfide, the formula unit is ZnS. The weight of a formula unit is called the *formula weight*. For zinc sulfide, the formula weight is the weight of a zinc atom plus the weight of a sulfur atom or 97.5 amu. One mole of ZnS weighs 97.5 grams.

3.15 Molecular weight and moles

■ EXERCISE  *Calculate the weight of a mole of each of the following substances:* (a) *The chief ore of lead, galena, or lead (II) sulfide*, $PbS$; (b) *Sucrose or sugar*, $C_{12}H_{22}O_{11}$; (c) *Cholesterol*, $C_{27}H_{45}OH$.

EXAMPLE 3  "Milk of magnesia" is a suspension of magnesium hydroxide $Mg(OH)_2$ in water. Calculate the weight of 1.00 mole of magnesium hydroxide.

*Given:* 1.00 mole $Mg(OH)_2$

*Find:* Weight

*Relationship:* The atomic weight of each element is: $Mg = 24.3$ amu, $O = 16.0$ amu, $H = 1.0$ amu

*Solution:* The formula $Mg(OH)_2$ indicates 1 atom of Mg, two atoms of O, and two atoms of H.

$$\begin{aligned} Mg &= 24.3 = 24.3 \text{ amu} \\ 2 \times O &= 2 \times 16.0 = 32.0 \text{ amu} \\ 2 \times H &= 2 \times 1.0 = \underline{\phantom{0}2.0 \text{ amu}} \\ \text{formula weight of } Mg(OH)_2 &= 58.3 \text{ amu} \end{aligned}$$

One mole of $Mg(OH)_2$ will weigh 58.3 grams.  ■

## 3.16  Writing balanced equations

A photographer weighed an unused photographic flash bulb and found it had a mass of 24.01 grams. After using the bulb (Fig. 3.7) and allowing it to cool, he reweighed it. Again the flash bulb had a mass of 24.01 grams. A flash bulb contains magnesium and oxygen sealed in a glass envelope. During the flash, these two substances react to form magnesium oxide, but as long as the glass envelope remains sealed no matter can enter or leave the bulb. It is a closed system. Evidently, even though two substances changed into a third substance, there was no noticeable mass change:

mass of the reactants = mass of the products

**Law of conservation of mass**

Many thousands of chemical reactions have been examined. In each case the sum of the masses of the reactants equals the sum of the masses of the products within the sensitivity of normal measuring instruments. This is known as the *Law of the Conservation of Mass.*

How can the Law of the Conservation of Mass be explained? John Dalton proposed that during a chemical change atoms are not created or destroyed but only rearranged. Thus the mass of these atoms will not change during the course of a reaction.

magnesium(s) + oxygen(g) $\rightleftharpoons$
    magnesium oxide(s) + energy

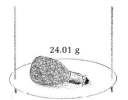

FIGURE 3.7
*A flash bulb was weighed before and after it was set off.*

In writing the formula equation for this reaction we must do two things: (1) Write the correct formula for each substance, and (2) maintain the same number of atoms of each element on each side of the equation. The formula for magnesium is Mg; the formula for oxygen, a diatomic gas, is $O_2$; the formula for magnesium oxide is MgO.

$$Mg(s) + O_2(g) \rightleftharpoons MgO(s) + energy$$

This equation satisfies the first requirement in that correct formulas are written for each substance. Are there equal numbers of atoms of each element indicated in the reactants and products? The symbol Mg represents one atom of free magnesium metal as a reactant. The symbol Mg in MgO represents one atom of combined magnesium in the products. The magnesium is balanced. Are there equal numbers of oxygen atoms in the reactants and products? The formula $O_2$ indicates two atoms of oxygen in the reactants. The O in MgO indicates only one oxygen atom combined in the product. The oxygen is not balanced. We can obtain two oxygen atoms on the right by indicating two formula units of magnesium oxide: 2MgO. The coefficient "2" means two units of magnesium oxide.

$$Mg(s) + O_2(g) \rightleftharpoons 2MgO(s) + energy$$

This balances the oxygen atoms. However, now we have one magnesium atom as a reactant and two in the products. We can balance the magnesium by placing a coefficient "2" in front of it.

$$2Mg(s) + O_2(g) \rightleftharpoons 2MgO(s) + energy$$

Notice that in balancing an equation, correct formulas, once written, are never changed. If the formula $H_2O$ is changed by placing a subscript 2 after O, the formula $H_2O_2$ is obtained. This formula represents the substance hydrogen peroxide, not water.

The magnesium oxide equation states that *two* atoms of magnesium react with *one* molecule of oxygen to give *two* formula units of magnesium oxide. An equation represents the *ratio* in which atoms, molecules, or formula units of a substance react.

What would we have if the equation:

two magnesium atoms + one oxygen molecule $\rightleftharpoons$
$$\text{two magnesium oxide formula units} \quad (3.6)$$

were multiplied by $6.02 \times 10^{23}$? We would have:

3.16 Writing balanced equations

$$12.04 \times 10^{23} \text{ magnesium atoms}$$
$$+ 6.02 \times 10^{23} \text{ oxygen molecules} \rightleftharpoons$$
$$12.04 \times 10^{23} \text{ magnesium oxide formula units} \quad (3.7)$$

Since $6.02 \times 10^{23}$ of a substance is a mole of a substance, we can write:

$$2.00 \text{ moles magnesium atoms}$$
$$+ 1.00 \text{ mole oxygen molecules} \rightleftharpoons$$
$$2.00 \text{ moles magnesium oxide formula units} \quad (3.8)$$

Equation 3.5 can also be interpreted as giving a *ratio* of moles of reactants and products. Since a mole of a substance has a definite weight, the weight ratios of reactants and products can be calculated using a balanced equation. This will be done in Chapter 6.

The following examples illustrate a method for balancing some simple equations.

**EXAMPLE 4** Write the equation for the reaction between hydrogen and chlorine, making sure that it is properly balanced.

*Word equation:* hydrogen(g) + chlorine(g) $\rightleftharpoons$ hydrogen chloride(g) + energy

*Formulas:* $H_2(g) + Cl_2(g) \rightleftharpoons HCl(g) +$ energy

*Shown in reactants:* 2 atoms of hydrogen in $H_2$ and 2 atoms of chlorine in $Cl_2$

*Shown in products:* 1 atom of hydrogen in HCl and 1 atom of chlorine in HCl

*Solution:* Placing a 2 in front of HCl gives two atoms of hydrogen and two atoms of chlorine in the products.

$$H_2(g) + Cl_2(g) \rightleftharpoons 2HCl(g) + \text{energy}$$

According to this equation, one mole of hydrogen molecules will react with one mole of chlorine molecules to produce two moles of hydrogen chloride.

**EXAMPLE 5** Balance the following equation.

*Word equation:* aluminum hydroxide(aq) + sulfuric acid (aq) $\rightleftharpoons$ aluminum sulfate(aq) + water(l)

*Formulas:* $Al(OH)_3(aq) + H_2SO_4 \rightleftharpoons Al_2(SO_4)_3(aq) + H_2O(l) +$ energy

This equation appears to be a bit more complicated. Notice, however, that there is a group of atoms ($SO_4$-the sulfate ion) that appears to act as a unit. It appears in both the reactants and products. Let us try to balance this group first.

*In the products:* 3($SO_4$) groups

*In the reactants:* 1($SO_4$) group

Placing a 3 in front of $H_2SO_4$ balances the ($SO_4$) group.

■ EXERCISE  Balance the following equations and state the mole ratios of each reactant to each product.
(a) hydrogen + oxygen ⇌ water

$$H_2(g) + O_2(g) \rightleftharpoons H_2O(l) + energy$$

(b) Hydrogen + nitrogen ⇌ ammonia

$$H_2(g) + N_2(g) \rightleftharpoons NH_3(g) + energy$$

(c) Magnesium oxide + water ⇌ magnesium hydroxide

$$MgO(s) + H_2O(l) \rightleftharpoons Mg(OH)_2(aq) + energy$$

(d) Sodium hydroxide + sulfuric acid ⇌ sodium sulfate + water

$$NaOH(aq) + H_2SO_4(aq) \rightleftharpoons Na_2SO_4(aq) + H_2O(l) + energy$$

$$Al(OH)_3(aq) + 3H_2SO_4(aq) \rightleftharpoons Al_2(SO_4)_3(aq) + H_2O(l) + energy$$

In the products: 2 Al atoms

In the reactants: 1 Al atom

Placing a 2 in front of $Al(OH)_3$ balances the Al atoms.

$$2Al(OH)_3(aq) + 3H_2SO_4(aq) \rightleftharpoons Al_2(SO_4)_3(aq) + H_2O(l) + energy$$

In the products: 2 H atoms

In the reactants: 12 H atoms, six in $2Al(OH)_3$ and six in $3H_2SO_4$.

Placing a 6 in front of $H_2O$ balances the H atoms.

$$2Al(OH)_3(aq) + 3H_2SO_4(aq) \rightleftharpoons Al_2(SO_4)_3(aq) + 6H_2O(l) + energy$$

Is this equation now balanced? How many Al's are on each side? How many O's? How many H's? How many S's? The equation is balanced. It is often useful to balance oxygen and *hydrogen last*. ■

## 3.17 Percent composition

Frequently, the results of examinations are expressed in terms of the percent of correct answers. For example, if a student answers thirty questions correctly on a forty-question exam, his percent correct would be calculated as follows:

$$\frac{\text{number correct}}{\text{total number}} \times 100 = \frac{30}{40} \times 100 = 75\% \text{ correct}$$

In the laboratory it is often useful to express the amount of an element in a compound as a percentage of the total weight. The percent weight of an element in a compound may be determined directly from a *formula* or it may be calculated from *data* obtained in the laboratory.

**Percent weight from formula**

Common table salt, sodium chloride, has the formula NaCl. What percent by weight of the compound is sodium and what percent by weight is chlorine? In order to determine what part of the formula weight is due to sodium we will have to calculate the formula weight.

$$\begin{aligned}1 \text{ Na atom} &= 23.0 \text{ amu} \\ \underline{1 \text{ Cl atom} }&\underline{= 35.5 \text{ amu}} \\ \text{Formula weight} &= 58.5 \text{ amu}\end{aligned}$$

$$\% \text{ weight of Na} = \frac{\text{weight of Na}}{\text{total weight}} \times 100$$

$$= \frac{23.0 \text{ amu}}{58.5 \text{ amu}} \times 100 = 39.3\%$$

Since there are only two elements in the compound, once we calculate the percent weight of sodium, we can determine the percent weight of chlorine by subtraction.

total weight % = 100.0
−weight % Na =  39.3
weight % Cl =  60.7%

EXAMPLE 6 Calculate the percent by weight of each element in iron(III) sulfate $Fe_2(SO_4)_3$.

*Given:* $Fe_2(SO_4)_3$

*Find:* The weight percent of each element in $Fe_2(SO_4)_3$

*Relationship:* % element = $\dfrac{\text{wt of element}}{\text{total wt}} \times 100$

*Solution:* The formula weight of $Fe_2(SO_4)_3$ is:

$$
\begin{aligned}
2\text{ Fe atoms} &= 2 \times 55.8 = 111.6 \text{ amu} \\
3\text{ S atoms} &= 3 \times 32.1 = \phantom{0}96.3 \text{ amu} \\
12\text{ O atoms} &= 12 \times 16.0 = 192.0 \text{ amu} \\
\text{formula weight} &= 399.9 \text{ or } 400 \text{ amu}
\end{aligned}
$$

% Fe = $\dfrac{111.6 \text{ amu}}{400 \text{ amu}} \times 100 = 27.9\%$ Fe

% S = $\dfrac{96.3 \text{ amu}}{400 \text{ amu}} \times 100 = 24.1\%$ S

% O = $\dfrac{192.0 \text{ amu}}{400 \text{ amu}} \times 100 = \underline{48.0\% \text{ O}}$
$\phantom{\% O = \dfrac{192.0 \text{ amu}}{400 \text{ amu}} \times 100 =}$ 100 %  ■

■ EXERCISE *Calculate the percent by weight of each element in the following compounds: Potassium chloride KCl; nitroglycerin $C_3H_5(NO_3)_3$; butane gas $C_4H_{10}$.*

**Percent weight from laboratory data**

In a laboratory experiment (Fig. 3.8), a mixture of copper and sulfur is heated, causing a compound of these elements to form. The copper is in the form of thin, flexible, metallic-like ribbons with a golden red color. These ribbons are weighed, placed in a crucible, and covered with some fine, yellow, powdered sulfur. The crucible is covered, placed in a hood, and the mixture slowly heated over a low flame. After the mixture has been heated for 20 minutes, the crucible is allowed to cool. When the top is removed, it is noted that the copper ribbon is much darker than before but still has its original shape. When the ribbon is touched with a clean glass rod, the ribbon crumbles into a dark powder. Evidently the copper and sulfur react to form a compound. After the crucible has cooled, it is weighed. The results of such an experiment by one student were:

1. Wt of copper ribbon = 2.54 g
2. Wt of copper-sulfur compound = 3.18 g

Based on these results, what is the percent composition of copper and sulfur in the new compound formed? If copper and sulfur combined to form a new compound we may write:

$$\text{copper(s)} + \text{sulfur(s)} \rightleftharpoons \text{a copper-sulfur compound(s)}$$

What weight of sulfur is there in the new compound? We know the weight of the compound is 3.18 grams. Of this, 2.54 grams are copper. The rest of the compound must be sulfur.

weight of compound = 3.18 g
−weight of copper = 2.54 g
weight of sulfur = 0.64 g

We now know the weight of copper and of sulfur and of the whole compound. The percentage composition of each element is:

■ EXERCISE  *What is the percent composition of each element in a sample of a compound that contains 4.80 grams carbon, 1.20 grams hydrogen, and 3.20 grams oxygen?*

$$\% \text{ Cu} = \frac{2.54 \text{ g Cu}}{3.18 \text{ g}} \times 100 = 79.9\% \text{ Cu}$$

$$\% \text{ S} = \frac{0.64 \text{ g S}}{3.18 \text{ g}} \times 100 = 20\% \text{ S}  \quad ■$$

## 3.18  Empirical formulas

How is it possible to determine the formula of a substance from weight data obtained in the laboratory? Consider the reaction mentioned in the previous section between copper and sulfur. What is the formula of the compound formed? We know the weight ratio between copper and sulfur is 2.54 grams to 0.64 gram. Does a formula represent a weight ratio? No, a formula represents an atom ratio. The formula $H_2O$ means *two* hydrogen atoms for *one* oxygen atom. Since a mole contains a given number of atoms, the formula $H_2O$ also means *two moles* of hydrogen atoms for *one mole* of oxygen atoms. We can determine the formula of the copper-sulfur compound if we can *change the weight ratio to mole ratios.*

How many moles are 2.54 grams of copper? The atomic weight of copper is 63.5. Therefore:

1 mole Cu = 63.5 g Cu

$$2.54 \text{ g Cu} \times \frac{1.00 \text{ mole Cu}}{63.5 \text{ g Cu}} = 0.0400 \text{ mole Cu}$$

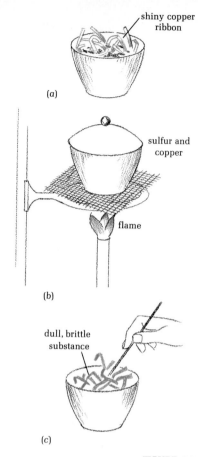

FIGURE 3.8

(a) Copper ribbon in a crucible. (b) The mixture of copper and sulfur is heated. (c) A substance with the shape of the copper ribbon crumbles when touched by a stirring rod.

How many moles is 0.64 gram S? The atomic weight of sulfur is 32.1. Therefore:

1 mole S = 32.1 g S

$$0.64 \text{ g S} \times \frac{1.00 \text{ mole S}}{32.1 \text{ g S}} = 0.020 \text{ mole}$$

There is a ratio of 0.0400 mole of copper to 0.020 mole of sulfur. We could write the formula as $Cu_{0.0400}S_{0.020}$, except that formulas are usually written in small, whole-number ratios. How can we change the ratio 0.0400 to 0.020 to small, whole numbers? First divide each number by the smallest number:

$$\frac{0.0400 \text{ mole Cu}}{0.020} = 2.0 \text{ moles Cu}$$

$$\frac{0.020 \text{ mole S}}{0.020} = 1.0 \text{ mole S}$$

The *mole ratio* of copper to sulfur is 2 to 1, therefore the *atom ratio* of copper to sulfur is also 2 to 1. The formula $Cu_2S$ is called an *empirical formula*. It is calculated from the weight ratios of elements in the compound. Often it is necessary to calculate the empirical formula when given the percentage composition. This is illustrated in Example 7. A scheme for determining empirical formulas is shown below.

$$\text{weight ratio} \xrightarrow{\text{divide by mole weight}} \text{mole ratio} \longrightarrow$$

convert to small, whole number

**EXAMPLE 7** Ethyl alcohol is 52.1% carbon, 13.0% hydrogen, and 34.8% oxygen. What is the empirical formula of ethyl alcohol?

*Given:* Ethyl alcohol is: 52.1% C, 13.0% H, 34.8% O

*Find:* Empirical formula of ethyl alcohol

*Relationships:*

$$\text{weight of element} \times \frac{1 \text{ mole of atoms}}{\text{weight of 1 mole of atoms}} = \text{moles of atoms}$$

*Solution:* We are given the percentage composition of carbon, hydrogen, and oxygen in the compound. If a compound is 52.1% carbon, then out of every 100 grams of the compound 52.1 grams are carbon. In this example we can arbitrarily assume that there are 100 grams of ethyl alcohol. Then the weight ratios are:

52.1% C in 100 grams gives 52.1 g C
13.0% H in 100 grams gives 13.0 g H
34.8% O in 100 grams gives 34.8 g O

We can change these weight ratios to mole ratios by multiplying by the unity term that relates one mole of atoms to the weight of a mole.

$$52.1 \text{ g C} \times \frac{1 \text{ mole C atoms}}{12.0 \text{ g C}} = 4.34 \text{ moles C atoms}$$

$$13.0 \text{ g H} \times \frac{1 \text{ mole H atoms}}{1.01 \text{ g H}} = 12.9 \text{ moles H atoms}$$

$$34.8 \text{ g O} \times \frac{1 \text{ mole O atoms}}{16.0 \text{ g O}} = 2.17 \text{ moles O atoms}$$

To change these ratios to small, whole numbers, divide each by the smallest number.

$$\frac{4.34}{2.17} \text{ moles C atoms} = 2.00 \text{ moles C atoms}$$

$$\frac{12.9}{2.17} \text{ moles H atoms} = 5.95 \text{ moles H atoms}$$

$$\frac{2.17}{2.17} \text{ moles O atoms} = 1.00 \text{ mole O atoms}$$

The number 5.95 is within 1 percent of 6.00. We will assume the ratio of atoms is 2 carbons to 6 hydrogens to 1 oxygen. The empirical formula is $C_2H_6O$.

EXAMPLE 8  A compound contains 1.79 grams of phosphorus and 2.29 grams of oxygen. What is the empirical formula of the compound?

*Given:* 1.79 grams of phosphorus, 2.29 grams of oxygen

*Find:* The empirical formula of the compound

*Relationships:*

$$\text{weight of element} \times \frac{1 \text{ mole of atoms}}{\text{weight of 1 mole of atoms}} = \text{moles of atoms}$$

*Solution:*

$$1.79 \text{ g P} \times \frac{1 \text{ mole P atoms}}{31.0 \text{ g P}} = 0.0577 \text{ mole P atoms}$$

$$2.29 \text{ g O} \times \frac{1 \text{ mole O atoms}}{16.0 \text{ g O}} = 0.143 \text{ mole O atoms}$$

Dividing by the smaller number of moles gives:

$$\frac{0.0577}{0.0577} \text{ mole P atoms} = 1.00 \text{ mole P atoms}$$

$$\frac{0.143}{0.0577} \text{ mole O atoms} = 2.48 \text{ moles O atoms}$$

The ratio is 1.00 phosphorus atom to 2.48 or about $2\frac{1}{2}$ oxygen atoms. If we double this ratio we have 2 phosphorus atoms to 5 oxygen atoms.

■ EXERCISE  *Calculate the empirical formula from the following data: A compound is 56.4% phosphorus and 43.6% oxygen.*

$$\frac{1.00 \text{ P atom}}{2.48 \text{ O atoms}} \times \frac{2}{2} = \frac{2.00 \text{ P atoms}}{4.96 \text{ O atoms}} = \frac{2 \text{ P atoms}}{5 \text{ O atoms}}$$

The empirical formula is $P_2O_5$.  ■

## Summary

Every type of matter is built up from the 105 known elements. Approximately 75 percent of the elements are solids that conduct heat and electricity and have a characteristic metallic luster. These elements are called *metals*. Elements that lack a metallic luster and are poor conductors of heat and electricity are called *nonmetals*.

Each element is represented by a *symbol*. Each symbol may represent an *atom* of the element or a mole of atoms of the element. Different elements combine to form compounds. The formulas of compounds indicate what elements are in the compound and what their atom *ratio* is.

An atom contains a positive *nucleus* surrounded by negative *electrons*. Although the nucleus represents a very small volume of the atom, most of the mass of the atom is contained in the nucleus. The positive *proton* and neutral *neutrons* make up the nucleus. The sum of protons and neutrons in an atom is called the *mass number*. The number of protons in an atom is called the *atomic number*. The atoms of different elements have different atomic numbers. A sample of a pure element may contain atoms with slightly different mass numbers. This variance in mass numbers is due to different numbers of neutrons in the nucleus. Atoms of an element with different mass numbers are called *isotopes*.

The average of the masses of all the naturally occurring isotopes of an element is called the *atomic weight* of that element. The unit of atomic weights is the *atomic mass unit* or *amu*. The standard for this scale is the carbon-12 isotope with a mass of exactly 12 amu. If a sample of an element has a mass equal to the atomic weight expressed in grams, the number of atoms in that sample is $6.02 \times 10^{23}$. This number is known as *Avogadro's number* and defines a *mole* of a substance.

The sum of the atomic weights of all the atoms in a molecule is known as the *molecular weight*. If a sample of a substance has a mass equal to the molecular weight expressed in grams, the number of molecules in that sample is $6.02 \times 10^{23}$.

During a chemical reaction, there is no noticeable mass change. We interpret this by assuming that there is no noticeable change in the number of atoms as reactants change to products. We indicate this by writing *balanced chemical equations*. The equation shows the ratios of atoms or molecules of each substance in the reactants and products.

The percentage composition of a compound shows the weight of each element in a compound on a percentage basis. The percentage composition of one element can be calculated from a molecular formula by dividing the total molecular weight of the compound into the weight of that element in the compound and then multiplying by 100.

If the percentage composition of a compound is known or if the weight of each element in a sample of a compound is known, the simplest ratio of atoms of each element in the compound may be found. This simplest ratio is known as the *empirical formula*. It is found by changing the weight ratios of elements to mole ratios.

## Objectives

By the end of this chapter, you should be able to do the following:

1. Given the name of a common element, write the symbol for that element. (Questions 2, 3)

2. Given samples of common elements, separate them into metals and nonmetals. (Question 10)

3. Restate the basic postulates of Dalton's atomic theory. (Question 9)

4. State the modifications that have been made in Dalton's atomic theory since his time. (Question 8)

5. Given the formula of a compound, state what elements make up that compound and what their ratios are. (Question 4)

6. Name the three fundamental particles needed to explain the chemical behavior of matter. List their charges and indicate their relative masses. (Question 6)

7. Given the symbolism showing the atomic number and mass number of an element, state how many electrons, protons, and neutrons the atom has. (Question 7)

8. State where in an atom each of the three fundamental particles may be found. (Question 13)

9. Cite the evidence that indicates that an atom is mostly empty space, that it has a small positive center, and that most of the mass of the atom is located in the center. (Question 11)

10. Given the percentage of the isotopes that make up an element, calculate the atomic weight of the element. (Question 12)

11. Given the weight of a sample of a substance, calculate: (a) the number of moles of the substance, (b) the number of atoms or molecules of the substance. (Questions 21, 22)

12 Given the formula of a compound, calculate the molecular weight. (Questions 15, 16)

13. Given the formula of a compound, calculate the percentage composition of each element in the compound. (Questions 17, 18)

14 Given the weight of each element in a sample of a compound, calculate the percent composition of each element in the compound. (Question 19)

15. Given the weight of each element in a sample of a compound, calculate the empirical formula of the compound. (Question 24)

16. Given the percentage composition of each element in a compound, calculate the empirical formula of the compound. (Question 23)

17. Given the formula of the reactants and the products in a reaction, write the balanced equation for the reaction. (Questions 25, 26, 27, 28, 29, 30, 31, 32)

## Questions

1. Explain the meaning of the following terms. Where possible give both operational and theoretical definitions: carbon-12; mole; metal; malleable; ductile; nonmetals; symbol; atomic weight; radioactive; theory; atoms; molecule; diatomic; subatomic particles; electron; proton; nuclear model; nucleus; atomic number; isotopes; mass number; formula; equation; neutron; atomic mass units; alpha particles; Avogadro's number.

2. Some of the major elements that occur in compounds present in sea water are hydrogen, oxygen, sodium, magnesium, sulfur, calcium, potassium, bromine, carbon, strontium, boron, silicon, and fluorine. Write the symbols for these elements.

3. Write the names of the following elements that occur in minor amounts in sea water: N, Al, Li, P, Ba, I, As, Mn, Cu, Zn, Pb, U, Ag, Ni, Hg, and Au.

4. What elements are present in each of the following compounds? What is the ratio of the atoms of each element in each compound? sodium sulfate $Na_2SO_4$; aluminum sulfate $Al_2(SO_4)_3$; magnesium chlorate $Mg(ClO_3)_2$; aluminum chlorate $Al(ClO_3)_3$.

5. Given the atom ratios for the following compounds, write their formulas: (a) iron(III) phosphate: one iron atom, one phosphorus atom, four oxygen atoms; (b) lead(II) chloride: one lead atom, two chlorine atoms; (c) lead(II) oxide: one lead atom, one oxygen atom; (d) calcium chloride: one calcium atom, two chlorine atoms; (e) calcium oxide: one calcium atom, one oxygen atom.

6. A lithium-7 isotope can be represented by $^{7}_{3}Li$. What subatomic particles make up a lithium-7 atom? What charges do each of these subatomic particles possess and what are their relative masses?

7. Contrary to Dalton's atomic model, some atoms possess the property of radioactivity; that is, their nuclei break up into other atomic nuclei and subatomic particles. The products of this decay may be detected with an instrument such as a Geiger counter. Many of these isotopes find wide use in medicine for detecting or curing diseases. Determine the atomic number, the mass number, the number of protons, the number of electrons, and the number of neutrons for each of the following isotopes: $^{131}_{53}I$; $^{24}_{11}Na$; $^{60}_{27}Co$; $^{99}_{43}Tc$; $^{51}_{24}Cr$; $^{130}_{54}Xe$; $^{197}_{80}Hg$; $^{198}_{79}Au$.

8. State how the following concepts have caused scientists to modify Dalton's model of the atom: isotopes, radioactivity, protons, electrons, and neutrons.

9. Under the proper environment, iron will react with sulfur to form iron(II) sulfide(FeS). In every case the weight ratios are 1.73 grams of

iron for 1.00 gram of sulfur. Using Dalton's atomic theory, explain what is happening when these substances react.

10. Consider the following substances: an aluminum frying pan, oxygen gas in a space vehicle, neon gas in neon light, a copper wire, graphite in a "lead" pencil, mercury in a thermometer, yellow sulfur from a druggist, and a silver spoon. Which of these substances are metals and which are nonmetals?

11. What evidence did Rutherford have that indicated that atoms consist of empty space? What evidence was there that atoms have a small positive nucleus? What evidence was there that this nucleus contains most of the mass of the atom?

12. The naturally occurring isotopes of lithium are $^6_3$Li and $^7_3$Li. Their distribution is: 7.40% lithium-6 and 92.6% lithium-7. Calculate the atomic weight of naturally occurring lithium.

13. An atom of oxygen-17 contains 8 protons, 8 electrons, and 9 neutrons. By means of a diagram show approximately where in the atom these subatomic particles occur.

14. Calculate the molecular or formula weight of each of the following substances: (a) $H_2$ hydrogen; (b) $CH_4$ methane; (c) $CCl_4$ carbon tetrachloride; (d) $Ca(OH)_2$ calcium hydroxide; (e) $Ba(OH)_2$ barium hydroxide; (f) $I_2$ iodine.

15. Calculate the weight of a mole of each of the following: (a) H hydrogen atom; (b) $H_2$ hydrogen molecules; (c) O oxygen atoms; (d) $O_2$ oxygen molecules; (e) F fluorine atom; (f) $F_2$ fluorine molecules; (g) Na sodium; (h) K potassium.

16. Calculate the weight of a mole of each of the following: (a) $H_2O$ water; (b) $H_2S$ hydrogen sulfide; (c) $Na_2S$ sodium sulfide; (d) $Na_2O$ sodium oxide; (e) $K_2SO_4$ potassium sulfate; (f) $Ba(NO_3)_2$ barium nitrate.

17. Geologists often use the property of hardness to identify minerals. An arbitrary scale from 1 to 10 has been established. A soft mineral such as talc has a hardness of 1. A hard mineral such as diamond has a hardness of 10. Calculate the percent composition of each element in the following minerals.

| Name | Hardness | Formula |
| --- | --- | --- |
| (a) diamond | 10 | C |
| (b) corundum | 9 | $Al_2O_3$ |
| (c) fluorite | 4 | $CaF_2$ |
| (d) gypsum | 2 | $CaSO_4$ |
| (e) talc | 1 | $Mg_3Si_4O_{10}(OH)_2$ |

18. Which of the following substances has the greatest percentage of sulfur? What has the lowest percentage of sulfur? (a) $SO_2$; (b) $SO_3$; (c) $Na_2S$; (d) $H_2SO_4$.

19. The weights of the component elements of some experimental samples of compounds are given here. Calculate the percent composition of each element in the compound. (a) 0.505 g H, 8.02 g S; (b) 0.168 g H, 2.33 g N, 8.00 g O; (c) 0.373 g H, 8.01 g C; (d) 0.642 g S, 0.640 g O.

20. How many atoms are present in each of the following? (a) 0.333 mole of helium atoms; (b) 5.00 moles of copper atoms; (c) 1.00 mole of sulfur atoms; (d) 0.195 mole of Pt atoms.

21. How many atoms are present in 5.000 grams of each of the following? B, C, Na, P, Ni.

22. How many moles of molecules are present in the following samples of matter? (a) 8.00 g hydrogen $H_2$; (b) 0.600 g iodine $I_2$; (c) 0.235 g water $H_2O$; (d) 1.00 g ammonia $NH_3$.

23. Calculate the empirical formulas of the following compounds: (a) 92.4% C, 7.76% H; (b) 85.6% C, 14.4% H; (c) 94.1% O, 5.94% H; (d) 88.8% O, 11.2% H; (e) 59.0% Na, 41.0% O; (f) 74.2% Na, 25.8% O; (g) 51.2% O, 34.9% Zn, 14.9% N; (h) 40.1% O, 39.8% Cu, 20.1% S.

24. Calculate the empirical formulas of the following compounds: (a) 0.103 g Pb, 0.00800 g O; (b) 9.45 g Pb, 0.533 g O; (c) 7.77 g Pb, 0.800 g O.

25. "Cherries jubilee" are prepared by igniting brandy in which the cherries have been soaked. Balance the equation for the burning of the alcohol in the brandy.

$$CH_3CH_2OH(l) + O_2(g) \rightleftharpoons CO_2(g) + H_2O(l) + energy$$

26. Minerals that contain calcium carbonate may be tested by putting some hydrochloric acid on them. Bubbles of carbon dioxide gas indicate that the mineral is a carbonate. Balance the equation for this reaction.

$$CaCO_3(s) + HCl(aq) \rightleftharpoons CaCl_2(aq) + H_2O(l) + CO_2(g)$$

27. Sulfur contributes to air pollution. It is formed when sulfur-containing fuels are burned. The sulfur dioxide $SO_2$ may be removed by passing it through calcium oxide CaO. The CaO and $SO_2$ react to form solid calcium sulfite $CaSO_3$. Write the balanced equation for this reaction.

28. Vinegar contains acetic acid $CH_3COOH$. Household ammonia contains a solution of $NH_3$ in water. These two substances will react to form ammonium acetate $NH_4CH_3COO$. Write the balanced equation for this reaction.

29. Balance the following equations.
(a) $Zn(s) + O_2(g) \rightleftharpoons ZnO(s)$
(b) $Mg(s) + O_2(g) \rightleftharpoons MgO(s)$
(c) $Fe(s) + S(s) \rightleftharpoons FeS(s)$

30. Balance the following equations.
(a) $CH_4(g) + O_2(g) \rightleftharpoons CO_2(g) + H_2O(l)$
(b) $C_2H_6(g) + O_2(g) \rightleftharpoons CO_2(g) + H_2O(l)$
(c) $C_6H_{14}(l) + O_2(g) \rightleftharpoons CO_2(g) + H_2O(l)$

31. Balance the following equations.
(a) $NH_3(g) + HF(g) \rightleftharpoons NH_4F(s)$
(b) $NH_3(g) + H_2SO_4(aq) \rightleftharpoons (NH_4)_2SO_4(aq)$
(c) $NH_3(g) + H_3PO_4(aq) \rightleftharpoons (NH_4)_3PO_4(aq)$

32. Balance the following equations.
(a) $SO_2(g) + H_2O(l) \rightleftharpoons H_2SO_3(aq)$
(b) $SO_3(g) + H_2O(l) \rightleftharpoons H_2SO_4(aq)$
(c) $Na_2O(s) + H_2O(l) \rightleftharpoons NaOH(aq)$

33. One mole of each of the following will remove one mole of carbon dioxide from the atmosphere. Which of these substances would you choose to use to remove carbon dioxide from the atmosphere of a space vehicle? (*Hint:* Weight is an important consideration for space vehicles.) LiOH, NaOH, KOH, CsOH, RbOH.

34. List the ways in which an atom of iron is different from an atom of gold. Use the isotopes $^{56}_{26}Fe$ and $^{197}_{79}Au$.

35. The mass of a proton is 1.007 amu; the mass of a neutron is 1.009 amu. A carbon-12 isotope contains six protons and six neutrons. Why is the mass of carbon-12 only 12 amu?

# Electronic structure

## 4.1 Introduction

Sodium and potassium are similar elements. They are both soft metals that react readily with many other substances. When added to water, both metals react with it to form a combustible gas and solutions that feel soapy. Experiments indicate that atoms of both sodium and potassium react in a one-to-one ratio with chlorine atoms. That is, one atom of chlorine can combine with one atom of sodium or with one atom of potassium. Table salt contains sodium combined with chlorine. "Sodium-free" salt contains potassium in place of sodium. Potassium salt is used by people who must limit their intake of sodium. In this chapter we will develop models for atoms which will help to explain the similarity between sodium and potassium atoms. These models are based on the amounts of energy which electrons in atoms are allowed to have. The arrangement of electrons according to the amount of energy they possess is known as the *electronic structure* of the atom.

Chemical changes involve a rearrangement of electrons in atoms. Each time that atoms *combine, rearrange, or break away* from one another, there is a rearrangement of electrons. The *types* of atoms that will react, the *ratio* in which they react, the *properties* of the products of a reaction, and the *energy* involved in a reaction depend upon the electronic structure of the atoms involved in the reaction. In Chapter 5 we will study a method for arranging the elements into groups with similar properties. This arrangement, known as the *periodic table,* is based on the electronic structure of the elements.

Before studying this chapter, review the concepts of kinetic and potential energy in Chapter 1.

## 4.2 Discrete packets—quanta

How many persons are in your family? Two, three, four . . . ? How many students are in your class? Twenty, fifty, three hundred? Would you expect an answer of 2.65 persons in your family or 48.38 students in your class? Why can't we speak of 1/10 of a person? People occur in packets or bundles of one and may be considered to be *quantized* because they occur in units of one, not as fractional parts of one.

The same idea of quantization was applied to matter by John Dalton. In his theory, he proposed that the smallest unit of matter was the atom. One can talk about one atom, or ten atoms, or ten million atoms, but a discussion of 4.79 or 16.41 atoms is questionable. Atoms are very small, but they still occur as packets or bundles of one.

**Energy**

In the early 1800s John Dalton put forth arguments for the quantization of matter (atoms). Experiments performed in the late 1800s indicated that electricity is quantized in the amount of charge on the electron. The smallest amount of charge yet found is that of the electron (or proton). Samples of matter have been found to have charges of one, two, three, or more times the charge of an electron. So far, no sample of matter has been found with $1/4$ or $2/3$ or any other fractional part of the charge of an electron.

At the end of the nineteenth century, light was believed to consist of waves of energy that traveled through a substance in space known as ether. In 1900 Max Planck proposed that light energy was also quantized. He developed a model for radiant energy to explain some "misunderstood" experimental data. According to Planck, light travels in bundles or packets called photons. Each photon possesses a definite amount of energy. The energy of a light ray or photon is directly proportional to the frequency of the ray.

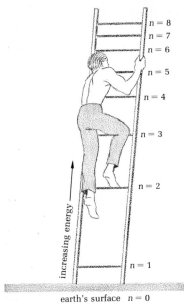

FIGURE 4.1
The allowed potential energy levels for a man on a ladder which has eight unevenly spaced rungs.

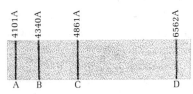

FIGURE 4.2
Frequencies of light absorbed by hydrogen atoms. Line A corresponds to an energy transition between energy shell 6 and energy shell 2. Line B corresponds to an energy transition between energy shell 5 and 2. Line C corresponds to an energy transition between energy shell 4 and 2. Line D corresponds to an energy transition between energy shell 3 and 2.

$$E = h\nu \qquad (4.1)$$

$E =$ energy of the ray (usually in units of ergs)
$h =$ a constant called Planck's constant (usually in units of ergs-sec)
$\nu =$ frequency of the ray (usually in units of sec$^{-1}$ or 1/sec; the symbol $\nu$ is the Greek letter nu)

If a purple ray has a greater frequency than a ray of yellow light, which ray has the greater energy? According to Planck's theory as expressed in Eq. 4.1, the purple ray has more energy because it has a greater frequency.

### Electron energy levels

Consider the potential energy of a man on a ladder. As the man goes up the ladder, he moves farther and farther away from the earth's surface. His potential energy, relative to the earth's surface, increases as his distance from the surface increases. The ladder in Fig. 4.1 has eight unevenly spaced rungs. The potential energy of the man is zero when he stands on the surface of the earth. When he stands on the first rung of the ladder, he has a potential energy which corresponds to $n = 1$. On the second rung, a man has a potential energy which corresponds to $n = 2$, and so on up to the eighth rung, $n = 8$. Can a man possess the energy which corresponds to a position somewhere between rung 1 and rung 2? Certainly a man can possess the energy which corresponds to rungs 1 to 8, but he *cannot* possess any of the energies in between these values. If a man tries to occupy a position other than rungs 1 to 8, his foot falls to the next lower rung or energy level. We say that the potential energy of the man is quantized. There are only eight *allowed* energy levels on this ladder corresponding to the eight rungs of the ladder.

Experiments have shown that atoms can absorb or emit photons of light of only certain energy levels. For instance, hydrogen atoms absorb frequencies of light with the energies shown in Fig. 4.2. Other energies of light pass through the hydrogen unabsorbed. In order to account for these observations, Nils Bohr proposed in 1913 that electrons traveling about the nucleus of an atom may possess only certain allowable amounts of energy. *Electrons may be on one energy level or another, but not in between.* The energy of an electron in an atom is analogous to the potential energy of a man on a ladder. *The energy of an electron in an atom is quantized.*

Let us consider a hydrogen atom as an example. The atom consists of one proton in the nucleus and one electron outside

FIGURE 4.3
The first four energy levels which can be occupied by the electron of hydrogen. The lowest possible energy state of electrons in an atom is called the ground state.

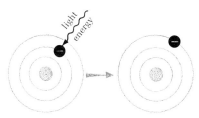

FIGURE 4.4
In order to move from a low energy level to a higher energy level, the electron must gain energy.

the nucleus. According to the Bohr theory, this one electron can exist only in certain allowed energy levels. The lowest allowed energy level is a region of space about 0.5 Angstroms from the nucleus. When an electron moves to energy levels which are farther from the nucleus, it is increasing its potential energy just as the man who moves up a ladder is increasing his potential energy. We can represent the *possible* energy levels in a hydrogen atom with the simple diagram shown in Fig. 4.3. This model will be modified later.

An electron in an atom may move from one energy level to another. Each change in the energy level of the electron is accompanied by the emission or absorption of energy in the form of a photon. In order for an electron to move away from the nucleus, that is, from a lower energy level to a higher energy level, the electron must gain or absorb energy (Fig. 4.4). The absorbed energy may come from light passing through the hydrogen. Bohr said that the absorption of light at a certain energy level corresponds to a particular change in the energy level of an electron in a hydrogen atom. In Fig. 4.2, the energies of light absorbed arise from an electron moving from the second level to a higher level. For example, absorbed light corresponding to the frequency of a point A in Fig. 4.2 causes an electron to move from the second energy level to the sixth energy level.

## 4.3  Energies of electrons in atoms

The energy of an electron in an atom is a combination of its kinetic energy and potential energy (see Chapter 1). The potential energy of a negative electron is caused by the force of attraction between it and the positive nucleus. The kinetic energy of an electron is due to its motion about the nucleus. In the following sections, the *total energy* of the electron—that is, the sum of its kinetic and potential energy—will be considered.

**Quantum numbers**

Quantum theory or quantum mechanical theory, which was developed during the twentieth century, helps to explain the allowable amounts of energy of electrons in atoms. Although the mathematical models used in this theory are beyond the scope of this book, the results of the theory are easier to understand. According to the quantum mechanical model of the atom, four separate quantum numbers are needed to specify the energy of an electron. Recall that two coordinates must be specified in order to locate a point on a graph. In the same way, four quantum numbers are required to locate an electron in energy space. The next four sections describe *how many* electrons each energy level can hold, the *order* in which the levels are filled, and the relative *energy difference* between levels.

## 4.4 Main energy level

The first quantum number, n, indicates the main energy level of the electron. It is the existence of the main energy levels that Bohr attempted to explain in 1913. A main energy level is also called a *shell*. In the hydrogen atom, which has one proton and one electron, the electron may be on any main energy level. These levels are indicated as $n = 1$, $n = 2$, $n = 3$, and so on. The lowest energy level is closest to the nucleus, and is called the first shell, $n = 1$ (Fig. 4.5). The second lowest energy level, $n = 2$, is called the second shell, the third lowest energy level, $n = 3$, is called the third shell, and so on.

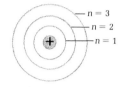

FIGURE 4.5
*The relative locations of the first three energy shells with respect to the nucleus.*

**Maximum number of electrons**

Each shell can contain a maximum number of electrons and no more. Experiments indicate that the first shell can hold a maximum of two electrons. The second shell can hold a maximum of eight electrons. According to the quantum mechanical model of the atom, the maximum number of electrons in any shell, n, is given by Eq. 4.2.

Maximum number of electrons in the nth shell $= 2(n^2)$  (4.2)

For example, in the second shell, $n = 2$. The maximum number of electrons in the second shell is:

$$2(n^2) = 2(2)^2 = 2(4) = 8$$

■ EXERCISE *Calculate the maximum number of electrons in the third and fourth shells. Check your answer using Table 4.1.*

The theoretical maximum number of electrons in a shell as given by Eq. 4.2 is in good agreement with the known behavior of atoms. ■

**Order of filling electron shells**

How are electrons distributed in the energy levels of atoms other than hydrogen? This arrangement of electrons is analagous to the arrangement of cars in a parking lot about a store in a shopping center. Usually the cars are as close as possible to

**TABLE 4.1**

*The relation between shell number and the maximum number of electrons in a given shell.*

| Shell | $2n^2$ |
|---|---|
| 1 | 2 |
| 2 | 8 |
| 3 | 18 |
| 4 | 32 |
| 5 | 50 |

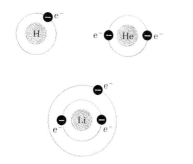

**FIGURE 4.6**

*The relative sizes of the hydrogen, helium, and lithium atoms and the number of electrons in each atom.*

the store. As the closer spaces fill up, cars must park farther away. The rapidly moving electrons normally fill the lowest possible energy level first. Then the next higher level begins to fill. An atom which has electrons distributed on the lowest possible energy level is said to be in the *ground state*. When one or more electrons of an atom are in energy levels higher than the ground state, the atom is said to be in an *excited state*. Only atoms in the ground state are discussed in this chapter.

Hydrogen, atomic number 1, has one proton and one electron. Each successive element has one more proton and one more electron. The electrons in atoms first fill up the $n = 1$ energy level. The one electron of hydrogen is in the $n = 1$, or first, shell. Helium, atomic number 2, has two electrons; both of these are on the $n = 1$ energy level. This fills the $n = 1$ energy level. The third element, lithium, has three electrons. Two of these electrons are on the $n = 1$ energy level and the third electron is on the $n = 2$ energy level. Figure 4.6 shows simplified diagrams of the hydrogen, helium, and lithium atoms. The second shell holds a maximum of eight electrons. The completion of this shell accounts for the electronic distribution of the first ten elements, which is shown in Table 4.2.

## 4.5 Energy sublevels—second quantum number

Additional observations of the radiant energy given off or absorbed by atoms led to a modification of Bohr's electron shell model of the atom. It became necessary to add the concept of energy sublevels to explain these observations. Each main

**TABLE 4.2**

*The number of electrons in the first and second shells for each of the first ten elements.*

| Element | Number of electrons first shell, $n = 1$ | Number of electrons second shell, $n = 2$ |
|---|---|---|
| H | 1 | 0 |
| He | 2 | 0 |
| Li | 2 | 1 |
| Be | 2 | 2 |
| B | 2 | 3 |
| C | 2 | 4 |
| N | 2 | 5 |
| O | 2 | 6 |
| F | 2 | 7 |
| Ne | 2 | 8 |

**TABLE 4.3**

The maximum number of electrons in each subshell and in each energy shell is indicated for the first four shells.

| n | Subshell | Number of electrons | Total e⁻ in shell |
|---|---|---|---|
| n = 1 | s | 2 | 2 |
| n = 2 | s | 2 | |
| | p | 6 | 8 |
| n = 3 | s | 2 | |
| | p | 6 | |
| | d | 10 | 18 |
| n = 4 | s | 2 | |
| | p | 6 | |
| | d | 10 | |
| | f | 14 | 32 |

energy level actually consists of one or more sublevels or *subshells*. A second quantum number, symbolized by the letter l, was introduced. This quantum number designates the shape of the region of space in which an electron is likely to be found.

According to the quantum mechanical model there is one subshell making up the $n = 1$ level. There are two subshells making up the $n = 2$ level. There are three subshells making up the $n = 3$ level. The number of subshells in each shell is equivalent to the shell number. Chemists use the letters s, p, d, f, etc. to designate the subshells in each shell.

The $n = 1$ shell is made up of an s subshell. Both the theory and experimental observations indicate that an s subshell can hold a maximum of *two* electrons.

The $n = 2$ shell is made up of an s subshell and a p subshell. The s subshell can hold a maximum of *two* electrons. The p subshell can hold a maximum of *six* electrons. This accounts for a maximum of eight electrons in the $n = 2$ shell.

The $n = 3$ shell, like the $n = 2$ shell, has an s subshell and a p subshell. It also has a d subshell which can hold a maximum of *ten* electrons. Thus the $n = 3$ shell can hold a maximum of *eighteen* electrons: two in the s, six in the p, and ten in the d.

The $n = 4$ shell has an s, p, d, and an f subshell. The f subshell can hold a maximum of fourteen electrons. If the s, p, and d subshells in the fourth shell are similar to those in the third shell, what is the maximum number of electrons in the fourth shell? (see Table 4.3)

**Energy level and order of filling subshells**

Experiments indicate that under certain conditions each subshell in a shell has a slightly different energy level than the other subshells. For instance, the third shell consists of three

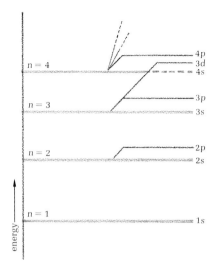

FIGURE 4.7
Energy shells and subshells arranged in order of increasing energy.

different subshells: an s subshell (written 3s), a p subshell (written 3p), and a d subshell (written 3d). Experiments show that the 3s sublevel will fill up before the 3p sublevel and the 3p sublevel will fill up before the 3d sublevel. The relative energy levels for the first eight subshells (1s to 4p) are shown in Fig. 4.7.

Electrons in ground state atoms occupy the lowest possible energy states. What is the order in which the subshells are occupied (Fig. 4.7)? Begin with the subshell of lowest energy. This is the 1s subshell; now, work up. The order is 1s, 2s, 2p, 3s, 3p, 4s, 3d, 4p. Notice that the 4s subshell is slightly less energetic than the 3d subshell; therefore, the 4s subshell fills before the 3d.

The symbolism that indicates the energy arrangement of electrons in an atom is called the *electronic configuration*. A hydrogen atom in the ground state, for example, has one electron in the s subshell of the first shell. The shell is indicated by its number; in this case a "1." The subshell is indicated by its letter; in this case an "s." The number of electrons in a subshell is indicated by a numerical superscript; in this case a "1." The electronic configuration of hydrogen is:

$_1$H     $1s^1$ — one electron
                — s subshell
                — first shell

The next element is helium, atomic number 2. Helium has two electrons both in the s subshell of the first shell. Its electronic configuration is:

$_2$He     $1s^2$

This fills the first shell. What is the electronic configuration of lithium, atomic number 3? Lithium has three electrons; two are in the 1s subshell, the third is in the 2s subshell.

$_3$Li     $1s^2$    $2s^1$

The electronic configuration of element number 4, beryllium is:

$_4$Be     $1s^2$    $2s^2$

This fills the 2s sublevel. Next comes the 2p subshell, which can hold six electrons. The electronic configuration of boron, element number 5 is:

$_5$B     $1s^2$    $2s^2$    $2p^1$

4.5 Energy sublevels—second quantum number

### TABLE 4.4

*The electronic configurations of the first 36 elements. The symbols (Ne) and (Ar) indicate the configurations of neon and argon; that is (Ne) = $1s^2\ 2s^2\ 2p^6$ and (Ar) = $1s^2\ 2s^2\ 2p^6\ 3s^2\ 3p^6$*

| | | | | | |
|---|---|---|---|---|---|
| 1. | **H** | $1s^1$ | 19. | **K** | (Ar) $4s^1$ |
| 2. | **He** | $1s^2$ | 20. | **Ca** | (Ar) $4s^2$ |
| 3. | **Li** | $1s^2\ 2s^1$ | 21. | **Sc** | (Ar) $3d^1\ 4s^2$ |
| 4. | **Be** | $1s^2\ 2s^2$ | 22. | **Ti** | (Ar) $3d^2\ 4s^2$ |
| 5. | **B** | $1s^2\ 2s^2\ 2p^1$ | 23. | **V** | (Ar) $3d^3\ 4s^2$ |
| 6. | **C** | $1s^2\ 2s^2\ 2p^2$ | 24. | **Cr** | (Ar) $3d^5\ 4s^1$ |
| 7. | **N** | $1s^2\ 2s^2\ 2p^3$ | 25. | **Mn** | (Ar) $3d^5\ 4s^2$ |
| 8. | **O** | $1s^2\ 2s^2\ 2p^4$ | 26. | **Fe** | (Ar) $3d^6\ 4s^2$ |
| 9. | **F** | $1s^2\ 2s^2\ 2p^5$ | 27. | **Co** | (Ar) $3d^7\ 4s^2$ |
| 10. | **Ne** | $1s^2\ 2s^2\ 2p^6$ | 28. | **Ni** | (Ar) $3d^8\ 4s^2$ |
| 11. | **Na** | (Ne) $3s^1$ | 29. | **Cu** | (Ar) $3d^{10}\ 4s^1$ |
| 12. | **Mg** | (Ne) $3s^2$ | 30. | **Zn** | (Ar) $3d^{10}\ 4s^2$ |
| 13. | **Al** | (Ne) $3s^2\ 3p^1$ | 31. | **Ga** | (Ar) $3d^{10}\ 4s^2\ 4p^1$ |
| 14. | **Si** | (Ne) $3s^2\ 3p^2$ | 32. | **Ge** | (Ar) $3d^{10}\ 4s^2\ 4p^2$ |
| 15. | **P** | (Ne) $3s^2\ 3p^3$ | 33. | **As** | (Ar) $3d^{10}\ 4s^2\ 4p^3$ |
| 16. | **S** | (Ne) $3s^2\ 3p^4$ | 34. | **Se** | (Ar) $3d^{10}\ 4s^2\ 4p^4$ |
| 17. | **Cl** | (Ne) $3s^2\ 3p^5$ | 35. | **Br** | (Ar) $3d^{10}\ 4s^2\ 4p^5$ |
| 18. | **Ar** | (Ne) $3s^2\ 3p^6$ | 36. | **Kr** | (Ar) $3d^{10}\ 4s^2\ 4p^6$ |

■ EXERCISE  Write the electronic configurations for the following elements: magnesium, neon, fluorine, chlorine.

What is the electron configuration of a sodium atom in the ground state? First, determine how many electrons sodium has. The atomic number of sodium is 11 and thus it must have eleven electrons. The 1s subshell can hold two electrons; this leaves nine electrons. The 2s subshell can hold two electrons; this leaves seven electrons. The 2p subshell can hold six electrons; this leaves one electron. The last electron can occupy the 3s subshell. The electronic configuration of sodium is:

$$_{11}\text{Na} \quad 1s^2 \quad 2s^2 \quad 2p^6 \quad 3s^1 \quad ■$$

When writing electronic configurations, remember that the lower energy levels fill first. An order-of-fill diagram which is easy to learn is shown in Fig. 4.8. The electronic configurations for the first thirty-six elements are shown in Table 4.4.

## 4.6  Orbitals

When the light emitted or absorbed by atoms in a strong magnetic field is analyzed, the frequencies emitted or absorbed are slightly different than those analyzed without the magnetic field. This experimental evidence has been interpreted to mean that electrons in atoms occupy regions of space that are

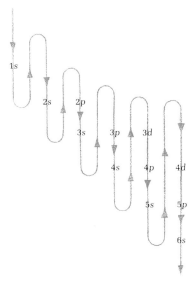

FIGURE 4.8
Order of filling the orbitals.

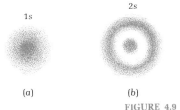

FIGURE 4.9
A diagram of a 1s and 2s orbital. Electrons are most likely to be found in darker areas.

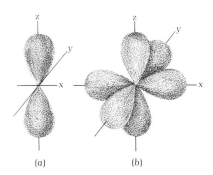

FIGURE 4.10
(a) A single p orbital, (b) three p orbitals.

oriented in different directions. According to the quantum mechanical model, these regions of space, known as orbitals, are indicated by the third quantum number, $m_l$. *Orbitals are regions of space where electrons are most likely to be found at any particular time. Each orbital can hold a maximum of two electrons.* Although the theory indicates the region of space in which the electron is most likely to be found, there is *nothing* in the theory that indicates the path electrons follow in traveling about the atom.

Each s subshell consists of an s orbital. The volumes of the orbitals in the 1s and 2s subshells are indicated in Fig. 4.9. One of the assumptions of the quantum mechanical model is that electrons do not travel in well-defined orbitals like the earth moving about the sun. Using this theory, one can only calculate the *probability* of finding an electron at any particular time. The darker regions in Fig. 4.9(a) indicate the regions of highest probability for finding an electron. Figure 4.9 shows only the volume of space occupied by electrons "most" of the time. Electrons spend some of their time in other regions of space.

**p Orbitals**

Every p subshell consists of three different p orbitals. Each of these orbitals can hold two electrons. The shape of the region of space which we call a p orbital is shown in Fig. 4.10(a). Each p orbital is at right angles to the others. They are often designated $p_x, p_y, p_z$. The three orbitals together are shown in Fig. 4.10(b). Remember, in the present model of the atoms there is no statement on the path electrons take in moving about the atom. One *cannot* conclude that since a p orbital resembles a figure "8," electrons in a p orbital travel in a figure "8."

## 4.7 Electron spin

Picture for a moment two electrons in the same orbital. Each electron is in the same atom, the same shell, the same subshell, and the same orbital. It appears that the two electrons are identical. However, according to the quantum theory the fourth quantum number, $m_s$ must be different. *One way* to interpret this number is to think of the electron as a spinning sphere. In a filled orbital, one electron spins in one direction, the other spins in the opposite direction. In this situation the electrons are said to be *paired*. Compare this

■ EXERCISE  Write out the electronic configurations for the following elements: boron, beryllium, fluorine, neon. In which case are there unpaired electrons?

with the arrangement of electrons in lithium, element number 3. Its electronic configuration is $1s^2\ 2s^1$. The two electrons in the 1s orbital are paired; however, the one electron in the 2s orbital is not paired. Its spin is not counterbalanced by a second electron. It is unpaired. ■

### Electrons and magnetism

Man has used magnets for thousands of years. Most students are already familiar with magnets and some of the aspects of magnetism. Chemists use magnetic properties of substances to help clarify the arrangement of electrons in atoms. Substances that are pushed out of a strong magnetic field are said to be *diamagnetic*. Substances exhibit diamagnetism when all their electrons are paired. Substances that are pulled into a strong magnetic field are said to be *paramagnetic*. Substances exhibit paramagnetism when they have one or more unpaired electrons. A spinning unpaired electron is believed to interact with the magnetic field and give a substance its paramagnetic properties.

The metals iron, nickel, and cobalt are easily magnetized and are called ferromagnetic. An explanation of their magnetic properties takes into consideration the presence of unpaired electrons. Thus the quantum mechanical model of the atom is a basis for the understanding of magnetism.

### Maximum unpairing

Let us examine the electronic configuration of carbon, element number 6. It has six electrons and its electronic configuration is: $_6C\ \ 1s^2\ 2s^2\ 2p^2$. The 1s and 2s orbitals are filled. Two electrons are in the 2p subshell. But there are three different p orbitals, each of which can hold two electrons. How are the two electrons in the 2p subshell of carbon distributed? Experimental evidence shows that individual carbon atoms have two unpaired electrons. Thus the configuration $_6C\ 1s^2\ 2s^2\ 2p_x^1\ 2p_y^1\ 2p_z^0$ must be correct.

A diagram of the p orbitals of a carbon atom is shown in Fig. 4.11. When any subshell is partially empty, the electrons distribute themselves to achieve *maximum unpairing* as in the case of carbon atoms. The reason is that a slight amount of energy is needed to overcome repulsion and place two negative particles in the same space region (same orbital). Within a subshell, electrons go into different orbitals before completely filling one orbital.

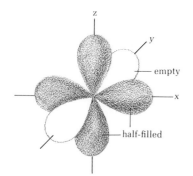

FIGURE 4.11
A diagram of the p orbitals of carbon showing that one of the orbitals is empty while each of the other two orbitals is half filled; that is, they contain one electron each.

EXAMPLE 1 Write the electronic configuration for an atom of nitrogen showing the detailed arrangement of electrons in orbitals.

Nitrogen is element number 7. Three electrons go into the 2p subshell. Because of the principle of maximum unpairing, these three electrons will each go into a different 2p orbital. The detailed electronic configuration is $_7$N $1s^2$ $2s^2$ $2p_x^1$ $2p_y^1$ $2p_z^1$.

EXAMPLE 2 Write the electronic configuration of an atom of sulfur showing the detailed arrangement of electrons in orbitals.

Sulfur is element number 16. Four electrons go into the 3p orbitals. Because of the principle of maximum unpairing, two electrons will fill one orbital and the other two electrons will go into two different orbitals. The detailed electronic configuration for sulfur is: $_{16}$S $1s^2$ $2s^2$ $2p^6$ $3s^2$ $3p_x^2$ $3p_y^1$ $3p_z^1$. ■

■ EXERCISE Write the electronic configuration for an atom of phosphorus and an atom of oxygen showing the detailed arrangement of electrons in orbitals.

## Atomic orbital diagrams

Now we will review much of what has been covered thus far in this chapter by *writing atomic orbital diagrams* for the elements. An atomic orbital diagram shows the electronic configuration of an atom by means of a diagram. An empty orbital is represented by ◯. An orbital with only one electron in it may be represented by ⊙. An orbital with two electrons in it may be represented by ⊙. The arrows are meant to represent the relative spin of the electrons. When two electrons are in the same orbital their spins oppose each other. This is indicated by writing the arrows pointed in opposite directions. The atomic orbital diagrams for the first 36 elements are shown below. Notice that only the outer shell is used in some of the diagrams. Compare these diagrams with the electronic configurations of the elements shown in Table 4.4.

The atomic orbital diagrams for the first ten elements are:

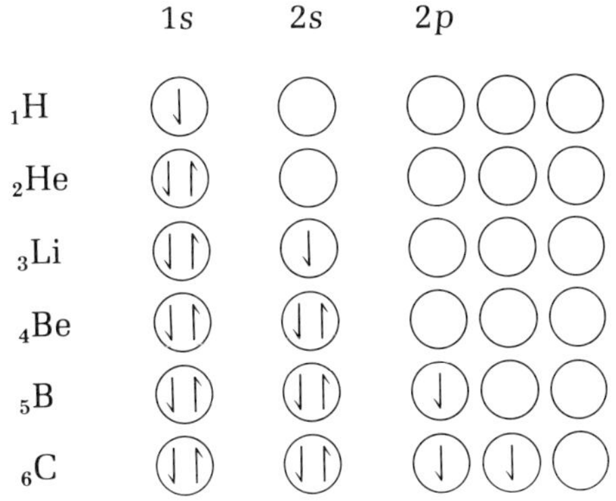

4.7 Electron spin

|   | 1s | 2s | 2p |
|---|---|---|---|
| ₇N | ⇅ | ⇅ | ↓ ↓ ↓ |
| ₈O | ⇅ | ⇅ | ⇅ ↓ ↓ |
| ₉F | ⇅ | ⇅ | ⇅ ⇅ ↓ |
| ₁₀Ne | ⇅ | ⇅ | ⇅ ⇅ ⇅ |

Note that each 2p orbital has one electron before any other has two electrons. The third shell consists of a 3s orbital, three different 3p orbitals, and five different 3d orbitals. However, as shown in Fig. 4.7, the 4s orbital is less energetic than the 3d orbitals. Consequently, the 4s orbital fills up before the 3d orbitals. The atomic orbitals for elements 11 to 20 are filled on the 1s, 2s, and 2p levels.

|   | 3s | 3p | 3d | 4s |
|---|---|---|---|---|
| ₁₁Na | ↓ | ○ ○ ○ | ○ ○ ○ ○ ○ | ○ |
| ₁₂Mg | ⇅ | ○ ○ ○ | ○ ○ ○ ○ ○ | ○ |
| ₁₃Al | ⇅ | ↓ ○ ○ | ○ ○ ○ ○ ○ | ○ |
| ₁₄Si | ⇅ | ↓ ↓ ○ | ○ ○ ○ ○ ○ | ○ |
| ₁₅P | ⇅ | ↓ ↓ ↓ | ○ ○ ○ ○ ○ | ○ |
| ₁₆S | ⇅ | ⇅ ↓ ↓ | ○ ○ ○ ○ ○ | ○ |
| ₁₇Cl | ⇅ | ⇅ ⇅ ↓ | ○ ○ ○ ○ ○ | ○ |
| ₁₈Ar | ⇅ | ⇅ ⇅ ⇅ | ○ ○ ○ ○ ○ | ○ |
| ₁₉K | ⇅ | ⇅ ⇅ ⇅ | ○ ○ ○ ○ ○ | ↓ |
| ₂₀Ca | ⇅ | ⇅ ⇅ ⇅ | ○ ○ ○ ○ ○ | ⇅ |

After the 4s orbital is filled, the 3d orbitals begin to fill. They are followed by the 4p orbitals. This accounts for elements 21 to 36.

|   | 3d | 4s | 4p |
|---|---|---|---|
| ₂₁Sc | ↓ ○ ○ ○ ○ | ⇅ | ○ ○ ○ |
| ₂₂Ti | ↓ ↓ ○ ○ ○ | ⇅ | ○ ○ ○ |
| ₂₃V | ↓ ↓ ↓ ○ ○ | ⇅ | ○ ○ ○ |

■ EXERCISE  Using the atomic orbital diagrams for elements from 1 to 36, answer the following questions.

1. List three elements whose individual atoms you would expect to be diamagnetic.
2. List three elements whose individual atoms you would expect to be paramagnetic.
3. List the elements that have only one electron in the following orbitals: 2s, 3s, 4s.
4. List the elements that have five electrons in each of the following subshells: 2p, 3p, 4p.

## 4.8 Valence electrons

Earlier in this chapter we saw that sodium and potassium were similar in chemical and physical properties. An examination of their atomic orbital diagrams provides an understanding of their similarities on the submicroscopic level.

What is the highest energy, occupied orbital in a sodium atom? In a potassium atom? How many electrons are in these orbitals? The occupied orbital of highest energy in a sodium atom is the 3s orbital and in a potassium atom the 4s orbital. In each case there is one electron in the outermost s orbital. Could this similarity in the outer electron orbitals be the reason why sodium and potassium are similar? In Chapter 6 we will discuss the role of electrons in determining chemical properties. What other elements resemble sodium and potassium? What properties does the element with one electron in the 2s orbital have? This element is lithium:

Experiments show that lithium behaves in many ways like sodium and potassium. It is a reactive metal. It liberates a combustible gas when added to water and the resulting solution feels soapy.

The number of electrons in the outermost shell is often useful in predicting and correlating properties of elements. The *outer shell electrons* are known as *valence electrons*. The outermost shell is called the *valence shell*. Lithium, sodium, and potassium each have one valence electron. The valence electron of sodium is in the third shell. In what shell is the valence electron of potassium?

In making comparisons, using the number of valence electrons to predict similar properties of elements does not work every time. For instance, what element has one electron in the first shell? It is hydrogen:

$$_1H \quad \underset{1s}{\downarrow}$$

Hydrogen, like lithium, sodium, and potassium, has one valence electron. Yet hydrogen is a gas that does not react readily at room temperature. It is not a solid reactive metal like lithium, sodium, and potassium.

Even though lithium, sodium, and potassium are similar, they are *not identical*. If all their properties were identical, how could we tell them apart? Lithium is harder to cut than sodium or potassium; potassium reacts with water much faster than lithium or sodium. Their densities are: lithium = 0.53 g/cm³, sodium = 0.97 g/cm³, potassium = 0.86 g/cm³. The macroscopic differences in lithium, sodium, and potassium depend upon submicroscopic differences. Atoms of lithium, sodium, and potassium are shown in Fig. 4.12. Using this figure, list some of the differences between the atoms of these elements.

Consider the elements tin and lead, which have similar chemical properties. Both are common metals and form compounds with similar formulas. They both unite with atoms of oxygen in the same ratios. One atom of tin combines with one

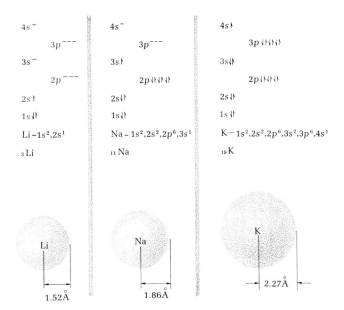

**FIGURE 4.12**
A comparison of the alkali metals, lithium, sodium, and potassium.

atom of oxygen and one atom of lead combines with one atom of oxygen. The resulting compounds are tin(II) oxide, SnO, and lead(II) oxide, PbO. Each metal will also combine with fluorine in the same ratios. One atom of tin will combine with four atoms of fluorine and one atom of lead will combine with four atoms of fluorine. The resulting compounds are tin(IV) fluoride, $SnF_4$, and lead(IV) fluoride, $PbF_4$. Do these two elements also have similar electronic configurations? Compare their valence shells. The atomic orbital diagrams for the valence shells of tin and lead are:

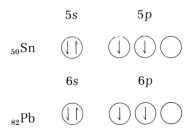

Note that the arrangement of elctrons in the valence shells of tin and lead are similar. We will see in Chapter 6 that the electrons in the p orbitals are used in forming SnO and PbO. All four electrons in the valence shells are used to form $SnF_4$ and $PbF_4$.

4.8 Valence electrons

■ EXERCISE  *Based on their electronic structure, which of the following elements would you expect to have similar chemical properties?*
(a) $_{13}$Al [Ne] $3s^2\ 3p^1$
   $_{15}$P  [Ne] $3s^2\ 3p^3$
   $_{33}$As [Ar] $4s^2\ 4p^3$
(b) $_{12}$Mg  $1s^2\ 2s^2\ 2p^6\ 3s^2$
   $_{17}$Cl   $1s^2\ 2s^2\ 2p^6\ 3s^2\ 3p^5$
   $_{20}$Ca   $1s^2\ 2s^2\ 2p^6\ 3s^2\ 3p^6\ 4s^2$

Tin and lead, in addition to their similarities, have many different properties. What are some of them? Remember that different properties are due to differences in atomic structure. Diagrams of atoms of both elements are shown in Fig. 4.13. Using this figure, list as many differences between atoms of tin and lead as you can. ■

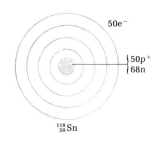

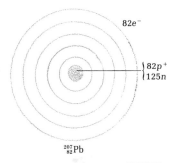

FIGURE 4.13
*Diagram of atoms of tin and lead.*

**Transition compounds**

Elements such as iron, with partially filled $3d$ orbitals, belong to the *first transition series*. Turn back to the atomic orbital diagrams for the first 36 elements. Which elements have partially filled $3d$ orbitals? As you can see they are elements 21 to 28. Copper and zinc, elements 29 and 30, are often included as members of the first transition series. There is a second and third transition series corresponding to the $4d$ and $5d$ orbitals. Even though the valence electrons of iron are in the fourth shell, iron atoms often use their $3d$ electrons when they combine with other atoms.

$$\quad\quad\quad 3s \quad\quad\quad 3p \quad\quad\quad\quad 3d \quad\quad\quad\quad 4s$$

$_{26}$Fe  (↓↑)  (↓↑)(↓↑)(↓↑)  (↓↑)(↓)(↓)(↓)(↓)  (↓↑)

The $3d$ and $4s$ electrons and orbitals are often used when iron forms compounds. In effect, iron can donate or share two or three electrons when it combines with other elements.

## 4.9 Electron-dot structures

Electronic configurations and atomic orbital diagrams are useful when you wish to analyze the valence shell of an element. Frequently, however, the *valence electrons* are symbolized by dots or crosses written around the symbol of the element. These symbols are known as *electron-dot structures*. For instance, the electron-dot structure for hydrogen, which has one valence electron, is:

H·

The dot structure for fluorine, which has seven valence electrons, is:

:F̈·

Usually the dots are written either directly above, below, to the left, or to the right of the symbol. Two electrons can go into each of these positions. The usual practice is to indicate the orbitals which are completely filled and those which are half-filled.

For nitrogen, write:

$\cdot \overset{..}{\underset{.}{N}} \cdot$

For boron, write:

$\overset{..}{B} \cdot$

These dot structures are very useful in explaining the ratios in which atoms combine together. Thus, they will be used repeatedly in Chapter 6. The following examples illustrate how to determine correct electron-dot structures.

EXAMPLE 3  Write the electron-dot structure for phosphorus, element number 15.

The electronic configuration of phosphorus is $1s^2\ 2s^2\ 2p^6\ 3s^2\ 3p^3$. There are five electrons in the valence shell. The dot structure is:

$\cdot \overset{..}{\underset{.}{P}} \cdot$

EXAMPLE 4  Write the electron-dot structure for iron, element number 26.

The electronic configuration for iron is $1s^2\ 2s^2\ 2p^6\ 3s^2\ 3p^6\ 3d^6\ 4s^2$. The valence shell is the fourth shell. It contains two electrons. The dot structure is:

$\cdot Fe \cdot$

Actually, Example 4 illustrates one of the inadequacies of electron-dot structures. According to the dot structure, iron has two electrons that it can use in forming compounds. We saw earlier, though, that in addition to these two electrons iron may also use electrons in its 3d energy level for forming compounds.

In Chapter 5 we will study the arrangement of elements into groups with similar electronic structures and similar properties.

## Summary

People, matter, and energy are all quantized; that is, they occur as bundles of one. The smallest possible bundle of energy is called a *quantum*. There are only certain allowed energy values; these values correspond to multiples of the smallest bundle of energy, the quantum.

There are only certain allowed energy values for the electrons in an atom. The allowed energy values for the electrons in an atom correspond to the allowed energy values for a man on a ladder. The particular wavelengths of light that an atom absorbs give experimental support to the hypothesis that the electrons in an atom have only certain allowed energy values. Electrons in atoms may have energy values which correspond to any one of the energy shells, but they cannot have any of the energies between these allowed values. It is impossible to locate positively the position of an electron at a given time. However, there is a high probability that an electron will be located in a certain region of space, called an *orbital*. Orbitals are classed as *s, p, d,* and *f*, according to their shapes. There is only one *s* orbital in a given energy shell. There are three *p* orbitals, five *d* orbitals, and seven *f* orbitals possible in a given energy shell.

The maximum number of electrons that a shell can hold is given by $2(n^2)$ where *n* is the shell number. The maximum number of electrons is 2 in the first shell, 8 in the second shell, 18 in the third shell, and 36 in the fourth shell.

The *electron configuration* of an atom is a symbolic representation of the arrangement of electrons in an atom. Quite a lot of information about an element and its properties can be obtained from its electron configuration: the number of valence electrons, the electron-dot structure, the number of unpaired electrons, and the similarity and differences between elements. The electron-dot structure for an element indicates the number of valence electrons.

The electronic configurations of elements can be used to predict their properties. A prediction can also be made about how one element will interact with another. In Chapter 5 we will show that the periodic table is really an arrangement of the elements according to their electronic configurations (electronic structure).

*Objectives*  By the end of this chapter you should be able to do the following:

1. State the simplest quanta that make up people, matter, electricity, and light. (Question 11)

2. State whether energy is absorbed or emitted as an electron changes energy levels in an atom. (Questions 12, 13)

3. Arrange the electron shells of an atom in order of increasing energy. (Question 14)

4. Determine the maximum number of electrons in each of the first five shells of an atom. (Questions 15, 16, 17)

5. Given the atomic number of an element from 1 to 38 indicate how many electrons are in each shell. (Question 18)

6. State the types of subshells in each of the first four shells. (Question 19)

7. State the maximum number of electrons possible in each s, p, d, or f subshell. (Question 20)

8. State the order of fill for the 1s to 4p subshells. (Questions 21)

9. Given the atomic number, write the electronic configuration of any of the first twenty elements. (Question 2)

10. Draw a contour of both an s orbital and a p orbital. (Questions 8, 22)

11. Given the atomic number, draw the atomic orbital diagram for any of the first twenty elements. (Question 23)

12. Given the electronic configuration of an element, classify it as having either diamagnetic properties or paramagnetic properties. (Questions 5, 24)

13. Given the electronic structure of an atom, identify the valence shell and the number of valence electrons. (Questions 3, 25, 26)

14. Given the atomic number and mass number of two different atoms, point out differences and similarities in mass, number of protons, number of neutrons, number of electrons, valence shell, and number of valence electrons. (Question 27)

15. Base on the electronic structure of a set of elements, which ones would you expect to have similar chemical properties? (Question 28)

16. Given the electronic structure of an element, write its electron-dot structure. (Question 6)

## Questions

1. Define the following terms: quanta; photon; 2p orbitals; ground state; excited state; electronic configuration; unpaired electrons; energy levels in an atom; electron shell; main energy level; electron subshell; electron orbital; electron spin; 1s orbital; 2s orbital; diamagnetic; paramagnetic; maximum unpairing; atomic orbital diagrams; valence electrons; valence shell; transition elements; electron-dot structures.

2. Write the electron configurations for: (a) $_4$Be; (b) $_{12}$Mg; (c) $_{20}$Ca; (d) $_8$O; (e) $_{16}$S; (f) $_9$F; (g) $_{17}$Cl; (h) $_{10}$Ne; (i) $_{18}$Ar.

3. How many valence electrons do each of the elements in Question 2 have?

4. How many unpaired electrons are there in each of the elements of Question 2?

5. Which of the elements of Question 2 are paramagnetic? diamagnetic?

6. Write electron-dot structures for the elements in Question 2.

7. In addition to s and p orbitals, there are also two other types of orbitals, d orbitals and f orbitals. If there are five d orbitals in a shell, and seven f orbitals in a shell, what is the maximum number of d electrons and f electrons which can exist in a single shell?

8. Draw a 1s orbital in such a manner as to indicate the probability of finding the electron of a hydrogen atom in this region of space. Discuss this probability.

9. Magnetic measurements indicate that silicon has two unpaired electrons. Use electron configurations to explain this phenomenon.

10. Calcium, magnesium, and barium are found to have similar properties. Can you find anything in their electronic structure (electronic configuration) which would cause this similarity? Explain.

11. The concept of quantum can be applied to people, matter, light, and electricity. What is a quantum of people, of matter (according to Dalton), of light, and of electricity?

12. When a compound containing potassium is heated in a hot flame, some electrons are raised from a lower energy level to a higher energy level. Does this process require energy or does it emit energy? Explain your answer.

13. Neon gas in a neon light emits an orange-red light when electricity is passed through it. The light represents the emission of energy from atoms where electrons are changing levels. Are the electrons that cause the emission of light moving from higher energy levels to lower energy levels or are they moving from lower energy levels to higher energy levels? Explain your answer.

14. A magnesium atom, such as might make up the magnesium in a flash bulb, contains electrons in three different shells: $n = 1$, $n = 2$, and $n = 3$. In which shell do the electrons possess the most energy? In which shell do the electrons possess the least energy? Explain your answer.

15. Calculate the maximum number of electrons in the first shell. What neutral atom has just this number of electrons?

16. Calculate the maximum number of electrons in the second shell. In what element is this shell just filled?

17. Calculate the maximum number of electrons in the third, fourth, and fifth shells.

18. How many electrons are in each shell of the following elements? (a) $_1$H; (b) $_3$Li; (c) $_{11}$Na; (d) $_{19}$K; (e) $_4$Be; (f) $_7$N; (g) $_{15}$P; (h) $_9$F; (i) $_{17}$Cl.

19. What subshells make up each of the following shells? (a) $n = 1$; (b) $n = 2$; (c) $n = 3$; (d) $n = 4$.

20. How many orbitals make up each of the following subshells? How many electrons are in each subshell? (a) s subshell; (b) p subshell; (c) d subshell; (d) f subshell.

21. Rearrange the following subshells in order of increasing energy: 2s, 3p, 1s, 4s, 3d, 2p, 3s, 4p.

22. Draw a 3p orbital in such a manner as to indicate the probability of finding an electron on this energy level.

23. Draw the atomic orbital diagram of the valence shell of each of the following elements: (a) $_3$Li; (b) $_{11}$Na; (c) $_{12}$Mg; (d) $_{20}$Ca; (e) $_6$C; (f) $_{14}$Si; (g) $_9$F; (h) $_{17}$Cl; (i) $_{10}$Ne.

24. Which of the following single atoms would you expect to be paramagnetic (one or more unpaired electrons)? (a) $_1$H; (b) $_2$He; (c) $_3$Li; (d) $_{11}$Na; (e) $_9$F; (f) $_{10}$Ne.

25. What is the valence shell of each of the following elements?
(a) $_3$Li    $1s^2\ 2s^1$
(b) $_{19}$K    $1s^2\ 2s^2\ 2p^6\ 3s^2\ 3p^6\ 4s^1$
(c) $_{38}$Sr    $1s^2\ 2s^2\ 2p^6\ 3s^2\ 3p^6\ 3d^{10}\ 4s^2\ 4p^6\ 5s^2$
(d) $_{35}$Br    $1s^2\ 2s^2\ 2p^6\ 3s^2\ 3p^6\ 3d^{10}\ 4s^2\ 4p^5$.

26. How many valence electrons do each of the elements in Question 24 have?

27. Compare the mass number, the number of protons, the number of neutrons, the number of electrons, the valence shell, and the number of valence electrons in each of the following pairs: (a) $^7_3$Li and $^{39}_{19}$K; (b) $^7_3$Li and $_4$Be; (c) $^{23}_{11}$Na and $^{24}_{12}$Mg; (d) $^{16}_8$O and $^{32}_{16}$S; (e) $^{14}_7$N and $^{16}_8$O; (f) $^{12}_6$C and $^{28}_{14}$Si.

28. Which of the elements in Question 27 would you expect to have similar physical and chemical properties?

# 5

# Periodicity

## 5.1 Introduction

What similarities exist between the elements? What differences? In Chapter 4 we saw that three elements—lithium, sodium, and potassium—have similar properties and one electron in their valence shells. Are there other groups of elements which have similarities in properties and valence electrons? Can the electronic structure of elements be used to explain similarities and differences in their properties? Answers to these and other questions will be considered in this chapter.

How do scientists and engineers make use of a knowledge of the similar properties of some elements? For example, consider sodium hydroxide, NaOH, a common laboratory substance. One of the properties of sodium hydroxide is its ability to purify air by chemically combining with and removing carbon dioxide, $CO_2$, from it. This property suggests that sodium hydroxide would be useful for purifying the atmosphere of closed vessels such as space ships or submarines. Since lithium metal and potassium metal show similarities to sodium metal, one might logically ask if lithium hydroxide, LiOH, and potassium hydroxide, KOH, have the same ability to remove carbon dioxide from the air as does NaOH. Experiments show that they do. Which of these three substances would you use to purify air in a space ship?

LiOH    mole weight = 23.9 g
NaOH    mole weight = 40.0 g
KOH    mole weight = 56.1 g

■ EXERCISE  Look at the periodic table to see which elements occur in the same vertical column as strontium, Sr. Although strontium is not a common element, it and calcium can readily be incorporated into bones in the human body. Strontium is produced in radioactive fallout. Why is fallout containing radioactive strontium so dangerous?

Since space vehicles must be as light as possible, lithium hydroxide would probably be preferred over sodium hydroxide or potassium hydroxide.

This chapter considers an arrangement of the elements known as the *periodic table*. This chart lists the 105 known elements in such a way that elements with similar properties can be recognized readily. Throughout this chapter you will want to refer to this chart, which is found on the inside back cover of this book. ■

## 5.2 Historical development

The periodic table or periodic model was developed over many years. By 1830 the German chemist Dobereiner had pointed out the existence of *triads*, or groups of three elements with similar properties. Three of these groups are shown in Table 5.1. The elements lithium, sodium, and potassium are soft, reactive metals; calcium, strontium, and barium are also reactive metals; chlorine, a gas, bromine, a liquid, and iodine, a solid, are reactive nonmetals.

TABLE 5.1

Three of the triads of Dobereiner.

| 1 | 2 | 3 |
|---|---|---|
| Li Lithium | Ca Calcium | Cl Chlorine |
| Na Sodium | Sr Strontium | Br Bromine |
| K Potassium | Ba Barium | I Iodine |

Newlands, a British chemist, reported his *Law of Octaves* in 1864. When he arranged the elements according to increasing *atomic weight*, he noted that every *eighth element* had similar properties. Table 5.2 shows how Newlands arranged the elements. Of course, many elements, such as the noble gases, had not yet been discovered and the numbers assigned by him applied only to his day. If we start with lithium, we find that sodium is the eight element from lithium. Potassium is the eighth element from sodium. According to Newland's model, lithium, sodium, and potassium should have similar proper-

**TABLE 5.2**

*Newland's Law of Octaves. Elements in horizontal rows have similar properties.*

| | | |
|---|---|---|
| H-1 | | |
| Li-2 | Na-9 | K-16 |
| Be-3 | Mg-10 | Ca-17 |
| B-4 | Al-11 | |
| C-5 | Si-12 | |
| N-6 | P-13 | |
| O-7 | S-14 | |
| F-8 | Cl-15 | |

ties. Why should every eighth element have similar properties? Later in this chapter we will use electronic configurations to help answer this question.

**Mendeleev and Meyer**

Meyer in Germany and Mendeleev in Russia independently developed a periodic arrangement or periodic model of the elements. Both arranged the elements according to *increasing atomic weight*. Mendeleev, working around 1870, left some blank spaces in his table. He predicted that there were elements, at that time undiscovered, that would be discovered and fill the blanks. One such blank was between silicon and tin on his periodic table. Mendeleev predicted that its properties and the properties of its compounds would be an "average" of the properties of silicon and tin. Fifteen years after Mendeleev's prediction, Winkler discovered germanium, an element with properties similar to those predicted by Mendeleev. (Check the properties of germanium in Table 5.7 and its position on the periodic table.

In addition to leaving blank spaces in his periodic table, Mendeleev reversed the expected order of elements in two places. In order to place elements with similar properties in the same vertical column or group, he reversed nickel and cobalt, and tellurium and iodine. In this case he modified his own order of *increasing atomic weight* in order to fit it to the observations. Forty years later, Moseley was able to show why these changes should be made.

**Moseley**

In 1914 Moseley bombarded various elements with electrons and studied the X-rays which were emitted from different target elements. Moseley concluded that the frequency of the X-rays produced depends on the number of *positive charges* in the nucleus of the atom. According to Moseley, *atomic numbers* can be assigned to elements on the basis of the frequency of the X-rays each element emits. He said that elements should be arranged in a periodic table in order of increasing *atomic number*.

Recall that Mendeleev placed cobalt ahead of nickel in his periodic table even though cobalt has an atomic weight of 58.9 and nickel has an atomic weight of 58.7. Moseley's data show that the atomic number of cobalt is 27 and that the atomic

■ **EXERCISE** *The atomic weight of tellurium is 127.6 and that of iodine is 126.9. The atomic number of tellurium is 52 and that of iodine is 53. Which of* these two elements should come first on the periodic chart? Why?

number of nickel is 28. Thus, on a basis of atomic numbers, cobalt should precede nickel on the periodic chart. ■

## 5.3 Structural basis of the periodic chart

The modern periodic chart was constructed by arranging the elements in horizontal rows and vertical columns. The elements in a particular vertical column are known as a *group* or *family* of elements. There are seven horizontal rows of elements in the chart, called *periods*. The periodic chart is formed by arranging the elements in order of *increasing atomic number* left to right across a period. ■

■ EXERCISE  Using the periodic chart, answer the following questions. What elements are in group IA? IIA? VIIA? VIIIA? period 1? period 2?

## 5.4 Periods

What structural relationship exists between the elements in a particular period? Consider the period 2 elements:

| Element | First shell, n = 1 | Second shell, n = 2 |
|---|---|---|
| $_3$Li | $1s^2$ | $2s^1$ |
| $_4$Be | $1s^2$ | $2s^2$ |
| $_5$B | $1s^2$ | $2s^2\ 2p^1$ |
| $_6$C | $1s^2$ | $2s^2\ 2p^2$ |
| $_7$N | $1s^2$ | $2s^2\ 2p^3$ |
| $_8$O | $1s^2$ | $2s^2\ 2p^4$ |
| $_9$F | $1s^2$ | $2s^2\ 2p^5$ |
| $_{10}$Ne | $1s^2$ | $2s^2\ 2p^6$ |

■ EXERCISE  Look up or write out the electronic configuration of the following elements: $_4$Be, $_{12}$Mg, $_{20}$Ca, $_8$O, $_{16}$S. What is the valence shell for each of these elements? In what period would you expect to find these elements? Check your answers in the periodic chart.

What is the valence shell of each of these elements? Note that the valence shell number 2 is the same as the period number. For all elements the *period number* and the *valence shell number* are identical. ■

## 5.5 Groups

What structural feature is common to the elements in a particular group? Consider the group IA elements and their electronic configurations:

| Element | | | | | | Valence shell |
|---|---|---|---|---|---|---|
| $_1$H | | | | | | $1s^1$ |
| $_3$Li | $1s^2$ | | | | | $2s^1$ |
| $_{11}$Na | $1s^2$ | $2s^2$ | $2p^6$ | | | $3s^1$ |
| $_{19}$K | $1s^2$ | $2s^2$ | $2p^6$ | $3s^2$ | $3p^6$ | $4s^1$ |
| $_{37}$Rb | $1s^2$ | $2s^2$ | . . . | $4s^2$ | $4p^6$ | $5s^1$ |
| $_{55}$Cs | $1s^2$ | $2s^2$ | . . . | $5s^2$ | $5p^2$ | $6s^1$ |
| $_{87}$Fr | $1s^2$ | $2s^2$ | . . . | $6s^2$ | $6p^6$ | $7s^1$ |

■ EXERCISE  On the basis of the number of valence electrons, in which group would you expect to find each of the following elements? Check your answer with the periodic chart.

(a) $_{13}$Al   $1s^2\ 2s^2\ 2p^6\ 3s^2\ 3p^1$
(b) $_7$N   $1s^2\ 2s^2\ 2p^3$
(c) $_{10}$Ne   $1s^2\ 2s^2\ 2p^6$

■ EXERCISE  Using the periodic chart to locate the group number, state how many electrons are in the valence shell of each of the following elements: phosphorus; iodine; krypton; silicon.

Each of these elements has *one electron* in its valence shell. This electron is located in an *s* orbital. *The group number I corresponds to the number of valence electrons.* Knowing what A group an element is in immediately tells us how many valence electrons the element has. This information will be useful in determining the formulas of compounds that elements form.

We saw in Chapter 4 that lithium, sodium, and potassium were similar. We will find that, throughout the periodic chart, *elements in the same groups have similar properties.* ■

## 5.6  Electron-dot structures for atoms

When elements react there is usually a gain, a loss, or a sharing of valence electrons. Before studying the formation of compounds by the combination of elements (Chapter 6), it will be useful to be able to write electron-dot structures quickly. Since the group number of an element indicates the number of valence electrons an atom of the element has, we need merely check the periodic chart for the group number and draw the corresponding number of dots about the symbol for the element. Each of the electrons is shown according to its placement in orbitals as in Fig. 5.1. ■

■ EXERCISE  Using a periodic chart, write electron-dot structures for the following elements: fluorine; chlorine; tin; lead.

## 5.7  Electronic configurations and the periodic chart

The periodic chart is formed by arranging the elements in order of increasing atomic number. When the elements are arranged

**FIGURE 5.1**

The electron dot structures of the A group elements in a periodic chart format. Electron dot structures are also called Lewis dot structures after G. N. Lewis, a famous chemist.

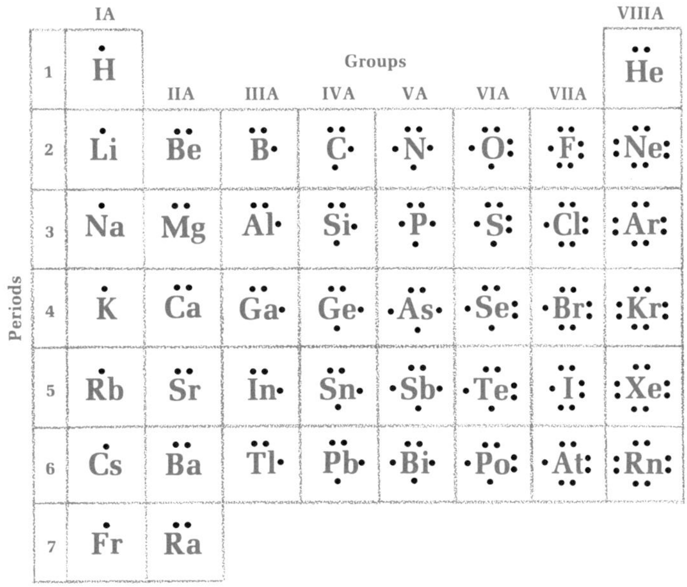

■ **EXERCISE** Using a periodic chart, classify the following elements as being either representative elements, transition elements, or inner transition elements: Cl; Na; U; Fe; Ag; Pb.

in this way they fall into groups which have the same number of valence electrons and similar properties.

As shown in Fig. 5.2, the periodic chart can be divided into four sections: an s orbital section, a p orbital section, a d orbital section, and an f orbital section. Each of these sections indicates the subshell to which the last differentiating electron has been added. This electron is added as one proceeds from element to element by increasing atomic number.

The s orbital section consists of two groups, IA and IIA. These groups consist of elements whose differentiating electrons are located in s orbitals.

The p orbital section consists of six groups: IIIA, IVA, VA, VIA, VIIA, and VIIIA. These groups consist of elements whose differentiating electrons are located in p orbitals.

Elements located in the s and p sections of the periodic chart (the A groups) are known as the *representative elements*. It is these elements that we will focus on in developing and applying various chemical models.

There are ten groups in the d section of the chart, corresponding to the ten d electrons per shell. Elements in this section of the chart are known as *transition elements*.

There are fourteen groups in the f section of the chart, corresponding to the fourteen f electrons per shell. Elements in this section of the chart are known as *inner transition elements*. ■

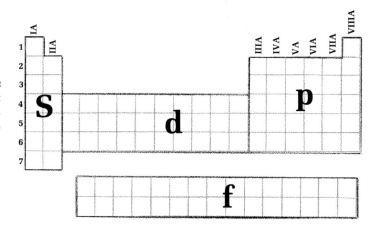

FIGURE 5.2
An outline of the periodic chart showing in which orbital the differentiating electron of each element is found.

## 5.8 The periodic law or periodic model

When the elements are arranged in order of increasing atomic number, they fall into groups with similar properties. Elements with similar properties keep reappearing; their properties are thus said to be periodic. Moseley presented evidence that the *properties of the elements are periodic functions of their atomic numbers*. This statement is known as the *periodic law* or *periodic model*.

### Trends in groups

Table 5.3 lists the boiling points, in Celsius degrees, of the group IA elements and the sizes of the atoms (in Angstroms) of each element. This information is plotted in Fig. 5.3.

What trend is there in the boiling points as you start with lithium and move down the group; that is, as the atomic mass increases? In general, there is a decrease in the boiling point of the elements in group IA. What trend is there in the sizes of

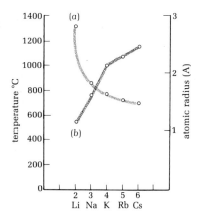

FIGURE 5.3
(a) The boiling points of the group IA elements plotted as a function of their period number. (b) The atomic radii of the group IA elements plotted as a function of their period number.

TABLE 5.3
The boiling points and atomic radii of the group IA elements.

| Element | Boiling point (°C) | Atomic radius (Å) |
|---|---|---|
| Lithium | 1326 | 1.23 |
| Sodium | 889 | 1.57 |
| Potassium | 757 | 2.03 |
| Rubidium | 679 | 2.16 |
| Cesium | 690 | 2.35 |
| Francium | — | — |

**TABLE 5.4**

The atomic radii and formulas for hydrogen compounds of the period two elements. Neon is omitted from this table since it forms no compound with hydrogen.

| Element | Atomic radius (Å) | Formulas for hydrogen compound |
|---|---|---|
| Lithium | 1.23 | LiH |
| Beryllium | 0.89 | $BeH_2$ |
| Boron | 0.80 | $BH_3$ |
| Carbon | 0.77 | $CH_4$ |
| Nitrogen | 0.74 | $NH_3$ |
| Oxygen | 0.74 | $H_2O$ |
| Fluorine | 0.72 | HF |

atoms as you start with lithium and move down the group; that is, as atomic mass increases? Both boiling points and atomic sizes indicate that properties of elements in a group tend to change in a gradual, orderly manner. Other properties which show similar, gradual changes with increasing atomic mass are melting points, size of ions, and acidity and basicity of oxides.

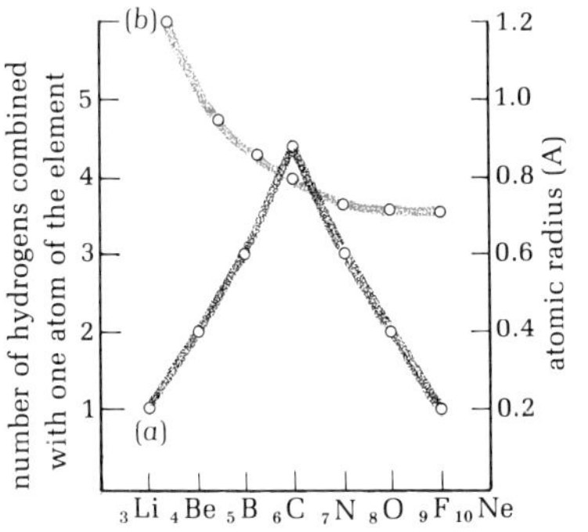

**Trends in periods**

Table 5.4 lists the sizes of atoms in the second period and the formulas for the compounds that each second period element forms with hydrogen. This information is plotted in Fig. 5.4.

What trend is there in the atomic sizes as you start with lithium and move across the second period? In general, there is a decrease in atomic size with increasing atomic numbers. What trend is there in the formulas of the compounds that the second period elements form with hydrogen? Notice that the number of hydrogen atoms increases from LiH to $CH_4$ and then decreases to HF. Both atomic sizes and formulas of hydrogen-containing compounds indicate that the properties of elements in a period change in a gradual, orderly manner.

**FIGURE 5.4**

The atomic radii and hydrogen compounds of period two elements. (a) Combined hydrogens, (b) atomic radius.

## 5.9 Ionization energy— a periodic property

If electrons of sufficient energy are shot through sodium vapor, a noticeable electronic current is detected. Some of the energy of the bombarding electron is transferred to the sodium atom on collision. When this energy equals or exceeds a certain minimum value, a positive sodium ion and a *free electron* are produced.

$$Na(g) \text{ (atom)} + \text{energy} \rightleftharpoons Na^+(g) + 1e^- \quad (5.1)$$

**TABLE 5.5**

The first ionization energies of the group IA elements.

| Element | First ionization energy (kcal/mole) |
|---|---|
| Lithium | 124 |
| Sodium | 118 |
| Potassium | 100 |
| Rubidium | 97.5 |
| Cesium | 90 |
| Francium | — |

■ EXERCISE  Which of the group VIIIA elements would you expect to have the lowest ionization energy?

■ EXERCISE  Which element among the following would you expect to have the lowest ionization energy? Which element would you expect to have the highest ionization energy? Sodium, magnesium, aluminum, silicon, phosphorus, sulfur, chlorine, and argon.

The process of forming an ion is called *ionization*. The energy required to remove the first electron from an atom is called the *first ionization energy* (I.E.). The energy required to remove the second electron is called the *second ionization energy*.

The first ionization energies for group IA elements are listed in Table 5.5 and plotted in Fig. 5.5. The values in Table 5.5 tell us how many kilocalories of energy are required to remove a mole of electrons from a mole of atoms.

What trend do you notice in the first ionization energies of the group IA elements? The ionization energy *decreases* in a gradual manner. This means that the elements at the *bottom* of the group tend to *lose electrons* more readily than elements at the *top* of the group.

Compare sodium (I. E. = 118 kcal/mole) with rubidium (I.E. = 97.5 kcal/mole). Why is there a decrease in ionization energy in going down a group? One reason for the difference in ionization energies is that the valence shell of rubidium is in the $n = 5$ shell. The valence electron in rubidium is in a higher energy level than that of sodium. It is also farther away from the nucleus. As a result, the valence electron of rubidium is removed from an atom more readily than a valence electron of sodium. For similar reasons, we find in the other A groups of the periodic chart that the *ionization energies* tend to *decrease in moving down the group*. ■

The first ionization energies of the second period elements are listed in Table 5.6 and are plotted in Fig. 5.6. What trend do you notice in the ionization energies of these elements in moving from left to right across the period?

Compare lithium (I.E. = 124 kcal/mole) with fluorine (I.E. = 401 kcal/mole). Why is there an increase in ionization energy as one moves across the period? The positive charge in the nucleus of the atom attracts the negative electrons. In going across the second period, the positive charges in the nucleus increase. This increases the atom's attraction for electrons, and consequently more work is required to overcome the force; the ionization energy *increases*. In moving across any *period from left to right, the ionization energy tends to increase*. ■

## 5.10 Stable configurations— the noble gases

The members of one group of the periodic chart do not readily form compounds. The group VIIIA elements, the *noble gases*,

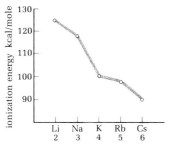

FIGURE 5.5

The first ionization energies of the group IA elements versus their period number.

■ EXERCISE Make a list of elements that have one less electron than each noble gas. In what group are these elements found? Make a list of elements that have one more electron than each noble gas. In what group are these elements found?

TABLE 5.6

The first ionization energies of the elements in period 2.

| Element | First ionization energy (kcal/mole) |
|---|---|
| Lithium | 124 |
| Beryllium | 215 |
| Boron | 191 |
| Carbon | 260 |
| Nitrogen | 336 |
| Oxygen | 314 |
| Fluorine | 401 |
| Neon | 497 |

usually will not combine with other elements to form compounds. Why is it that the elements of this group are relatively inert?

We have suggested that when elements combine to form compounds, *valence electrons may be lost, gained, or shared between atoms.* Evidently atoms of the noble gases do not readily *lose, gain, or share* electrons. Let us examine their valence shell configurations to see if we can find a reason for this behavior.

$_2$He   $1s^2$
$_{10}$Ne . . . $2s^2\ 2p^6$
$_{18}$Ar . . . $3s^2\ 3p^6$
$_{36}$Kr . . . $4s^2\ 4p^6$
$_{54}$Xe . . . $5s^2\ 5p^6$
$_{86}$Rn . . . $6s^2\ 6p^6$

Note that the noble gases have completely filled s and p orbitals in the valence shell. The noble gases have no room for additional electrons in the s or p valence orbitals and thus *do not tend to gain or share* electrons from other atoms. (Xenon does form some compounds. A model will be developed in later courses to explain this property.)

Because the noble gases are at the end of a period, the nuclei of their atoms have a strong attraction for their valence electrons. This means that large amounts of energy must be supplied if they are to lose the valence electrons. For this reason, *the noble gases do not lose their valence electrons readily to other elements.* ■

## 5.11 A classification for the elements

As defined in Chapter 2, metals are elements which usually have a characteristic luster, are good conductors of heat and electricity, can be drawn into wires, and hammered into different shapes. Using ionization energy data, we can also define metals as those elements with *low ionization energies.* Since atoms of elements with low ionization energies tend, when they react, to lose electrons more easily than elements with high ionization energies, we can say that atoms of *metals tend to lose electrons.*

As defined in Chapter 2, nonmetals are usually poor con-

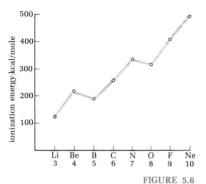

FIGURE 5.6
The first ionization energies of the elements in period two versus the atomic number of the element.

ductors of heat and electricity. Using ionization energy as a model, we can also define nonmetals as those elements with *high ionization energies*. Since the atoms of elements with high ionization energies tend to gain electrons when they react with other elements, we can say that atoms of *nonmetals tend to gain electrons*.

There is no clear-cut division between metals and nonmetals. Some elements have properties that are partly metallic and partly nonmetallic. These substances are called *metalloids*.

Figure 5.7 illustrates, in outline form, the grouping of the elements into these classifications in the periodic chart.

## 5.12 Prediction of properties

The periodic chart lists the elements in order of increasing atomic number. *Since the electronic configurations of an element vary in a periodic manner, any property that is a function of the electronic configuration of an atom will also vary in a periodic manner.* Some of these properties are: atomic size, density, boiling point, volume of a mole, melting point, ionization energy, electrical conductivity, acid-base properties, and thermal conductivity.

Let us examine the predictions that Mendeleev made for what he called eka-silicon, the element that is below silicon on the periodic chart. They are found in Table 5.7. Listed alongside these predictions are the actual properties of eka-silicon as measured by Winkler when in 1886 he discovered the element, now called germanium.

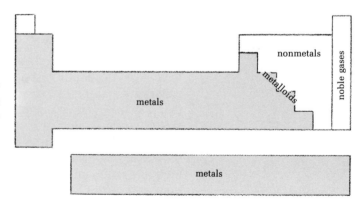

FIGURE 5.7
Outline of the periodic chart, showing position of metals, nonmetals, metalloids, and noble gases.

TABLE 5.7

Predictions of properties by Mendeleev compared to measurements by Winkler.

| Property | Eka-Silicon, predicted | Germanium, measured |
|---|---|---|
| Atomic weight | 72 | 72.59 |
| Density (g/ml) | 5.5 | 5.47 |
| Formula of chlorine compound | EsCl$_4$ | GeCl$_4$ |
| Density of chlorine compound (g/ml) | 4.7 | 4.70 |
| Melting point of chlorine compound | 100 °C | 86 °C |

### Groups

How could the properties of an unknown element be predicted? For example, what should be the formula of the compound formed between germanium and hydrogen? Germanium is in group IVA. The group IVA hydrogen compounds are: $CH_4$; $SiH_4$, Ge?, $SnH_4$, $PbH_4$. Since all the other members of the group have *one* group atom to *four* hydrogen atoms, it seems reasonable to expect the formula to be $GeH_4$. This is the correct formula.

When oxygen combines with the group IVA elements, the formulas are: $CO_2$, $SiO_2$, Ge?, $SnO_2$, $PbO_2$, and SnO, PbO. What is the formula of the compound formed between germanium and oxygen?

Can we predict the melting point of germanium if we know the melting points of the other elements in the group?

| C | Si | Ge | Sn | Pb |
|---|---|---|---|---|
| 3570°C | 1414°C | ? | 232°C | 328°C |

We might expect the melting point of germanium to be somewhere between the melting points of the elements above and below it on the periodic chart. The melting point of silicon is 1414°C, and of tin, 232°C. Their average is 823°C. The melting point of germanium is 958°C. ■

■ EXERCISE *From the data given, predict the formula of the compound that germanium forms with fluorine:* $CF_4$, $SiF_4$, Ge?, $SnF_4$, $PbF_4$.

■ EXERCISE *From the boiling point data given, estimate the boiling point of germanium. Check your answer in The Handbook of Chemistry and Physics or other reference source.*

| C | Si | Ge | Sn | Pb |
|---|---|---|---|---|
| 3470°C | 2355°C | ? | 2362°C | 1755°C |

### Periods

Just as trends in *groups* help us to predict properties, trends in *periods* also help us to predict properties. For example, Mendeleev predicted the atomic weight of germanium. The elements before and after germanium are:

| Cu | Zn | Ga | Ge | As | Se |
|---|---|---|---|---|---|
| 63.5 | 65.4 | 69.7 | ? | 74.9 | 79.0 |

What would you predict is the atomic weight of germanium? The actual value is 72.6.

From the following list of ionization potentials (kcal/mole) of representative elements in the fourth period, can you estimate the ionization potential of germanium?

| K | Ca | Ga | Ge | As | Se | Br | Kr |
|---|---|---|---|---|---|---|---|
| 100 | 141 | 139 | ? | 236 | 225 | 273 | 323 |

The ionization potential of germanium should be somewhere between 139 and 236 (kcal/mole). The actual value is 180 kcal/mole. ■

■ EXERCISE  From the following atomic radii (Å), estimate the radius of a germanium atom: K, 2.03; Ca, 1.74; Ga, 1.25; As, 1.21; Se, 1.17; Br, 1.14.

We have attempted to show that knowing the properties of some elements in a group or in a period helps one to predict or estimate the properties of other elements in the same group or period. Trends also occur along diagonals. Study the location of germanium in the periodic chart in the back of this book. Many properties of germanium can be inferred from the properties of: silicon and tin in group IVA; gallium and arsenic in period 4; aluminum and antimony (diagonal); and phosphorus and indium (diagonal).

## Summary

In 1870 Mendeleev found that when he arranged elements in order of increasing atomic *weights*, elements fell into groups with similar properties. This model was later improved by Moseley, who arranged elements in order of increasing atomic *number*.

Vertical columns in the periodic chart are called groups. The elements in any one group have similar properties. Horizontal rows in the periodic chart are called periods. There is usually a gradual change in properties of elements as one goes across a period. The elements in a group have a similar number of valence electrons. This is one reason why they have similar properties.

In going across a period from left to right, the number of protons and the number of valence electrons in an atom increases in a gradual manner. This is one reason why properties of elements change gradually in going across a period.

The energy required to pull an electron from an atom is called the ionization potential. This is a periodic property and is a measure of the attraction of an atom for electrons.

In going down a group, the energy level of the valence electrons increases. This makes it easier for electrons to be pulled off an atom, and the ionization potential decreases. In going across a period, the number of protons in the nucleus increases and the main energy level for electrons remains the same. The attraction of the nucleus for electrons increases and the ionization potential increases.

By realizing that elements in a group have similar properties and that there is a gradual change in properties across a period, you should be able to estimate or predict the properties of elements about which you may otherwise have little information.

## Objectives

*By the end of this chapter you should be able to do the following:*

1. State the criteria Mendeleev used to arrange elements in his periodic chart. (Question 2)

2. Given a periodic chart, identify groups I–VIIIA and periods 1 to 7. (Questions 3, 4)

3. State the criteria Moseley used to modify Mendeleev's model of the periodic chart. (Question 5)

4. Given a periodic chart and the atomic number of a representative element, state the group number and period number of the element. (Question 6)

5. Given the group number of a representative element, state how many valence electrons the element has. (Question 7)

6. Given the period number of an element, state what the valence shell of the element is. (Question 9)

7. Given a periodic chart, write the electron-dot structure of a representative element. (Question 10)

8. Given a periodic chart, identify the elements according to the following classifications: representative elements, transition elements, and inner transition elements. (Question 11)

9. Given a periodic chart, identify the elements according to the following classifications: metals, metalloids, nonmetals, and noble gases. (Question 12)

10. State the general trend of the ionization potential in moving across a period from left to right. (Question 13)

11. State the general trend of the ionization potential in moving down a group from top to bottom. (Question 13)

12. State what effect an increasing number of protons in the nucleus has on the ionization potential. (Question 14)

13. State what effect an increasing valence shell number has on the ionization potential. (Question 15)

14. Identify the group of representative elements that does not easily gain, lose, or share electrons. (Question 16)

15. Given data on a number of elements in the periodic chart, estimate the properties of another element. (Questions 20, 21)

**Questions**

1. Explain each of the following terms (where possible, give examples): *triads; Law of Octaves; periodic chart; group of elements; periods; electron-dot structures; transition elements; inner transition elements; representative elements; periodic model; metals; nonmetals; metalloids; noble gases; ionization energy; periodic trends.*

2. What criteria did Mendeleev use to arrange elements in his periodic chart? How many exceptions to Mendeleev's rule can you find in a modern periodic chart?

3. Identify each group in the periodic chart by listing the first element in each group.

4. Identify each period in the periodic chart by listing the *first* element in each period.

5. How did Moseley modify Mendeleev's model of the periodic chart?

6. Using a periodic chart as a reference, give the group number and period number of each of the following elements: (a) helium; (b) tin; (c) krypton; (d) nitrogen; (e) magnesium; (f) bromine; (g) phosphorus; (h) aluminum; (i) calcium.

7. How many valence electrons does each of the following elements have? (a) Group IIA: calcium, strontium, barium; (b) Group IVA: carbon, silicon, tin; (c) Group VIA: sulfur, tellurium; (d) Group VIIIA: neon, krypton, helium.

8. How many valence electrons do each of the following elements have? Refer to a periodic chart for the group number of each element. (a) nitrogen; (b) bismuth; (c) oxygen; (d) sodium; (e) boron; (f) hydrogen.

9. Using a periodic chart, determine the period in which each of the following elements is located. What is the number of the valence shell of each of these elements? (a) nitrogen; (b) bismuth; (c) uranium; (d) calcium; (e) germanium; (f) randon.

10. Draw electron-dot structures for each element listed in Questions 7 and 8.

11. Using a periodic chart, classify the following as either representative elements, transition elements, or inner transition elements. (a) carbon; (b) neon; (c) iron; (d) thorium; (e) chromium; (f) manganese; (g) curium; (h) berkelium; (i) nickel.

12. Place each element listed in Questions 7, 8, and 11 into one of the following categories: (a) metal; (b) nonmetal; (c) metalloid; (d) noble gas.

13. What is the general trend in ionization energy in going across a period from left to right? What is the general trend in ionization energy in going down a group from top to bottom?

14. What effect would an increasing number of protons in the nucleus of an atom have on the atom's attraction for electrons? How would this increasing number of protons affect the ionization energy of the atom?

15. What effect does an increasing valence shell have on the attraction of an atom for electrons? How does the valence shell number affect the ionization energy of an atom?

16. Which group of elements in the periodic chart does not readily gain, lose, or share electrons?

17. Why is energy required to remove an electron from a neutral atom? Would more energy be required to remove an electron from a positive ion or from a corresponding neutral atom? Explain in terms of electron interaction.

18. Using a periodic chart, decide which element of the following pairs has the lower ionization energy: (a) K or Na; (b) F or Cl; (c) K or Br; (d) Cl or Ar; (e) B or Al; (f) O or F; (g) Mg or Ba; (h) H or Li; (i) Se or Br.

19. Determine the electronic configurations of the following ions. Do the electronic configurations of these ions bear any relationship to those of other, neutral atoms? If so, list which ones. (a) Li$^+$; (b) Be$^{2+}$; (c) H$^+$; (d) Na$^+$; (e) Mg$^{2+}$; (f) F$^-$; (g) O$^{2-}$; (h) N$^{3-}$; (i) K$^+$; (j) Ca$^{2+}$; (k) Sc$^{3+}$; (l) Ti$^{4+}$; (m) P$^{3-}$; (n) S$^{2-}$; (o) Cl$^-$.

20. From the location of sulfur in the periodic chart and the following information, estimate some properties of sulfur. Check your answers in a reference book.

|  | Oxygen | Selenium | Phosphorus | Chlorine | Sulfur |
|---|---|---|---|---|---|
| Atomic number | 8 | 34 | 15 | 17 | |
| Atomic weight | 16.0 | 79.0 | 31.0 | 35.5 | |
| Electron structure | $1s^2\ 2s^2\ 2p^4$ | $\ldots 4s^2\ 4p^4$ | $\ldots 3s^2\ 3p^3$ | $\ldots 3s^2\ 3p^5$ | |
| Boiling point | $-183°C$ | $685°C$ | $280°C$ | $-35°C$ | |
| Melting point | $-218°C$ | $217°C$ | $44°C$ | $-101°C$ | |
| Density (g/ml) | 1.14 | 4.79 | 1.82 | 1.56 | |
| Atomic radius (Å) | 0.73 | 1.16 | 1.06 | 0.99 | |
| Ionization energy (kcal/mole) | 314 | 225 | 254 | 300 | |
| Volume of mole (ml) | 14 | 16.5 | 17.0 | 18.7 | |

21. Astatine is the last element in group VIIA. Because the element is uncommon and because it is highly radioactive, not many of its properties are well-known. From the following data can you predict some of the properties of astatine?

|  | Fluorine | Chlorine | Bromine | Iodine | Astatine |
|---|---|---|---|---|---|
| Boiling point (°C) | $-188$ | $-35$ | 58 | 183 | |
| Melting point (°C) | $-220$ | $-101$ | $-7.2$ | 114 | |
| Density (g/ml) | 1.11 | 1.56 | 3.12 | 4.94 | |
| Electron structure | $1s^2\ 2p^2\ 2p^5$ | $\ldots 3s^2\ 3p^5$ | $\ldots 4s^2\ 4p^5$ | $\ldots 5s^2\ 5p^5$ | |
| Hydrogen compound | HF | HCl | HBr | HI | |
| Magnesium compound | MgF$_2$ | MgCl$_2$ | MgBr$_2$ | MgI$_2$ | |
| Aluminum compound | AlF$_3$ | AlCl$_3$ | AlBr$_3$ | AlI$_3$ | |
| Ionization energy (kcal/mole) | 402 | 300 | 273 | 241 | |
| Atomic radius (Å) | 0.72 | 0.99 | 1.14 | 1.33 | |
| Volume of a mole (ml) | 17.1 | 18.7 | 23.5 | 25.7 | |

# Bonding

## 6.1 Introduction

Why are water and table salt different? In this chapter we will consider some of the factors that determine the properties of substances. We will study the forces that hold atoms together in *chemical aggregates* or *molecules*. These forces of attraction are called *bonds*.

Why is the formula for water $H_2O$? Why is the formula for table salt NaCl? We will use our model of electronic orbitals to help explain the ratios in which atoms combine to form aggregates.

The atomic models developed in Chapters 3, 4, and 5 will be used to develop a model for bonding. In Chapter 7 we will survey the system used for naming these compounds.

## 6.2 Chemical bonding

*The force which results from the simultaneous attraction of electrons to two nuclei is known as a chemical bond.* The element hydrogen can be used to develop a model of chemical bonding. It has been experimentally observed that when two hydrogen atoms approach each other closely enough, they stick together to form a molecule. A diagram of this process is shown in Fig. 6.1. Can a model be developed which will explain the

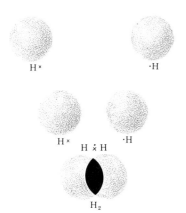

**FIGURE 6.1**
*If two hydrogen atoms approach each other closely enough, they may bind themselves together to form $H_2$.*

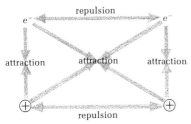

**FIGURE 6.2**
*The attractive and repulsive forces within the hydrogen molecule.*

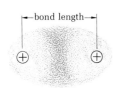

**FIGURE 6.3**
*The average distance between the two nuclei in a molecule is the bond length.*

forces which hold the molecule together? The bond in the hydrogen molecule could result from unlike charges attracting each other. The positive nucleus of each atom could attract both negative electrons since unlike charges attract each other. In like manner each electron could attract both nuclei. These forces bind the atoms together into the molecule.

Are there repulsive forces within the hydrogen molecule? Since we know that like charges repulse each other, we should expect a proton to repulse another proton just as an electron should be expected to repulse another electron (Fig. 6.2). *The attractive forces must overcome the repulsive forces since the atoms are held together in the molecule.*

A model similar to the one used for hydrogen can be used to explain the formation of *all chemical bonds.* Atoms join together to form molecules because of the attractive forces between electrons and protons.

The electrons spend most of their time between the two nuclei which are involved in the bond formation. This is indicated by the darker area between the two hydrogen nuclei in Fig. 6.3. The average distance between the two nuclei is called the *bond length.*

The hydrogen molecule is viewed as two nuclei sharing a pair of electrons and can be represented by the electron dot formula, H⋮H. The dot and cross (⋮) represent a shared electron pair with one electron donated by each atom.

When hydrogen and oxygen are mixed together in a reinforced test tube, and the resulting mixture subjected to a spark, an explosion results. (Caution: A large quantity of heat is given off during the explosion.) A cold beaker placed above the tube is found to have droplets of a colorless, odorless liquid formed on it during the explosion (Fig. 6.4).

Is water the colorless, odorless liquid formed on the beaker? Cobalt chloride test paper changes from blue to pink in water, and therefore can be used to test for water. When the liquid on the beaker is tested with cobalt chloride test paper, the same color change (blue to pink) is observed. We conclude that water must be formed when a mixture of oxygen and hydrogen explodes. The equation for the reaction is:

$$2H_2(g) + O_2(g) + \text{spark} \rightleftharpoons 2H_2O(g) + \text{energy} \qquad (6.1)$$

When hydrogen reacts with oxygen, bonds between hydrogen atoms in H—H and bonds between oxygen atoms in O—O must be broken. These processes require energy—in this case, the initial spark. When these atoms come together to form

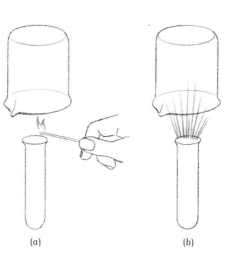

FIGURE 6.4
(a) A mixture of hydrogen and oxygen in a reinforced test tube. (b) The mixture of hydrogen and oxygen is ignited, with a resulting explosion. (b) After the explosion, a colorless, odorless liquid forms on the bottom of the cold beaker.

bonds in water, H—O—H, energy is released. Since the overall reaction releases 68 kcal of energy per mole of water, we know that more energy is released than is used.

### Three general classifications of chemical bonds

Although nuclei share electrons in a bond, they may not share them equally. In most cases one of the nuclei has a greater attraction for electrons than the other. Bonds can be classified according to whether the electrons are shared equally or unequally. For convenience, we separate chemical bonds into three general classifications.

### Nonpolar covalent

Consider the case in which an atom, A, is bonded to an atom, B.

A∶B

If the atoms A and B have the same attraction for electrons, the electron pair will be shared equally between the two nuclei. The shared electrons spend about as much time around one nucleus as around the other. When the electron pair is shared equally, we call this a *nonpolar covalent bond*. Molecules which contain nonpolar covalent bonds are $H_2$, $O_2$, $N_2$, $F_2$, IBr, and $CH_4$ (Figs. 6.5 and 6.6).

FIGURE 6.5
When atoms of the same element share electrons evenly, the bond is nonpolar covalent.

### Polar covalent

Consider the case in which the two atoms that are bonded together have quite different attractions for electrons. The bonding electrons spend most of their time near the atom which has the highest attraction for electrons; the bonding electrons are therefore unequally shared. Consider a molecule formed by atom A bonding to atom B, where B has a considerably greater attraction for electrons than A.

$$A \overset{\times}{\cdot} B$$

Since the bonding electrons spend more time near atom B, the space around B is relatively negative while the space around A is relatively positive. This bond is called a *polar covalent bond*. The separation of charge is indicated by:

$$A \overset{\times}{\cdot} B \quad \text{or} \quad A^{\delta+} \overset{\times}{\cdot} B^{\delta-}$$

where $\delta+$ means slightly positive, $\delta-$, slightly negative.

The aggregate AB as a whole is electrically neutral since it contains as many positive charges as negative charges. Some examples of molecules which contain polar covalent bonds are: HCl, HBr, $NH_3$, and $H_2O$ (Fig. 6.7).

### Ionic

Consider the case in which an atom, A, which has a quite low attraction for electrons, is bonded to an atom, B, which has a quite high attraction for electrons. The shared electron pair will spend essentially all of its time around B and almost no time around A.

$$\overset{+}{A} \overset{\times}{\cdot} \overset{-}{B}$$

Atom A has essentially given up or lost an electron to atom B. Recall that an ion is a charged particle. A is a positive ion represented by $A^+$. B is a negative ion represented by $B^-$.

The positive ion $A^+$ has a strong attraction in all directions for negative ions. The resulting force between the charged particles holds them together and is called an *ionic bond*. An example of a chemical aggregate which contains ionic bonds is sodium chloride, $Na^+Cl^-$.

The sodium chloride molecule consists of two charged particles, $Na^+$ and $Cl^-$. The sodium chloride crystal represented in Fig. 6.8 is an array in space of positive sodium ions, $Na^+$, and negative chloride ions, $Cl^-$. There is nothing which we can properly call a molecule of sodium chloride. Sodium chloride is an *ionic substance*.

It should be noted that there is no sharp distinction among the three classifications of bonds. When atom A is bonded to

IBr

$CH_4$

**FIGURE 6.6**
*When atoms share electrons almost evenly, the bond is essentially nonpolar covalent.*

**FIGURE 6.7**
*Three representations of the water molecule, $H_2O$, which consists of two atoms of hydrogen and one atom of oxygen.*

6.2 Chemical bonding

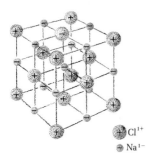

+ − + − + − +
− + − + − + −
+ − + − + − +
− + − + − + −

**FIGURE 6.8**
*The sodium chloride crystal consists of an array of positive sodium ions, $Na^+$, and negative chloride ions, $Cl^-$.*

**FIGURE 6.9**
*The periodic change of both size and electronegativity is shown here. Note the direction of change in each case.*

atom B, we classify the bond between them as *nonpolar covalent, polar covalent,* or *ionic,* depending upon the nature of A and B. We shall now investigate the factors which influence the attraction an atom has for electrons.

## 6.3 Electronegativity

*Electronegativity* is a measure of the ability of an *atom to attract the electrons in a bond* which it forms with another element. Numerical values of electronegativity were first suggested by Linus Pauling in 1932. The most electronegative element, fluorine, is assigned an electronegativity of 4.0 on the Pauling scale. Those elements, the nonmetals, which have a strong attraction for electrons are highly electronegative. The nonmetals have electronegativity values roughly between 2.0 and 4.0. The metals have fairly weak attraction for electrons; thus they have relatively low electronegativities. The metals have electronegativity values roughly between 0.7 and 2.2. Since the noble gases do not readily form bonds, no electronegativity values are currently given for them.

As might be expected, electronegativity is influenced by the same factors which influence ionization energies. These factors include:

1. the size of the nuclear charge
2. the valence shell number, that is, the number of the shell in which the valence electrons are located
3. the shielding effect produced by the inner electrons
4. the availability of an orbital for the bonding electrons

| IA | | | | | | | | | | | | | | | | | VIIIA |
|---|---|---|---|---|---|---|---|---|---|---|---|---|---|---|---|---|---|
| 0.371 H 2.1 | IIA | Atomic radii (angstroms) Electronegativity | | | | | | | | | | IIIA | IVA | VA | VIA | VIIA | H |
| 1.520 Li 1.0 | 1.113 Be 1.5 | | | | Transition elements | | | | | | | 0.794 B 2.0 | 0.772 C 2.5 | 0.74 N 3.0 | 0.74 O 3.5 | 0.709 F 4.0 | Ne |
| 1.858 Na 0.9 | 1.598 Mg 1.2 | IIIB | IVB | VB | Groups VIB | VIIB | VIII | | | IB | IIB | 1.432 Al 1.5 | 1.176 Si 1.8 | 1.10 P 2.1 | 1.02 S 2.5 | 1.00 Cl 3.0 | Ar |
| 2.272 K 0.8 | 1.974 Ca 1.0 | 1.606 Sc 1.3 | 1.448 Ti 1.5 | 1.311 V 1.6 | 1.249 Cr 1.6 | 1.366 Mn 1.5 | 1.241 Fe 1.8 | 1.253 Co 1.8 | 1.246 Ni 1.8 | 1.278 Cn 1.9 | 1.332 Zn 1.6 | 1.221 Ga 1.6 | 1.225 Ge 1.8 | 1.24 As 2.0 | 1.16 Se 2.4 | 1.142 Br 2.8 | Kr |
| 2.48 Rb 0.8 | 2.152 Sr 1.0 | 1.776 Y 1.2 | 1.590 Zr 1.4 | 1.429 Nb 1.6 | 1.362 Mo 1.8 | 1.325 Tc 1.5 | 1.325 Ru 2.2 | 1.345 Rh 2.2 | 1.376 Pd 2.2 | 1.444 Ag 1.9 | 1.490 Cd 1.7 | 1.626 In 1.7 | 1.405 Sn 1.8 | 1.45 Sb 1.9 | 1.432 Te 2.1 | 1.333 I 2.5 | Xe |
| 2.654 Cs 0.7 | 2.174 Ba 0.9 | 1.870 La 1.1 | 1.563 Hf 1.3 | 1.43 Ta 1.5 | 1.370 W 1.7 | 1.370 Re 1.9 | 1.338 Os 2.2 | 1.358 Ir 2.2 | 1.388 Pt 2.2 | 1.442 Au 2.4 | 1.502 Hg 1.9 | 1.704 Tl 1.8 | 1.750 Pb 1.8 | 1.548 Bi 1.9 | 1.67 Po 2.0 | At 2.2 | Rn |
| Fr 0.7 | Ra 0.9 | 1.878 Ac 1.1 | | | | | | | | | | | | | | | |

118    6: Bonding

**Nuclear charge**

Negatively charged electrons are attracted to atoms by the positively charged protons in the nucleus. The greater the number of protons in the nucleus, the greater will be the attraction for electrons. Note that in period 2 of Fig. 6.9 the electronegativity values increase steadily from 1.0 for lithium to 4.0 for fluorine. The nuclear charge also increases from left to right across the period; thus the attraction which an atom has for electrons increases in the same direction. In general, *electronegativities increase across a period from left to right.*

**Valence shell number–atomic size**

Note in Fig. 6.9 that sodium in period 3 has a lower electronegativity than lithium. The smaller an atom is, the closer its nucleus can get to the electrons of another atom and the greater its electronegativity will be. The size of an atom depends on its valence shell. These relationships are indicated in Fig. 6.9. Examine the sizes of the atoms in group IA in the same figure. There is an increase in atomic size down the group as the number of valence shells increases. Starting with lithium and moving down the group, we note that the electronegativity values decrease steadily from 1.0 for lithium to 0.7 for francium. *In general, electronegativities decrease down a group; the larger atoms of a group have the lower electronegativity values.*

**Shielding**

The effective pull which an electron experiences depends upon the atom's nuclear charge and the distance of the electron from the nucleus. The effective pull on an electron also depends upon the number of inner shell electrons. This "shielding effect" or "insulating effect" of the inner shell electrons is most notable in the transition elements.

**Orbital availability**

Electronegativity values increase from left to right in the second period, reaching a maximum of 4.0 at fluorine. However, neon, which lies to the right of fluorine, has no electronegativity value listed. Why is this? Because of its lack of available orbitals and its high ionization energy, neon forms no known compounds. Since no bond involving neon has ever been observed, there is no way to measure the attraction neon has for the electrons in a bond.

**Electronegativity and ionization energy**

A measure of the attraction an atom has for electrons is included in both its ionization energy and its electronegativity. Ionization energies can be accurately measured in the laboratory but *electronegativity values must be derived* from other measurements, since there is no accurate way to measure the attraction an atom has for the electrons in a bond. In fact, the electronegativity of an element may change slightly from compound to compound.

## 6.4 Covalent bonding

It has been experimentally observed that when a stream of individual hydrogen atoms is passed over a piece of platinum, Pt, metal, the metal becomes hot and glows. Measurements indicate that the number of individual particles is reduced to one-half the original number as a result of passing over the platinum. Chemical analysis of the gas which results from the interaction of hydrogen atoms, H, with the platinum indicates that it consists of hydrogen molecules, $H_2$ (Fig. 6.10). Can you explain the results of this experiment?

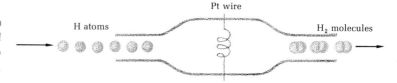

FIGURE 6.10
When hydrogen atoms, H, are passed over a platinum wire, the wire heats up and has a reddish glow.

Energy is evidently given off when hydrogen atoms collect on the surface of the platinum and combine into pairs to form hydrogen molecules, $H_2$.

$$H\cdot + \cdot H \longrightarrow H\!:\!H + 108.8 \text{ kcal/mole}$$

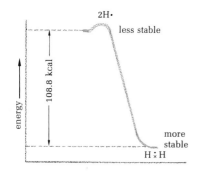

FIGURE 6.11
Energy, 108.8 kilocalories, is given off when two moles of hydrogen atoms combine to form a mole of hydrogen molecules.

When the chemical bond is formed between the two hydrogen atoms, energy is released. Energy must be absorbed if this bond is to be broken. The same quantity of energy is released in the bond formation as is absorbed in the bond breaking. The chemical system has gone from a state of higher energy to a state of lower energy with a net release of energy (Fig. 6.11). Chemical changes generally occur if the change leads to a lower energy state and thus a more stable structure.

■ EXERCISE  *What is the bond type in the hydrogen molecule? Does one atom have a higher electronegativity than the other?*

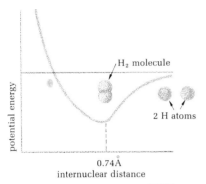

FIGURE 6.12

*As two hydrogen atoms move closer, the attraction between them increases. The attraction between the two hydrogen atoms will reach a maximum when the two nuclei are at a distance of 0.74 Angstroms from each other. This is the bond length of the hydrogen molecule.*

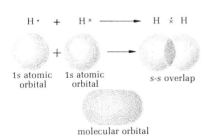

FIGURE 6.13

*When one spherical s orbital of a hydrogen atom overlaps with the spherical s orbital of another hydrogen, a molecular orbital is formed.*

As two hydrogen atoms approach each other, the potential energy between them decreases and reaches a minimum when the two nuclei are 0.74 angstrom ($7.4 \times 10^{-9}$ Å) apart. There is an increase in potential energy if the two nuclei move closer or farther apart (Fig. 6.12). The distance 0.74 Å is the average distance between the two nuclei in a hydrogen molecule and is called the hydrogen bond length. ■

### Atomic orbitals and combining ratios

Why does the hydrogen molecule contain two atoms of hydrogen instead of three or four? The ratio in which atoms combine is explained by the concept of overlapping atomic orbitals. The atomic orbital diagram of a hydrogen atom is:

$_1$H    1s  (↓)

According to the concept of overlapping atomic orbitals, a half-filled orbital of one atom can overlap with a half-filled orbital of another atom to form a molecular orbital which is completely filled. The term *molecular orbital* indicates an orbital which is formed by the overlap of two atomic orbitals (Fig. 6.13). When the two atomic orbitals overlap to form the molecular orbital, the proton in each nucleus is strongly attracted to two electrons instead of one. This attraction holds the molecule together.

In most cases, formation of a chemical bond results in a more nearly filled valence shell. In the case of hydrogen, the 1s orbital of one atom overlaps the 1s orbital of another atom to form a molecular orbital which contains two electrons, the maximum number of electrons in the first shell. In effect, then, hydrogen shares a pair of electrons with itself to effectively fill the 1s orbital of each.

Can the concept of overlap of atomic orbitals be used to explain why the fluorine molecule consists of two atoms? The atomic orbital diagram for fluorine is:

$_9$F    2s (↑↓)    2p (↑↓)(↑↓)(↓)

Each fluorine atom has one valence orbital which is half-filled. In this respect, it is similar to hydrogen. Two fluorine atoms can bond together by overlap of their half-filled *p* orbitals

6.4 Covalent bonding

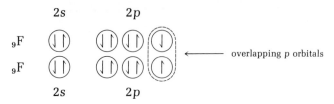

A diagram of the overlapping p orbitals in the fluorine molecule, $F_2$, is shown in Fig. 6.14.

The bond between the two fluorine atoms consists of a shared electron pair. The fluorine molecule is represented by:

:F̈ ⋮ F̈:

There is no difference in the electrons; we have used · and × merely for the purpose of bookkeeping. Note that each of the fluorine atoms effectively has eight electrons in its outermost shell. This is one example of the "rule of eight," or octet rule, which states that many elements tend either to accept electrons, give up electrons, or share electrons in order to attain the noble gas configuration of eight electrons in the outermost shell. We have already observed that noble gases are quite unreactive, and we therefore conclude that the noble gas configuration, eight electrons in the outermost shell, is a stable configuration. ■

■ EXERCISE *Draw an atomic orbital diagram for the valence shell of chlorine. Describe the orbitals which overlap to form the $Cl_2$ molecule. Write the electron-dot structure for the chlorine molecule. Repeat for the $Br_2$ molecule.*

Nitrogen gas consists of diatomic molecules, $N_2$. The $N_2$ molecule is broken into nitrogen atoms, N, only with great difficulty.

$$N_2(g) + 171{,}000 \text{ cal} \rightleftharpoons 2N(g) \quad (6.2)$$

This equation indicates that the two nitrogen atoms are strongly bonded to each other. Although nitrogen is a highly electronegative element (only oxygen and fluorine are more so), it reacts with the active metals only at high temperatures,

FIGURE 6.14

*The valence electrons of the fluorine atom are located in the mutually perpendicular p orbitals. The half-filled p orbital of one fluorine atom can overlap with the half-filled p orbital of another fluorine atom to form the molecular orbital of the fluorine molecule, $F_2$.*

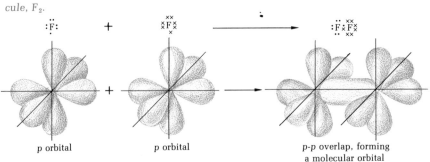

forming compounds called nitrides. Nitrogen will *not* support combustion as do $O_2$ and $F_2$. What model for the bond between the two N atoms will explain the properties of the $N_2$ molecule?

The atomic orbital diagram for nitrogen is:

Each nitrogen atom has three half-filled p orbitals. Each of the p orbitals on one nitrogen atom could overlap with one p orbital on the other nitrogen atom to form *three* bonds:

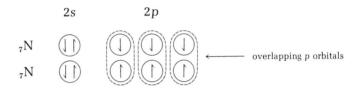

The electron-dot structure of the nitrogen molecule is:

:N::××N××

The triple bond between the two N atoms makes it quite difficult to break the $N_2$ molecule apart into atoms.

Since nitrogen in the combined state is an important component of fertilizers, chemists have been looking for economical ways to split the $N_2$ molecule apart and form compounds which can be used commercially.

Hydrogen sulfide, $H_2S$, is the foul-smelling gas which is primarily responsible for the odor of rotten eggs. Each sulfur atom bonds to *two* hydrogen atoms to form $H_2S$. The atomic orbitals involved in forming the $H_2S$ molecule are:

The electron dot structure for $H_2S$ is:

H:S:
  ××
  ··
  ××
H

■ EXERCISE  *Write the electron configuration for oxygen. Indicate which orbitals are used in forming $H_2O$. Write the electron-dot structure for $H_2O$.*

123   6.4 Covalent bonding

■ EXERCISE  What is the general formula of the compounds of the halogens, group VIIA, with hydrogen? Write the electron-dot structure for each member of this series. What orbitals are involved in each case?

The sulfur, in effect, shares two electrons in order to achieve the noble gas configuration of eight electrons in its outermost shell. All of the members of group VIA, O, S, selenium (Se), tellurium (Te), and polonium (Po), form compounds analogous to $H_2S$, that is, $H_2O$, $H_2S$, $H_2Se$, $H_2Te$, and $H_2Po$. ■

Helium does not burn or support combustion. It does not react with the active metals even at very high temperatures. Why is it so unreactive? Helium has one completely filled atomic orbital located in the first energy shell. These completely filled orbitals show no tendency to overlap, and consequently no bond forms between two helium atoms.

$1s$

He  (↓↑)

He  (↓↑)

### Covalent bonds and electronegativity

How can one determine whether the bonds in a certain molecule should be classified as primarily covalent, polar covalent, or primarily ionic? The relationship between the electronegativities of the bonding elements and the type of bond formed between them is indicated in Table 6.1. For our purposes, we will consider a bond to be largely nonpolar covalent if the electronegativity difference between the two bonding elements is less than 0.5. If the difference is greater than 1.7, we will consider the bond to be largely ionic in character. Bonds between elements with electronegativity differences between 0.5 and 1.7 will be considered as polar covalent. When we classify bonds in this way, we should keep in mind that all bonds are a result of the sharing of one or more pairs of electrons. ■

■ EXERCISE  Use Fig. 6.9 to determine the electronegativity difference between the two elements in each of the following: $NH_3$, $H_2S$, $SCl_2$, $SrCl_2$, NaCl, $SrF_2$, $PCl_3$, BrCl, LiH, and $NO_2$. Indicate the bond type in each case.

■ EXERCISE  Make a plot of electronegativity difference versus percent ionic character using the data in Table 6.1.

TABLE 6.1

The relationship between the electronegativity difference between the two bonding elements and the percentage of ionic character of the bond formed between the two elements.

| Electronegativity difference | Percentage of ionic character | | Electronegativity difference | Percentage of ionic character | |
|---|---|---|---|---|---|
| 0.2 | 1 | nonpolar covalent | 1.8 | 55 | |
| 0.4 | 4 | | 2.0 | 63 | |
| 0.6 | 9 | | 2.2 | 70 | |
| 0.8 | 15 | | 2.4 | 76 | primarily ionic |
| 1.0 | 22 | polar covalent | 2.6 | 82 | |
| 1.2 | 30 | | 2.8 | 86 | |
| 1.4 | 39 | | 3.0 | 89 | |
| 1.6 | 47 | | 3.2 | 92 | |
| | | | 3.4 | 96 | |

## 6.5 Formulas and names of covalent compounds

A formula indicates the number and kind of atoms bonded together in a molecule. A molecule of hydrogen iodide contains one atom of hydrogen and one atom of iodine. HI is the only formula which can be written to represent hydrogen iodide. Note that in the name for hydrogen iodide:

1. The name of the less electronegative element is written first.
2. The name of the more electronegative element is written last and has a modified ending of -ide.

Compounds containing two different elements are called *binary compounds*. These same rules can be applied to the naming of most binary compounds. For example, $H_2S$ is called hydrogen sulfide. The more electronegative element is written last and the ending is changed to -ide. ■

■ EXERCISE  *Use the rules for naming compounds to name the following:* HCl, HBr, $H_2Se$, $BeI_2$, *and* HAt.

## 6.6 Ionic bonding

There is no noticeable effect when sodium metal is placed in contact with chlorine gas. This is demonstrated by placing some sodium metal in a tube attached to a vacuum pump. The pump removes all of the gases from the tube, and chlorine gas is then allowed to fill the tube. No visible change occurs in the sodium after a period of five minutes. When heated gently, however, the sodium melts and a vigorous reaction occurs, releasing both heat and light. A white solid is formed. The white solid is soluble in water and its water solution conducts an electrical current, although the solid itself does not. It causes a flame to burn with a yellow color. This white solid has all of the properties, including taste (*caution:* highly dangerous), of table salt. The chemical equation is:

$$2Na(s) + Cl_2(g) \rightleftharpoons 2NaCl(s) + energy$$

The electronegativity difference between Na and Cl is 2.1. According to our classification, bonds between elements which have an electronegativity difference of 1.7 or greater are considered to be ionic. Thus, NaCl is considered as *an ionic compound*.

Magnesium burns in oxygen to form a white solid, magnesium oxide, MgO.

$$2Mg(s) + O_2(g) \rightleftharpoons 2MgO(s) + energy$$

Magnesium oxide is also an ionic compound. The metal magnesium is bonded to the nonmetal oxygen in magnesium oxide. *Ionic compounds are generally formed by the bonding of a metallic element with a nonmetallic element.*

**Electron transfer**

What happens when a sodium atom collides with a chlorine atom? Recall that sodium has a low electronegativity and that chlorine has a high electronegativity. Chlorine has an orbital which could accept an electron since it is only half-filled:

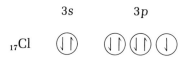

When a chlorine atom approaches a sodium atom closely enough, an electron from the sodium is transferred to the chlorine. Such a transfer can be illustrated using electron-dot structures.

$$\text{Na} + \cdot\ddot{\text{Cl}}: \rightleftharpoons \text{Na}^+ + :\ddot{\text{Cl}}:^-$$

Since Cl gained an electron, the Cl⁻ ion has eight electrons in its outermost shell. This is the stable configuration of the noble gas, argon. The Na⁺ has the configuration of neon.

The reaction of a magnesium atom with an oxygen atom to form the ionic compound MgO can be represented with electron-dot structures. This is another example of the rule of eight.

$$\ddot{\text{Mg}} + {}_\times^\times\!\!\overset{\times\times}{\underset{\times}{\text{O}}} \rightleftharpoons \text{Mg}^{2+} + {}_\times^\times\!\!\overset{\times\times}{\underset{\cdot\times}{\text{O}}}{}^{2-}$$

**Structure and properties**

Sodium chloride and magnesium oxide are representative ionic compounds. At room temperature they are both moderately hard solids. Both have high melting points as well as high boiling points. Sodium chloride melts at about 800°C and boils at about 1400°C. Magnesium oxide melts at about 2800°C and boils at about 3600°C. Neither solid conducts electricity well, but the molten liquids of both are good conductors. What model will explain these properties?

The bonding in NaCl is viewed as a positive sodium ion, Na⁺, being attracted to a negative chloride ion, Cl⁻. The elec-

**FIGURE 6.15**
A model of the sodium chloride crystal. Only seven ions are shown here to emphasize that each chloride ion is surrounded by six sodium ions. For a more complete model of the NaCl crystal, see Fig. 6.8.

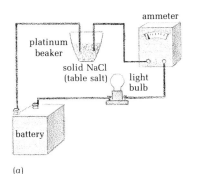

(a)

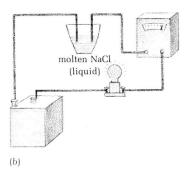

(b)

**FIGURE 6.16**
(a) When solid sodium chloride is placed in the platinum beaker of this conductivity apparatus, the light bulb does not light. (b) When molten NaCl is tested, the light bulb burns brightly and the ammeter indicates a flow of current through the system.

trostatic force between ions depends on (1) the *size of the charges* of the two ions and (2) the *size of the ions*, since the size determines how close the ions can get to each other.

In ionic compounds, the ions are arranged in an orderly array in space which we call a crystal (Fig. 6.15). In sodium chloride each chloride ion is surrounded by six sodium ions. In the same way, each sodium ion is surrounded by six chloride ions. The number of neighboring ions, six in each of these cases, is the *coordination number*. Each ion in the crystal is strongly bound to six neighboring ions. In this way, the entire crystal is held together tightly as one unit.

Because of the strong attractive forces between ions throughout the crystal, it is difficult to break or pull the ions away from each other. We should expect that quite a lot of thermal energy would be required to provide each ion with enough energy to overcome the attractive forces and break away from the crystal. This accounts for the high melting and boiling points of ionic compounds such as NaCl and MgO and for the fact that ionic compounds are moderately hard and do not scratch easily.

We shall now describe an experiment which investigates the properties of ionic compounds. A light bulb conductivity apparatus shown in Fig. 6.16 was assembled. When plugged into an electrical source, the bulb will light only when the two probes are placed in a substance which will conduct a current. For instance, if the two probes are placed in some liquid mercury, the bulb burns brightly; thus we know that mercury is a good conductor. When the probes are placed in a beaker containing finely divided iron, again the bulb burns brightly. The light bulb does not light, however, when the probes are placed in a beaker containing solid NaCl. After the NaCl is melted, the bulb lights when the probes are placed in the molten NaCl. Why should molten NaCl conduct a current and cause the bulb to light when solid NaCl does not?

In both solid and molten NaCl ions are present which could conduct a current and cause the bulb to light. In the solid NaCl, however, the ions are rigidly held and are not free to move about. Melting the solid increases the energy of the ions, breaks down the rigid crystalline structure, and allows the ions to become mobile. The positive sodium ions are free to move to the negative electrode; the negative chloride ions are free to move to the positive electrode. In this way a current is conducted through molten NaCl.

## Atomic orbital diagrams

Review the formation of NaCl by the reaction of sodium with chlorine. We can represent this transfer of electrons using atomic orbital diagrams:

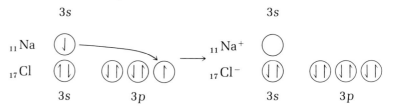

Sodium achieves the noble gas configuration of neon by losing an electron. Chlorine achieves the noble gas configuration of argon by gaining an electron. Metals tend to achieve the nearest noble gas configuration by losing one or more electrons. Nonmetals tend to gain one or more electrons to achieve the nearest noble gas configuration.

## Combining Ratios

The ratio of sodium ions to chloride ions in NaCl is one to one, or 1:1. This is so because in the formation of NaCl each sodium atom lost one electron while each chlorine atom gained this same number of electrons. The ratio of magnesium ions to oxide ions in MgO is also 1:1. Each oxygen atom gains the same number of electrons, two, lost by each magnesium atom when MgO is formed. What would be the ratio between magnesium and chlorine in the compound magnesium chloride? Each atom of magnesium which reacts can lose *two electrons*. Each chlorine atom which reacts can accept *only one electron*. How many chlorine atoms are needed to accept the two electrons given off by one magnesium atom?

$$\text{Mg} \begin{matrix} \nearrow :\ddot{\underset{..}{\text{Cl}}}: \\ \searrow :\ddot{\underset{..}{\text{Cl}}}: \end{matrix} \longrightarrow \text{Mg}^{2+} + \begin{matrix} :\ddot{\underset{..}{\text{Cl}}}:^- \\ :\ddot{\underset{..}{\text{Cl}}}:^- \end{matrix}$$

The formula for magnesium chloride is $MgCl_2$.

What is the formula for sodium oxide? In the formation of sodium oxide, each sodium atom can lose *one electron*. Each oxygen atom can gain *two electrons*. The formula for sodium oxide is $Na_2O$.

$$\begin{matrix} \text{Na}\times \searrow \\ \phantom{xx} + \cdot\ddot{\underset{..}{\text{O}}}: \\ \text{Na}\times \nearrow \end{matrix} \longrightarrow \begin{matrix} \text{Na}^+ \\ \phantom{xx} + :\ddot{\underset{..}{\text{O}}}:^{2-} \\ \text{Na}^+ \end{matrix}$$

■ EXERCISE *Aluminum atoms have three valence electrons which they lose easily. Write electron-dot structures to represent the combination of aluminum with both chlorine and oxygen. Write the formulas for aluminum oxide and aluminum chloride. What are the combining ratios in each case?*

The combining ratios of elements which form ionic compounds depend upon the number of electrons gained and lost by each element. When two elements combine to form an ionic compound, the number of electrons lost by one element must equal the number gained by the other element (Table 6.2). ■

## 6.7 Formulas and names of binary ionic compounds

Binary ionic compounds are named according to the same general rules developed earlier for covalent compounds.

1. Name the metal first.
2. Name the nonmetal second but change the ending of the nonmetal to -ide.

Some typical examples are potassium fluoride (KF); calcium fluoride ($CaF_2$); aluminum oxide ($Al_2O_3$); and cesium oxide ($Cs_2O$). Later we will develop this system of nomenclature to cover a greater number of compounds.

Formulas for ionic compounds are written according to the same general rules developed for covalent compounds. For ionic compounds which contain one metal and one nonmetal:

1. Write the symbol for the metal first and for the nonmetal second.
2. Indicate the combining ratios of atoms by means of subscripts.

The electron configurations of the elements and the octet rule are valuable aids in determining combining ratios. The metals in group IA have the general outer electron configuration $ns^1$. These metals easily lose one electron to form $M^+$ ions. Lithium, $1s^2, 2s^1$, which forms $Li^+$, is an example. The metals of group IIA have the general outer configuration $ns^2$. These metals readily lose two electrons to form $M^{2+}$ ions. Beryllium, $1s^2, 2s^2$, which forms $Be^{2+}$, is an example. Aluminum which is in group IIIA has the outer configuration $3s^2, 3p^1$. Aluminum readily

TABLE 6.2

*The electron transfer involved in the formation of NaCl, $Na_2O$, $MgCl_2$, and MgO.*

| Metal | Group number of metal | Number of electrons lost by each metal atom | Nonmetal | Group number of nonmetal | Number of electrons gained by each atom of nonmetal | Formula of compound |
|---|---|---|---|---|---|---|
| Na | IA | 1 | Cl | VIIA | 1 | NaCl |
| Na | IA | 1 | O | VIA | 2 | $Na_2O$ |
| Mg | IIA | 2 | Cl | VIIA | 1 | $MgCl_2$ |
| Mg | IIA | 2 | O | VIA | 2 | MgO |

loses three electrons to form the Al$^{3+}$ ion. The metals in each of the groups considered so far lose one or more electrons to form an ion which has the nearest rare gas configuration.

The nonmetals of group VIIA have the general outer configuration $ns^2$, $np^5$. These nonmetals readily gain one electron to form X$^-$ ions. Chlorine, whose outer configuration is $3s^2$, $3p^5$ forms the Cl$^-$ ion. The group VIA nonmetals have the general outer configuration of $ns^2$, $np^4$. These nonmetals can gain two electrons and form X$^{2-}$ ions. Oxygen, $2s^2$, $2p^4$, which forms O$^{2-}$, is an example. The outer configuration of nitrogen is $2s^2$, $2p^3$. Nitrogen can combine with the more reactive metals to form the ion, N$^{3-}$. In each of these cases, the nonmetal gains one or more electrons to form a negative ion with the nearest noble gas configuration. Note that the number of the group plus the number of electrons gained is equal to eight in each case.

We shall now demonstrate the use of the periodic table and the octet rule in determining combining ratios.

■ EXERCISE  Write the formulas for the elements which occur in the free state as diatomic molecules.

■ EXERCISE  Write balanced equations for the following:
(a) calcium + oxygen ⇌ calcium oxide + energy
(b) aluminum + chlorine ⇌ aluminum chloride + energy

EXAMPLE 1  Write the formulas of each of the following compounds. What is the combining ratio in each case?

(a) Barium bromide. Barium, one of the group IIA metals, tends to lose two electrons. Two bromine atoms are required to accept the 2 electrons lost by one barium atom. The combining ratio is 1:2. The formula of barium bromide is BaBr$_2$.

(b) Barium nitride. Barium tends to lose 2 electrons. Nitrogen, group VA, tends to gain 3 electrons. One barium atom loses 2 electrons and one nitrogen atom can gain 3 electrons. Three barium atoms are required to donate 6 electrons to two nitrogen atoms. The combining ratio is 3:2. The formula of barium nitride is Ba$_3$N$_2$. ■

## 6.8 The polar covalent bond

Unlike the covalent molecules of H$_2$ and Cl$_2$, hydrogen chloride is very soluble in H$_2$O. Unlike the solid ionic compounds NaCl and MgO, hydrogen chloride is a gas at room temperature. In both its properties and its bond type, hydrogen chloride lies somewhere between covalent compounds and ionic compounds. The bond between H and Cl in HCl is classified as a polar covalent bond. The bond has partial ionic character and partial covalent character (Fig. 6.17).

The water molecule is another example of a molecule which contains polar covalent bonds.

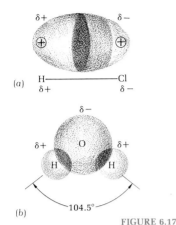

(a)

(b) 104.5°

**FIGURE 6.17**
*In HCl and in $H_2O$, the bounding electrons spend more time near the more electronegative element---causing that region to be relatively negatively charged. Dark areas indicate highest electron density.*

■ **EXERCISE** *Indicate which orbitals are used in forming HF, $H_2S$, and $NH_3$. Write the electron-dot structure for each of these molecules.*

**Atomic orbital diagrams**

Which orbitals of the chlorine atom are available for bonding? Does hydrogen have an orbital which can be used in bonding? Each of these atoms has a half-filled orbital, and these half-filled orbitals can overlap to form a bond.

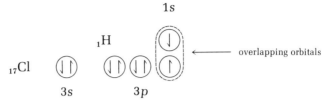

The 1s orbital with one electron overlaps with the 3p orbital of the chlorine that has only one electron. The electron-dot structure for the HCl molecule is H.

$$\overset{\times\times}{\underset{\times\times}{:HCl\overset{\times}{\phantom{.}}}}$$

What orbitals are used in forming the water molecule? An oxygen atom has two p orbitals which are only half-filled. Each of these p orbitals can overlap with a 1s orbital of a hydrogen atom.

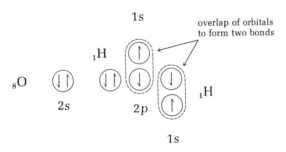

The electron-dot structure for water is:

$$H\overset{\times}{\underset{\overset{\times}{H}}{:}}\overset{..}{O}:$$

The shared electrons in HCl spend most of their time somewhere between the two nuclei, but do they spend as much time around one nucleus as around the other? Chlorine has the higher electronegativity, therefore the shared electrons will spend more time around the chlorine nucleus. The vicinity around chlorine is negatively charged and the vicinity around the hydrogen nucleus is positively charged (Fig. 6.17(a)).

The water molecule is another example of a molecule which

6.8 *The polar covalent bond*

contains polar covalent bonding. Experiments indicate that the two hydrogen nuclei are bonded at an angle of 104.5° as shown in Fig. 6.17(b). The electrons shared between H and O spend most of their time around the oxygen nucleus since oxygen is more electronegative. This causes the region around oxygen to be relatively negatively charged. The vicinity of the hydrogen must then be relatively positively charged. If the *s* orbital of each hydrogen atom overlaps a *p* orbital of oxygen, we would expect the angle between to be 90°. Because the positive hydrogens repulse each other, however, the angle is larger—that is, 104.5°.

**Polar molecules**

Hydrogen and chlorine are held together in the hydrogen chloride molecule by a polar bond. The hydrogen end of the molecule is relatively positively charged and the chlorine end is negatively charged. *When the center of positive charge and the center of negative charge do not exactly coincide, the molecule is polar.* A polar molecule must have at least one *polar bond*.

Dipole moment is the property which we measure experimentally to determine whether a molecule is polar. Molecules which are significantly polar have a measureable dipole moment. Experiments show that HCl has a measureable dipole moment; thus HCl is a polar molecule. Water is also found by experiment to possess a dipole moment. The water molecule is polar since it contains two equivalent polar bonds directed at an angle of 104.5° from each other (see Fig. 6.17). The oxygen end of the molecule is negative, and the hydrogen end of the molecule is positive.

Carbon dioxide, which contains two polar bonds, is found experimentally to have no measureable dipole moment; the $CO_2$ molecule is nonpolar. How can this be? Each of the C—O bonds in $CO_2$ is polar since oxygen is more electronegative than carbon.

$$\ddot{\underset{..}{O}}:\!\!:C\!\!:\!\!:\ddot{\underset{..}{O}}$$

The molecule can be nonpolar only if these two polarities exactly cancel each other. What geometric shape of the molecule would allow the two polarities to cancel each other?

$$O=C=O \qquad O \longleftarrow\!\!+\, C \,+\!\!\longrightarrow O$$

The two polar bonds will exactly oppose each other only if the three nuclei lie on a straight line. A linear molecule is the only structure which will explain why the $CO_2$ molecule is nonpolar even though it contains two polar bonds.

In the water molecule the center of negative charge is centered at the oxygen. The center of positive charge is midway between each of the relatively positive hydrogens.

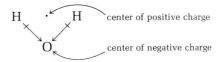

The water molecule is polar because the centers of positive and negative charge do not exactly coincide. In carbon dioxide, however, the center of positive charge is centered at the carbon nucleus, whereas the center of negative charge is the midpoint between the two oxygen atoms.

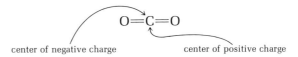

The centers of positive and negative charge coincide exactly, and therefore $CO_2$ is a nonpolar molecule. Whenever the polar bonds in a molecule are arranged in an opposing symmetric fashion, the molecule will still be nonpolar because the centers of charge coincide.

## 6.9 Intramolecular and intermolecular forces

There is a strong attractive force between the two nitrogen atoms in a nitrogen molecule, and a large quantity of energy is required to pull them apart. We designate this force of attraction as a nonpolar covalent bond. There is also a strong bond between the two hydrogen atoms in a hydrogen molecule and between the two fluorine atoms in a fluorine molecule. We have already studied the force of attraction or chemical bond between two atoms in a molecule, but could there also be a force of attraction *between two molecules?* The chemical bonds which hold molecules together are *intramolecular forces*, or forces acting within the molecule. Any forces of attraction or repulsion between molecules are called *intermolecular forces*.

Why are some substances gases, some liquids, and some solids at room temperature? The phase in which a substance exists at 20°C is a reflection of the strength of the forces that hold the atoms, molecules, or ions in the substance together. Substances that are solids at 20 °C have stronger forces of attraction between particles than substances that are liquids. Substances that are gases at room temperature have weak forces of attraction between particles.

Hydrogen exists as a gas at room temperature. Hydrogen gas must be cooled to a very low temperature, −235 °C, before it liquefies. This temperature, −235 °C, is the boiling point of hydrogen. Even at this extremely low temperature, hydrogen molecules have sufficient energy, and therefore motion, to break away from each other. If very little energy is required for the hydrogen molecules to break away from each other, the attractive forces between the *molecules* must be very weak compared to forces between the atoms within the molecule. These weak attractions are called *van der Waals forces*. (Fig. 6.18) The intermolecular forces are very weak compared to the intramolecular forces in this case. Are the intermolecular forces between other covalent molecules comparably weak? The boiling points of other covalent substances ($N_2 = -196$ °C, $O_2 = -183$ °C, and $HI = -35$ °C), are sufficiently low to indicate weak intermolecular forces in each of these cases. The solids formed by these light covalent substances are easily scratched, which indicates that the molecules are easily pulled apart even in the solid.

FIGURE 6.18
Van der Waals forces. The hydrogen molecule is nonpolar. However, it is possible that at any one instant both electrons will be on one side of the molecule. This makes one end of the molecule positive and one end negative.

*Nonpolar covalent substances* are poor conductors of electricity. Since electricity is due to the motion of charged particles, the charged particles in covalent molecules must not be able to move freely. The motion of the electrons and protons which make up the atoms of a molecule must be restricted to the volume of the molecule. They cannot move from molecule to molecule through the substance.

The forces between *polar molecules* are due to the attraction of the positive end of one molecule for the negative end of another molecule (Fig. 6.19). These intermolecular forces are called *dipole-dipole interactions* and are greater than van der Waals forces but smaller than the forces which exist between the ions in an ionic compound.

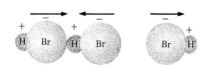

FIGURE 6.19
Dipole-dipole forces. Polar molecules such as HBr have a dipole; that is, they have a positive and negative end. Polar molecules also have van der Waals forces of attraction.

Since the attractive forces between polar molecules are greater than those between nonpolar molecules, polar molecules generally have higher melting and boiling points than nonpolar compounds. Even though there is a separation of charge in polar molecules, there are evidently no mobile, charged particles available for conducting electricity; polar compounds are, in general, poor electrical conductors.

Attractive forces between particles depend on the locations of the electrons and protons that make up these particles. These attractive forces are conveniently classified as (1) van der Waals forces, (2) dipole-dipole attractions, (3) hydrogen bonds, (4) ionic bonds, (5) metallic bonds, and (6) network bonds (Table 6.3).

### Hydrogen bonds

When a hydrogen atom in a molecule is bonded to a highly electronegative atom such as nitrogen, oxygen, or fluorine, it takes on a relatively high positive charge. The positive charge results because the bonding pair of electrons spend most of their time around the more electronegative atom. This hydrogen atom exhibits a strong attraction for relatively negative atoms in other molecules (Fig. 6.20). This attractive force is called *hydrogen bonding*. It is stronger than ordinary dipole-dipole attractions. Hydrogen bonding causes many lightweight molecules to liquefy at lower temperatures than would otherwise be expected.

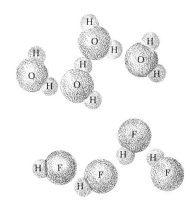

FIGURE 6.20

*When hydrogen is bonded to a highly electronegative element such as nitrogen, oxygen, or fluorine, it becomes highly positive. As such it has a strong attraction for other partially negative atoms in the same molecule or other molecules.*

### Ionic bonds

The positive ions within a crystal lattice are strongly attracted to surrounding negative ions. These attractive forces are much greater than van der Waals forces, dipole-dipole attractions, and hydrogen bonding. Ionic compounds are solids at 20°C.

### Metallic bonds

What are the properties of metals? Metals generally can be hammered into shapes and drawn into wires. Freshly cut metal sur-

TABLE 6.3

*The relationship of bonding types and states of matter.*

|  | van der Waals forces | Dipole-dipole attractions | Hydrogen bonding | Ionic bonding | Metallic bonding | Network bonding |
|---|---|---|---|---|---|---|
| Attractive forces | Very weak | Weak | Weak to moderate | Strong | Moderate to strong | Strong |
| Melting point | Low | Low | Medium | High | Medium to high | High |
| Boiling point | Low | Low | Medium | High | Medium to high | High |
| Hardness | Soft | Soft | Moderate | Moderate | Soft to moderate | High |
| Typical examples | $H_2$, $N_2$, $F_2$, He, Ne, Ar, $CH_4$, $SiF_4$ | CO, HI, $PH_3$, $H_2S$ | HF, $NH_3$, $H_2O$, $H_2SO_4$ | NaCl, $CaF_2$, KBr | Na, Hg, Fe, W | $SiO_2$, C (Diamond), $Al_4C_3$ |

faces have a shiny luster. Metals are good conductors of heat and electricity.

Since a copper wire is made up of uncharged copper atoms, how can we explain the fact that a copper wire is an excellent electrical conductor? Since all copper atoms have the same electronegativity, what type of bonding might we expect between them? Copper atoms should form nonpolar covalent bonds, but unlike the nonpolar covalent nonmetals such as hydrogen and chlorine, copper has a fairly low ionization energy. Electrons can be relatively easily removed from copper atoms. Copper conducts electricity, thus the electrons must be free to move throughout the metal. Copper behaves as if it consists of closely packed ions surrounded by "a sea of electrons" (Fig. 6.21). The attraction of the negative electrons for the positive ions is the force which is called a *metallic bond*. Metallic bonds hold the metal together.

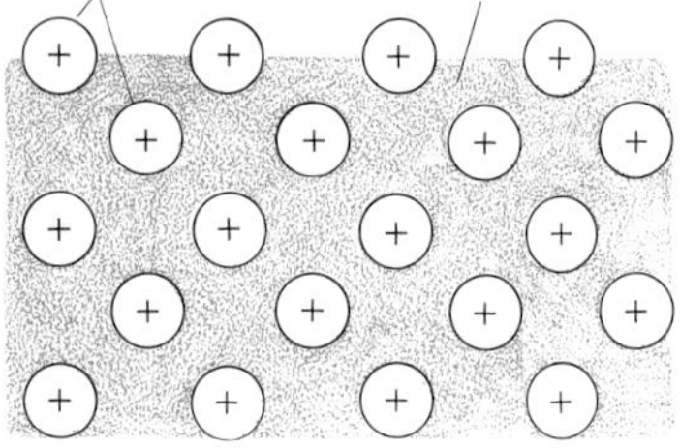

The attractive forces or metallic bonds which hold metals together are intermediate to strong. Evidence for these forces is the fact that metals have intermediate to high melting and boiling points compared to covalent and ionic substances. Metals are good electrical conductors since the negatively charged electrons are fairly loosely held and are free to move throughout the metal. When metals are heated, the electrons and ions of the metal are set into rapid notion. The ions are only free to vibrate since they are rigidly fixed in certain locations. As the temperature of a metal is increased, the electrons move more rapidly and the ions vibrate more rapidly. The mobile electrons can strike other electrons or neighboring atoms with a resultant transfer of motion. We sense this motion of particles as heat. The ability of electrons to move through metals makes metals good conductors of heat. The shiny surfaces of metals can also be explained by the mobile electrons. The light that strikes a metal surface is absorbed by the electrons which quickly re-emit the light, thus giving the metal its characteristic luster.

Many of the properties of metals can be explained by using the model of rigidly held metal ions in a "sea of mobile electrons."

FIGURE 6.21
Metallic bonding. The attraction of the positive ions for negative electrons holds the metal together.

## 6.10 Moles of atoms, molecules, and ions

The use of the mole concept was first demonstrated in Chapter 3. At that time, we were concerned primarily with covalent molecules. The mole concept applies equally well to ionic, polar covalent, and covalent substances. The example which follows should serve as a brief review of the mole concept. For a further review, refer to Chapter 3.

EXAMPLE 2  A student carefully weighed a sample of NaCl. He found that the sample had a mass of 5.24 grams. After some thought, the student wondered how many chloride, $Cl^-$, ions he had weighed. Can you determine the number of chloride ions in 5.24 grams of NaCl?

Given: 5.24 g of NaCl

Find: Number of $Cl^-$ ions

Relationship: 1.00 mole NaCl = 58.5 g NaCl = $6.02 \times 10^{23}$ NaCl particles

Solution: In order to determine the number of $Cl^-$ in 5.24 grams of NaCl, we first calculate the number of moles of NaCl in 5.24 grams. To do this, we need a unity term with moles NaCl in the numerator and grams NaCl in the denominator.

$5.24 \text{ g} \times \boxed{\phantom{xxxxx}} = ?$ moles

$5.24 \text{ g} \times \dfrac{1.00 \text{ mole NaCl}}{58.5 \text{ g NaCl}} = 0.0895$ mole NaCl

There is 0.0895 mole of NaCl in 5.24 g. Since 1.00 mole of NaCl contains 1.00 mole of $Na^+$ and 1.00 mole of $Cl^-$, 0.0895 mole of NaCl contains 0.0895 mole of $Na^+$ and 0.0895 mole of $Cl^-$.

$0.0895 \text{ mole } Cl^- \times \boxed{\phantom{xxxxx}} = ? \ Cl^-$ ions

$0.0895 \text{ mole } Cl^- \times \dfrac{6.02 \times 10^{23} \ Cl^- \text{ ions}}{1.00 \text{ mole of } Cl^- \text{ ions}}$

$\qquad\qquad = 0.538 \times 10^{23} \ Cl^-$ ions or $5.38 \times 10^{22} \ Cl^-$ ions

There are $5.38 \times 10^{22} \ Cl^-$ ions in 0.0895 mole of $Cl^-$ or in 5.24 grams of NaCl. There is the same number of $Na^+$ ions in the sample as there are $Cl^-$ ions.

## Summary

A chemical bond is formed when two electrons are simultaneously attracted to two nuclei. Bonds may be classified as covalent, polar covalent, or ionic, depending upon how equally or unequally the electrons are shared between the two bonding nuclei. If there is little difference in the electronegativities of the bonding elements, a nonpolar covalent bond forms. Bonds between elements whose electronegativities differ by more than 0.5 but less than 1.7 are classified as polar covalent bonds. When the electronegativity difference between the bonding elements is greater than 1.7, the bond is classified as ionic. Even though we classify bonds as covalent, polar covalent, or ionic, depending on how equally or unequally the bonding electrons are shared, we realize that all bonds result from the same electrostatic attraction of electrons to two nuclei. The classifications are arbitrary and are used only to help us in understanding chemical bonding.

The forces which exist between atoms in molecules are called intramolecular forces. The forces which exist between molecules are intermolecular forces. In general, intermolecular forces are smaller in magnitude than are intramolecular forces, but they are both a result of electrostatic attractions.

The octet rule states that elements generally tend to accept, give up, or share electrons in order to achieve the noble gas configuration of eight electrons in the outermost energy shell. Since exceptions to the octet rule are few in a beginning course, it can be conveniently used to determine combining ratios.

A certain quantity of energy is released when a particular bond is formed; this same quantity of energy is the minimum energy which must be absorbed if the same bond is to be broken. Since energy is given off when a bond is formed, a chemical system may go to a more stable state by forming bonds. Generally chemical reactions occur in such a way as to go to a more stable state, that is, the state in which there are the greatest number of chemical bonds.

If a molecule contains only one polar covalent bond, there is a separation of charge within the molecule; the molecule is polar. Polar molecules are those molecules whose positive and negative centers do not exactly coincide. If a molecule does contain polar bonds, but is found by experimental measurements to be nonpolar, it can be concluded that the polar bonds are distributed symmetrically. In this case the polar bonds oppose each other causing net cancellation of the separations of charge.

The properties of substances depend upon both the number and kind of atoms in the substance and upon the bonds which hold the atoms together.

## Objectives

By the end of this chapter you should be able to do the following:

1. State and show by means of a diagram what attractive forces and what repulsive forces exist in a hydrogen molecule. (Question 14)

2. Given a binary compound, use electronegativity values to classify the bonding as nonpolar covalent, polar covalent, or ionic. (Question 2)

3. State the effect of nuclear charge, valence shell number, atomic size, and shielding on the electronegativity of an element. (Questions 15, 16, 17)

4. Use atomic orbital diagrams to indicate electrons shared between atoms in given molecules. (Question 18)

5. Use atomic orbital diagrams to indicate electrons transferred during the formation of a binary ionic compound. (Question 18)

6. Draw electron-dot structures for various binary compounds. (Questions 3)

7. Write formulas for compounds formed between representative metals and nonmetals. (Question 6)

8. State the relative magnitude of some of the physical properties of some classes of compounds (covalent molecules, ionic compounds, metals). (Question 7)

9. Given the geometry of a molecule, state whether the molecule is polar. (Question 19)

10. State the model used to explain the properties of metals. Use the model to explain these properties. (Question 20)

11. Given the weight of a sample of a substance, calculate the number of molecules, atoms, or ions in that sample. (Questions 11, 12)

## Questions

1. Discuss or explain each of the following terms in your own words: *chemical bond; electrostatic forces; bond length; nonpolar covalent bond; polar covalent bond; ionic bond; electronegativity; shielding effect; orbital availability; orbital overlap; dot structure; octet rule; double bond; triple bond; unpaired electron; molecular orbital; intermolecular forces; ion; crystal; coordination number; noble gas configuration; polar molecule; symmetric molecule; metallic bonding; molecular weight.*

2. Determine the electronegativity difference between the pairs of elements in each of the following compounds. Use this electronegativity difference to classify the bond in each case. Note any periodic trends. (a) KF; (b) KCl; (c) KBr; (d) KI; (e) NaCl; (f) $MgCl_2$; (g) $AlCl_3$; (h) $SiCl_4$; (i) $BeH_2$; (j) $BF_3$; (k) $CH_4$; (l) $NH_3$; (m) $H_2O$; (n) $SO_2$; (o) $SeO_2$.

3. Draw electron-dot structures for each of the compounds in Question 2.

4. What would be the charge on the metal ion formed by the reaction of each of the following metals with the highly electronegative element, fluorine? (a) Al; (b) Li; (c) K; (d) Rb; (e) Mg; (f) Ca; (g) Ba; (h) Ra.

5. What would be the charge on a negative ion formed by the reaction of a very reactive metal, potassium, with each of the following nonmetals? (a) F; (b) Cl; (c) Br; (d) I; (e) As; (f) S; (g) Se; (h) Te; (i) N.

6. Write balanced equations for the following word equations.
(a) hydrogen + fluorine ⇌ hydrogen fluoride
(b) hydrogen + sulfur ⇌ hydrogen sulfide
(c) aluminum + iodine ⇌ aluminum iodide
(d) sodium + bromine ⇌ sodium bromide
(e) strontium + fluorine ⇌ strontium fluoride
(f) barium + fluorine ⇌ barium fluoride
(g) strontium + oxygen ⇌ strontium oxide
(h) barium + oxygen ⇌ barium oxide
(i) barium + sulfur ⇌ barium sulfide
(j) ammonia + hydrogen bromide ⇌ ammonium bromide

7. Which of the following are likely to be (a) gases at room temperature; (b) solids at room temperature; (c) good conductors of electricity as solids; (d) good conductors of electricity as liquids? (1) nitrogen; (2) hydrogen; (3) tin; (4) barium; (5) potassium fluoride; (6) fluorine; (7) sodium oxide; (8) hydrogen bromide; (9) hydrogen sulfide; (10) chlorine dioxide, $ClO_2$; (11) cobalt; (12) strontium chloride; (13) mercury; (14) sodium bromide; (15) methane, $CH_4$.

8. Add the atomic weights to calculate the molecular or formula weights of each of the following compounds or molecules: (a) sodium nitride, $Na_3N$; (b) ethyl alcohol, $C_2H_6O$; (c) ammonium sulfide, $(NH_4)_2S$; (d) asbestos, $H_4Mg_3Si_2O_9$; (e) cholesterol, $C_{22}H_{45}OH$; (f) glucose, $C_6H_{12}O_6$; (g) washing soda, $Na_2CO_3 \cdot 10H_2O$; (h) baking soda, $NaHCO_3$.

9. How many grams are there in the given number of moles of each of the following cases? (a) 5.00 moles $H_2$; (b) 8.12 moles $O_2$; (c) 8.12 moles $I_2$; (d) 0.0146 mole $H_2SO_4$; (e) 0.0146 mole $C_{22}H_{45}OH$; (f) 23.0 moles $NH_4Br$.

10. Calculate the number of moles in each of the following samples of substances: (a) 5.00 grams $Cl_2$; (b) 50.0 grams $Cl_2$; (c) 2.65 grams $Na_2S$; (d) 2.65 grams $Na_2SO_4$; (e) 2.65 grams $H_4Mg_3Si_2O_9$; (f) 100.0 grams $CS_2$.

11. How many molecules or formula units are there in each of the following samples? (a) 7.23 moles $XeF_4$; (b) 7.23 moles $NO_2$; (c) 200.0 moles $CH_4$; (d) 5.00 grams Xe; (e) 7.23 grams $XeF_4$; (f) 7.23 grams $NO_2$.

12. Calculate the number of *atoms* of each of the elements in the samples of compounds listed in Question 11.

13. Calculate the number of moles in each of the following samples: (a) $6.02 \times 10^{23}$ molecules $Cl_2$; (b) $1.20 \times 10^{22}$ molecules HCl; (c) $9.11 \times 10^{10}$ molecules $SO_2$.

14. Consider the protons and electrons in a hydrogen molecule. Between what particles are there attractive forces? Between what particles are there repulsive forces?

15. What factor causes the electronegativity to increase in moving across a period from left to right?

16. What factor causes the electronegativity to decrease in going down a group in the periodic chart?

17. Why doesn't the electronegativity increase more rapidly from element 21 to 30?

18. Use atomic orbital diagrams to show which orbitals overlap in the following compounds. If the compound is ionic, from which orbitals are electrons transferred? (a) LiF; (b) $NF_3$; (c) $OF_2$; (d) $F_2$; (e) ClF; (f) NaCl; (g) $MgCl_2$; (h) $AlCl_3$; (i) $SiCl_4$; (j) $PCl_3$; (k) $SCl_2$; (l) $Cl_2$; (m) BrCl; (n) BaO; (o) LiH; (p) $PH_3$.

19. Which of the following molecules are polar?

(a) H—Be—H  (b) 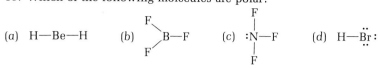B—F with F's  (c) :N—F with F above and F below  (d) H—B̈r:

20. Use a model to explain why metals (a) conduct heat; (b) conduct electricity; (c) reflect light; (d) are usually solids.

# Oxidation numbers and nomenclature

## 7.1 Introduction

The previous chapter dealt with compounds and the forces that hold atoms together in a compound. Before considering additional properties of compounds, we will learn how to name them. The naming of compounds is called *chemical nomenclature*.

You already know some chemical nomenclature, since you have learned the names and symbols of many elements. The names of most compounds include part of the name of each element that is present in the compound. The naming of compounds is thus systematic, and makes it easier to name several thousand compounds than to name several thousand people.

The naming of compounds depends in part on the type of bonds between atoms and on the number of electrons that atoms have gained or lost. We will consider a system for keeping track of electron loss and gain before developing the system of nomenclature.

## 7.2 Oxidation numbers

The bookkeeping system for electrons is known as the *oxidation number* system. The oxidation number, or oxidation state, of an atom is the apparent charge which an atom bears, or appears to bear, in a substance. The oxidation number system is useful in naming compounds and in tracing the electron change that an atom may undergo during a reaction. Because oxidation numbers depend on the electronic structure of atoms, oxidation numbers are *a periodic property*.

### Elements

What is the oxidation number of an aluminum atom in a piece of aluminum metal (Fig. 7.1)? In addition to neutrons, each aluminum atom contains 13 protons and 13 electrons. Each atom is electrically neutral since the charge on each atom is zero. Consider the two hydrogen atoms in the hydrogen molecule shown in Fig. 7.2. If we assume that each electron in the bond is being shared equally, then each atom has one proton and one electron; that is, each atom is neutral. The oxidation number of atoms in elemental form is assumed to be zero. This applies to all 105 elements.

FIGURE 7.1

*Each aluminum atom in a piece of aluminum metal is neutral.*

**Rule 1** *The oxidation number of atoms in elemental form is zero.*

### Monoatomic ions

FIGURE 7.2

*Hydrogen atoms in a hydrogen molecule. Each atom contains one proton and one electron and therefore is neutral.*

The ions that exist in sodium chloride are $Na^+$ and $Cl^-$. Each of these ions are examples of monoatomic ions. A *monoatomic ion* is an ion containing only one atom. The oxidation number of an atom in the form of a monoatomic ion is the same as the charge on the ion. The oxidation number of sodium in NaCl is positive one (1+). The oxidation number of chlorine in NaCl is negative one (1−).

**Rule 2** *The oxidation number of a monoatomic ion is the same as the charge on the ion.*

We saw in Chapter 6 that since sodium has a low-ionization energy, a low electronegativity, and one valence electron, it usually loses one electron when forming compounds. Sodium usually has a 1+ oxidation number. For the same reason, the other elements in group IA usually have a 1+ oxidation number (Table 7.1).

What is the oxidation number of magnesium in magnesium chloride, $MgCl_2$? This substance contains $Mg^{2+}$ ions and $Cl^-$ ions. The oxidation number of the magnesium is 2+. In forming compounds, magnesium and other group IIA elements usually lose their two valence electrons. Their oxidation number in compounds is usually 2+ (Table 7.1).

When chlorine reacts with a metal, each chlorine atom gains one electron to form a $Cl^-$ ion. Chlorine gains an electron because it has a high electronegativity and room for one more electron in its valence shell. When bonded to metals, its usual oxidation number is 1−. For the same reasons, the other elements in group VIIA have a 1− oxidation number when bonded to metals (Table 7.1).

What is the oxidation number of oxygen in a sample of magnesium oxide, MgO? Each oxygen atom gains two electrons when it reacts with a metal. The oxidation number of oxygen and the other group VIA elements in compounds is usually 2− (Table 7.1).

**TABLE 7.1**

The oxidation number of the element in compounds is a function of the group number.

| Common oxidation number of metals in group IA | | Common oxidation number of metals in group IIA | | Common oxidation number of metals in group IIIA | | Common oxidation number of nonmetals in group VIA | | Common oxidation number of nonmetals in group VIIA | |
|---|---|---|---|---|---|---|---|---|---|
| 1+ | | 2+ | | 3+ | | 2− | | 1− | |
| LiCl | $Na_2O$ | $BeF_2$ | $SrCl_2$ | $AlF_3$ | $Al_2O_3$ | MgO | $Na_2Se$ | LiF | $BaI_2$ |
| KF | $K_2S$ | $MgI_2$ | BaS | $AlBr_3$ | $Al_2S_3$ | MgS | $K_2Se$ | LiCl | $AlI_3$ |
| CsI | $Li_3N$ | $CaBr_2$ | CaO | $GaI_3$ | $Ga_2O_3$ | $K_2O$ | BaO | $CaBr_2$ | KF |

■ **EXERCISE** Using a periodic chart, determine the oxidation number of each atom in the following substances: (a) $Na_2S$; (b) $Li_3N$; (c) BaS; (d) $O_2$; (e) S; (f) $AlF_3$.

**Hydrogen and oxygen**

In the previous section we used the periodic chart to relate the common oxidation numbers of some of the elements. Because hydrogen and oxygen are two common elements and because they occur in many compounds, it is useful to learn their common oxidation numbers.

Rule 3   The usual oxidation number of hydrogen in compounds is 1+. (Exception: hydrogen has an oxidation number of 1− in hydrides which it forms with less electronegative elements, for example, sodium and potassium.)

Rule 4   The usual oxidation number of oxygen in compounds is 2−. (Exception: peroxides, when oxygen is bonded to oxygen or when oxygen is bonded to the more electronegative fluorine.) ■

**Summing oxidation numbers**

The oxidation number system is meant to help keep track of electrons. For instance, when a magnesium atom reacts with an oxygen atom, magnesium oxide is formed.

$$\overset{..}{\text{Mg}} + \cdot\overset{..}{\text{O}}: \rightleftharpoons \text{Mg} \ :\overset{..}{\underset{..}{\text{O}}}: + \text{energy} \qquad (7.1)$$

What is the oxidation number of magnesium in MgO? What is the oxidation number of oxygen in MgO? Magnesium is 2+; oxygen is 2−. Each magnesium atom has lost two electrons to the oxygen atom. The sum of the oxidation numbers in this compound is zero, or $(2+) + (2-) = 0$. Since a compound bears no electrical charge, the sum of the oxidation numbers in a compound is zero.

Rule 5  *The sum of oxidation numbers of all the atoms in a compound is zero.*

**Variable oxidation numbers**

Copper forms two different compounds with chlorine as shown in Table 7.2. Note how different some of the properties of CuCl and $CuCl_2$ are. How can we determine the oxidation number of copper? First, assume that the oxidation number of each chloride ion is 1−. Since the sum of the oxidation numbers of the atoms of copper and chlorine in a formula unit must equal zero, we have:

For CuCl:

| oxidation number of copper | + | oxidation number of chlorine | = 0 |
|---|---|---|---|
| (Cu) | + | (1−) | = 0 |
|  |  | Cu | = 1+ |

For $CuCl_2$:

| oxidation number of copper | + | 2 [oxidation number of chlorine] | = 0 |
|---|---|---|---|
| (Cu) | + | 2(1−) | = 0 |
|  |  | Cu | = 2+ |

In CuCl the oxidation number of copper is 1+. This copper ion is named the copper(I) ion. In $CuCl_2$ the oxidation number of copper is 2+. This copper ion is named the copper(II) ion. Copper is an example of one of the elements that exhibit a variable oxidation number. The ions of elements with a variable oxidation number are named by placing a Roman numeral, equivalent to the amount of charge, after the name of the element.

**TABLE 7.2**

*Some of the properties of the two compounds that copper forms with chlorine.*

| Substance | Copper(I) chloride | Copper(II) chloride |
|---|---|---|
| Formula | CuCl | CuCl$_2$ |
| Mole weight (g) | 99.0 | 134 |
| Color | White | Yellow |
| Water soluble | Slightly soluble | Soluble |
| Color in dilute water solution | Colorless | Blue |

Let us examine another element with more than one oxidation state. Manganese forms several different compounds with oxygen. Two of these compounds are MnO and Mn$_2$O$_7$. What is the oxidation number of manganese in each of these compounds? We will assume in each case that oxygen has a 2− oxidation number.

For MnO:

$$\begin{pmatrix}\text{oxidation number}\\ \text{of manganese}\end{pmatrix} + \begin{pmatrix}\text{oxidation number}\\ \text{of oxygen}\end{pmatrix} = 0$$

$$(\text{Mn}) + (2-) = 0$$

$$\text{Mn} = 2+$$

The name of this maganese ion is manganese (II).

For Mn$_2$O$_7$:

$$2\begin{bmatrix}\text{oxidation number}\\ \text{of manganese}\end{bmatrix} + 7\begin{bmatrix}\text{oxidation number}\\ \text{of oxygen}\end{bmatrix} = 0$$

$$2(\text{Mn}) + 7(2-) = 0$$

$$2\,\text{Mn} = 14+$$

$$\text{Mn} = 7+$$

Manganese has an oxidation number of 7+ in this compound. We name it manganese(VII).

The more *common metals* with a variable oxidation number are the transition elements in the fourth period, the transition elements in the group IB and IIB, and some representative elements. These elements and their more common oxidation numbers are listed in Table 7.3. ■

■ **EXERCISE** Determine the oxidation number of each element in the following compounds. Name the metal ions. (a) FeCl$_2$; (b) FeCl$_3$; (c) PbO$_2$; (d) SnS; (e) SnS$_2$; (f) HgCl$_2$; (g) FeO; (h) Fe$_2$O$_3$; (i) Hg$_2$Cl$_2$.

**TABLE 7.3**

Some metallic elements have a variable oxidation number. The more common cases are listed here.

| | VIB | VIIB | VIIIB | | | IB | IIB | VA | IVA |
|---|---|---|---|---|---|---|---|---|---|
| | $Cr^{6+}_{3+,2+}$ | $Mn^{7+}_{4+,2+}$ | $Fe^{3+}_{2+}$ | $Co^{3+}_{2+}$ | $Ni^{3+}_{2+}$ | $Cu^{2+}_{1+}$ | | $As^{5+}_{3+}$ | |
| | | | | | | | | $Sn^{4+}_{2+}$ | $Sb^{5+}_{3+}$ |
| | | | | | $Au^{3+}_{+}$ | $Hg^{2+}_{+}$ | $Pb^{4+}_{2+}$ | $Bi^{5+}_{3+}$ | |

$$H \overset{\cdot\times}{\underset{\times}{:}} O:$$

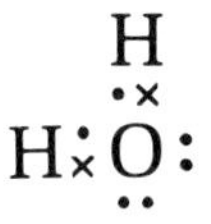

**FIGURE 7.3**

The electron-dot structure of a water molecule. The shared pair of electrons are assigned to the oxygen atom because it is more electronegative.

■ **EXERCISE** What is the oxidation number of each of the elements in the following substances? You may have to check the electronegativity values of some elements. (a) HCl; (b) HF; (c) $SO_2$; (d) $SO_3$; (e) $NO_2$; (f) $N_2O_4$.

**Covalent compounds**

What are the oxidation numbers of hydrogen and oxygen in a molecule of water (Fig. 7.3)? Each hydrogen atom is sharing a pair of electrons with an oxygen atom. For the purpose of calculating oxidation numbers, electrons in a bond are assigned to the atom with the higher electronegativity. This means that the one electron of hydrogen is assigned to oxygen, giving each hydrogen a 1+ oxidation number. Each oxygen has been assigned two negative electrons from the hydrogen atoms, giving the oxygen atom a 2− oxidation number.

Keep in mind that the oxidation number system is an *arbitrary system* used to keep track of electrons. The oxidation number of hydrogen in water is 1+ and the oxidation number of oxygen is 2−. However, the water molecule is still covalent and *no* hydrogen ions ($H^+$) or oxide ions ($O^{2-}$) exist in a water molecule. ■

**Ternary compounds**

Many compounds contain three different elements. Some examples are sodium carbonate, $Na_2CO_3$, copper(II) sulfate, $CuSO_4$, and nitric acid, $HNO_3$. Compounds with three different elements in them are known as *ternary compounds*. What are the oxidation numbers of the elements in nitric acid?

$$\begin{bmatrix}\text{oxidation} \\ \text{number} \\ \text{of hydrogen}\end{bmatrix} + \begin{bmatrix}\text{oxidation} \\ \text{number} \\ \text{of nitrogen}\end{bmatrix} + 3\begin{bmatrix}\text{oxidation} \\ \text{number} \\ \text{of oxygen}\end{bmatrix} = 0$$

We will assume that hydrogen is 1+ and that oxygen is 2−.

$$1+ \quad + \quad (N) \quad + \quad 3(2-) = 0$$
$$(N) \quad + \quad (5-) = 0$$
$$(N) = 5+$$

The oxidation number of nitrogen in nitric acid is 5+.

What are the oxidation numbers of the elements in $Na_2CO_3$?

$$2\begin{bmatrix}\text{oxidation}\\\text{number}\\\text{of sodium}\end{bmatrix} + \begin{matrix}\text{oxidation}\\\text{number}\\\text{of carbon}\end{matrix} + 3\begin{bmatrix}\text{oxidation}\\\text{number}\\\text{of oxygen}\end{bmatrix} = 0$$

We will assume that sodium, being in group IA, is 1+, and that oxygen is 2−.

$$2(1+) \quad + \quad (C) \quad + \quad 3(2-) \quad = 0$$
$$(C) \quad + \quad (4-) \quad = 0$$
$$C \quad = 4+$$

■ **EXERCISE** What are the oxidation numbers of the elements in the following substances? (a) $H_2SO_4$; (b) $H_3PO_4$; (c) $KNO_3$; (d) $Li_2CO_3$; (e) $Ba(OH)_2$; (f) $NaClO_3$.

**Polyatomic ions**

Chemists have found that under certain conditions some chemical aggregates carry an electric charge. An aggregate of two or more atoms that carries a charge is called a *polyatomic ion* (*poly* means "many"). Some of these polyatomic ions are common chemical species (Fig. 7.4).

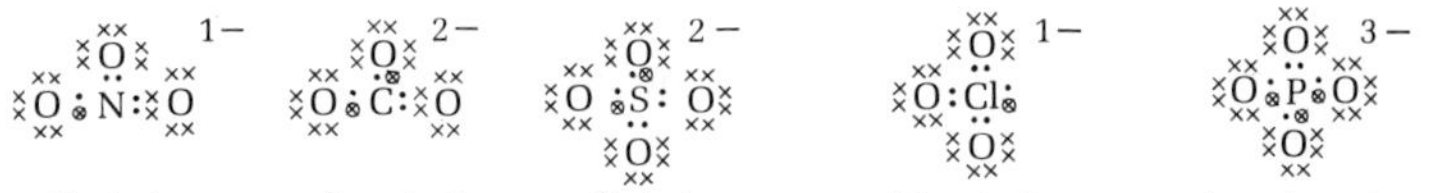

**FIGURE 7.4**
Electron-dot structures of five polyatomic ions. Electrons shared between any two atoms are assigned to the more electronegative atom, oxygen in each case.

nitrate ion $NO_3^-$ (5+) 3(2−)

carbonate ion $CO_3^{2-}$ (4+) 3(2−)

sulfate ion $SO_4^{2-}$ (6+) 4(2−)

chlorate ion $ClO_3^-$ (5+) 3(2−)

phosphate ion $PO_4^{3-}$ (5+) 4(2−)

What is the oxidation number of elements in a polyatomic ion? Let us modify Rule 5 to read:

*Rule 5* The sum of the oxidation numbers of the atoms in a molecule (or ion) is equal to the charge on the molecule (or ion).

What is the oxidation number of P in $PO_4^{3-}$?

$$\begin{matrix}\text{oxidation number}\\\text{of phosphorus}\end{matrix} + 4\begin{bmatrix}\text{oxidation number}\\\text{of oxygen}\end{bmatrix} = \begin{matrix}\text{charge on}\\\text{the ion}\end{matrix}$$

$$(P) \quad + \quad 4(2-) \quad = \quad 3-$$
$$(P) \quad + \quad (8-) \quad = \quad 3-$$
$$(P) \quad = \quad 5+$$

The oxidation number of phosphorus in the phosphate ion is 5+. How can a nonmetal such as phosphorus have a positive oxidation number? Remember that in this case the phosphorus is bonded to oxygen, a highly electronegative element. The

■ EXERCISE *Determine the oxidation number of elements in the following species:* (a) $SO_3^{2-}$; (b) $ClO_4^-$; (c) $PO_3^{3-}$.

electrons shared in bonds are assigned to the oxygen atoms since oxygen is more electronegative than phosphorus (Fig. 7.4). ■

## 7.3 Naming binary compounds

In this section we will discuss the naming of binary compounds. *Binary compounds are those compounds that contain two different elements.* For ease in naming them, we will divide binary compounds into the following classifications.

1. Metals with nonmetals
2. Variable oxidation number metals with nonmetals
3. Nonmetals with nonmetals
4. Compounds containing $NH_4^+$, $OH^-$, and $CN^-$
5. Hydrogen with nonmetals (compounds and aqueous solutions)
6. Hydrogen with metals

**Metals with nonmetals**

What system is used to name NaCl and MgO? In naming compounds containing a metal and a nonmetal we follow these rules:

Rule 1   *Name the positive element (metal) first.*
Rule 2   *Name the negative element (nonmetal) second and modify the ending of the nonmetal to -ide.*

We will first study compounds where the metal has only one common positive oxidation state. These metal ions, their names, and their charges are listed in Table 7.4. The common nonmetal ions, their names, and their charges are listed in Table 7.5.

What is the name of $BaCl_2$? Note that this compound is binary and contains the metal barium and the nonmetal chlorine. We name the metal first (barium), and then the nonmetal with an -*ide* ending (chloride). $BaCl_2$ is called barium chloride.

TABLE 7.4

*Ions of metallic elements which exhibit only one common oxidation number.*

| + | 2+ | 3+ |
|---|----|----|
| Lithium, $Li^+$ | Magnesium, $Mg^{2+}$ | Aluminum, $Al^{3+}$ |
| Sodium, $Na^+$ | Calcium, $Ca^{2+}$ | |
| Potassium, $K^+$ | Strontium, $Sr^{2+}$ | |
| Rubidium, $Rb^+$ | Barium, $Ba^{2+}$ | |
| Cesium, $Cs^+$ | Zinc, $Zn^{2+}$ | |
| Silver, $Ag^+$ | Cadmium, $Cd^{2+}$ | |

| TABLE 7.5 | 3⁻ | 2⁻ | 1⁻ |
|---|---|---|---|
| Common monoatomic nonmetallic ions. Note that the names of these ions all end in -ide. | Nitride, $N^{3-}$<br>Phosphide, $P^{3-}$ | Oxide, $O^{2-}$<br>Sulfide, $S^{2-}$ | Fluoride, $F^-$<br>Chloride, $Cl^-$<br>Bromide, $Br^-$<br>Iodide, $I^-$ |

■ EXERCISE  Name the following compounds: (a) LiBr; (b) $Na_3N$; (c) SrS; (d) $Ba_3N_2$; (e) AlP; (f) AgBr.

The following are examples for naming binary compounds:

KBr, potassium bromide   $K_3N$, potassium nitride
$K_2S$, potassium sulfide   $Al_2O_3$, aluminum oxide ■

**EXAMPLE 1**   What is the formula of magnesium iodide? The name tells us that the compound: (a) contains magnesium ions ($Mg^{2+}$); (b) is binary (ide ending); and (c) contains iodide ions ($I^-$). Since the oxidation numbers of a molecule or formula unit must add up to zero, how many iodide ions ($I^-$) are necessary to balance the 2+ of magnesium ($Mg^{2+}$)? Two iodide ions will balance the charge of magnesium. The formula of magnesium iodide is $MgI_2$.

**EXAMPLE 2**   What is the formula of aluminum sulfide? The name tells us that the compound: (a) contains aluminum ions ($Al^{3+}$), (b) is binary (ide ending); and (c) contains sulfide ions ($S^{2-}$). How many sulfide ions will balance out the 3+ oxidation number of aluminum? One sulfide is not enough and two sulfides give a total charge of 4− versus 3+ for aluminum. If we use two aluminum ions and three sulfide ions, we will have a net charge of zero (6+) + (6−) = 0. The formula for aluminum sulfide is $Al_2S_3$. ■

■ EXERCISE  Write the formulas of the following compounds: (a) lithium bromide; (b) potassium oxide; (c) barium fluoride; (d) zinc oxide; (e) aluminum oxide; (f) cadmium sulfide.

**Variable oxidation number metals with nonmetals**

What is the name of CuCl? What is the name of $CuCl_2$? Both of these compounds contain copper and chlorine, yet they have different formulas and different properties (Table 7.2). We must therefore give them different names. Notice that the oxidation number of copper is different in each compound. We have $Cu^+$ in CuCl and $Cu^{2+}$ in $CuCl_2$. $Cu^+$ is named copper(I) and $Cu^{2+}$ is named copper(II). *The Roman numeral indicates the oxidation number* of each particular copper ion. The rest of the compound is named as before.

CuCl, copper(I) chloride
$CuCl_2$, copper(II) chloride

What are the names of FeS and $Fe_2S_3$? Evidently the iron has two different oxidation numbers. For FeS,

$$\begin{array}{ccc}
\text{oxidation number of iron} + \text{oxidation number of sulfur} &=& 0 \\
(\text{Fe}) + (2-) &=& 0 \\
\text{Fe} &=& 2+
\end{array}$$

The name of the $Fe^{2+}$ ion is iron(II). The name of FeS is iron(II) sulfide. For $Fe_2S_3$,

$$\begin{array}{ccc}
2\left[\text{oxidation number of iron}\right] + 3\left[\text{oxidation number of sulfur}\right] &=& 0 \\
2(\text{Fe}) + 3(2-) &=& 0 \\
\text{Fe} &=& 3+
\end{array}$$

The name of the $Fe^{3+}$ ion is iron(III). The name of $Fe_2S_3$ is iron(III) sulfide. The common metals with variable oxidation numbers are listed in Table 7.3.

The following are some examples of this system:

$CoCl_2$, cobalt(II) chloride   $PbO_2$, lead(IV) oxide
$CoO$, cobalt(II) oxide   $HgI_2$, mercury(II) iodide ■

■ EXERCISE  Name the following substances: (a) CuO; (b) $Cu_2O$; (c) CrO; (d) $Cr_2O_3$; (e) $SnS_2$; (f) HgO.

■ EXERCISE  Write the formulas for the following substances: (a) manganese(IV) oxide; (b) cobalt(II) sulfide; (c) mercury(II) sulfide; (d) mercury(I) chloride (merecury(I) is diatomic).

**Nonmetals with nonmetals**

The system of nomenclature in which Roman numerals are used to indicate the oxidation number of an element is called the IUPAC system of nomenclature. This same system can be applied to naming compounds containing two nonmetals. However, an older, classical system of nomenclature can also be used for naming such compounds.

In the classical system of nomenclature, *Greek prefixes are used to indicate the number of atoms of each element present in a molecule.* These prefixes are listed here. The prefix mono is often omitted if it applies to the first element in the compound.

mono - one      hexa - six
di - two        hepta - seven
tri - three     octa - eight
tetra - four    nona - nine
penta - five    deca - ten

Carbon dioxide is the name of $CO_2$. The prefix "di" tells us that there are two oxygen atoms per molecule. In naming binary compounds containing two nonmetals:

1. Name the more positive element first and the more negative element second.

7.3 Naming binary compounds

2. Use the prefixes "di," "tri," "tetra," etc., to indicate the number of atoms of each type in the molecule.
3. Use the -ide ending.

The following are examples for naming these compounds.

$N_2O$, dinitrogen monoxide   $N_2O_3$, dinitrogen trioxide
NO, nitrogen monoxide   $N_2O_4$, dinitrogen tetraoxide
$NO_2$, nitrogen dioxide

One exception to these rules is ammonia, $NH_3$. ■

■ EXERCISE  Name the following substances: (a) $XeF_2$; (b) $NCl_3$; (c) $PBr_5$; (d) $P_4O_6$; (e) $P_4O_{10}$; (f) $N_2O_5$.

■ EXERCISE  Write the formulas of the following compounds: (a) boron trichloride; (b) carbon disulfide; (c) silicon tetrachloride; (d) dinitrogen trioxide.

### Compounds containing $NH_4^+$, $OH^-$, and $CN^-$

Three common polyatomic ions are the ammonium ion $NH_4^+$ the hydroxide ion $OH^-$, and the cyanide ion $CN^-$. The ammonium ion has many properties that are similar to the properties of sodium or potassium ions. In naming compounds containing the ammonium ion, we treat the $NH_4^+$ ion as if it were a monoatomic ion such as $Na^+$ or $K^+$. For example,

$NH_4Cl$; ammonium chloride   $(NH_4)_2S$; ammonium sulfide

Both the hydroxide ion, $OH^-$, and the cyanide ion, $CN^-$, are named as if they were monoatomic like $Cl^-$ or $S^{2-}$. The names of both $OH^-$ and $CN^-$ have an -ide ending. For example,

NaOH, sodium hydroxide   NaCN, sodium cyanide
$Al(OH)_3$, aluminum hydroxide   $Al(CN)_3$, aluminum cyanide. ■

■ EXERCISE  Name the following compounds: (a) $NH_4F$; (b) $(NH_4)_2Se$; (c) $NH_4CN$; (d) LiOH; (e) LiCN; (f) $Ca(CN)_2$.

■ EXERCISE  Write the formulas of the following compounds: (a) ammonium iodide; (b) ammonium sulfide; (c) lithium hydroxide; (d) barium hydroxide; (e) barium cyanide; (f) cesium cyanide.

### Hydrogen with nonmetals

Hydrogen forms covalent compounds with most nonmetals. You are probably familiar with some of these compounds already. The substance hydrogen sulfide, $H_2S$, is responsible for the odor of rotten eggs. Ammonia, $NH_3$, is used in water solutions as a cleaning agent. The most common of all compounds is water, $H_2O$.

When the gases HF, HCl, HBr, HI, and $H_2S$ are dissolved in water, the resulting solutions have properties of acids. These water solutions have characteristic names. Compare the names of these water solutions with the names of the corresponding gases.

*Water Solutions*
HF(aq), hydrofluoric acid
HCl(aq), hydrochloric acid
HBr(aq), hydrobromic acid
HI(aq), hydroiodic acid
$H_2S$(aq), hydrosulfuric acid

*Gaseous Compound*
HF(g), hydrogen fluoride
HCl(g), hydrogen chloride
HBr(g), hydrogen bromide
HI(g), hydrogen iodide
$H_2S$(g), Hydrogen sulfide

■ EXERCISE  Write the formulas of the following compounds: (a) hydrogen fluoride; (b) hydroiodic acid; (c) hydrogen iodide; (d) magnesium hydride.

**Hydrogen with metals**

Hydrogen reacts with many metals to form compounds called hydrides. In these compounds, the hydrogen is more electronegative than the metal and it is assigned an oxidation number of 1−. For example,

LiH, lithium hydride    $AlH_3$, aluminum hydride. ■

## 7.4 Naming ternary compounds

Ternary compounds are those compounds that contain three different elements. For ease in naming them, we will divide ternary compounds into the following classifications.

1. -ate compounds and -ic acids
2. -ite compounds and -ous acids
3. salts with more than one type of positive ion

**-ate compounds and -ic acids**

Certain polyatomic negative ions are commonly found in compounds. The most common of these negative ions are:

carbonate, $CO_3^{2-}$      sulfate, $SO_4^{2-}$
nitrate, $NO_3^-$           chlorate, $ClO_3^-$
phosphate, $PO_4^{3-}$      perchlorate, $ClO_4^-$

Figure 7.5 gives a more complete list in periodic format.
In a compound, each of these ions is surrounded by an appropriate number of positive ions. For example, washing soda is

FIGURE 7.5
Polyatomic ions ending in -ate (a periodic format).

| IIIA | IVA | VA | VIA | VIIA |
|---|---|---|---|---|
| $BO_3^{3-}$ borate | $CO_3^{2-}$ carbonate | $NO_3^-$ nitrate | | |
| | $SiO_3^{2-}$ silicate | $PO_4^{3-}$ phosphate | $SO_4^{2-}$ sulfate | $ClO_4^-$ perchlorate $ClO_3^-$ chlorate |
| | | $AsO_4^{3-}$ arsenate | $SeO_4^{2-}$ selenate | $BrO_4^-$ perbromate $BrO_3^-$ bromate |
| | | | $TeO_4^{2-}$ tellurate | $IO_4^-$ periodate $IO_3^-$ iodate |

mainly sodium carbonate, $Na_2CO_3$. Each sodium ion has a 1+ charge. Two of these sodium ions are necessary to balance the 2− charge of the carbonate ion.

In naming compounds containing these ions, we name the metal or positive ion first and the negative ion last. Note that each of these ions has an ending of -ate.

$K_2CO_3$, potassium carbonate  $FeCO_3$, iron(II) carbonate
$KNO_3$, potassium nitrate   $Fe(NO_3)_2$, iron(II) nitrate
$K_3PO_4$, potassium phosphate  $Fe(ClO_4)_2$, iron(II) perchlorate

How do we know that the oxidation number of iron in the above compounds is 2+ if we are only given the formulas? Examine $FeCO_3$; in this case you must be able to recognize the $(CO_3)$ group as the carbonate ion $(CO_3^{2-})$. Knowing that the carbonate ion has a 2− charge tells us that the lone iron ion must be 2+. We could have made use of the charges to write an algebraic equation. For $FeCO_3$:

$$\text{charge on the iron ion} + \text{charge on the carbonate ion} = 0$$

$$(\text{Fe}) + (2-) = 0$$

$$\text{Fe} = 2+$$

The charge on the iron ion is 2+: the name of the compound is iron(II) carbonate.

■ EXERCISE  Name the following compounds: (a) $Li_2CO_3$; (b) $LiNO_3$; (c) $BaSO_4$; (d) $Ba(ClO_3)_2$; (e) $Fe(NO_3)_3$; (f) $Sn_3(PO_4)_2$.

■ EXERCISE  Write the formulas of the following substances: (a) copper (I) sulfate; (b) sodium perchlorate; (c) silver perchlorate; (d) lead (II) carbonate; (e) magnesium phosphate; (f) zinc phosphate.

EXAMPLE 3  What is the formula for lithium sulfate? The lithium ion has a 1+ charge; it is in group IA. The sulfate ion $(SO_4^{2-})$ has a 2− charge. There must be two lithium ions for each sulfate ion, or $Li_2SO_4$. ■

Figure 7.6 shows hydrogen ions covalently bonded to several different chemical aggregates. Aqueous solutions of substances such as these have the properties of acids. For this reason, we name aqueous solutions of these substances as acids.

Is there a consistent relationship between the name of an acid such as carbonic acid, $H_2CO_3$, and a salt such as sodium carbonate, $Na_2CO_3$? Notice that in the acid, two hydrogen ions have replaced the two sodium ions in the salt. In naming acids such as this we modify the -ate ending of the salt to -ic. Compare the following list.

$Na_2CO_3$, sodium *carbonate*   $Na_2SO_4$, sodium *sulfate*
$NaNO_3$, sodium *nitrate*    $NaClO_3$, sodium *chlorate*
$Na_3PO_4$, sodium *phosphate*   $NaClO_4$, sodium *perchlorate*

H$_2$CO$_3$(aq), *carbonic* acid  
HNO$_3$(aq), *nitric* acid  
H$_3$PO$_4$(aq), *phosphoric* acid  
H$_2$SO$_4$(aq), *sulfuric* acid  
HClO$_3$(aq), *chloric* acid  
HClO$_4$(aq) *perchloric* acid  

### -ite compounds and -ous acids

What is different about the two compounds NaNO$_3$ and NaNO$_2$? NaNO$_3$ is called sodium nitrate, NaNO$_2$ is called sodium nitrite. Note that the nitrite ion, NO$_2^-$, has one less oxygen than the nitrate ion, NO$_3^-$. Other ions that have -*ite* endings are compared to their *ate* counterparts in the following list.

NO$_2^-$, nitrite  
SO$_3^{2-}$, sulfite  
PO$_3^{3-}$, phosphite  
ClO$_2^-$, chlorite  
ClO$^-$, hypochlorite  

NO$_3^-$, nitrate  
SO$_4^{2-}$, sulfate  
PO$_4^{3-}$, phosphate  
ClO$_3^-$, chlorate  

The following are examples of the names of these compounds.

KClO$_2$, potassium chlorite  
KClO, potassium hypochlorite  
Ag$_3$PO$_3$, silver phosphite  
Zn(ClO$_2$)$_2$, zinc chlorite  

■ **EXERCISE** *Name the following compounds: (a) H$_2$SO$_4$; (b) H$_2$SO$_3$; (c) Li$_2$SO$_4$; (d) Li$_2$SO$_3$; (e) CuSO$_4$, (f) PbSO$_3$.*

When the metal ion in a compound ending in -*ite* is replaced by a hydrogen ion, the new compound is named as an acid. The ending -*ite* changes to -*ous*. For example,

■ **EXERCISE** *Write the formulas of the following substances: (a) magnesium chlorite; (b) sulfurous acid; (c) iron(III) sulfite; (d) nitrous acid.*

KNO$_2$, potassium nitrite  
K$_2$SO$_3$, potassium sulfite  
KClO, potassium hypochlorite  
HNO$_2$, nitrous acid  
H$_2$SO$_3$, sulfurous acid  
HClO, hypochlorous acid ■  

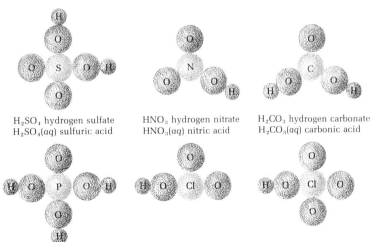

**FIGURE 7.6**  
*The structure of several different hydrogen compounds whose water solutions are acids.*

H$_2$SO$_4$ hydrogen sulfate  
H$_2$SO$_4$(aq) sulfuric acid  

HNO$_3$ hydrogen nitrate  
HNO$_3$(aq) nitric acid  

H$_2$CO$_3$ hydrogen carbonate  
H$_2$CO$_3$(aq) carbonic acid  

H$_3$PO$_4$ hydrogen phosphate  
H$_3$PO$_4$(aq) phosphoric acid  

HClO$_3$ hydrogen chlorate  
HClO$_3$(aq) chloric acid  

HClO$_4$ hydrogen perchlorate  
HClO$_4$(aq) perchloric acid

**Compounds with more than one positive ion**

An important industrial compound commonly called alum is KAl(SO$_4$)$_2$. What name can we assign to this compound? You should recognize K$^+$ as the potassium ion, Al$^{3+}$ as the aluminum ion, and SO$_4^{2-}$ as the sulfate ion. The name of KAl(SO$_4$)$_2$ includes the name of each ion: potassium aluminum sulfate. In this particular compound there are two positive ions. The correct name of a compound like this simply gives the name of each ion present.

Sometimes it is necessary to state the number of positive ions present in a compound. For instance, how would you name each of these compounds? NaH$_2$PO$_4$, Na$_2$HPO$_4$. In these cases the prefixes di- or mono- are used to indicate the number of hydrogens present in the compound. This gives: NaH$_2$PO$_4$, sodium dihydrogen phosphate and Na$_2$HPO$_4$, sodium monohydrogen phosphate.

## Summary

The oxidation number of an element is the charge that an atom of the element bears or seems to bear in a compound. In the elemental state, each atom has an oxidation number of zero. For monoatomic ions such as K$^+$ and Cl$^-$ in KCl, the oxidation number is the same as the charge. In most compounds, the oxidation number of hydrogen is 1+ and the oxidation number of oxygen is 2−. The sum of the oxidation numbers of atoms in a molecule or ion is equal to the charge on the molecule or ion.

Many metal atoms can have a variable oxidation number. For example, in different compounds copper ions may be 1+ or 2+. Roman numerals must be used to name these ions. Cu$^+$ is called copper(I) and Cu$^{2+}$ is called copper(II).

The names of binary compounds end in -ide. For a binary compound, the more positive element is named first, and the less positive element is named second. Greek prefixes are often used in naming binary compounds containing two nonmetals. For purposes of nomenclature the ammonium ion, NH$_4^+$, the hydroxide ion, OH$^-$, and the cyanide ion, CN$^-$, are treated as monoatomic ions.

Some compounds that contain hydrogen and a nonmetal dissolved in water form acid solutions. These solutions have special names. For example, a solution of hydrogen chloride is called hydrochloric acid. Compounds containing hydrogen and a metal are called hydrides.

Ternary compounds contain three different elements. One group of ternary compounds contains positive metal ions and negative ions whose names have an -ate ending; an example is potassium sulfate, K$_2$SO$_4$. When the positive metal ions are replaced by hydrogen ions, the compound is named as an acid with an -ic ending; an example is sulfuric acid, H$_2$SO$_4$.

Another group of ternary compounds contains positive metal ions and negative ions whose names have an -ite ending; an example is potassium sulfite, $K_2SO_3$. When the positive metal ions are replaced by hydrogen ions, the compound is named as an acid with an -ous ending; an example is sulfurous acid, $H_2SO_3$.

## Objectives

By the end of this chapter, you should be able to do the following:

1. Given the formula of a substance, determine the oxidation numbers of each element in the substance. (Questions 4, 5)

2. Given the formula of a binary compound, write the name of the compound. (Questions 2, 3)

3. Given the name of a binary compound, determine the formula of the compound. (Questions 6, 7, 8)

4. Given the name of a binary compound, determine the oxidation number of each element. (Question 12)

5. Given the formula of a ternary compound, write the name of the compound. (Question 5)

6. Given the name of a ternary compound, determine the formula of the compound. (Questions 9, 10, 11)

7. Given the name of a ternary compound, determine the oxidation number of each element. (Question 12)

8. Given a positive ion and a negative ion, write the formula and give the name of the compound the two ions can form. (Question 13)

## Questions

1. Using your own words, define each of the following terms; give an example of its use: (a) oxidation number; (b) binary compounds; (c) ternary compounds; (d) salts; (e) formula; (f) nomenclature; (g) variable oxidation number; (h) Greek prefixes.

2. Determine the oxidation number of each atom in the following substances; name each substance: (a) $H_2$; (b) Pb; (c) Ag; (d) KCl; (e) $NH_4I$; (f) $Ba(OH)_2$; (g) $Hg_2Cl_2$; (h) $XeF_4$; (i) $P_4O_{10}$.

3. Determine the oxidation number of each atom in the following substances; name each substance: (a) NiO; (b) $NiO_2$; (c) $PbO_2$; (d) $SnO_2$; (e) $TiO_2$; (f) $V_2O_3$; (g) $MnO_2$; (h) $Cu_2O$; (i) CuO.

4. Determine the oxidation number of each atom in the following substances; name each substance: (a) $KNO_3$; (b) $KNO_2$; (c) $NO_2$; (d) $Rb_2SO_4$; (e) $Rb_2SO_3$; (f) $SO_3$; (g) $Na_3AsO_4$; (h) $Na_3AsO_3$; (i) $As_2O_3$.

5. Determine the oxidation number of each atom in the following substances; name each substance: (a) $Pb(ClO_3)_2$; (b) $Pb(ClO_2)_2$; (c) $Fe(OH)_3$; (d) CuCN; (e) $Hg_3(PO_4)_2$; (f) $CrPO_4$; (g) $Cr_3(PO_4)_2$; (h) $Bi_2(SO_4)_3$; (i) $Sb(NO_3)_3$.

6. Write the correct formula for the following compounds: (a) sodium iodide; (b) ammonium sulfide; (c) rubidium nitride; (d) magnesium nitride; (e) zinc hydroxide; (f) cadmium oxide.

7. Write the correct formula for each of the following compounds: (a) iron(III) chloride; (b) copper(I) oxide; (c) iron(II) bromide; (d) copper(II) sulfide; (e) lead(IV) fluoride; (f) lead(II) oxide; (g) chromium(III) sulfide; (h) manganese(II) iodide; (i) gold(III) chloride; (j) tin(II) bromide.

8. Write the correct formula for each of the following compounds: (a) ammonia; (b) hydrobromic acid; (c) nitrogen trifluoride; (d) carbon tetrafluoride; (e) dichlorine heptaoxide; (f) oxygen difluoride.

9. Write the formula for the following compounds. (a) carbonic acid; (b) nitric acid; (c) chlorous acid; (d) hydrobromic acid; (e) chloric acid; (f) nitrous acid.

10. Write the formulas for the following compounds: (a) ammonium sulfate; (b) ammonium nitrite; (c) silver hydroxide; (d) sodium cyanide; (e) aluminum phosphite; (f) aluminum carbonate.

11. Write formulas for the following compounds: (a) lead(II) sulfate; (b) tin(II) nitrate; (c) gold(III) sulfite; (d) bismuth(III) chlorite; (e) manganese(II) nitrite; (f) mercury(II) carbonate.

12. Determine the oxidation number of each element in the compounds named in Questions 7, 8, 9, 10, and 11.

13. Write correct names and formulas for the compounds formed from the combinations of the following ions.

|          | $Cl^-$ | $O^{2-}$ | $NO_3^-$ | $NO_2^-$ | $SO_3^{2-}$ | $SO_4^{2-}$ | $CO_3^{2-}$ | $PO_4^{3-}$ | $OH^-$ |
|----------|--------|----------|----------|----------|-------------|-------------|-------------|-------------|--------|
| $K^+$    | KCl    | $K_2O$   | $KNO_3$  | $KNO_2$  | $K_2SO_3$   | $K_2SO_4$   | $K_2CO_3$   | $K_3PO_4$   | KOH    |
| $Ba^{2+}$| $BaCl_2$ | BaO    |          |          |             |             |             |             |        |
| $Al^{3+}$| $AlCl_3$ | $Al_2O_3$ |        |          |             |             |             |             |        |
| $Zn^{2+}$|        |          |          |          |             |             |             |             |        |
| $Cr^{3+}$|        |          |          |          |             |             |             |             |        |
| $Mn^{2+}$|        |          |          |          |             |             |             |             |        |
| $Fe^{2+}$|        |          |          |          |             |             |             |             |        |
| $Fe^{3+}$|        |          |          |          |             |             |             |             |        |
| $Co^{3+}$|        |          |          |          |             |             |             |             |        |
| $Hg^{2+}$|        |          |          |          |             |             |             |             |        |
| $Pb^{2+}$|        |          |          |          |             |             |             |             |        |
| $Au^{3+}$|        |          |          |          |             |             |             |             |        |
| $Ag^{1+}$|        |          |          |          |             |             |             |             |        |

14. We have listed the formulas of a number of familiar compounds along with their common names. Supply the chemical name in each case.

| Common name | Formula | Common name | Formula |
|---|---|---|---|
| Muriatic acid | HCl(aq) | Baking soda | NaHCO$_3$ |
| Lye | NaOH | Oil of vitriol | H$_2$SO$_4$ |
| Table salt | NaCl | Laughing gas | N$_2$O |
| Lime | CaO | Fool's gold | FeS$_2$ |
| Limestone | CaCO$_3$ | Alumina | Al$_2$O$_3$ |
| Sal ammoniac | NH$_4$Cl | Milk of magnesia | Mg(OH)$_2$ |
| Saltpeter | NaNO$_3$ | Galena | PbS |
| Litharge | PbO | | |

15. The names of analogous compounds in a group are similar. For example, both phosphorus and arsenic are in group VA. Phosphoric acid is H$_3$PO$_4$ and arsenic acid is H$_3$AsO$_4$. The phosphate ion is PO$_4^{3-}$ and the arsenate ion is AsO$_4^{3-}$. Fill in the blanks:

(a) lithium phosphate, Li$_3$PO$_4$
    lithium arsenate, _____
(b) sodium phosphite, Li$_3$PO$_3$
    sodium arsenite, _____
(c) potassium chlorate, KClO$_3$
    potassium bromate, _____
    potassium iodate, _____
(d) hydrogen sulfide, H$_2$S
    hydrogen telluride, _____
    hydrogen selenide, _____
(e) ammonium sulfate, (NH$_4$)$_2$SO$_4$
    ammonium selenate, _____
(f) chloric acid, HClO$_3$
    bromic acid, _____
    iodic acid, _____

# States of matter

## 8.1 Introduction

Matter may exist in either the *solid, liquid, gas,* or *plasma* state. A solid is a sample of matter with a definite shape and a definite volume. A liquid is a sample of matter with a definite volume and an indefinite shape. A gas is a sample of matter with an indefinite shape and an indefinite volume. A plasma consists of positively and negatively charged particles at high temperatures. The plasma state is not commonly found in the chemistry laboratory.

The state in which a given sample of matter will exist depends upon the following factors: (1) pressure, (2) temperature, (3) particles, and (4) bonding. Particles and the measurement of pressure and temperature will be discussed in this chapter.

The existence of different states of matter can be explained by a model called the *kinetic molecular theory* (KMT). The basic assumption of this model is that matter consists of particles in motion. Kinetic molecular theory predictions agree quite well with the observed quantitative behavior of gases. The theory of liquids and solids is a subject of current research.

The state in which substances exist is of current interest to scientists, engineers, and the informed layman. For example, one powerful fuel for use in rocket engines is liquid hydrogen, but hydrogen is a gas at temperatures above $-252\,°C$. Since it is difficult to handle substances at such low temperatures, other fuels are often used in place of hydrogen.

What factors determine the state of a substance? We will examine the kinetic molecular theory first, then pressure, and then temperature.

## 8.2 Kinetic molecular theory

### Particles

■ EXERCISE *Name two elements that are monoatomic and two elements that are polyatomic. Name two compounds that are polyatomic and two compounds that are ionic.*

Matter consists of particles. Particles of matter may be either atoms, molecules, or ions, The simplest substances are *monoatomic*; their particles consist of only one atom. Other substances are *polyatomic*; their particles consist of two or more atoms bonded together. Still other substances such as sodium chloride, NaCl, consist of ions. ■

### Motion and kinetic energy

According to the kinetic molecular theory, all particles are in motion. This motion may be *rotational, vibrational,* or *translational* as shown in Fig. 8.1. Given sufficient motion, atoms, molecules, or ions that are held together by attractive forces can break away from one another and form separate atoms, molecules, or ions.

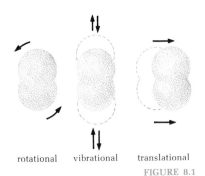

rotational   vibrational   translational

FIGURE 8.1

*Three types of kinetic energy.*

Kinetic energy is the energy which a particle possesses because of its motion. The faster a particle moves, the more kinetic energy it possesses. In addition to velocity, the kinetic energy of a particle also depends upon the mass of the particle. Consider a marble and a bowling ball rolling down a bowling alley at the same velocity; the object with the greater mass has the greater kinetic energy. The relationship between translational kinetic energy, mass, and velocity is:

$$\text{Kinetic Energy} = \text{K.E.} = \tfrac{1}{2}mv^2 \tag{8.1}$$

where $m =$ mass of particle and $v =$ velocity of particle.

A basic assumption of the kinetic molecular theory is that the relationship between kinetic energy, mass, and velocity as given in Eq. 8.1 applies to submicroscopic particles as well as to macroscopic particles. If we know the mass and velocity of a molecule or atom we can calculate its kinetic energy.

EXAMPLE 1   At 0.00°C the *average velocity* of an oxygen molecule is 42,500 cm/sec. If the mass of the molecule is $5.32 \times 10^{-23}$ g, what is the kinetic energy of the particle?

Given: $m = 5.32 \times 10^{-23}$ g
$v = 42{,}500$ cm/sec $= 4.25 \times 10^4$ cm/sec

*Find:* the kinetic energy

*Relationship:* K.E. $= \frac{1}{2} mv^2$

*Solution:* K.E. $= \frac{1}{2} mv^2 = \frac{1}{2} \times 5.32 \times 10^{-23}$ g $\times (4.25 \times 10^4$ g cm/sec$)^2$
$= \frac{1}{2} \times 5.32 \times 10^{-23}$ g $\times (1.80 \times 10^9$ cm$^2$/sec$^2)$
$= 2.66 \times 10^{-23}$ g $\times (1.80 \times 10^9$ cm$^2$/sec$^2)$
$= 4.79 \times 10^{-14}$ g cm$^2$/sec$^2$

The unit g cm$^2$/sec$^2$ is a unit of energy. It is more commonly written as an erg.

■ **EXERCISE** At 0.00°C, the average velocity of a hydrogen molecule is $1.70 \times 10^5$ cm/sec. If the mass of the molecule is $3.33 \times 10^{-24}$ g, what is the kinetic energy of the particle? Compare this value with the kinetic energy of the oxygen molecule in Example 1.

$$1.00 \text{ g cm}^2/\text{sec}^2 = 1.00 \text{ erg} \tag{8.2}$$
K.E. $= 4.79 \times 10^{-14}$ erg

The average kinetic energy of an oxygen molecule in a sample of gas at 0.00°C is $4.79 \times 10^{-14}$ erg. ■

### Overcoming attractive forces

In Chapter 6 we considered the attractive and repulsive forces that exist between particles. These forces range in magnitude from very weak forces between molecules such as $H_2$ to very strong forces between ions in compounds such as NaCl. These forces determine whether a substance is solid, liquid, or gas. How are the attractive forces between particles overcome?

Consider a sample of solid lead. The atoms of lead are attracted to one another and vibrate about fixed positions in the solid. When heat energy is added to the sample, the kinetic energy of the atoms increases. This means that the vibrational motion of the atoms increases. At sufficiently high kinetic energies, the lead atoms break away from the solid phase and form a liquid (Fig. 8.2).

Some substances melt at low temperatures because of weak attractive forces between their particles. Hydrogen, for example, melts at $-252°C$. Other substances melt at high temperatures because of strong attractive forces—for example, tungsten melts at 3380°C.

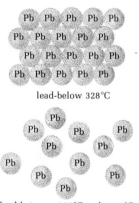

lead-below 328°C

lead-between 328°C and 1750°C

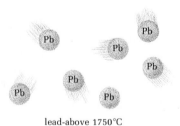

lead-above 1750°C

**FIGURE 8.2** Atoms in solid, liquid, and gaseous lead.

### Kinetic molecular model for gases

In the mid-1800s a number of scientists contributed to the development of a model which satisfactorily explained the known behavior of gases. This model, known as the kinetic molecular theory, has been modified somewhat over the years, but the basic assumptions of this model remain:

1. The molecules in a gas are far apart.

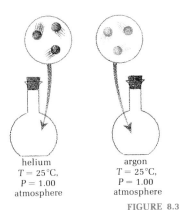

FIGURE 8.3
When the temperature of a sample of helium equals the temperature of a sample of argon, the average kinetic energy of helium atoms is equal to the average kinetic energy of argon atoms.

helium
T = 25°C,
P = 1.00 atmosphere

argon
T = 25°C,
P = 1.00 atmosphere

2. The attractive forces between molecules in a gas are very small because the particles are far apart.
3. The molecules travel in straight lines.
4. There is no loss in energy when a molecule collides with another molecule or with the walls of a container. Such collisions are said to be *elastic*.
5. The average kinetic energy of particles is directly proportional to the temperature (K.E. $\propto$ T).
6. At the same temperature, molecules of two different gases have the same average kinetic energy (Fig. 8.3). (Compare Example 1 and the exercise that follows it.)
7. A gas in a container exerts a pressure on the walls of a container because of the collisions of the molecules with the walls.

Before studying solids, liquids, and gases further, we must examine the concepts of pressure and temperature more closely.

## 8.3 Pressure

**Force per unit area**

Consider the collision of a particle with its container wall. At the instant of impact, the particle exerts a force on the wall over the area of impact. The force exerted per unit area is defined as pressure.

$$\text{pressure} = \frac{\text{force}}{\text{area}} \quad \text{or} \quad P = \frac{F}{A} \tag{8.3}$$

According to Eq. 8.3, the greater the force exerted on a given area, the greater will be the pressure (Fig. 8.4). Likewise, the

FIGURE 8.4
The pressure of a system depends upon the force and area.

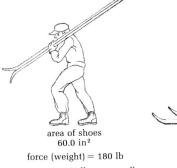

area of shoes
60.0 in²
force (weight) = 180 lb

$$\text{pressure} = \frac{180 \text{ lb}}{60.0 \text{ in}^2} = 3.00 \frac{\text{lb}}{\text{in}^2}$$

area of skis
400 in²
force (weight) = 180 lb

$$\text{pressure} = \frac{180 \text{ lb}}{400 \text{ in}^2} = 0.450 \frac{\text{lb}}{\text{in}^2}$$

smaller the area over which a given force is exerted, the greater will be the pressure. Why are hypodermic needles designed so that the point is quite small?

### The barometer

How can we measure the pressure which molecules exert when they collide with the walls of their container? After a series of experiments in the 1640s, Torricelli developed an instrument called a *barometer*, which is used to measure the pressure caused by molecules in the atmosphere. As shown in Fig. 8.5, a simple barometer is fairly easy to construct. A glass tube, about 900 mm long, is filled with mercury. (*Caution:* Remember that mercury is slightly volatile and extremely toxic; it should be handled carefully. Any spillage should be cleaned up immediately and placed in a waste mercury bottle.) The filled tube is then inverted into a dish of mercury. Under normal atmospheric conditions, the level of the mercury in the tube will remain about 760 mm above the level of the mercury in the dish.

What keeps the mercury level in the tube above that in the dish? Gravity acts on the mercury, tending to pull it down. If this were the only force acting on the mercury, then the mercury level in the dish and the tube would be equal. One model, proposed in 1661, considered that an invisible cord always held the mercury in the tube at a height of 760 mm. This model was disproved when it was shown that the mercury level changes with altitude. A second model, which is still accepted today, assumes that particles of air collide with the mercury in the dish and exert a pressure on it. This pressure is transferred by mercury atoms to the column of mercury in the tube. If the air pressure is increased, the mercury will rise in the tube. ■

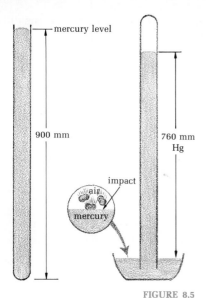

FIGURE 8.5
*Construction of the mercury barometer.*

■ **EXERCISE** Suppose you were an astronaut traveling from planet to planet. If you made a barometer on each, what would be the difference in height between the mercury in the dish and the mercury in the column under the following circumstances? Assume that gravity on each planet is the same as on the earth: (a) an extremely thin atmosphere; (b) no atmosphere; (c) a very dense atmosphere; (d) an atmosphere very similar to that of the earth.

### Units of pressure

Normal atmospheric pressure will cause the level of a column of mercury to remain at a height of 760 mm at sea level. The level changes slightly with a change in atmospheric conditions. At higher elevations, fewer air particles strike the mercury; thus the level decreases with increasing altitude.

Since pressure may be measured by the height of a column of mercury, one common pressure unit of measurement is *mm Hg*, also known as a *Torr*. Since the pressure of the atmosphere is equivalent to 760 *mm Hg*, *atmosphere* (atm) is another common unit of pressure measurement.

$$760 \text{ mm Hg} = 1.00 \text{ atm} \tag{8.4}$$

Pressure may also be expressed as a force per unit area; common units are *pounds per square inch*.

$$760 \text{ mm Hg} = 1.00 \text{ atm} = 14.7 \frac{\text{lb}}{\text{in}^2} \tag{8.5}$$

EXAMPLE 2  At an altitude of 6.21 miles above sea level, the gases in the atmosphere support a column of mercury 210 mm. (a) What is the pressure in units of atmospheres? (b) What is the pressure in units of lb/in²?

(a)

Given: pressure = 210 mm Hg

Find: the pressure in atmospheres

Relationship: 760 mm Hg = 1.00 atm

Solution: 210 mm Hg × unity term = ? atm

$$210 \text{ mm Hg} \times \frac{1.00 \text{ atm}}{760 \text{ mm Hg}} = 0.276 \text{ atm}$$

Therefore, 210 mm Hg is equivalent to 0.276 atm.

(b)

Given: pressure = 210 mm Hg

Find: the pressure in lb/in²

Relationship: 760 mm Hg = 14.7 lb/in²

Solution: $210 \text{ mm Hg} \times \dfrac{14.7 \text{ lb/in}^2}{760 \text{ mm Hg}} = 4.07 \text{ lb/in}^2$

Therefore, 210 mm Hg is equivalent to 4.07 lb/in². ■

■ EXERCISE  If the mercury column in a barometer is at a height of 760 mm, what is its height in inches?

■ EXERCISE  The pressure inside a space capsule is 0.200 atmosphere. What is the pressure (a) in mm Hg? (b) in lb/in²?

### The manometer

Another simple instrument used for pressure measurement is the *manometer*, illustrated in Fig. 8.6. One arm of the manometer is attached to a chemical system and the other end is open to the atmosphere. When the mercury levels in each arm are at equal heights as shown in Fig. 8.6(a), the pressure of the chemical system equals the pressure of the atmosphere. If the pressure in the chemical system is increased by, for example, a chemical reaction, the increased pressure will force more of the mercury up into the open tube (see Fig. 8.6(b)). The *difference* between the height of the two levels is a measure of the *difference* in pressure between the chemical system and the atmosphere. The pressure in the chemical system can be calculated by:

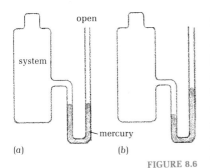

FIGURE 8.6

Construction of the manometer. (a) Pressure in the system equals the pressure of the environment. (b) Pressure in the system is higher than in the environment.

■ EXERCISE  A manometer was used to measure the pressure of a liquid flowing in a pipe. The mercury level in the atmospheric side of the manometer was 53 mm higher than the system side. If atmospheric pressure was 758 mm Hg, what was the pressure of the liquid in the pipe?

$P =$ atmospheric pressure $+ (h_a - h_s)$ (8.6)

where atmospheric pressure is measured in mm Hg, $h_s =$ height of mercury column on system side in mm, and $h_a =$ height of mercury column on atmospheric side in mm.

What is the pressure in the chemical system shown in Fig. 8.7? The mercury level on the system side is 287 mm above the table. The mercury level on the atmospheric side is 143 mm above the table. If atmospheric pressure is 760 mm Hg, then the pressure in the chemical system is:

$P =$ 760 mm Hg $+$ (143 mm Hg $-$ 287 mm Hg)
$=$ 760 mm Hg $+ (-144$ mm Hg$)$
$=$ 616 mm Hg  ■

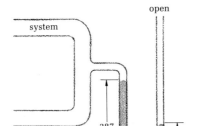

FIGURE 8.7
The pressure of a chemical system as measured by a manometer.

## 8.4 Temperature

**Velocity distribution**

Consider 1.00 liter of nitrogen gas at 0°C. Are all the molecules moving at the same velocity? By allowing some molecules to escape from their container and measuring how long it takes for the molecules to travel a certain distance, chemists have found that:

1. Some molecules travel at above average velocities.
2. Some molecules travel at or near average velocities.
3. Some molecules travel below average velocities.

The greatest number of molecules have a velocity somewhere near the average velocity. The number of nitrogen molecules in a sample of gas at 0°C having a particular velocity is plotted versus the velocity in Fig. 8.8(a). The left-hand side of the graph shows that a few molecules have very low velocities. The middle section of the graph shows that many molecules have near average velocities. The right-hand section of the graph shows that a few molecules have above average velocities.

What happens to the velocities when the temperature of the sample is increased (Fig. 8.8(b))? This graph shows that the average velocity of the molecules has increased. Notice that the velocities become spread out over a greater range.

What is temperature? One theoretical definition is that the *temperature of a system is a measure of the average random kinetic energy of the particles in the system.* Therefore, heating a system to a *higher temperature increases* the average kinetic energy of the particles in the system; cooling a system to a *lower temperature decreases* the average kinetic energy of the particles.

## Temperature measurement

Why does the mercury in a thermometer expand when the thermometer is placed in contact with an object which has a higher temperature? The particles of high kinetic energy in the object transfer some of their kinetic energy to the particles of glass in the thermometer. This kinetic energy is in turn transferred to the atoms of mercury. In other words, the mercury becomes hotter. As the motion of the mercury atoms increases, these atoms push each other farther apart, and the liquid expands (Fig. 8.9).

In the process just described, note that kinetic energy is transferred from the "hotter" object to the "colder" object. Energy that flows from hot to cold is called *thermal energy, heat energy,* or just plain *heat. When two objects of different temperatures are placed in contact, heat energy flows spontaneously from the hot object to the cold object.*

## Temperature scales

A scale is needed to measure the expansion of mercury in a thermometer as it is heated. Three different temperature scales are in common use today. They are the Fahrenheit scale, Celsius scale, and the Kelvin scale (Fig. 8.10).

The *Celsius* scale is most commonly used in the laboratory. On this scale the freezing point of water is 0.00 °C. The boiling point of water is 100 °C. Note that there are 100 Celsius degrees (100 °C) between the freezing point and boiling point of water.

The *Kelvin* scale is an absolute scale. It has been theorized that the coldest possible temperature is the temperature at which molecular motion ceases. This temperature is called *absolute zero.* This point is 0 °K. The freezing point of water is 273 °K, and the boiling point is 373 °K. Note that there are 100 Kelvin degrees (100 K°) between the freezing point and boiling point of water.

The *Fahrenheit* scale is used in the United States for nonscience purposes such as cooking and weather reporting. The freezing point of water on this scale is 32 °F. The boiling point of water is 212 °F. Note that there are 180 Fahrenheit degrees (180 F°) between the freezing point and boiling point of water.

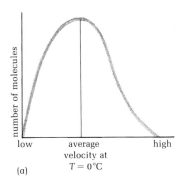
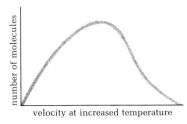
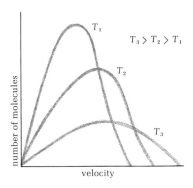

FIGURE 8.8

(a) *The curve represents the velocity distribution of molecules in a sample of gas at a given temperature.* (b) *When the temperature is increased, the curve becomes spread out. The average velocity of the molecules increases.* (c) *A comparison of the velocity distributions of molecules in a sample of gas at three different temperatures.*

8.4 Temperature

## Temperature conversions

The *size of a degree on the Kelvin and Celsius scales is the same, but they have different zero points or starting points. At the freezing point of water the celsius scale reads 0°C but the Kelvin scale says 273°K. To convert a Celsius reading to a Kelvin reading, add 273 to the Celsius reading. To convert a Kelvin reading to a Celsius reading, subtract 273 from the Kelvin reading.*

■ EXERCISE  Mercury melts at −39°C. What is the melting point of mercury on the Kelvin scale? Would a mercury thermometer be more useful for measuring winter temperatures in Miami, Florida, or in Fairbanks, Alaska?

■ EXERCISE  Absolute zero is 0°K. What is this temperature on the Celsius scale?

**EXAMPLE 3**  Potassium metal melts at 63°C. What is the melting point of potassium on the Kelvin scale?

Given: melting point of potassium = 63°C

Find: melting point of potassium on Kelvin scale

Relationship: °K = °C + 273

Solution: °K = 63°C + 273 = 336°K  ■

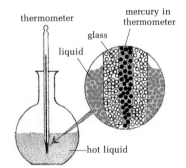

FIGURE 8.9

*Some of the kinetic energy of the liquid particles is transferred to the glass particles in the thermometer. The glass particles then transfer some of their kinetic energy to the mercury particles, causing the mercury to expand.*

An equation that expresses the relationship between degrees Celsius and degrees Farenheit is:

$$°C = \frac{(°F - 32)}{1.80} \tag{8.7}$$

or, solving this equation for °F,

$$°F = 1.80°C + 32 \tag{8.8}$$

The value 1.80 is also commonly written as 9/5.

**EXAMPLE 4**  At one point during re-entry, the heat shield on a space vehicle reached 1200 °C. What was the temperature in °F?

Given: temperature = 1200 °C

Find: temperature in °F

Relationship: °F = 1.80 °C + 32

Solution: °F = 1.80°C + 32

= 1.80 × 1200 + 32

= 2192°F or 2190°F (to three significant figures)

**EXAMPLE 5**  Acetic acid boils at 245°F. What is the boiling point on the Celsius scale?

Given: temperature = 245°F

Find: temperature in °C

Relationship: °C = (°F − 32)/1.80

■ EXERCISE  How many degrees Fahrenheit is a room that has a temperature of 20.0°C?

■ **EXERCISE** An American student traveling in France developed a fever of 103°F. (a) What was the reading on the Celsius thermometer used to measure his temperature? (b) What is the temperature in °K?

Solution: 
$$°C = \frac{(°F - 32)}{1.80}$$
$$= \frac{(245 - 32)}{1.80}$$
$$= \frac{(213)}{1.80}$$
$$= 118°C \quad ■$$

## 8.5 Specific heat

Heat energy may be used to raise the temperature or to bring about a phase change in a chemical system. For example, the application of heat energy to an iron bar at 20°C will raise the temperature of the bar. On the other hand, the application of heat energy to a sample of ice at 0°C causes the ice to change into the liquid state, but the temperature remains at 0°C until all the ice melts.

A common unit used for heat energy is the calorie. The *calorie* is defined as the amount of energy necessary to raise the temperature of 1.00 gram of *water* at 14.5°C to 15.5°C. A *kilocalorie* is equivalent to 1000 calories.

If 1.00 gram of water is heated from 0.00°C to 1.00°C, 1.00 calorie of heat energy is needed. If the sample is heated another degree to 2.00°C, another 1.00 calorie of heat energy is needed. The amount of heat energy needed to raise the temperature of 1.00 gram of a substance by 1.00 Celsius degree is called the *specific heat* of that substance. The specific heat of water is 1.00 calorie/gram °C. The specific heats of a number of other substances are listed in Table 8.1.

How many calories are needed to heat 15.0 grams of water from 10.0°C to 43.0°C? To raise 1.00 gram of water by 1.00 degree Celsius requires 1.00 calorie. We have 15.0 grams of water that must be heated by 33.0 degrees Celsius. To calculate the number of calories required, we multiply the specific heat by the mass in grams and the change in temperature.

$$\left(\frac{1.00 \text{ cal}}{\text{g °C}}\right) \times (15.0 \text{ g}) \times (33.0°C) = 49.5 \text{ cal}$$

- heat required per g per degree Celsius for water
- mass of sample
- temperature change

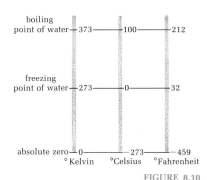

FIGURE 8.10
The three common temperature scales.

**TABLE 8.1**

*The specific heats of a number of substances. This is the amount of energy in calories required to raise the temperature of 1.00 gram of a given substance by 1.00°C*

| Substance | Specific heat (cal/g °C) |
|---|---|
| Aluminum, Al | 0.212 |
| Carbon, C | 0.127 |
| Ice, $H_2O$ | 0.5 |
| Water, $H_2O$ | 1.00 |
| Steam, $H_2O$ | 0.5 |
| Alcohol, $C_2H_5OH$ | 0.581 |

■ EXERCISE  How many calories are necessary to raise the temperature of a pond of water by 2.00°C if the pond holds $2.41 \times 10^8$ grams of water?

**EXAMPLE 6**  What will be the temperature of a 9.35 gram sample of aluminum at 20.0°C after 50.0 calories of heat energy have been added to it?

Given: 9.35 g aluminum
20.0°C
50.0 cal of heat energy

Find: the final temperature of sample

Relationship: From Table 8.1, 0.212 cal is required to raise the temperature of 1.00 g of Al by 1.00°C. The specific heat can be used as a unity term. It relates heat, mass, and temperature change.

Solution: How many calories are available per 1.00 gram of aluminum?

$$\frac{50.0 \text{ cal}}{9.35 \text{ g}} = 5.35 \text{ cal/g}$$

How many degrees Celsius will 5.35 cal/g raise the temperature of aluminum? We can use the specific heat relationship:

$$5.35 \text{ cal/g} \times \frac{1.00 \text{ g} \times °C}{0.212 \text{ cal}} = 25.2 \text{ °C temperature rise}$$

The original temperature was 20.0°C. The new temperature is 20.0°C + 25.2°C = 45.2°C. The problem could have been solved in one step.

$$\frac{50.0 \text{ cal}}{9.35 \text{ g}} \times \frac{1.00 \text{ g} \times °C}{0.212 \text{ ca.}} = 25.2 °C \text{ temperature rise} \quad ■$$

## 8.6  Solids

We will now consider some properties of solids in terms of the kinetic molecular model.

### Rigidity

Solids are *rigid bodies* with definite shapes and volumes; they are difficult to *compress*. Solids consist of atoms, molecules, or ions. The difficulty in compressing solids suggests that these particles are packed closely together. Strong attractive forces between the particles would prevent them from moving past one another easily; this would account for the rigidity of solids.

### Crystallinity

Many solids consist of crystals. A *crystal* is a homogeneous sample of solid matter with smooth surfaces and definite angles between these surfaces. Sodium chloride is an example of a crystalline substance that has 90° angles between surfaces. According to the kinetic molecular model, the particles in solids are arranged in definite, repeating geometric patterns. The angles and surfaces in a solid depend upon the geometric arrangement of particles in the solid.

**Definite melting point**

Most solids have definite melting points. In fact, the melting point of a solid can aid in the identification of the substance. How can the kinetic molecular theory (KMT) model explain this property? The particles in a solid vibrate about fixed positions. If a solid is heated, the particles vibrate faster. Their kinetic energy increases. Eventually, the particles gain sufficient kinetic energy to break the bonds holding them together, and the solid liquefies. The definite melting point of a particular solid is due to the definite kinetic energy necessary to break the bonds between the particles in the solid. For example, bonds between $H_2O$ molecules in ice break at 0°C at 1.00 atm.

**Sublimation**

Naphthalene is a white solid commonly sold as moth balls. Left in a closet, a small sample of naphthalene eventually disappears. The solid also has a noticeable odor. The odor and eventual disappearance of this solid indicate that it slowly vaporizes. The process of a solid changing directly to a vapor is called *sublimation*. For solids that undergo sublimation we assume that many particles in the solid possess sufficient vibrational energy to break away from the solid and enter the surrounding gaseous state. Two other solids that readily sublime are solid carbon dioxide (dry ice) and $H_2O$ (in the form of snow below 0°C).

**Heat of fusion**

What happens when heat energy is added to a sample of ice at 0°C? It begins to melt. At this temperature the $H_2O$ molecules possess sufficient energy of motion to overcome the forces holding them together in the solid. The heat energy absorbed by the molecules is used to break bonds rather than increase the temperature of the ice. The melting of ice is an example of a *phase change* from a solid phase to a liquid phase. At a given pressure, the *phase change* of a pure substance is a *constant temperature process*.

The heat energy necessary to melt 1.00 gram of ice at 0°C is 80.0 calories. The amount of heat energy necessary to melt 1.00 gram of a substance is known as the *heat of fusion*, $H_F$. For water,

$$H_F = +80.0 \text{ cal/g}$$

The positive sign indicates that the process is endothermic.

Since heat energy absorbed by a substance during the melting process is used to break bonds between the particles in the solid, the heat of fusion gives an estimate of the strength of the bonds between particles in the solid. If the structure of the liquid formed is similar to the structure of the solid, then the heat of fusion will be low.

When a liquid such as water freezes, it liberates the same amount of energy necessary to melt an equal quantity of ice.

heat energy $(H_F) + H_2O(s) \rightleftharpoons H_2O(l)$

where $T = 0°C$, $P = 1.00$ atm, $H_F = 80.0$ cal/g.

**EXAMPLE 7** How many calories are necessary to melt 1.00 mole of ice at 0°C?

Given: 1.00 mole of ice

Find: the number of calories necessary for melting

Relationship: $H_F = +80.0$ cal/g or 80.0 cal melt (equivalent to) 1.00 g

Solution: 1.00 mole ice × $\dfrac{18.0 \text{ g}}{1.00 \text{ mole}}$ × 80.0 cal/g = 1440 cal   ■

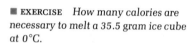
changes grams to calories

changes moles to grams

■ **EXERCISE** How many calories are necessary to melt a 35.5 gram ice cube at 0°C.

■ **EXERCISE** How many calories will 5000 grams of water give off when it freezes to solid snow at an altitude of 15,000 feet?

### Solid-liquid equilibria

Consider an insulated closed system containing 50.0 grams of water and 30.0 grams of ice. The pressure on the system is 1.00 atmosphere and the temperature is 0.00°C. As long as no energy is allowed to enter or leave the system, the amount of water and the amount of ice remains constant. The macroscopic properties of the system do not change. The shape of the ice may change with time, however. Molecules are continually leaving the ice phase and entering the liquid phase. At the same time, an equal number of molecules are leaving the liquid phase and entering the solid phase. This may cause the ice to change shape; however, we can detect no change in the amount of ice or water (Fig. 8.11). When no macroscopic changes occur in a

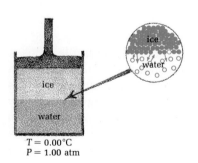

FIGURE 8.11
Ice and water in equilibrium.

| TABLE 8.2 | Substance | Melting point (°C) | Substance | Melting point (°C) |
|---|---|---|---|---|
| The melting points of a number of substances. | Lithium, Li | 180 | Helium, He | −269 |
| | Sodium, Na | 98 | Neon, Ne | −249 |
| | Magnesium, Mg | 650 | Tungsten, W | 3380 |
| | Aluminum, Al | 660 | Iron, Fe | 1539 |
| | Silicon, Si | 1423 | Copper, Cu | 1083 |

chemical system, we say the system has reached a state of *dynamic equilibrium*.

The temperature at which a liquid and a solid are in equilibrium is called the *melting point*, mp, of the substance. The melting point varies with pressure. The *normal melting point* of a substance is the temperature at which liquid and solid are in equilibrium *at 1.00 atmosphere pressure* (Table 8.2).

Since the melting point of a substance is the temperature at which the particles in the solid have sufficient kinetic energy to overcome the forces holding them together, the melting point is an indication of the strength of the forces holding the particles together in a solid. ■

■ EXERCISE  Which of the metals in Table 8.2 has the highest melting point? In which of the metals in Table 8.2 are the forces between particles in the solid probably the strongest?

**Solid-gas equilibrium**

Below −78.5 °C carbon dioxide is a solid at 1.00 atmosphere pressure. If a sample of solid carbon dioxide (dry ice) is warmed to −78.5 °C in a closed system it begins to sublime (Fig. 8.12). If no more heat is added, the amount of solid carbon dioxide and gaseous carbon dioxide remains constant. The solid phase and gaseous phase are in equilibrium:

heat energy + $CO_2(s) \rightleftharpoons CO_2(g)$

where $T = -78.5\,°C$ and $P = 1.00$ atm.

If more heat energy is added to the system, the temperature remains constant but some of the solid carbon dioxide changes to gaseous carbon dioxide. If heat energy is removed from the system, the temperature remains constant but some gas changes into the solid phase. ■

■ EXERCISE  Tin melts at 232 °C. If a sample of solid tin at 232 °C is slowly heated, what would the temperature be when the final amount of solid melts?

## 8.7 Liquids

We will now examine some of the properties of liquids and compare them with the kinetic molecular model.

## Fluidity

Liquids are *fluid*. They take the shape of their container. They maintain their volume and are difficult to compress. The difficulty in compressing liquids leads us to suppose that the particles in liquids are already close together. Strong forces of attraction must keep these particles together. However, the ability of liquids to undergo change of shape easily and to flow indicates that their particles can move past one another readily.

When poured from a container, maple syrup flows slowly. Liquids that flow slowly are said to have a high *viscosity*. The constituent particles of such liquids are large, and therefore have difficulty flowing past one another. This is due to the large size of such particles. Carbon tetrachloride flows readily. It has a low viscosity because its particles are small, and therefore flow past one another readily (Fig. 8.13). ■

■ EXERCISE  Name two substances or mixtures, other than syrup, that have a high viscosity. Name two substances or mixtures, other than water, that have a low viscosity.

## Diffusion

Several drops of purple potassium permanganate solution were carefully placed on top of a layer of water without mixing. Slowly the purple color spread out into the originally clear colorless water. As it mixed, its color became less intense. When two liquids slowly spread through one another as potassium permanganate solution and water do, they are said to diffuse.

How does the kinetic molecular model explain diffusion? The particles that make up a liquid are in rapid random motion, moving about and striking one another. Diffusion is caused by the mixing of the rapidly moving particles in each liquid.

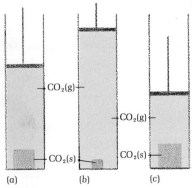

FIGURE 8.12

(a) At $-78.5\,°C$ solid carbon dioxide is in equilibrium with gaseous carbon dioxide. If no heat energy is added, the amount of solid and the amount of gas remain constant. (b) If heat energy is added, the amount of solid carbon dioxide decreases and the amount of gas increases. (c) If heat energy is removed, the amount of solid carbon dioxide increases and the amount of gas decreases. (Temperature, $-78.5\,°C$, and pressure, 1.00 atmosphere, are held constant.)

## Surface tension

Carefully place a clean, dry razor blade on top of some water in a beaker. The razor blade floats, even though the density of steel is greater than the density of water. The ability of a liquid such as water to support a razor or a pin is due to *surface tension*.

Surface tension is due to attractive forces exerted on surface molecules by molecules below it. A molecule deep within a liquid is attracted in all directions by the surrounding molecules (Fig. 8.14). A molecule on the surface is attracted downward by the molecules below it. The molecules which experience this downward attraction form a thin surface film which keeps the razor blade afloat. Once the blade pierces this film, it will sink to the bottom. ■

■ EXERCISE  Does surface tension keep a piece of wood afloat? Does surface tension keep a battleship afloat?

174   8: States of matter

(a) molecules in a liquid with a low viscosity

(b) molecules in a liquid with a high viscosity

FIGURE 8.13

*The viscosity of a liquid depends upon the ability of the molecules to flow past one another. Long-chain molecules may entangle one another and make the liquid viscous.*

### Freezing

Pure liquids freeze at the same temperature at which their solids melt. When 1.00 gram of liquid freezes, it gives off heat energy equivalent to the heat of fusion.

When a liquid is cooled, its particles move more and more slowly. At the freezing point, the particles are moving slowly enough for the attractive forces to hold them in rigid positions. Cooled enough, *all liquids freeze*.

### Vapor pressure

Have you ever noticed that water left in an open container at room temperature slowly disappears? We call this process *evaporation*.

We can use the apparatus shown in Fig. 8.15 to learn more about vaporization and its dependence on temperature. The apparatus consists of a sealed jar of dry air at 25 °C. There are slots available for attaching a manometer and a medicine dropper. Initially, the manometer reading indicates that the pressure inside the jar is equal to the atmospheric pressure.

What happens when a few drops of water are introduced into the jar? In a short time the manometer registers a pressure increase of 23.8 mm Hg. Addition of a few more drops of water has no more effect on the pressure. What causes the pressure to increase? The volume of liquid water added to the jar is negligible and could not account for the pressure change. Perhaps some of the molecules in the liquid enter the vapor state. This increases the number of gaseous molecules and therefore also increases the number of collisions of gaseous particles with the walls of the jar. The increased number of collisions causes the pressure inside the container to rise. The pressure due to the vapor from the liquid is known as the *vapor pressure*.

How does a change in temperature affect the vapor pressure? The apparatus for measuring vapor pressure was placed in a water bath and readings were made at various temperatures. The data are recorded in Table 8.3. Notice that the vapor pressure increases with the temperature.

Why does an increase in temperature raise the vapor pressure? As the temperature is increased, the average kinetic energy of particles in the liquid increases. If a particle in the liquid possesses sufficient kinetic energy, it can overcome the attractive forces of other particles in the liquid and escape into the surrounding atmosphere. Increasing the temperature

TABLE 8.3

*The vapor pressure of water as a function of temperature.*

| Temperature (°C) | Vapor pressure (mm Hg) |
| --- | --- |
| 0 (Ice) | 4.6 |
| 0 (Water) | 4.6 |
| 10 | 9.2 |
| 20 | 17.5 |
| 30 | 31.8 |
| 40 | 55.3 |
| 50 | 92.5 |
| 60 | 149 |
| 70 | 234 |
| 80 | 355 |
| 90 | 525 |
| 100 | 760 |

■ **EXERCISE** What is the vapor pressure of water at 0.00°C? What is the vapor pressure of ice at 0.00°C? Why are the two values the same?

■ **EXERCISE** At what temperature is

the vapor pressure of water equal to 1.00 atmosphere?

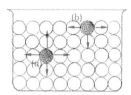

FIGURE 8.14
Water molecules attract one another. (a) A molecule within the liquid is attracted in all directions by surrounding molecules. However, (b) a molecule at the surface is attracted downward and sideward by molecules beneath it and to the sides.

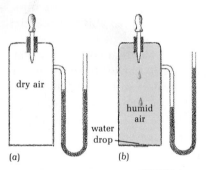

FIGURE 8.15
(a) The manometer shows that the pressure of the dry air in the system is equal to atmospheric pressure. (b) When a few drops of water are added to the system, the manometer shows that the pressure of the system is greater than atmospheric pressure.

■ EXERCISE How many calories are necessary to vaporize 6.00 moles of $H_2O(l)$?

■ EXERCISE When air rises it cools. This cooling may cause $H_2O(v)$ to

enables more particles to enter the vapor state and therefore raises the vapor pressure. ■

### Liquid-gaseous equilibrium

When water boils, pockets of vapor appear within the liquid. For water this occurs at 100°C when the pressure is 1.00 atmosphere. Note in Table 8.3 that at 100°C the vapor pressure of water is 1.00 atmosphere. When the vapor pressure of a liquid reaches atmospheric pressure, the liquid *boils*. In other words, the *normal boiling point* of a liquid is the temperature at which the vapor pressure equals 1.00 atmosphere (760 mm Hg).

Consider a closed system containing only water at 100°C and 1.00 atmosphere. As *more heat energy* is added, the temperature remains at 100°C, but more water changes to steam. After all of the liquid has been converted to gas, further heating causes the temperature to rise again.

If *no heat energy* enters or leaves a closed system of liquid and gaseous $H_2O$, the amount of liquid and the amount of gas remain constant. Molecules are continually leaving the liquid phase and entering the gaseous phase. At the same time an equal number of molecules are leaving the gaseous phase and entering the liquid phase. The system is in a state of *dynamic equilibrium*:

heat energy + $H_2O(l) \rightleftharpoons H_2O(g)$

where $T = 100°C$ and $P = 1.00$ atm.

The temperature at which the liquid and gaseous states of a substance are in equilibrium at 1.00 atmospheres is called the *normal boiling point*.

### Heat of vaporization

When heat energy is added to water at 100°C and 1.00 atmosphere, some liquid changes to gas. The amount of heat energy needed to change 1.00 gram of liquid to 1.00 gram of gas is called the *heat of vaporization*, $H_v$. The heat of vaporization for water is

$H_v = +540$ cal/g

EXAMPLE 8  How many calories are necessary to change 5.00 grams of liquid $H_2O$ to gaseous $H_2O$ at 100°C?

*Given:* 5.00 g $H_2O(l)$

*Find:* the number of calories necessary to vaporize the water

change to $H_2O(l)$. How many calories will be liberated by the condensation of 10,000 grams of $H_2O(v)$?

Relationship: $H_v = +540$ cal/g or 540 cal vaporize 1.00 g $H_2O(l)$

Solution: 5.00 g × 540 cal/g = 2700 cal

              ↑
           | converts grams to calories

EXAMPLE 9  A sample of ice at $-13.0\,°C$ weighed 2.90 grams. How many calories are necessary to change the ice to steam at 120°C?

Given: initial temperature = $-13\,°C$
       final temperature   = 120°C
       mass of ice          = 2.90 g

Find: calories necessary to change ice at initial temperature to steam at final temperature

Relationship:
1. Specific heat of ice = 0.5 cal/g °C
2. Specific heat of water = 1.00 cal/g °C
3. Specific heat of steam = 0.5 cal/g °C
4. Heat of fusion = +80 cal/g
5. Heat of vaporization = +540 cal/g

Solution: To find the number of calories necessary we will break this problem up into five separate steps.
1. Heat necessary to change ice at $-13.0°$ to ice at 0.0°C
2. Heat necessary to change ice at 0.0°C to water at 0.0°C
3. Heat necessary to change water at 0.0°C to water at 100°C
4. Heat necessary to change water at 100°C to steam at 100°C
5. Heat necessary to change steam at 100°C to steam at 120°C.

In outline form we have:

| 1 | 2 | 3 | 4 | 5 |
|---|---|---|---|---|
| Ice ⟶ | Ice ⟶ | Water ⟶ | Water ⟶ | Steam ⟶ Steam |
| $-13.0\,°C$ | 0°C | 0°C | 100°C | 100°C     120°C |

Step 1: 2.90 g × 13.0°C × 0.5 cal/g°C =   19 cal ≈   20 cal
Step 2: 2.90 g × 80.0 cal/g           = 232 cal ≈ 232 cal
Step 3: 2.90 g × 100°C × 1.00 cal/g°C = 290 cal ≈ 290 cal
Step 4: 2.90 g × 540 cal/g           = 1570 cal ≈ 1570 cal
Step 5: 2.90 g × 20°C × 0.5 cal/g°C  =   29 cal ≈   30 cal
Total of five steps is                    = 2142 cal ≈ 2140 cal

■ EXERCISE  How many calories are necessary to change 20.0 grams of ice from $-25.0\,°C$ to steam at 135°C?

Note: The specific heats of ice and of steam used here contain only one significant figure because the values change with temperature. Therefore, calculations using these figures will contain only one significant figure. ■

8.7 Liquids

## 8.8 Gases

**Compressibility**

A gas takes the size and shape of its container. A given volume of gas can be compressed into a smaller volume or expanded into a larger volume. These observations lead us to believe that particles in a gas must be far apart. When we compress a gas, the particles are pushed closer together. When we allow a gas to expand, the particles move farther apart.

The particles in a gas are in rapid random motion. This motion causes them to fill a container in which they are placed. The ease with which a gas fills its container leads us to believe that the forces between particles in a gas are weak. This is due to relatively large distances between particles.

**Diffusion**

If the gas jet to a bunsen burner is accidently turned on but not ignited, the odor of gas is quickly detected by those standing nearby. The methane gas and the additive gases which are responsible for the odor readily diffuse with the gases in the environment.

What causes diffusion of gases? The rapid motion of particles in a gas enables them to move into the spaces between particles of another gas (Fig. 8.16). The diffusion of gases occurs without the aid of any mixing or agitation.

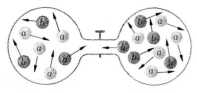

FIGURE 8.16
*The atoms or molecules that make up a gas are well-spaced and in rapid motion. The motion and spacing of these particles allow two different gases to mix readily.*

**Condensation**

When cooled sufficiently, all gases become either liquid or solid. When a gas changes from the gaseous state to a liquid state it is said to *condense*.

As a gas is cooled, heat energy is removed from the system. As the motion of the particles slows down, collisions may result in attractive forces holding two particles together. When cooled to the boiling point, attractive forces are great enough to hold the particles together in the liquid state.

As a gas condenses to a liquid, new bonds are formed. The formation of these bonds results in the liberation of energy equivalent to the heat of vaporization. When 1.00 gram of gaseous $H_2O$ condenses, it liberates 540 calories. For this reason, steam at 100°C causes more severe burns than water at 100°C.

## Summary

The properties of solids, liquids, and gases can be understood in terms of the kinetic molecular theory. According to this model, matter consists of particles: either atoms, molecules, or ions. These particles possess kinetic energy due to their rotational, vibrational, and translational motion. Particles are held closely together in solids and liquids by attractive forces. When particles are given sufficient kinetic energy, they can overcome these attractive forces: solids melt, liquids boil. The particles in a gas are far apart and attractive forces between them are weak. The pressure of a gas is due to the elastic collisions of the particles with the walls of the container.

Pressure is a force exerted over a given area. In discussing the properties of a solid, liquid, or gas, we must specify or assume a certain pressure in the system. Atmospheric pressure can be measured with a barometer and the pressure of a chemical system can be measured with a manometer. The common units used for pressure measurements are mm Hg, or Torr, atmospheres, or pounds per square inch.

Temperature is a measure of the average kinetic energy of the particles in a system. At a given temperature, particles have a range of velocities. Some particles move slower than average, many move near average, some faster than average. When the temperature of a system is increased, the average velocity of the particles increases. Temperatures are usually recorded on one of three scales: the Kelvin, the Celsius, or the Fahrenheit scale.

When a warm object is placed in contact with a cold object, the warm object cools and the cold object warms. We say that heat energy has been transferred from the warm object to the cold object. The unit of heat energy is the calorie. The amount of heat necessary to raise 1.00 gram of a substance by 1.00 Celsius degree is called the specific heat.

Solids are samples of matter with definite shapes and definite volumes. Particles in a solid are close together and for the most part vibrate about fixed positions. At sufficiently high temperatures, solids melt. The heat required to melt 1.00 gram of solid is called the *heat of fusion*. The temperature at which solid and liquid exist in equilibrium at 1.00 atmosphere is called the normal melting point (or freezing point). Instead of melting, some solids sublime; that is, they go directly from the solid state to gaseous state.

Liquids are samples of matter with definite volumes and indefinite shapes. Particles in a liquid are close together but can move past one another readily. At sufficiently high temperatures liquids boil. The heat required to boil 1.00 gram of liquid is called the heat of vaporization. The temperature at which liquid and gas exist in equilibrium at 1.00 atmosphere is called the normal boiling point. Many liquids and some solids produce significant amounts of vapor even below the normal boiling point. The pressure produced by particles in the vapor state in equilibrium with the liquid is called the vapor pressure.

A gas is a sample of matter that takes the size and shape of its container. Because particles in a gas are in constant motion and far apart, gases diffuse into one another readily. Cooled sufficiently, all gases will condense into a liquid or solid.

*Objectives*  By the end of this chapter, you should be able to do the following:

1. List the assumptions of the kinetic molecular theory as it applies to gases. (Question 2)

2. Given the mass and velocity of a particle, calculate its kinetic energy. (Question 4)

3. Given a pressure reading in one set of units (mm Hg, atmospheres, or in lbs/in²) calculate the pressure in a different set of units. (Questions 6, 7, 8, 12)

4. Given the heights of the mercury levels in a manometer, determine the pressure in a chemical system. (Question 9)

5. Draw a graph depicting the approximate velocity distribution of molecules at a given temperature. (Question 13)

6. Given the temperature of a system on one temperature scale (°K, °C, or °F) calculate the temperature on another scale. (Questions 10, 11, 12, 14)

7. Given the specific heat, mass, and temperature change for a chemical system, calculate the energy change. (Questions 15, 16)

8. List four assumptions of the kinetic molecular theory as applied to solids. (Question 2)

9. Given the heat of fusion or heat of vaporization and mass of a sample of matter, calculate the heat required for the sample to undergo a phase change. (Question 17)

10. List four assumptions of kinetic molecular theory as applied to liquids. (Question 2)

11. List the phases that may be present at the normal melting point of a substance and at the normal boiling point of a substance. (Question 19)

*Questions*  1. Define the following terms. Indicate how each term fits into the behavior and study of solids, liquids, and gases. Where possible, give theoretical and operational definitions: solid; liquid; gas; kinetic molecular theory; rotational motion; vibrational motion; translational motion; velocity; kinetic energy; elastic collisions; pressure; barometer; mm Hg; atmospheres; lb/in²; manometer; average kinetic energy; temperature; thermal energy; Fahrenheit scale; Celsius scale; Kelvin scale; absolute zero; calorie; specific heat; compressibility; crystal; normal melting point; sublimation; heat of fusion; fluid; viscosity; diffuse; surface tension; evaporation; vapor pressure; normal boiling point; heat of vaporization; condense; phase; phase change.

2. Without referring to the text, list the assumptions of the kinetic molecular theory as it applies to gases. What modifications would you have to make in order to apply the theory to liquids? What modifications would you have to make in order to apply the theory to solids?

3. Often the assumptions of a model are oversimplified in order to allow us to calculate an approximate answer to a question. For example, the particles in some solids may move about within the solid quite a bit. Examine the assumptions of the kinetic molecular theory as it applies to gases and to liquids. Which assumptions do you think might be oversimplifications?

4. At 273°K the average velocity of a sulfur dioxide molecule is 32,300 cm/sec. What is the average kinetic energy of a sulfur dioxide molecule?

5. If a molecule has a velocity of 32,300 cm/sec, what is its velocity in miles per hour?

6. During stormy weather the pressure over an area usually drops because air with a large amount of water vapor has a lower density than dry air. If the height of a mercury barometer during a hurricane is 28.93 inches, calculate: (a) the height of the column in millimeters; (b) the pressure in atmospheres; and (c) the pressure in $lb/in^2$.

7. The air pressure inside a car tire was 34.0 $lb/in^2$ *greater* than atmospheric pressure. Atmospheric pressure on that day was 14.5 $lb/in^2$. (a) What was the total pressure of the gas in the tire in $lb/in^2$? (b) Convert your answer to mm Hg; (c) Convert your answer to atmospheres.

8. A nurse found that a pressure of 115 mm Hg was sufficient to cut off the flow of blood in a patient's arm. (a) What was the pressure reading in atmospheres? (b) What was the pressure reading in $lb/in^2$?

9. A mercury manometer was used to measure the pressure in a small steam boiler. The atmospheric side of the manometer was 370 mm higher than the system side of the manometer. If atmospheric pressure was 758 mm Hg, what was the pressure in the boiler? Draw a diagram of the system.

10. The estimated temperature of lightning is 18,000°F. What is this temperature in degrees Celsius? In degrees Kelvin?

11. In devising a temperature scale in the early 1700s, Daniel Fahrenheit thought that 0°F was the coldest possible temperature. What is 0°F on the Celsius scale? What is 0°F on the Kelvin scale? How many Kelvin degrees above absolute zero was Fahrenheit's coldest possible temperature? If Mr. Fahrenheit planned to make body temperature equal to 100°F, what might you conclude about his health at this time?

12. The planet just closer to the sun than the earth is Venus. Mars is just farther than the earth from the sun. The estimated surface temperatures and pressures of these planets are:

|  | Temperature (°K) | Pressure |
|---|---|---|
| **Venus** | 700 | 100 atm |
| **Earth** | 290 | 1.00 atm |
| **Mars** | 200 | 0.01 atm |

(a) Convert the average temperature of each planet to degrees Fahrenheit. (b) Convert the estimated pressure of each planet to lb/in$^2$.

13. Draw a graph depicting the velocity distributions of gaseous molecules at 700°K. Repeat for a higher temperature and for a lower temperature.

14. The activity and hunting ability of fish are dependent on the temperature of the water. For example, at 48°F the brook trout exhibits a peak of activity. At 63°F the brook trout can only catch minnows; at 70°F they cannot catch minnows, and at 77°F they die. Convert each of these temperatures to degrees Celsius. (Because of the temperature restrictions on most life forms, thermal pollution is of great concern.)

15. The specific heat of hydrogen gas is approximately 3.4 cal/g. How many calories are necessary to heat 12.0 g of hydrogen from 300°K to 400°K?

16. The specific heat of carbon (graphite) is approximately 0.27 cal/g. How many calories are necessary to heat 17 g of graphite from 5°C to 1500°C? Watch your significant figures.

17. The heat of vaporization of carbon tetrachloride is 46.3 cal/g. How many calories are needed to: (a) vaporize 10.0 g of $CCl_4$; (b) vaporize 1.00 mole of $CCl_4$; (c) vaporize 1.00 pound of $CCl_4$?

18. The vaporization of 1.00 mole of mercury requires 14,200 cal. How many calories are necessary to vaporize 1.00 g of mercury?

19. Consider a number of chemical systems at a pressure of 1.00 atmosphere. If one of the following substances were in each system, what phases could be present in each system at the temperature indicated?

| Substance | Boiling point (°K) | System temperature (°K) |
|---|---|---|
| **Nitrogen** | 77.4° | 80° |
| **Ammonia** | 240° | 240° |
| **Hexane** | 342° | 343° |
| **Acetic acid** | 391° | 390° |
| **Ethyl alcohol** | 351° | 351° |

# Water and its elements

## 9.1 Introduction

Have you ever considered water to be an unusual substance? Most chemists do. What other substance has a solid phase that will float on its liquid phase as ice floats on liquid water? What other low molecular weight substance has such a high boiling point? Can you think of a common liquid that will dissolve substances such as sugar, table salt, hydrogen chloride, and ammonia as water does? In this chapter we will study some of the properties of water and attempt to explain these properties using the structure of water as our model.

The previous chapter presented the concepts of phases and phase changes. We will apply these concepts and other concepts such as bonding, electronegativity, polar molecules, oxidation number, and nomenclature to the study of water.

## 9.2 Physical properties of water

**Phase changes**

Pure water is a colorless, odorless, tasteless liquid. Natural waters contain dissolved carbon dioxide, oxygen, and various salts, which impart a taste to water.

At one atmosphere of pressure, water boils at 100°C. This temperature is called the *normal boiling point* of water. At this same temperature, steam condenses into water when heat is removed. As shown in Fig. 9.1, the boiling point of water changes as the applied pressure changes. At *pressures higher* than 1.00 atmosphere, the boiling point is *above* 100°C. At *pressures below* 1.00 atmosphere, the boiling point is *below* 100°C. If the pressure is lowered sufficiently, it is possible to boil water at 0.10°C.

At one atmosphere of pressure, water freezes into ice at 0°C. This temperature is called the *normal freezing point* of water. At this same temperature, ice melts into water when heat is added to it. As shown in Fig. 9.2, the freezing point of water changes as the pressure changes. At *pressures above* 1.00 atmosphere, the freezing point is *less than* 0°C. At *pressures below* 1.00 atmosphere, the freezing point is *greater than* 0°C. ■

■ **EXERCISE** *Under the conditions described below, state how you think the boiling point and freezing point of water would compare with the normal boiling point and normal freezing point of water: (a) In an elevated city such as Denver, where the pressure is less than one atmosphere; (b) in a space vehicle where the pressure is 0.20 atmosphere; (c) in a deep-water diving vessel, where the pressure is 12 atmospheres; (d) under the blade of a person who is ice skating (freezing point change only).*

Figure 9.3 shows the temperature changes undergone by $H_2O$ as it is heated. The heat required to melt a mole of ice (1.44 kcal) and the heat required to boil a mole of water (9.72 kcal) is found to be high compared to other substances. Why is water so unusual?

### Solvent properties of water

When a substance is mixed with water to form a homogeneous phase, the substance is said to be soluble in water. Which substances are soluble in water? Note in Table 9.1 that the soluble substances are either ionic or polar. The nonpolar substances are not soluble in water. Why does water seem to be a good solvent for ionic and polar substances but not for nonpolar substances? Let us look at our model.

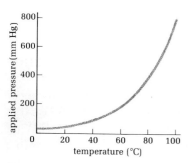

FIGURE 9.1
*The boiling point of water as a function of pressure.*

## 9.3 The water molecule, a model

### The composition of water

Accurate determinations indicate that the weight ratio of hydrogen to oxygen in water is 1.008 grams to 8.000 grams. This corresponds to a ratio of 2 hydrogen atoms to 1 oxygen atom, or $H_2O$.

### The H—O—H bond angle

The two hydrogen atoms in water are attached to the oxygen atom. Since each hydrogen atom has a half-filled *s* orbital, these

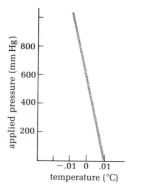

FIGURE 9.2
*The freezing point of water as a function of pressure.*

184    9: Water and its elements

**TABLE 9.1**

Examples of the solvent properties of water. Most nonpolar substances are only slightly soluble in water.

| Substance | Formula | Class | Soluble |
|---|---|---|---|
| Sodium sulfate | $Na_2SO_4$ | ionic | yes |
| Sodium hydroxide | NaOH | ionic | yes |
| Ammonia | $NH_3$ | polar | yes |
| Ethyl alcohol | $CH_3CH_2OH$ | polar | yes |
| Sucrose | $C_{12}H_{22}O_{11}$ | polar | yes |
| Hydrogen | $H_2$ | nonpolar | no |
| Carbon tetrachloride | $CCl_4$ | nonpolar | no |
| Silver chloride | AgCl | ionic | no |

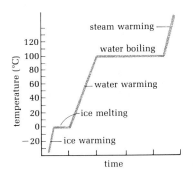

FIGURE 9.3
The temperature change over a period of time as a mole of ice is heated from $-35°C$.

atoms can overlap each of the two half-filled p orbitals of oxygen.

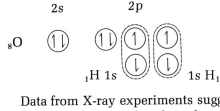

Data from X-ray experiments suggest that the angle between the hydrogen-oxygen bonds is about 104.5° (Fig. 9.4). There are also two pairs of unshared electrons in the water molecule. Because the vicinity around the oxygen atom is slightly negative and the vicinity around the hydrogen atoms is slightly positive, the water molecule is polar (Fig. 9.5). (See section on polar molecules in Chapter 6.)

**Hydrogen bonding**

What effect does the polar nature of the water molecule have on its properties? Let us compare the boiling point of water with the other compounds that hydrogen forms with the group VIA elements. Figure 9.6 shows that the boiling point of $H_2O$ is much higher than the boiling point of $H_2S$, $H_2Se$, or $H_2Te$. Why is this?

Because water is highly polar, the positive end of one molecule has a strong attraction for the negative end of another molecule. In a water solution the positive hydrogen atoms in one molecule are attracted to the negative oxygen atoms of other molecules (Fig. 9.7(a)). These attractive forces make it difficult to separate the molecules. Before molecules in the liquid will fly away from one another to form steam, they must be given kinetic energy equivalent to 100°C.

The attraction between water molecules is greater than that

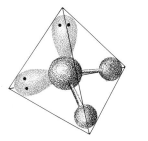

FIGURE 9.4
The water molecule can be imagined to fit the shape of a tetrahedron. The oxygen atom is in the center and the two hydrogen atoms are pointed toward two of the corners. The two unshared electron pairs are pointed toward the other corners.

9.3 The water molecular, a model

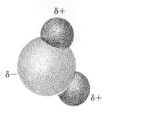

FIGURE 9.5
The water molecule is polar.

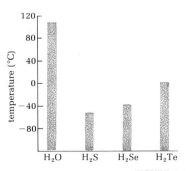

FIGURE 9.6
The unusually high boiling point of water is due to the attractive forces between water molecules.

normally found between polar molecules. It is given the special name of *hydrogen bonding*. Hydrogen bonding is the attractive force between a positive hydrogen atom in one molecule and a highly electronegative atom in another molecule. Hydrogen bonding usually occurs when the hydrogen atom is attached to an atom of either nitrogen, oxygen, or fluorine.

Why does ice float on water? Normally when a liquid freezes, the molecules slow down, come closer together, and form a solid which sinks to the bottom of the liquid. When water freezes, the molecules slow down. The hydrogen atom of one molecule becomes strongly attracted to the oxygen atom of another molecule. In the ice that forms, the hydrogen bonds cause the molecules to be oriented as shown in Fig. 9.7 (b). In the crystal there are many empty spaces, thus giving ice a lower density than water.

### Solvent properties

The hydrogen bonds in water provide a useful model to explain the high boiling point and high heat of vaporization of water and the low density of ice. How can we explain the ability of water to dissolve polar and ionic substances but not nonpolar substances?

Sodium chloride is soluble in water (37 g per 100 ml $H_2O$). Notice how the water molecules are arranged about the positive sodium ions in Fig. 9.8. The negative part of the water molecule is directed toward the sodium ions and tends to pull these ions off of the crystal and into the solution. Once a sodium ion is pulled away from the crystal it is surrounded by the negative end of the water molecules. In a similar manner chloride ions are dissolved by being pulled away from the crystal by the positive ends of water molecules (Fig. 9.8(b)).

What happens if the forces between ions or atoms in a crystal are stronger than the forces from water molecules? In this case

FIGURE 9.7
(a) Water molecules. (b) Molecules of $H_2O$ in ice.

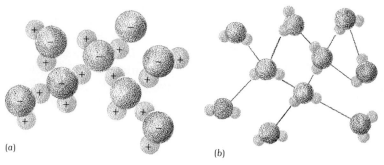

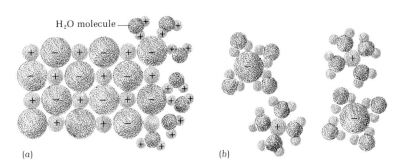

**FIGURE 9.8**
(a) Sodium ions (+) and chloride ions (−) in a crystal of sodium chloride. The polar water molecules pull on the charged ions. (b) If the force of water molecules on the ions is strong enough, the crystal will dissolve.

■ EXERCISE Hydrogen bonding can occur in molecules that contain a hydrogen atom attached to a nitrogen, oxygen, or fluorine atom. Put the following substances into one of two groups: (1) with hydrogen bonds, and (2) without hydrogen bonds.

(a) sulfuric acid, $H_2SO_4$,

$$H-O-\underset{\underset{O}{\overset{O}{\|}}}{S}-O-H$$

(b) sodium hydroxide, NaOH,

$Na^+\ O-H^-$

(c) propane, $C_3H_8$,

$$H-\underset{\underset{H}{|}}{\overset{\overset{H}{|}}{C}}-\underset{\underset{H}{|}}{\overset{\overset{H}{|}}{C}}-\underset{\underset{H}{|}}{\overset{\overset{H}{|}}{C}}-H$$

(d) ethyl alcohol, $CH_3CH_2OH$,

$$H-\underset{\underset{H}{|}}{\overset{\overset{H}{|}}{C}}-\underset{\underset{H}{|}}{\overset{\overset{H}{|}}{C}}-O-H$$

the crystal does not dissolve to any appreciable extent, and the substance is said to be only slightly soluble. Such is the case for silver chloride ($1.4 \times 10^{-4}$ g dissolves per 100 ml $H_2O$).

Polar molecules such as ammonia and ethyl alcohol dissolve readily in water. As in the case of sodium chloride, the positive end of the water molecule attracts the negative end of the polar molecule, pulling it into solution.

What happens when a nonpolar liquid such as carbon tetrachloride is added to water? As shown in Chapter 1, these two do not mix. Since the carbon tetrachloride molecule is nonpolar, it has neither a positive end nor a negative end to attract a water molecule. Water molecules attract other water molecules, and the carbon tetrachloride molecules exert no force capable of pushing the water molecules apart and mixing with them. ■

## 9.4 Chemical properties of water

In this section we will study the reaction of water with various substances. In the next section we shall develop the model of the water molecule further to help explain these reactions.

**Reaction with elements**

*Metals* Although water reacts slowly with zinc, it reacts rapidly with other metals. For instance, if a piece of sodium metal the size of a grain of rice is placed in a beaker of water, a rapid reaction occurs (Fig. 9.9). The small piece of metal is seen to float on the water and move rapidly about. It makes a hissing noise and rapidly disappears. The water remains clear and colorless. If the reaction is carried out in a test tube and a burning splint is inserted, a "pop" occurs. Evidently the reaction pro-

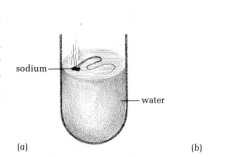

**FIGURE 9.9**
(a) A small piece of sodium metal moves about rapidly on a water surface. It hisses, grows smaller, and then disappears (b) If a flame is held over the test tube while the sodium reacts, a small "pop" is produced.

duces hydrogen gas. If the dye phenolphthalein is added to the clear, colorless solution, it turns red. Past experiments have shown that when phenolphthalein turns red there are many hydroxide ions in the solution. When sodium reacts with water, the products are sodium hydroxide and hydrogen.

$$2Na(s) + 2H_2O(l) \rightleftharpoons 2NaOH(aq) + H_2(g) \tag{9.1}$$

The other metals in group IA and barium, strontium, and calcium in group IIA also react with water to produce hydroxides and hydrogen.

Some metals are not as active as sodium or barium. They may react rapidly only with hot water or steam. For example, when a burning piece of magnesium ribbon is placed into a flask of steam, the magnesium continues to burn (Fig. 9.10). A white solid is formed and hydrogen gas is released. Magnesium reacts with steam to form magnesium oxide and hydrogen gas.

$$Mg(s) + H_2O(g) \rightleftharpoons MgO(s) + H_2(g) \tag{9.2}$$

Other metals such as zinc and iron also react with steam to form oxides and hydrogen.

$$Zn(s) + H_2O(g) \rightleftharpoons ZnO(s) + H_2(g) \tag{9.3}$$

Some metals are less active than magnesium or zinc. Copper, mercury, silver, and gold will not react in water or in steam. Table 9.2 lists the common metals in decreasing order of ease with which they react with water; it shows also the relative ease of decomposition of the metal oxides.

*Nonmetals* The only nonmetal that we will discuss is chlorine. Chlorine gas is slightly soluble in water. Some of the molecules that dissolve react with the water to form hypochlorous acid, HClO, and hydrochloric acid.

$$Cl_2(aq) + H_2O(l) \longrightarrow HClO(aq) + HCl(aq) \tag{9.4}$$

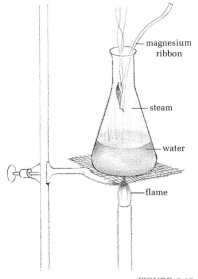

**FIGURE 9.10**
Burning magnesium ribbon continues to burn when it is placed in steam.

| TABLE 9.2 | | Symbol | Reaction | Decomposition |
|---|---|---|---|---|
| Reaction of metals with water; decomposition of metal oxides. | More active metals | K<br>Ba<br>Sr<br>Ca<br>Na | React with cold water to give hydroxide + hydrogen | Metal oxide + hydrogen + heat gives no reaction |
| | | Mg<br>Al<br>Mn<br>Zn<br>Cr | React with steam to give oxide + hydrogen | |
| | | Fe | | |
| | | Cd<br>Co<br>Ni<br>Sn<br>Pb | React with very hot steam to give oxide + hydrogen | Metal oxide + hydrogen + heat gives metal + water |
| | | H | Common reference point | |
| | | Cu | | |
| | Less active metals | Hg<br>Ag<br>Pt<br>Au | No reaction with steam | Metal oxide + heat gives metal + oxygen |

■ EXERCISE  Using Table 9.2 as a guide, write completed balanced equations for the following. If there is no reaction, write "no reaction." (a) potassium + water → ; (b) calcium + water → ; (c) silver + water ⇌ ; (d) mercury + steam ⇌ .

■ EXERCISE  Bromine and iodine are partially soluble in water. They are in the same group as chlorine and react with water in the same way that chlorine does. Write balanced equations for the reaction of bromine and iodine in water. Name the products.

Only a small number of the chlorine molecules react at any one time. When some HClO and HCl have formed, they begin to react with each other to re-form $Cl_2$ and $H_2O$.

$$Cl_2(aq) + H_2O(l) \longleftarrow HClO(aq) + HCl(aq) \qquad (9.5)$$

In a solution of chlorine and water, both reactions 9.4 and 9.5 are occurring at the same time and at equal rates. Consequently the amount of $Cl_2$, $H_2O$, HClO, and HCl does not change. The system is said to be in equilibrium. This equilibrium can be expressed by using a double arrow.

$$Cl_2(aq) + H_2O(l) \rightleftharpoons HClO(aq) + HCl(aq) \qquad ■ \qquad (9.6)$$

### Reaction with oxides

*Metal oxides*  When a few pieces of white calcium oxide are placed on a watch glass and a few drops of water are added, the

9.4 Chemical properties of water

FIGURE 9.11
When a few drops of water are added to lumps of calcium oxide, some calcium oxide dissolves in excess water to form a solution that turns phenolphthalein red.

solid calcium oxide seems to expand and crack where the water touches it (Fig. 9.11). Steam forms, indicating that the mixture is hot. When a large amount of water is added, a white solid and a milky solution are left. Phenolphthalein turns red when mixed with some of the solution.

The red color of the phenolphthalein indicates that hydroxide ions have formed. When calcium oxide mixes with water, calcium hydroxide and heat are produced. The calcium hydroxide is only partially soluble in water.

$$CaO(s) + H_2O(l) \rightleftharpoons Ca(OH)_2(aq) + \text{heat} \qquad (9.7)$$

Other metal oxides of the group IA and IIA elements react with water to produce hydroxides.

$$Na_2O(s) + H_2O(l) \rightleftharpoons 2NaOH(aq) + \text{heat} \qquad (9.8)$$

**Nonmetal oxides** Many fuels such as coal and gasoline contain sulfur. When these fuels burn, one of the products is sulfur dioxide, $SO_2$. In the presence of sunlight, sulfur dioxide reacts with substances in the atmosphere to produce sulfur trioxide, $SO_3$. Sulfur trioxide reacts readily with water vapor to form small droplets of sulfuric acid in the air; it is especially dangerous when inhaled by children and elderly people.

$$S \text{ (in fuels)} + O_2(g) \rightleftharpoons SO_2(g) \qquad (9.9)$$
$$2SO_2(g) + O_2(g) \text{ (in air)} \rightleftharpoons 2SO_3(g) \qquad (9.10)$$
$$SO_3(g) + H_2O(g) \rightleftharpoons H_2SO_4(l) \qquad (9.11)$$

The behavior of $SO_3$ in the presence of water is typical of nonmetal oxides. These oxides usually react with water to form acids. Usually during the course of the reaction the nonmetal forms an acid in which the oxidation number of the nonmetal does not change. Compare $SO_2$ and $SO_3$:

$$SO_2(g) + H_2O(l) \rightleftharpoons H_2SO_3(aq) \qquad (9.12)$$
$$CO_2(g) + H_2O(l) \rightleftharpoons H_2CO_3(aq) \qquad (9.13)$$
$$P_4O_{10}(s) + 6H_2O(l) \rightleftharpoons 4H_3PO_4(aq) \qquad (9.14)$$

■ **EXERCISE** Write balanced equations for the reactions between:
(a) $BaO(s) + H_2O(l)$
(b) $SrO(s) + H_2O(l)$
(c) $Li_2O(s) + H_2O(l)$
(d) $SeO_2(g) + H_2O(l)$ (Se is similar to S).

**Reaction with hydrogen compounds**

*Metal hydrides* Metal hydrides react with water to produce hydrogen gas and hydroxides. For example, lithium hydride reacts with water to produce lithium hydroxide and hydrogen.

$$LiH(s) + H_2O(l) \rightleftharpoons LiOH(aq) + H_2(g) \qquad (9.15)$$

*Ammonia* Ammonia gas dissolves readily in water to produce a solution that turns phenolphthalein solution red (Fig.

9.12). The ammonia molecules react with water molecules to form ammonium ions and hydroxide ions.

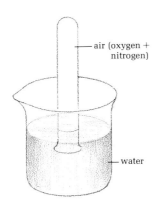

$$H:\overset{H}{\underset{H}{N}}: + \boxed{H}:\overset{..}{\underset{..}{O}}: \rightleftharpoons H:\overset{H}{\underset{H}{N}}:H^+ + :\overset{..}{\underset{..}{O}}:H^-$$

Electrical measurements indicate that at any one time, only about 5 percent of the total number of ammonia molecules react. Once some $NH_4^+$ ions and $OH^-$ ions have formed, they start to react and re-form $NH_3$ and $H_2O$. All four species are present at any one time. The system is in equilibrium.

$$NH_3(g) + H_2O(l) \rightleftharpoons NH_4^+(aq) + OH^+(aq) \quad (9.16)$$

*Hydrogen halides* What happens if a test tube of hydrogen chloride is inverted and placed in a beaker of water (Fig. 9.12)? The water rises rapidly into the test tube, indicating that hydrogen chloride is soluble in water.

Does hydrogen chloride react with water as it dissolves? When litmus dye is added to the solution it turns red. Past experiments have shown that litmus turns red in solutions that contain many *hydronium ions*, $H_3O^+$. Such solutions are called acids. When hydrogen chloride dissolves in water it produces hydronium ions and chloride ions.

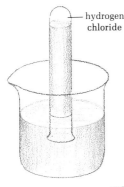

$$HCl(g) + H_2O(aq) \rightleftharpoons H_3O^+(aq) + Cl^-(aq) \quad (9.17)$$

$$\boxed{H:\overset{..}{\underset{..}{Cl}}: + }:\overset{..}{\underset{H}{O}}:H \rightleftharpoons :\overset{..}{\underset{..}{Cl}}:^- + H:\overset{..}{\underset{H}{O}}:H^+$$

This solution is called hydrochloric acid.

The hydronium ion is actually a hydrogen ion, $H^+$, attached to a water molecule. Frequently, to simplify matters, chemists do not write the water molecule in the equation. In this case Eq. 9.17 becomes

$$HCl(g) \xrightleftharpoons{H_2O} H^+(aq) + Cl^-(aq)$$

The other hydrogen halides behave the same way as hydrogen chloride when they are added to water.

*Ternary acids* When a few milliliters of concentrated sulfuric acid are slowly and cautiously added to a beaker of water, the clear colorless mixture becomes hot. If too much acid is added too quickly, the mixture may boil violently and splatter hot acid into the air. (*Caution:* Never add water to concentrated acid because splattering occurs.) The mixture turns litmus dye

FIGURE 9.12

(a) When an inverted test tube of air is lowered into a beaker of water, little or no water rises into the test tube. (b) When an inverted test tube of hydrogen chloride gas is lowered into a beaker of water, the water rapidly rises into and fills the test tube.

red, indicating the presence of hydronium ions. These observations indicate that a new substance has formed when sulfuric acid mixes with water. Sulfuric acid reacts with water to form hydronium ions and sulfate ions.

$$H_2SO_4(l) + 2H_2O(l) \rightleftharpoons 2H_3O^+(aq) + SO_4^{2-}(aq) \qquad (9.19)$$

The reaction is believed to occur in two steps. First one hydrogen ion comes off the $H_2SO_4$ molecule. Then a second hydrogen ion comes off the $HSO_4^-$ ion.

Step 1 $\quad H_2SO_4 + H_2O \rightleftharpoons H_3O^+ + HSO_4^- \qquad (9.20)$
Step 2 $\quad HSO_4^- + H_2O \rightleftharpoons H_3O^+ + SO_4^{2-} \qquad (9.21)$

Other ternary acids react with water in a way similar to sulfuric acid. They form hydronium ions and negative ions.

$$HNO_3(l) + H_2O(l) \rightleftharpoons H_3O^+(aq) + NO_3^-(aq) \qquad (9.22)$$

■ EXERCISE  Complete and balance the following equations.
(a) $CaH_2(s) + H_2O(l) \rightarrow$
(b) $HI(g) + H_2O(l) \rightarrow \quad;$
(c) $H_2CO_3(aq) + H_2O(l) \rightleftharpoons \quad;$
(d) $HClO_4(aq) + H_2O(l) \rightleftharpoons \quad.$

**Reaction with certain common ions**

*Ions that form acid solutions*  When certain salts are dissolved in water, the resulting solutions turn litmus red. This indicates that the solutions contain an excess of hydronium ions and are acidic. A number of salts with this behavior are listed in Table 9.3. How do these salts produce hydronium ions in a water solution? Let us see what happens when one such salt, ammonium chloride, is placed in water.

$$NH_4Cl(s) + H_2O(l) \rightarrow \;?$$

The ionic salt dissolves readily in water; thus we can assume that aqueous ammonium ions and chloride ions form.

$$NH_4Cl \xrightarrow{H_2O} NH_4^+(aq) + Cl^-(aq) \qquad (9.23)$$

The ammonion ion then reacts further with water to give the ammonia molecule and hydronium ion.

$$NH_4^+(aq) + H_2O(l) \rightleftharpoons NH_3(aq) + H_3O^+(aq) \qquad (9.24)$$

$$\overset{H}{\underset{H}{H:\ddot{N}:}}(H\; +)\; \overset{..}{\underset{H}{:\ddot{O}:H}} \rightleftharpoons \overset{H}{\underset{H}{H:\ddot{N}:}} + \overset{+}{\underset{H}{H:\ddot{O}:H}}$$

Consider the salt sodium hydrogen sulfate. How does this salt produce hydronium ions in solution? First, sodium ions and hydrogen sulfate ions are formed.

**TABLE 9.3**

Some salts that produce either $H_3O^+$ ions or $OH^-$ ions when added to water.

| | Produce $H_3O^+$ in solution (litmus red) | | Produce $OH^-$ in solution (litmus blue) | |
|---|---|---|---|---|
| Salt | Reacting ion | | Salt | Reacting ion |
| $NH_4Cl$ | $NH_4^+$ | | $Na_2CO_3$ | $CO_3^{2-}$ |
| $NH_4NO_3$ | $NH_4^+$ | | $Na_2S$ | $S^{2-}$ |
| $NH_4ClO_3$ | $NH_4^+$ | | $Na_3PO_4$ | $PO_4^{2-}$ |
| $NaHSO_4$ | $HSO_4^-$ | | $NaCH_3COO$ | $CH_3COO^-$ |
| $KHSO_4$ | $HSO_4^-$ | | $Na_2SO_3$ | $SO_3^{2-}$ |
| $NaH_2PO_4$ | $H_2PO_4^-$ | | $NaNO_2$ | $NO_2^-$ |
| $KH_2PO_4$ | $H_2PO_4^-$ | | $Na_2O$ | $O^{2-}$ |

$$NaHSO_4 \text{ (s)} \xrightleftharpoons{H_2O} Na^+ \text{(aq)} + HSO_4^- \text{(aq)} \tag{9.25}$$

The hydrogen sulfate ions then react with water to give hydronium ions and sulfate ions.

$$HSO_4^- \text{(aq)} + H_2O\text{(l)} \longrightarrow H_3O^+\text{(aq)} + SO_4^{2-}\text{(aq)} \tag{9.26}$$

*Ions that form basic solutions* When certain salts are dissolved in water, the resulting solutions turn litmus blue. This indicates that the solutions contain excess hydroxide ions and are basic. A number of salts with this behavior are listed in Table 9.3. How do these salts produce hydroxide ions in solutions?

When sodium carbonate ($Na_2CO_3$) is placed in water it dissociates into sodium ions and carbonate ions.

$$Na_2CO_3(s) \rightleftharpoons 2Na^+\text{(aq)} + CO_3^{2-}\text{(aq)} \tag{9.27}$$

The carbonate ion reacts further with the water molecule: A hydrogen ion is transferred from $H_2O$ to $CO_3^{2-}$ to form a hydrogen carbonate ion and a hydroxide ion.

$$CO_3^{2-}\text{(aq)} + H_2O\text{(l)} \rightleftharpoons HCO_3^-\text{(aq)} + OH^-\text{(aq)} \tag{9.28}$$

Many other negative ions have the ability to accept a hydrogen ion from a water molecule.

$$S^{2-} + H_2O\text{(l)} \rightleftharpoons HS^-\text{(aq)} + OH^-\text{(aq)} \tag{9.29}$$

$$CH_3COO^-\text{(aq)} + H_2O\text{(l)} \rightleftharpoons CH_3COOH\text{(aq)} + OH^-\text{(aq)} \tag{9.30}$$

■ EXERCISE *State whether the following salts will produce solutions containing excess hydronium ions or excess hydroxide ions. Write an equation showing the ions that react to form these solutions.*
*(a) $NH_4Br$; (b) $K_2CO_3$; (c) $CaO$; (d) $K_2S$; (e) $NaH_2PO_4$; (f) $LiNO_2$.*

The salts listed in Table 9.3 form either acidic or basic solutions. Many salts such as KCl, $KNO_3$, NaBr, and $NaClO_3$ do not produce hydronium ions or hydroxide ions in solution. Ions that do not react with water to form $OH^-$ ions or $H_3O^+$ ions are shown in Table 9.4. ■

**TABLE 9.4**

Some common ions that do not react with water to produce either excess $OH^-$ or excess $H_3O^+$ ions.

| | | | |
|---|---|---|---|
| $Li^+$ | $Mg^{2+}$ | $NO_3^-$ | $Cl^-$ |
| $Na^+$ | $Ca^{2+}$ | $SO_4^{2-}$ | $Br^-$ |
| $K^+$ | $Sr^{2+}$ | $ClO_3^-$ | $I^-$ |
| $Ca^+$ | $Ba^{2+}$ | | |
| $Rb^+$ | | | |

■ EXERCISE  Write formulas for (a) barium hydroxide octahydrate; (b) calcium chloride hexahydrate.

## Formation of hydrates

When the blue solid cobalt(II) chloride, $CoCl_2$, is placed on a watch glass and left in the open air, it slowly changes color from blue to red. Weight measurements indicate that the solid gains weight as it changes in color. When the red solid is gently heated in an oven, it loses weight and changes back to the blue color. What happens to the cobalt(II) chloride? It is believed that both the first weight and color changes are brought about by the solid picking up water from the atmosphere.

$$6H_2O(g) + CoCl_2(s) \rightleftharpoons CoCl_2 \cdot 6H_2O(s) \qquad (9.31)$$
$$\text{blue} \qquad\qquad \text{red}$$

The dot between the $CoCl_2$ and $6H_2O$ indicates that the $6H_2O$ are covalently bonded to the $CoCl_2$.

The reaction between a metal salt such as $CoCl_2$ and water to form a compound such as $CoCl_2 \cdot 6H_2O$ is termed a *hydration reaction*. The salt without water is said to be *anhydrous*. Salts containing $H_2O$, such as $CoCl_2 \cdot 6H_2O$, are called hydrates. Anhydrous metal salts that remove water from the atmosphere are said to be hydroscopic. One of these anhydrous salts, $CaCl_2$, is frequently used by chemists to maintain a dry atmosphere in chemical systems. Anhydrous salts can be used in the home to keep cookie jars dry.

Many hydrates do not feel or look wet. Have you ever seen or touched washing soda or Epsom salts? Washing soda is a white solid, sodium carbonate decahydrate, $Na_2CO_3 \cdot 10H_2O$. Epsom salt is also a white solid, magnesium sulfate heptahydrate, $MgSO_4 \cdot 7H_2O$.

The amount of water that a hydrate contains depends on the temperature and the amount of water vapor in the atmosphere. Depending on pressure and temperature, copper(II) sulfate forms hydrates of $CuSO_4 \cdot H_2O$, $CuSO_4 \cdot 3H_2O$, and $CuSO_4 \cdot 5H_2O$. ■

## Ionization of water

Aqueous solutions often are good conductors of electricity. If distilled water is boiled to remove any dissolved carbon dioxide, the sample conducts electricity, but only slightly (Fig. 9.13)

Substances will conduct electricity only if *charged particles can move through the substance*. What charged particles are in a sample of pure water? The electrons and protons that make

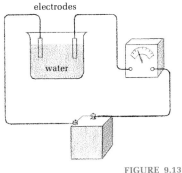

FIGURE 9.13
*Pure water conducts electricity only slightly.*

up the elements hydrogen and oxygen are part of a neutral water molecule. Chemists believe that water molecules react with themselves to form ions by transferring a hydrogen ion. One water molecule reacts with a second water molecule to form a hydronium ion and a hydroxide ion.

$$H_2O(l) + H_2O(l) \longrightarrow H_3O^+(aq) + OH^-(aq) \qquad (9.32)$$

These ions are continually reacting with themselves to form more water.

$$H_2O + H_2O \longleftarrow H_3O^+ + OH^- \qquad (9.33)$$

Both reactions 9.32 and 9.33 occur at the same rate. The amount of water and of ions does not change with time; the system is in equilibrium.

$$H_2O + H_2O \rightleftharpoons H_3O^+ + OH^- \qquad (9.34)$$

At 25°C there is $10^{-7}$ mole of hydronium ions and of hydroxide ions in each liter of water. This same volume of water contains $5.5 \times 10^1$ moles of water molecules.

## 9.5 Chemical properties and structure—a model

In Section 9.3 the water molecule was said to consist of a relatively negative oxygen atom and two positive hydrogen atoms. Will this model allow us to explain the chemical properties of water? Let us examine the water molecule from two different aspects: (1) the active hydrogen and (2) the available electron pair of oxygen.

**The active hydrogen**

The hydrogen nucleus in a water molecule is attached to a water molecule because of its attraction to a shared pair of electrons. This hydrogen nucleus or proton can be transferred to other molecules or ions that will make a pair of electrons available for sharing. Usually the proton will be attracted to a pair of electrons that are in a negative species—that is, a negatively charged ion, atom, or molecule.

$$:\ddot{X}:^- \overset{\delta+ \;\; \delta-}{H) :\ddot{O}:} \longrightarrow :\ddot{X}:H + :\ddot{O}:^-$$
$$\phantom{:\ddot{X}:^- H)} H_{\delta+} \phantom{\longrightarrow :\ddot{X}:H +} H$$

What examples that we have studied are in this category?

*Metallic oxides + water*

$$Na_2O(s) + H_2O(l) \rightleftharpoons 2NaOH(aq)$$

The proton from a water molecule is transferred to an unshared pair of electrons on a negative oxide ion.

$$:\ddot{\underset{..}{O}}:^{2-}(aq) + H):\ddot{\underset{H}{O}}:(l) \rightleftharpoons :\ddot{\underset{..}{O}}:H^-(aq) + :\ddot{\underset{..}{O}}:H^-(aq)$$

*Ammonia* The proton is transferred to an unshared pair of electrons on a partially negative nitrogen atom.

$$H:\overset{H}{\underset{H}{\ddot{N}}}:^{\delta-}(aq) + H)\overset{\delta+}{:\ddot{\underset{H}{O}}}:(l) \rightleftharpoons H:\overset{H}{\underset{H}{N}}:H^+(aq) + :\ddot{\underset{..}{O}}:H^-(aq)$$

*Metal hydrides*

$$LiH(s) + H_2O(l) \rightleftharpoons LiOH(aq) + H_2(g)$$

The proton from water is transferred to the hydride ion.

$$Li^+(aq) + H:^-(aq) + H):\ddot{\underset{H}{O}}:(l) \rightleftharpoons Li^+(aq) + H:H(g) + :\ddot{\underset{..}{O}}:H^-(aq)$$

*Common ions* The proton from water is transferred to an unshared pair of electrons on a negative ion.

$$:\ddot{\underset{..}{S}}:^{2-}(aq) + H):\ddot{\underset{H}{O}}:(l) \rightleftharpoons :\ddot{\underset{..}{S}}:H^-(aq) + :\ddot{\underset{..}{O}}:H^-(aq)$$

*Ionization of water* Water molecules can react with themselves to form hydronium ions and hydroxide ions. A proton from one water molecule is transferred to the unshared pair of electrons on an oxygen in a second water molecule.

$$H-\overset{\delta-}{\underset{H}{\ddot{O}}}:(l) + H)\overset{\delta+}{:\ddot{\underset{H}{O}}}:(l) \rightleftharpoons H:\overset{+}{\underset{H}{\ddot{O}}}:H(aq) + :\ddot{\underset{..}{O}}:H^-(aq)$$

*Active metals—electron transfer* When an active metal such as sodium is added to water, sodium hydroxide and hydrogen form.

$$2Na(s) + 2H_2O(l) \rightleftharpoons 2NaOH(aq) + H_2(g)$$

Active metals have a low ionization energy and electrons are easily transferred from the metal atom to the protons on a water molecule.

$$\begin{array}{c} \text{Na}\cdot \\ \text{Na}\cdot \end{array} \begin{array}{c} \text{H}:\overset{..}{\underset{..}{\text{O}}}: \\ \text{H}:\overset{..}{\underset{..}{\text{O}}}: \\ \text{H} \end{array} \longrightarrow \begin{array}{c} \text{Na}^+(\text{aq}), \\ \text{Na}^+(\text{aq}), \end{array} \;\; (\text{H}:\text{H}) \;\; \begin{array}{c} :\overset{..}{\underset{..}{\text{O}}}:\text{H}^-(\text{aq}) \\ :\overset{..}{\underset{..}{\text{O}}}:\text{H}^-(\text{aq}) \end{array}$$

**The available electron pair of oxygen**

We have just seen that when water reacts with itself, the unshared pair of electrons of the oxygen atom are made available to a proton for sharing. In many of the chemical reactions of water the relatively negative oxygen atom in the water molecule makes available its unshared electron pair to some species that can gain a share in an electron pair.

$$-\underset{|}{\overset{|}{\text{X}}} + :\overset{..}{\underset{..}{\text{O}}}:\text{H} \rightleftharpoons -\underset{|}{\overset{|}{\text{X}}}:\overset{..}{\underset{..}{\text{O}}}:\text{H}$$
$$\hspace{2em}\text{H}\hspace{5em}\text{H}$$

*Hydrogen halides and ternary acids* Hydrogen halides and ternary acids dissolve in water to form hydronium ions.

$$\text{HCl}(g) + \text{H}_2\text{O}(l) \rightleftharpoons \text{H}_3\text{O}^+(aq) + \text{Cl}^-(aq)$$

A proton in HCl is transferred to a pair of electrons on the oxygen atom in water.

$$\text{H}:\overset{..}{\underset{..}{\text{O}}}{}^{\delta-}:(l) + \text{H}{}^{\delta+}:\overset{..}{\underset{..}{\text{Cl}}}:(g) \longrightarrow \text{H}:\overset{..}{\underset{..}{\text{O}}}:\text{H}^+(aq) + :\overset{..}{\underset{..}{\text{Cl}}}:^-(aq)$$
$$\hspace{1em}\text{H}\hspace{12em}\text{H}$$

*Common ions* The ammonium ion reacts with water to form hydronium ions.

$$\text{H}:\overset{..}{\underset{..}{\text{O}}}:(l) + \text{H}:\overset{\text{H}}{\underset{\text{H}}{\text{N}}}:\text{H}^+(aq) \longrightarrow \text{H}:\overset{..}{\underset{..}{\text{O}}}:\text{H}^+(aq) + :\overset{\text{H}}{\underset{\text{H}}{\text{N}}}:\text{H}(aq)$$
$$\hspace{1em}\text{H}$$

*Nonmetallic oxides* Most nonmetallic oxides react with water. The relatively negative oxygen in the water molecule donates a share in a pair of electrons to the positive nonmetal atom:

$$\begin{array}{c} \overset{\times\times}{\underset{\times\times}{\text{O}}} \\ \text{O}{\times\times}:\text{S}^{\delta+} + :\overset{..}{\underset{..}{\text{O}}}:\text{H} \longrightarrow \overset{\times\times}{\underset{\times\times}{\text{O}}}:\text{S}:\overset{..}{\underset{..}{\text{O}}}:\text{H} \\ \text{O}\hspace{2em}\text{H} \hspace{4em} \overset{\times\times}{\underset{\times\times}{\text{O}}} \\ \hspace{10em}\text{H} \end{array}$$

■ EXERCISE *Show by means of electron-dot structures whether water molecules in the following reactions*

9.5 Chemical properties and structure—a model

are (1) donating protons or (2) providing a pair of unshared electrons:

(a) $BaO(s) + H_2O(l) \rightleftharpoons Ba(OH)_2(aq)$
(b) $CO_2(g) + H_2O(l) \rightleftharpoons H_2CO_3(aq)$
(c) $H_2SO_4(l) + 2H_2O(l) \rightleftharpoons 2H_3O^+(aq) + SO_4^{2-}(aq)$
(d) $KH(s) + H_2O(l) \rightleftharpoons KOH(aq) + H_2(g)$
(e) $Na_3PO_4(s) + 2H_2O(l) \rightleftharpoons 3NaOH(aq) + H_3PO_4(aq)$
(f) $NiSO_4(s) + 2H_2O(l) \rightleftharpoons NiSO_4 \cdot 2H_2O(s)$

or $SO_3(g) + H_2O(l) \rightleftharpoons H_2SO_4(aq)$

**Hydrates** A salt such as $CoCl_2$ can pick up water from the atmosphere to form a hydrate. We can imagine this reaction as an attraction of positive cobalt ions for the unshared pair of electrons of oxygen atoms in water (Fig. 9.14).

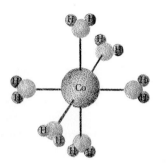

FIGURE 9.14
The geometry of $Co(H_2O)_6^{2+}$. The six oxygen atoms share electron pairs with the cobalt(II) ion.

## 9.6 Decomposition of water

**Reaction and equation**

As discussed in Chapter 3, water can be decomposed into hydrogen and oxygen by passing an electric current through an appropriate sample. At one electrode hydrogen is produced; at another electrode oxygen is produced. What reactions are occurring at each electrode?

Into one electrode, excess electrons are being pumped by an electrical generator or battery (Fig. 9.15). The excess electrons make the electrode negative. The positive part of the water molecule is attracted to the electrode and reacts with it. Relatively positive hydrogen atoms in the water molecule gain electrons from the electrode. Hydrogen gas and hydroxide ions are formed.

$$\text{energy} + 2H_2O + 2e^- \rightleftharpoons H_2(g) + 2OH^-(aq) \quad (9.35)$$

From the other electrode, electrons are being removed by the electrical generator. The removal of electrons makes the electrode positive. The negative part of the water molecule is attracted to the positive electrode and reacts with it. Relatively negative oxygen atoms in the water lose electrons to the electrode. Oxygen molecules and hydronium ions are formed.

$$\text{energy} + 6H_2O \rightleftharpoons O_2 + 4H_3O^+ + 4e^- \quad (9.36)$$

Because as many electrons must be used up as are generated, reaction 9.35 occurs twice as often as reaction 9.36. If we double the coefficients in reaction 9.35 and then add it to 9.36 we obtain the total equation.

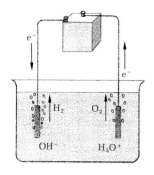

$4H_2O + 4e^- \longrightarrow 2H_2 + 4OH^-$

$4H_2O \longrightarrow O_2 + 4H_3O^+ + 4e^-$

**FIGURE 9.15**
An electrical generator pumps electrons from one electrode to another.

$$\begin{aligned}\text{energy} + 4H_2O + 4e^- &\rightleftharpoons 2H_2 + 4OH^- \\ \text{energy} + 6H_2O &\rightleftharpoons O_2 + 4H_3O^+ + 4e^- \\ \hline \text{energy} + 10H_2O + 4e^- &\rightleftharpoons 2H_2 + O_2 + 4OH^- \\ &\qquad + 4H_3O^+ + 4e^- \quad (9.37)\end{aligned}$$

The term $4e^-$ can be removed from Eq. 9.37 since it appears on each side.

Note the terms $4H_3O^+$ and $4OH^-$. These ions would react in solution to form $8H_2O$. We can remove the term $8H_2O$ from the left side of Eq. 9.37 and $4H_3O^+$, $4OH^-$ from the right side. The amount of energy required to decompose 2 moles of water is 137 kcal. This gives us

$$2H_2O(l) + 137 \text{ kcal} \rightleftharpoons 2H_2(g) + O_2(g) \quad (9.38)$$

**Oxidation and reduction**

What elements change oxidation number when water is decomposed? Hydrogen is 1+ in water; oxygen is 2−.

$$\begin{array}{cccc}2H_2O & \longrightarrow & 2H_2 & + & O_2 \\ (1+,\, 2-) & & (0) & & (0)\end{array} \quad (9.39)$$

The oxidation number of hydrogen goes from 1+ in $H_2O$ to 0 in $H_2$. A process whereby an element has its oxidation number lowered is called *reduction*. In this reaction the hydrogen is reduced from +1 to 0. The electrode at which a reduction takes place is called the *cathode*. As in the case of hydrogen, *reduction occurs when electrons are gained*.

The oxidation number of oxygen goes from 2− in $H_2O$ to 0 in $O_2$. A process whereby an element has its oxidation number increased is called *oxidation*. In this reaction the oxygen is oxidized from 2− to 0. The electrode at which an oxidation takes place is called the *anode*. As in the case of oxygen, *oxidation occurs when electrons are lost*.

Many reactions that we have studied so far involve a transfer of electrons from one atom to another. One atom is reduced, another is oxidized. These reactions are called *redox reactions*. ■

■ **EXERCISE** Each of the following equations represents a reaction in which one substance has been oxidized and another reduced. Use oxidation numbers to identify the elements which have been oxidized or reduced.
(a) $2Mg(s) + O_2(g) \rightleftharpoons 2MgO(s) + \text{energy}$
(b) $2HgO(s) + \text{heat} \rightleftharpoons 2Hg(l) + O_2(g)$
(c) $2KClO_3(s) + \text{heat} \rightleftharpoons 2KCl(s) + 3O_2(g)$.

## 9.7 Mole ratios

Consider again Eq. 9.38 for the decomposition of water by electrical energy.

$$2H_2O(l) + 137 \text{ kcal} \rightleftharpoons 2H_2(g) + O_2(g)$$

According to this equation, 2 moles of water can be decomposed by 137 kcal of energy into 2 moles of hydrogen and 1 mole of oxygen. The balanced equation gives the mole ratio in which substances react. Each term ($H_2O$, kcal, $H_2$, $O_2$) in this equation is numerically related to the other terms by the coefficients in the equation. Therefore, 2 moles of water are chemically equivalent to 2 moles of hydrogen and to 1 mole of oxygen and to 137 kcal of energy for this reaction. When this equation applies, then:

2 moles $H_2O$ ≈ 2 moles $H_2$ ≈ 1 mole $O_2$ ≈ 137 kcal

How many moles of hydrogen can be produced when 15 moles of $H_2O$ are decomposed? From the balanced equation we see that 2 moles of water are chemically equivalent to 2 moles of hydrogen. Let us use unity terms to solve the problem.

$$15 \text{ moles } H_2O \times \frac{2 \text{ moles } H_2}{2 \text{ moles } H_2O} = 15 \text{ moles } H_2$$

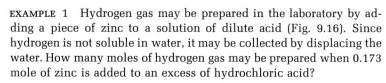
Ratio from equation

The unity term 2 moles $H_2$/2 moles $H_2O$ comes directly from the balanced equation.

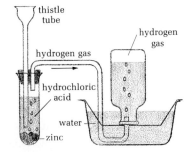

FIGURE 9.16
Apparatus for the preparation and collection of hydrogen gas.

EXAMPLE 1  Hydrogen gas may be prepared in the laboratory by adding a piece of zinc to a solution of dilute acid (Fig. 9.16). Since hydrogen is not soluble in water, it may be collected by displacing the water. How many moles of hydrogen gas may be prepared when 0.173 mole of zinc is added to an excess of hydrochloric acid?

$Zn(s) + 2HCl(aq) \rightleftharpoons ZnCl_2(aq) + H_2(g)$

*Given:* 0.173 mole of zinc; balanced equation

*Find:* moles of hydrogen produced

*Relationship:* From the balanced equation we see that 1 mole of zinc is chemically equivalent to 1 mole of hydrogen.

*Solution:* $0.173 \text{ mole Zn} \times \dfrac{1 \text{ mole } H_2}{1 \text{ mole Zn}} = 0.173 \text{ mole } H_2$

ratio from balanced equation

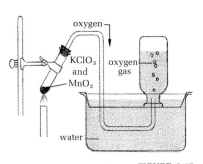

FIGURE 9.17
Apparatus for the preparation and collection of oxygen gas. (Caution: Using large amounts of $KClO_3$ or using a dirty test tube may cause an explosion.)

EXAMPLE 2  Oxygen gas may be produced in the laboratory by cautiously heating potassium chlorate and collecting the oxygen by displacement of water (Fig. 9.17). The reaction occurs faster and at a lower temperature when a small amount of a substance such as $MnO_2$ or $Fe_2O_3$ is mixed with the potassium chlorate. Substances that alter the speed of reactions are called *catalysts* (for further discussion, see Section 10.3).

$$2KClO_3(s) \xrightarrow{MnO_2} 2KCl(s) + 3O_2(g)$$

(a) How many moles of oxygen can be produced when 0.0549 mole of $KClO_3$ is used? (b) How many moles of KCl will be produced?

Given: 0.0549 mole of $KClO_3$

Find: moles of oxygen produced

Relationship: The balanced equation tells us that 2 moles of $KClO_3$ will produce 3 moles of $O_2$.

(a) Solution: 0.0549 moles $KClO_3 \times \dfrac{3 \text{ moles } O_2}{2 \text{ moles } KClO_3}$

$$= 0.0822 \text{ mole } O_2$$

(b) Solution: 0.0549 mole $KClO_3 \times \dfrac{2 \text{ moles } KCl}{2 \text{ moles } KClO_3}$

$$= 0.0549 \text{ mole KCl} \quad \blacksquare$$

■ EXERCISE  *One convenient method of preparing hydrogen gas is to add a metal hydride to water. How many moles of hydrogen can be prepared from the addition of 2.07 moles of barium hydride to an excess of water?*
$BaH_2(s) + 2H_2O(l) \rightleftharpoons Ba(OH)_2(aq) + 2H_2(g)$

■ EXERCISE  *Pure oxygen was first prepared by heating mercury(II) oxide. How many moles of mercury(II) oxide would be needed to prepare 0.500 mole of oxygen?* $2HgO(s) + heat \rightleftharpoons 2Hg(l) + O_2(g)$

## 9.8  Hydrogen

**Physical properties**

Hydrogen is a colorless, odorless, tasteless gas. Each diatomic molecule of hydrogen contains two electrons shared equally by two nuclei. Because the hydrogen molecule is nonpolar, there is very little attractive force between different molecules. For this reason the boiling point, freezing point, heat of vaporization, and solubility of $H_2$ in $H_2O$ are all relatively low (Table 9.5.).

**Preparation**

The common methods of preparing hydrogen in the laboratory have already been discussed.

(a) *Metal + acid* (Example 1)

$$Zn(s) + 2HCl(aq) \rightleftharpoons ZnCl_2(aq) + H_2(g)$$

(b) *Metal hydride + water* (Section 9.7)

$$BaH_2(s) + 2H_2O(l) \rightleftharpoons Ba(OH)_2(aq) + H_2(g)$$

(c) *Active metal + water* (Section 9.4)

$$Ba(s) + 2H_2O(l) \rightleftharpoons Ba(OH)_2(aq) + H_2(g)$$

(d) *Decomposition of Water by Electricity* (Section 9.6)

$$2H_2O(l) + energy \rightleftharpoons 2H_2(g) + O_2(g)$$

TABLE 9.5

Some of the physical properties of hydrogen.

| | |
|---|---|
| Formula | $H_2$ |
| Density | 0.0899 g/liter |
| Solubility in $H_2O$ | $\dfrac{9 \times 10^{-5} \text{ mole } H_2}{100 \text{ ml } H_2O}$ |
| Boiling point | $-253\,°C$ |
| Heat of vaporization | 0.216 kcal/mole |
| Freezing point | $-259\,°C$ |
| Heat of fusion | 0.028 kcal/mole |
| First ionization energy | 313 kcal/mole |
| Electronegativity | 2.1 |
| Atomic radius | 0.37 Å |
| Isotopes | $^1_1H, ^2_1H, ^3_1H$ |

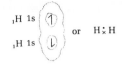

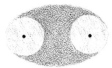

FIGURE 9.18
The bond between two hydrogen atoms is strong because the nuclei are strongly attracted to the electron pair.

## Chemical properties

At room temperature the hydrogen molecule is relatively unreactive. This is because the two nuclei are strongly attracted to the electron pair; the bond between the two hydrogen atoms is strong (Fig. 9.18). At elevated temperatures or in intense light, hydrogen reacts with most elements, both metals and nonmetals.

*Reaction with oxygen* One method of testing for *small* amounts of hydrogen that may be formed in a test tube is to hold a burning splint near the test tube. A sharp "pop" indicates hydrogen.

$$2H_2 + O_2 \longrightarrow 2H_2O + 137 \text{ kcal}$$

The energy from this reaction is used to run rocket engines and fuel cells.

*Reaction with nitrogen* Nitrogen is an important component of fertilizers. Before plants can use nitrogen, however, it must be combined with other elements to form compounds. One large scale method of preparation is to take nitrogen from the air and react it with hydrogen at elevated temperature and pressure.

$$N_2(g) + 3H_2(g) \rightleftharpoons 2NH_3(g) + \text{energy}$$

*Reaction with other elements*

$$H_2(g) + Cl_2(g) \rightleftharpoons 2HCl(g) + \text{energy}$$
$$H_2(g) + S(s) \rightleftharpoons H_2S(g) + \text{energy}$$
$$2Li(s) + H_2(g) \rightleftharpoons 2LiH(s) + \text{energy}$$

*Reaction with metal oxides* Hydrogen is often used to remove oxygen from metal oxides, forming pure metal and water.

$$CuO(s) + H_2(g) \rightleftharpoons Cu(s) + H_2O(l)$$

The ease with which oxygen can be removed from metal oxides varies with the metal. In some cases, simple heating liberates oxygen. In other cases heating with hydrogen removes the oxygen. In still other cases, $H_2$ will not remove the oxygen (Table 9.2).

## 9.9 Oxygen

### Physical properties

Oxygen is a colorless, odorless, tasteless gas. It forms nonpolar diatomic molecules. Some of the electrons in the molecule are

**TABLE 9.6**

*Some of the physical properties of oxygen.*

| | |
|---|---|
| Formula | $O_2$ |
| Density | 1.43 g/liter |
| Solubility in $H_2O$ | $\dfrac{1.3 \times 10^{-4} \text{ mole } O_2}{100 \text{ ml } H_2O}$ |
| Boiling point | $-183\,°C$ |
| Heat of vaporization | 1.62 kcal/mole |
| Freezing point | $-218\,°C$ |
| Heat of fusion | 0.112 kcal/mole |
| First ionization energy | 314 kcal/mole |
| Electronegativity | 3.5 |
| Atomic radius | 0.73 Å |
| Isotopes | $^{16}_{8}O$, $^{17}_{8}O$, $^{18}_{8}O$ |

unpaired, thus giving oxygen paramagnetic properties. Because the oxygen molecule is nonpolar, there is little attractive force between different molecules. For this reason the freezing point, boiling point, heat of vaporization, and solubility of $O_2$ in $H_2O$ are all relatively low (Table 9.6).

**Preparation**

The common methods of preparing oxygen in the laboratory have already been discussed.

(a) From $KClO_3$ (Example 2)
$$2KClO_3(s) + \text{heat} \rightleftharpoons 2KCl(s) + 3O_2(g)$$
(b) From HgO (Section 9.7)
$$2HgO(s) + \text{heat} \rightleftharpoons 2Hg(l) + O_2(g)$$
(c) Decomposition of water (Section 9.6)
$$2H_2O(l) + 137 \text{ kcal} \rightleftharpoons 2H_2(g) + O_2(g)$$
(d) From the air

Approximately 20 percent (by volume or by moles) of the atmosphere is oxygen. Large quantities of oxygen are obtained economically for industrial uses by cooling the air until the oxygen liquefies.

**Reactions**

At high enough temperatures, oxygen reacts with all elements to form oxides.

At room temperature, oxygen reacts slowly with many metals. It reacts with a clean surface of aluminum to form aluminum oxide.

$$4Al(s) + 3O_2(g) \rightleftharpoons 2Al_2O_3(s)$$

The $Al_2O_3$ adheres to the reactive metal and provides a dull but protective coating.

Compounds that contain the elements carbon, hydrogen, and sulfur usually react with oxygen to form $CO_2$, $H_2O$, or $SO_2$.

$$2C_8H_{18}(l) + 25O_2(g) \rightleftharpoons 16CO_2(g) + 18H_2O(g)$$
$$2H_2S(g) + 3O_2(g) \rightleftharpoons 2H_2O(g) + 2SO_2(g)$$
$$CH_4(g) + 2O_2(g) \rightleftharpoons CO_2(g) + 2H_2O(g)$$

When an electric discharge is passed through oxygen, some atoms may rearrange and form an *ozone molecule*, $O_3$.

$$3O_2(g) + \text{energy} \longrightarrow 2O_3(g)$$

The characteristic odor of electric motors is due to traces of ozone.

■ EXERCISE  *Dirigibles are filled with the unreactive, lighter than air gas, helium. Why isn't hydrogen, a cheaper and lighter gas, used? Support your reason with a balanced equation.*

Because most oxides are energetically stable, energy is produced when most substances combine with oxygen to form oxides. This energy may supply heat, as when fuels are burned; or light, as when magnesium in a flash bulb is used; or power, as when hydrogen in a rocket is burned. The energy required to maintain life in animals is supplied by the reaction of oxygen in the carefully controlled systems in living cells. ■

## Summary

Because the water molecule is polar, the positive end of one molecule has a high attraction for the negative end of another molecule. This attraction involves a hydrogen atom and is called *hydrogen bonding*. Significant amounts of energy are required to pull water molecules away from one another. This is why water has a high freezing point, high boiling point, and high heat of vaporization. When ice forms, hydrogen bonding causes $H_2O$ molecules to be oriented in such a way that the crystal contains many vacant spaces. This makes ice less dense than water.

Water is a good solvent for salts and polar substances. The positive or negative ends of the water molecule can attract ions or polar molecules and dissolve them. Water molecules and nonpolar substances have little attraction between one another. Nonpolar molecules cannot push themselves between water molecules and therefore they do not dissolve in water.

Many of the chemical reactions of water are characterized by either (1) the donation of a hydrogen ion from water to another substance, or (2) the availability of the electron pair of oxygen for sharing with another substance. Metallic hydrides, oxides, and some negative ions can remove a hydrogen ion from a water molecule and leave behind $OH^-$ ions. Nonmetallic oxides, hydrogen halides, ternary acids, and some ions can donate a hydrogen ion to the unshared electron pair on an oxygen atom and form $H_3O^+$ ions.

The electrical decomposition of water is an oxidation reduction reaction that produces two elements, hydrogen and oxygen. Both of these elements consist of nonpolar, diatomic molecules that are only slightly soluble in water. At sufficiently high temperatures, both elements will react with most other elements.

Balanced equations give the mole ratio in which substances react. The ratio can be used to calculate the amount of another substance used or produced in a reaction.

## Objectives

By the end of this chapter you should be able to do the following:
1. State the conditions of pressure and temperature at which water will undergo a phase change. (Question 1)

2. State how the freezing point and boiling point of water are affected by changes in pressure. (Questions 2, 4)

3. Draw an electron-dot structure for the water molecule. Locate positive and negative areas of charge. (Question 3)

4. Use the concept of hydrogen bonding to predict the properties a substance might have. (Questions 7, 8)

5. Write completed balanced equations for reactions that water undergoes with: (a) active metals; (b) chlorine; (c) oxides; (d) hydrogen compounds; (e) certain common ions; (f) itself. (Questions 9, 10, 11, 12)

6. Given the name of a hydrate, write its formula. (Question 13)

7. Given the formula of a hydrate, write its name. (Question 14)

8. Given a balanced equation and the number of moles of the reactants, calculate the number of moles of product formed. (Questions 19–24)

9. Write balanced equations for methods of preparing hydrogen. (Question 15)

10. Write balanced equations for the reaction of hydrogen with certain representative elements. (Question 16)

11. Write balanced equations for the reaction of oxygen with other representative elements and some compounds. (Question 16)

12. Given a balanced equation, identify any elements that may have undergone oxidation or reduction. (Question 17)

## Questions

1. Explain each of the following terms: *normal boiling point of water; normal freezing point of water; solution; soluble; shared electron pair; unshared electron pair; hydrogen bonding; slightly soluble; equilibrium; hydroxide ions; hydronium ions; proton transfer reaction; electron transfer reaction; oxidation; anode; reduction; cathode; hydrate; anhydrous; hydroscopic; redox reaction; ozone.*

2. The normal boiling point of water is 100 °C and the normal freezing point is 0 °C. Explain how it is possible for the temperature of water in a car engine to fall below 0 °C on a cold night and yet not freeze. How is it possible for the temperature of water in a car engine to rise above 100 °C and not boil?

3. Draw an electron-dot structure for the water molecule. Account for the original eight electrons in the oxygen atom and the one electron in each hydrogen atom.

4. The moon has little atmosphere and thus little atmospheric pressure. What do you think would happen to an open container of water on the moon at room temperature? What would happen to an open container of ice on the moon at $-10°C$?

5. Geologists believe that water in tiny cracks in rocks helps break up these rocks when it freezes. Explain what *molecular* forces are breaking up rocks in this process.

6. Explain why steam at 100 °C causes more severe burns than water at 100°C. (*Hint:* Which one possesses more energy on a molecular level?)

7. What forces make methyl alcohol a liquid at room temperature. Why is methane a gas?

$$\begin{array}{cc} \text{H} & \text{H} \\ | & | \\ \text{H—C—O—H} & \text{H—C—H} \\ | & | \\ \text{H} & \text{H} \\ \text{Methyl alcohol} & \text{Methane} \end{array}$$

8. Which of the two substances in Question 7 would you expect to be more soluble in water? Why?

9. None of the following elements are found "free" in nature. Write the equation for what would happen if water came in contact with one of these elements. (a) Li; (b) Na; (c) K; (d) Rb; (e) Cs; (f) Ca; (g) Sr; (h) Ba; (i) Ra.

10. Chlorine or bromine can be used as a disinfectant in swimming pools. Write the equations for the reaction of chlorine and bromine in water.

11. Although sodium metal readily reacts with oxygen in the air, no deposits of sodium oxide have ever been found on the earth's surface. Write the equation for the reaction that would occur if water did come in contact with deposits of the following substances: (a) $Na_2O$; (b) $Li_2O$; (c) KH; (d) MgO; (e) $P_4O_6$; (f) $P_4O_{10}$; (g) $SO_3$; (h) $SO_2$; (i) $NH_4I$; (j) $Na_2S$; (k) RaO; (l) $RaH_2$; (m) HCl; (n) $CaCl_2$.

12. Carbonated beverages contain water, sugar, carbon dioxide, and flavoring. What possible reaction do water and carbon dioxide undergo?

13. Write formulas for the following compounds: (a) cadmium nitrate tetrahydrate; (b) (gypsum) calcium sulfate dihydrate; (c) copper(II) sulfate monohydrate; (d) calcium sulfate monohydrate.

14. Name the following compounds: (a) $CuSO_4 \cdot 3H_2O$; (b) $Na_3PO_4 \cdot 12H_2O$ (dodeca = twelve); (c) $FeCl_3 \cdot 6H_2O$; (d) $Cr_2(SO_4)_3 \cdot 8H_2O$; (e) $ZnSO_4 \cdot 7H_2O$.

15. Write a balanced equation for the production of hydrogen using (a) sodium hydride, (b) sodium metal, (c) zinc.

16. Complete the following equations:
(a) $H_2(g) + I_2(s) \rightarrow$
(b) $H_2(g) + S(s) \rightarrow$
(c) $H_2(g) + Li(s) \rightarrow$
(d) $H_2(g) + Ba(s) \rightarrow$
(e) $C(s) + O_2(g) \rightarrow$
(f) $C_2H_6(g) + O_2(g) \rightarrow$
(g) $S(s) + O_2(g) \rightarrow$
(h) $Zn(s) + O_2(g) \rightarrow$

17. In which of the following reactions does electron transfer occur? Which element is oxidized? Which element is reduced?
(a) $2H_2(g) + O_2(g) \rightleftharpoons 2H_2O(l)$
(b) $Li_2O(s) + H_2O(l) \rightleftharpoons 2LiOH(aq)$
(c) $SO_3(s) + H_2O(l) \rightleftharpoons H_2SO_4(aq)$
(d) $H_2(g) + Br_2(g) \rightleftharpoons 2HBr(g)$
(e) $Na_2S(s) + H_2O(l) \rightleftharpoons 2NaOH(aq) + H_2S(g)$
(f) $2KClO_3(s) \rightleftharpoons 2KCl(s) + 3O_2(g)$
(g) $2Ag_2O(s) \rightleftharpoons 4Ag(s) + O_2(g)$

18. A fog, smog, haze, mist may occur frequently in some industrialized regions of the country. Sometimes this event is triggered by small particles of sulfuric acid in the atmosphere. Write a series of reactions showing how sulfur in gasoline reacts with oxygen and then water vapor to form sulfuric acid.

19. Hydrofluoric acid cannot be stored in glass containers because it reacts with silicon dioxide in the glass.

$$SiO_2(s) + 4HF(g) \rightleftharpoons SiF_4(g) + 2H_2O(g) + energy$$

(a) How many moles of $SiO_2$ will 3.25 moles of HF dissolve? (b) How many moles of $SiF_4$ and $H_2O$ will be produced if 3.25 moles of HF react?

20. Silicon carbide is nearly as hard as diamond. It is formed by heating sand, $SiO_2$, and carbon in an electric furnace.

$$SiO_2(s) + 3C(s) \rightleftharpoons SiC(s) + 2CO(g)$$

If 175 moles of $SiO_2$ are reacted, (a) How many moles of C are required? (b) How many moles of SiC and CO are produced?

21. Large quantities of carbon disulfide are used in making rayon and cellophane. The carbon disulfide can be made from methane and sulfur.

$$CH_4(g) + 4S(g) \rightleftharpoons CS_2(g) + 2H_2S(g)$$

(a) How many moles of sulfur are needed to produce 7000 moles of $CS_2$? (b) If 7000 moles of $CS_2$ are made per day, how many moles of sulfur are needed per year? (c) How many moles of $CH_4$ are needed to produce 7000 moles $CS_2$? What is the volume of $CH_4$ if one mole has a volume of 22.4 liters?

22. How many moles of water are in one mole of washing soda, $Na_2CO_3 \cdot 10H_2O$?

23. What is the weight percent of water in washing soda, $Na_2CO_3 \cdot 10H_2O$?

24. Peroxides contain the peroxide ion, $O_2^{2-}$. Each oxygen atom has a 1− oxidation number. When peroxides are left open to the air, or when water is added to them, they react to form hydroxides plus oxygen.

$$2Na_2O_2(s) + 2H_2O(l) \rightleftharpoons 4NaOH(aq) + O_2(g)$$

If 0.0277 mole of sodium peroxide ($Na_2O_2$) were placed in a test tube, (a) How many moles of sodium hydroxide and of oxygen would be produced? (b) How many moles of water would be required to react with the peroxide? (c) From you answer to (b), calculate the weight of water necessary and the volume of water necessary. Could the reaction be carried out in a test tube?

# 10

# Chemical changes and equations

## 10.1 Introduction

Many large rockets contain liquid hydrogen and liquid oxygen. Hydrogen burns in oxygen liberating 68.3 kcal of energy per mole of water formed. This released energy provides propulsion for the rocket. Can you write a balanced equation for this reaction and calculate the weight of oxygen needed to react with 2.00 moles of hydrogen? Can you explain this reaction using a model based on the rearrangement of atoms, or with a model based on electron transfer? What factors might speed up or slow down the reaction?

In this chapter we will review the information that an equation contains and discuss how this information is useful. The factors that control the speed of a reaction and two methods of classifying reactions will also be discussed.

## 10.2 Interpretation of a balanced equation

We said previously that hydrogen will burn in oxygen, forming water and liberating 68.3 kcal of energy per mole of product. The balanced equation for this reaction is:

$$2H_2(g) + O_2(g) \rightleftharpoons 2H_2O(l) + 137 \text{ kcal} \tag{10.1}$$

What information does this equation give us? First, it tells us what substances are the reactants, $H_2$ and $O_2$, and what substances are the products, $H_2O$. The balanced equation also gives information about (1) *mole ratios* of reactants and products; (2) relationship of *heat* to moles of reactants or products; (3) *molecular ratios* of reactants and products, and (4) the relationship of *weights* of reactants and *weights* of products (from mole ratios).

**Mole ratios of reactants and products**

The coefficients in a balanced equation give the mole ratio in which each of the substances in the reaction are chemically related. Consider Eq. 10.1, for example. According to the coefficients in the equation, *two moles* of hydrogen molecules will react with *one mole* of oxygen molecules to form *two moles* of water molecules. We can write the following ratios, each of which is a chemical unity term.

$$\frac{2 \text{ moles } H_2}{1 \text{ mole } O_2} = 1 \qquad \frac{2 \text{ moles } H_2O}{1 \text{ mole } O_2} = 1 \qquad \frac{2 \text{ moles } H_2}{2 \text{ moles } H_2O} = 1$$

The reciprocal of each of these are also chemical unity terms.

$$\frac{1 \text{ mole } O_2}{2 \text{ moles } H_2} = 1 \qquad \frac{1 \text{ mole } O_2}{2 \text{ moles } H_2O} = 1 \qquad \frac{2 \text{ moles } H_2O}{2 \text{ moles } H_2} = 1$$

**Relationship of heat to moles of reactants and products**

Equation 10.1 indicates that *137 kcal* of energy are released when *two moles* of water are formed from hydrogen and oxygen. If only one mole of water was formed in a reaction, only 68.3 kcal of energy would be released. The unity term that indicates the ratio of moles of water formed to kcal of energy released is:

$$\frac{1.00 \text{ mole } H_2O}{68.3 \text{ kcal}} = 1 \quad \text{or} \quad \frac{68.3 \text{ kcal}}{1.00 \text{ mole } H_2O} = 1$$

**Molecule ratio of reactants and products**

In addition to giving information about the mole ratio of each substance in the reaction, the *coefficients* in an equation give a *particle ratio*. In Eq. 10.1, the particle ratios are *two molecules* of hydrogen to *one molecule* of oxygen to *two molecules* of water. Possible unity terms are:

$$\frac{2 \text{ molecules } H_2}{1 \text{ molecule } O_2} = 1 \qquad \frac{1 \text{ molecule } O_2}{2 \text{ molecules } H_2} = 1$$

$$\frac{2 \text{ molecules } H_2}{2 \text{ molecules } H_2O} = 1 \qquad \frac{2 \text{ molecules } H_2O}{2 \text{ molecules } H_2} = 1$$

$$\frac{2 \text{ molecules } H_2O}{1 \text{ molecule } O_2} = 1 \qquad \frac{1 \text{ molecule } O_2}{2 \text{ molecules } H_2O} = 1$$

**Relationship of weights of reactants and weights of products**

The coefficients in a balanced equation give the *mole ratio* of reactants to products. How can we use this information to *calculate the weight* of a substance that will be produced in a given reaction? We can calculate the weight of a mole of any particular substance and then convert *moles* of a sample to the *weight* of the sample.

EXAMPLE 1  What weight of water will be produced when 12.0 moles of hydrogen are completely burned in oxygen?

Given: 12.0 moles of $H_2$ react

Find: weight of $H_2O$ produced

Relationship: Equation 10.1 gives us the mole ratio

$$2H_2 + O_2 \rightleftharpoons 2H_2O$$

Solution: (1) How many moles of $H_2O$ will be produced from 12.0 moles $H_2$?

$$12.0 \text{ moles } H_2 \times \frac{2 \text{ moles } H_2O}{2 \text{ moles } H_2} = 12.0 \text{ moles } H_2O$$

(2) How many grams is 12.0 moles of $H_2O$?

$$12.0 \text{ moles } H_2O \times \frac{18.0 \text{ g } H_2O}{1 \text{ mole } H_2O} = 216 \text{ g } H_2O$$

We could have arranged the problem in one expression:

$$12.0 \text{ mole } H_2 \times \underbrace{\frac{2 \text{ moles } H_2O}{2 \text{ moles } H_2}}_{\text{changes moles of } H_2 \text{ to moles of } H_2O} \times \underbrace{\frac{18.0 \text{ g } H_2O}{1 \text{ mole } H_2O}}_{\text{changes moles of } H_2O \text{ to grams of } H_2O} = 216 \text{g } H_2O$$

■ EXERCISE  How many grams of sodium sulfide are formed when 6.00 moles of sodium react with a sufficient amount of sulfur?

$2Na(s) + S(s) \rightleftharpoons Na_2S(s)$

Hint: (1) How many moles of sodium sulfide are formed? (2) What is the weight in grams of this many moles of sodium sulfide?

## 10.3  Collision theory

Hydrogen iodide decomposes at high temperatures to produce iodine and hydrogen.

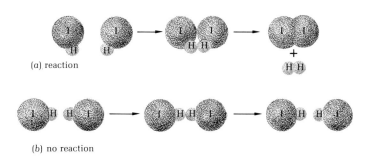

FIGURE 10.1
Molecules of hydrogen iodide could collide with many different orientations. In (a) it is possible for the collision to produce hydrogen and iodine. In (b) because of the orientation of the molecules, the collision may not result in these products.

(a) reaction

(b) no reaction

$$\text{energy} + 2\text{HI(g)} \rightleftharpoons \text{H}_2\text{(g)} + \text{I}_2\text{(g)} \tag{10.2}$$

What model can be proposed to explain this reaction? What role does temperature have in the speed at which the hydrogen iodide reacts?

### Orientation

A useful model for reactions is the *collision theory model*. In this model, we assume that molecules must collide before reacting, but not every collision will result in a reaction. When two hydrogen iodide molecules collide, they may produce a hydrogen molecule and an iodine molecule.

$$\text{HI} + \text{HI (upon colliding)} \rightleftharpoons \text{H}_2 + \text{I}_2 \tag{10.3}$$

Does every collision produce a reaction? In Fig. 10.1(a) the atoms are oriented in such a manner that a reaction is possible. If the molecules approach each other at other angles, a reaction may not occur (Fig. 10.1(b)).

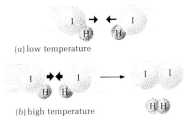

(a) low temperature

(b) high temperature

FIGURE 10.2
(a) At low temperatures, most collisions do not result in a reaction because the particles do not collide with sufficient kinetic energy to break the bond. (b) At high temperatures, many molecules collide with sufficient kinetic energy to break the H—I bond, and a reaction occurs.

### Temperature

Experimentally it has been determined that an increase in temperature causes hydrogen iodide to react faster. The speed at which a reaction occurs is known as the *rate of the reaction*. Why does an increase in temperature cause an increase in the rate of the reaction?

When two hydrogen iodide molecules collide and produce hydrogen and iodine, bonds are broken between hydrogen atoms and iodine atoms. The energy required to break these bonds comes from the force or kinetic energy with which these molecules strike one another. At low temperatures, most molecules are traveling at a relatively low velocity. They do not collide with sufficient energy to break the hydrogen–iodine bond (Fig. 10.2). As the temperature increases, the molecular

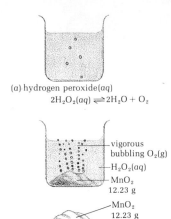

(a) A solution of hydrogen peroxide undergoes slow decomposition to form water and oxygen. (b) Manganese (IV) oxide speeds up the reaction even though it does not undergo change itself.

velocities increase; more molecules collide at sufficiently high velocities to cause a reaction.

## Catalysts

When manganese(IV) oxide is placed in a solution of hydrogen peroxide, vigorous bubbling occurs (Fig. 10.3). However, the manganese(IV) oxide does not appear to undergo any change. What is happening?

When hydrogen peroxide solution is left in a clear glass bottle it decomposes slowly, with no noticeable reaction. The products are water and oxygen.

$$2H_2O_2(aq) \rightleftharpoons 2H_2O(l) + O_2(g) \tag{10.4}$$

When manganese(IV) oxide is placed in the solution, the reaction occurs at a much greater rate. Oxygen is liberated so rapidly that bubbles form and rise out of the solution. Substances that change the rate of a reaction without undergoing a permanent change themselves are called *catalysts*.

It is believed that catalysts take part in the reaction and are temporarily changed but later re-form. *They are both reactants and products.*

$$2H_2O_2(aq) + MnO_2(s) \rightleftharpoons 2H_2O(l) + O_2(g) + MnO_2(s) \tag{10.5}$$

A catalyst such as $MnO_2$ may combine with peroxide molecules.

$$2H_2O_2(aq) + MnO_2(s) \rightleftharpoons 2H_2O_2 \cdot MnO_2(s) \tag{10.6}$$

This product may then decompose to regenerate $MnO_2$, and produce water and oxygen.

$$2H_2O_2 \cdot MnO_2(s) \rightleftharpoons MnO_2(s) + 2H_2O(l) + O_2(g) \tag{10.7}$$

Chemists believe that some catalysts may *lower the energy required for a molecule to react* (Fig. 10.4). By providing a new pathway for reactants to form products, molecules with a low energy may react. Other catalysts help orient the angles at which molecules approach each other so that more collisions

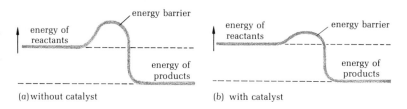

FIGURE 10.4
A catalyst increases the rate of a reaction by lowering the energy barrier between reactants and products.

■ EXERCISE  Use the concepts of collision, orientation, temperature, and catalysts to explain the following observations. (a) When hydrogen is mixed with oxygen at room temperature no reaction is noticeable. (b) At high temperatures hydrogen and oxygen react to produce water (steam). (c) At room temperature a stream of hydrogen gas will react with oxygen gas in the presence of platinum metal.

will produce a reaction. This may be how the more than 100,000 different enzymes in the body catalyze chemical reactions. ■

## Concentration

To test for the presence of oxygen in a test tube, place a glowing splint in the tube. If the splint bursts into flame, oxygen is probably present. Why does the oxygen in the test tube cause the splint to burst into flame?

The glowing splint is actually burning with oxygen in the air at a slow rate. Air is approximately 20 percent oxygen (Fig. 10.5). Therefore, one out of five molecules in the air can react with the splint as follows:

carbon and hydrogen (in splint) + oxygen(g) ⟶
    carbon dioxide(g) + water(g)

If the test tube contains a higher percent of oxygen—say about 75 percent oxygen—then three out of four molecules in the air that collide with the splint can react with it. The rate at which the splint burns increases sharply and the glowing splint bursts into flame.

The *amount of a substance per unit volume* is called the *concentration*. In general, an increase in the concentration of a reactant will increase the rate at which the reaction occurs.

## Surface area

How long does it take for the magnesium in a flash bulb to burn? Perhaps it takes less than 0.10 second. An equal weight of magnesium ribbon, however, may take one or two seconds to

FIGURE 10.5
(a) In air, only one molecule in five is oxygen. (b) When the concentration of oxygen is increased, a greater number of molecules can react and the splint burns faster.

(a) air (20% $O_2$)

(b) increased concentration of oxygen (75%)

■ **EXERCISE** *Usually a lump of coal or a piece of grain will burn slowly. Occasionally, though, a mine full of coal dust or a storage bin full of grain dust will explode. Explain these rapid reactions in terms of particle size.*

■ **EXERCISE** *Oxygen that is taken up by the lungs undergoes a chemical reaction. Explain, in terms of the factors that affect the rate of a reaction, why many athletes had difficulty performing in the 1968 Mexico City Olympics. (Elevation of Mexico City is about 1½ miles above sea level.)*

burn (Fig. 10.6). Why does the magnesium in the bulb burn so fast? If we examine the contents of an unburned bulb, we see many small, thin pieces of magnesium ribbon. Cutting the magnesium into many small pieces greatly increases the surface area. In the case of the uncut ribbon, only magnesium atoms at the surface of the metal can collide with oxygen molecules; therefore, only the surface atoms can react with oxygen. Increasing this surface area increases the number of magnesium atoms that can react in a given time. This increases the rate at which the magnesium burns.

In general, increasing the surface area of a reactant will increase its rate of reaction. The surface area may be increased by cutting or grinding the substance into many small pieces. ■

## 10.4 Reactions—a rearrangement of atoms

John Dalton, in 1805, described a chemical reaction as the combining, disunion, or rearrangement of atoms. Simple chemical reactions can be classified into four basic categories: (1) combination, (2) decomposition, (3) replacement, and (4) displacement (see Table 10.1).

magnesium ribbon burns in one second

fine magnesium wire burns in less than 0.10 second

**FIGURE 10.6**
*When a substance is cut into small pieces, its surface area increases and the substance reacts faster.*

### Combination

When heated, zinc and sulfur will combine to form zinc sulfide.

$$Zn(s) + S(s) \rightleftharpoons ZnS(s) + \text{energy} \tag{10.8}$$

This is an example of the combination of atoms from one element with atoms of another element to form a compound.

**TABLE 10.1** *Classification of reactions according to the rearrangement of atoms.*

| Category | Description | Reaction |
|---|---|---|
| Combination | Bond is formed between atoms A and B | $A + B \rightleftharpoons AB$ |
| Decomposition | Bond is broken between atoms A and B | $AB \rightleftharpoons A + B$ |
| Replacement | Bond between B and C is broken; bond between A and C is formed | $A + BC \rightleftharpoons AC + B$ |
| Displacement | Bonds broken between A and B and C and D; bonds formed between A and D and C and B | $AB + CD \rightleftharpoons AD + CB$ |

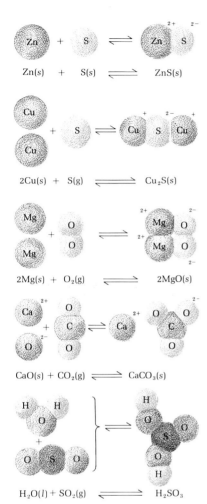

FIGURE 10.7
Five examples of combination reactions. Atoms of two different substances combine to form one substance.

During the reaction bonds are formed between A and B. Some examples are illustrated in Fig. 10.7.

### Decomposition

When mercury(II) oxide is heated, mercury and oxygen are produced.

$$2HgO(s) + \text{energy} \rightleftharpoons 2Hg(l) + O_2(g) \tag{10.9}$$

This reaction is an example of decomposition. When atoms that are bound together in a single substance are broken apart (as mercury and oxygen atoms are) we say that decomposition has occurred. During this process bonds are broken (Fig. 10.8).

### Replacement

Since solutions containing copper(II) ions are blue, the blue color is believed to be due to these ions. When a piece of clean zinc metal is added to a beaker of blue copper(II) sulfate solution, the zinc falls to the bottom of the beaker and begins to turn black. When the solution is stirred, dark, solid particles can be seen in the solution. The zinc appears to get smaller, and the blue color becomes less intense. In a few more minutes the solution is colorless. If the zinc is removed and washed off, it is seen to be considerably smaller (Fig. 10.9). What has happened?

The dark solid has the properties of copper. The solution contains zinc sulfate. Evidently, zinc has reacted with aqueous copper(II) sulfate to form copper and aqueous zinc sulfate.

$$Zn(s) + CuSO_4(aq) \rightleftharpoons Cu(s) + ZnSO_4(aq) \tag{10.10}$$

In this reaction zinc has replaced copper in the sulfate compound. Reactions where atoms of one element replace atoms of another element in a compound are called replacement reactions (Fig. 10.10).

### Displacement

When a drop of hydrochloric acid is added to a solution of silver nitrate, a white solid forms, indicating that a chemical reaction has occurred (Fig. 10.11). This solid soon settles to the bottom. The solid formed when two solutions are mixed is called a *precipitate*. What reaction is occurring?

$$HCl(aq) + AgNO_3(aq) \longrightarrow ?$$

10.4 Reactions—a rearrangement of atoms

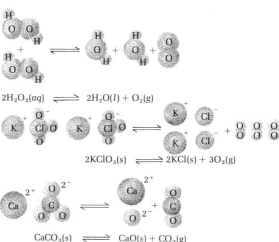

**FIGURE 10.8**
Five examples of decomposition reactions. Atoms of one substance disunite to form two or more simpler substances.

$2HgO(s) \rightleftharpoons 2Hg(l) + O_2(g)$

$2H_2O(l) \rightleftharpoons 2H_2(g) + O_2(g)$

$2H_2O_2(aq) \rightleftharpoons 2H_2O(l) + O_2(g)$

$2KClO_3(s) \rightleftharpoons 2KCl(s) + 3O_2(g)$

$CaCO_3(s) \rightleftharpoons CaO(s) + CO_2(g)$

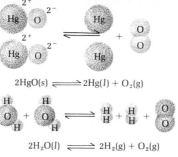

(a)

(b)

CuSO$_4$(aq) blue solution
zinc
blue color disappears
dark solid particles form

**FIGURE 10.9**
(a) When a piece of zinc is added to an aqueous solution of copper(II) sulfate, the blue color of the solution fades. (b) The zinc grows smaller and a dark solid forms. After several minutes the blue color disappears.

Analysis of the white solid formed shows that it is silver chloride. The remaining clear solution has the properties of nitric acid. Silver nitrate and hydrochloric acid react to produce silver chloride and nitric acid.

$$AgNO_3(aq) + HCl(aq) \rightleftharpoons AgCl(s) + HNO_3(aq) \qquad (10.11)$$

Note that in this reaction, the Ag$^+$ and H$^+$(aq) (actually H$_3$O$^+$) have exchanged places. When an atom in one compound exchanges places with an atom in a second compound, we classify the reaction as a displacement (Fig. 10.12). (This kind of reation is also known as a double displacement or methathesis reaction.)

*Neutralization*  Solutions of strong bases contain many hydroxide ions in solution. These bases turn litmus blue, taste bitter, and feel slippery. Solutions of strong acids contain many hydronium ions in solution. These acids turn litmus red, taste sour, and dissolve zinc, releasing hydrogen gas. When appropriate amounts of an acid and a base are mixed, a solution with

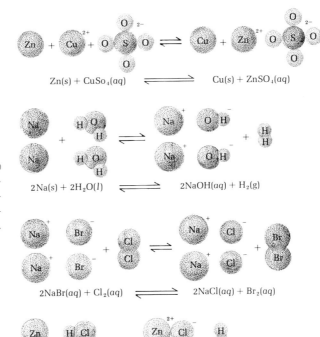

FIGURE 10.10
Four examples of replacement reactions. In each case, an atom of one element replaces the atom of another element in a substance.

a new set of properties is formed. This indicates that a reaction has occurred. The reaction between an acid and a base is called neutralization.

When a dilute solution of hydrochloric acid is mixed with a dilute solution of sodium hydroxide (a base), no visible change is observed. A temperature increase in the system indicates that some heat is released. The resulting mixture tastes salty and does not have the characteristic properties of acids or bases. The explanation is that water and sodium chloride are formed.

$$NaOH(aq) + HCl(aq) \rightleftharpoons NaCl(aq) + H_2O(l) \qquad (10.12)$$

In this reaction the sodium and hydrogen exchange places. The bond formed between $H^+$ and $OH^-$ is stronger than bonds between $Na^+$ and $OH^-$ or $H^+$ and $Cl^-$. This neutralization reaction is an example of a displacement reaction. Other reactions between acids and bases are also displacement reactions. In each case the products are water and a salt.

Base + Acid $\rightleftharpoons$ Salt + Water
$$Ba(OH)_2(aq) + H_2SO_4(aq) \rightleftharpoons BaSO_4(s) + 2H_2O(l) \qquad (10.13)$$

■ EXERCISE  Classify the following as combination, decomposition, replacement, or displacement reactions.
(a) $3Mg + 2FeCl_3 \rightleftharpoons 3MgCl_2 + 2Fe$
(b) $6Li + N_2 \rightleftharpoons 2Li_3N$
(c) $Ba(OH)_2 + MgSO_4 \rightleftharpoons BaSO_4 + Mg(OH)_2$
(d) $2Ag_2O \rightleftharpoons 4Ag + O_2$

217   10.4 Reactions—a rearrangement of atoms

## 10.5 Reactions—classification by mechanism

John Dalton's model of a reaction was proposed before scientists had developed the electron–proton model of the atom. We can begin to interpret what happens in a reaction by studying what happens to the electrons of an atom in a reaction. Four possible methods for classifying reactions are: (1) electron transfer reactions, (2) proton transfer reactions, (3) electron pair donation reactions, and (4) precipitation reactions.

### Electron transfer reactions

Most metals will react with and combine with many nonmetals. We have mentioned in Section 10.4 the reactions between zinc and sulfur, copper and sulfur, magnesium and oxygen, and sodium and chlorine. In each of these reactions, metal atoms that have a low ionization energy lose electrons to nonmetal atoms that have a high attraction for electrons. *Electrons are transferred from the metal atom to the nonmetal atom.*

The change in oxidation state for zinc is shown in the following *half-reaction*. (In a half-reaction, only the oxidation or the reduction step is shown.)

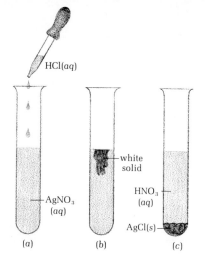

**FIGURE 10.11**
(a) Both aqueous hydrochloric acid and aqueous silver nitrate are colorless, clear liquids. (b) When a drop of hydrochloric acid is added to a few milliliters of silver nitrate, a white solid forms. (c) This slowly sinks to the bottom of the test tube. A clear liquid remains above the solid. The solid is silver chloride.

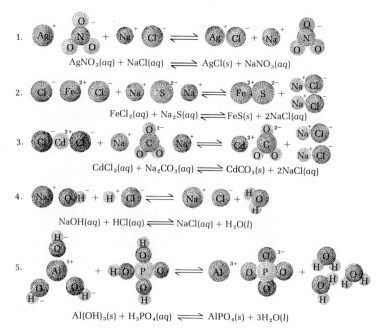

**FIGURE 10.12**
Five examples of displacement reactions.

1. $AgNO_3(aq) + NaCl(aq) \rightleftharpoons AgCl(s) + NaNO_3(aq)$

2. $FeCl_2(aq) + Na_2S(aq) \rightleftharpoons FeS(s) + 2NaCl(aq)$

3. $CdCl_2(aq) + Na_2CO_3(aq) \rightleftharpoons CdCO_3(s) + 2NaCl(aq)$

4. $NaOH(aq) + HCl(aq) \rightleftharpoons NaCl(aq) + H_2O(l)$

5. $Al(OH)_3(s) + H_3PO_4(aq) \rightleftharpoons AlPO_4(s) + 3H_2O(l)$

$$Zn^0 \rightleftharpoons Zn^{2+} + 2e^- \quad \text{(oxidation)}$$

At the same time, sulfur undergoes the following change.

$$S^0 + 2e^- \rightleftharpoons S^{2-} \quad \text{(reduction)}$$

The overall reaction is the sum of these two half-reactions

$$Zn + S \rightleftharpoons ZnS \quad (10.14)$$
$$(0) \quad (0) \quad (2+)(2-)$$

where the zero represents the original oxidation state.

The reaction between magnesium and oxygen is:

$$Mg + O_2 \rightleftharpoons 2MgO \quad (10.15)$$
$$(0) \quad (0) \quad (2+)(2-)$$

(Electrons are transferred from magnesium to oxygen.)

The reaction between sodium and chlorine is:

$$2Na + Cl_2 \rightleftharpoons 2NaCl \quad (10.16)$$
$$(0) \quad (0) \quad (1+)(1-)$$

(Electrons are transferred from sodium to chlorine.)

Electron transfer reactions may be recognized by a change in oxidation state of at least one element as reactants form products. The decomposition of potassium chlorate is an electron transfer reaction.

$$2KClO_3 \xrightarrow{\text{heating}} 2KCl + 3O_2 \quad (10.17)$$
$$(1+)(5+)(2-) \quad (1+)(1-) \quad (0)$$

(Electrons are transferred from oxygen to chlorine.)

All of the replacement reactions discussed in Section 10.4 involve electron transfer. For example:

$$Zn(s) + CuSO_4(aq) \rightleftharpoons Cu(s) + ZnSO_4(aq) \quad (10.18)$$
$$(0) \quad (2+)(6+)(2-) \quad (0) \quad (2+)(6+)(2-)$$

The zinc atom transfers electrons to the copper(II) ion.

$$Zn^0 + Cu^{2+} \rightleftharpoons Cu^0 + Zn^{2+} \quad (10.19)$$

Evidently, since a zinc atom is donating electrons to a copper(II) ion, a zinc atom tends to lose electrons more readily than a copper atom. The atoms of many other metals will lose electrons to copper ions. The higher up a metal is in Table 10.2, the more readily it will lose electrons.

An atom of a metal listed in Table 10.2 *will replace an ion of a metal* listed below it from solution. Thus, since magnesium is

**TABLE 10.2**

*The tendency of metals to lose electrons. Metals at the top of the table lose electrons or undergo oxidation most readily.*

| | | |
|---|---|---|
| K(s) | $\rightleftharpoons$ K$^+$(aq) | $+ e^-$ |
| Ba(s) | $\rightleftharpoons$ Ba$^{2+}$(aq) | $+ 2e^-$ |
| Sr(s) | $\rightleftharpoons$ Sr$^{2+}$(aq) | $+ 2e^-$ |
| Ca(s) | $\rightleftharpoons$ Ca$^{2+}$(aq) | $+ 2e^-$ |
| Na(s) | $\rightleftharpoons$ Na$^+$(aq) | $+ e^-$ |
| Mg(s) | $\rightleftharpoons$ Mg$^{2+}$(aq) | $+ 2e^-$ |
| Al(s) | $\rightleftharpoons$ Al$^{3+}$(aq) | $+ 3e^-$ |
| Mn(s) | $\rightleftharpoons$ Mn$^{2+}$(aq) | $+ 2e^-$ |
| Zn(s) | $\rightleftharpoons$ Zn$^{2+}$(aq) | $+ 2e^-$ |
| Cr(s) | $\rightleftharpoons$ Cr$^{2+}$(aq) | $+ 2e^-$ |
| Fe(s) | $\rightleftharpoons$ Fe$^{2+}$(aq) | $+ 2e^-$ |
| Cd(s) | $\rightleftharpoons$ Cd$^{2+}$(aq) | $+ 2e^-$ |
| Co(s) | $\rightleftharpoons$ Co$^{2+}$(aq) | $+ 2e^-$ |
| Ni(s) | $\rightleftharpoons$ Ni$^{2+}$(aq) | $+ 2e^-$ |
| Sn(s) | $\rightleftharpoons$ Sn$^{2+}$(aq) | $+ 2e^-$ |
| Pb(s) | $\rightleftharpoons$ Pb$^{2+}$(aq) | $+ 2e^-$ |
| H$_2$(g) + 2H$_2$O(l) | $\rightleftharpoons$ 2H$_3$O$^+$(aq) | $+ 2e^-$ |
| Cu(s) | $\rightleftharpoons$ Cu$^{2+}$(aq) | $+ 2e^-$ |
| Hg(l) | $\rightleftharpoons$ Hg$^{2+}$(aq) | $+ 2e^-$ |
| Ag(s) | $\rightleftharpoons$ Ag$^+$(aq) | $+ 1e^-$ |
| Pt(s) | $\rightleftharpoons$ Pt$^{4+}$(aq) | $+ 4e^-$ |
| Au(s) | $\rightleftharpoons$ Au$^{3+}$(aq) | $+ 3e^-$ |

above tin, if we added magnesium metal to a solution of tin(II) nitrate we would have:

$$Mg(s) + Sn(NO_3)_2(aq) \rightleftharpoons Sn(s) + Mg(NO_3)_2(aq) \quad (10.20)$$

or ignoring $NO_3^-$,

$$Mg(s) + Sn^{2+}(aq) \rightleftharpoons Sn(s) + Mg^{2+}(aq) \quad (10.21)$$

(Electrons are transferred from magnesium to tin.)

However, since silver is below copper, if we added silver metal to a solution of copper(II) nitrate, we would have:

$$Ag(s) + Cu(NO_3)_2 \longrightarrow \text{No reaction}$$

Silver atoms do not lose electrons readily. However, we could force the reaction to go by means of electrical energy. ■

■ EXERCISE  The following reactions involve electron transfer. Identify the atom that loses electrons and the atom that gains electrons.
(a) $Mg(s) + H_2O(l) \rightleftharpoons MgO(s) + H_2(g)$
(b) $2Ag_2O(s) \xrightleftharpoons[]{heating} 4Ag(s) + O_2(g)$
(c) $Fe(s) + 2AgNO_3(aq) \rightleftharpoons Fe(NO_3)_2(aq) + 2Ag(s)$

**Proton transfer reactions**

When the two invisible gases ammonia and hydrogen chloride mix, a white smoke forms. The reaction is:

$$NH_3(g) + HCl(g) \rightleftharpoons NH_4Cl(s) \quad (10.22)$$

In this reaction, the *hydrogen nucleus* or *bare proton is transferred* from the hydrogen chloride to the ammonia molecule. The HCl is a *proton donor*. $NH_3$ is a *proton acceptor*. Chemical

$$\underset{\underset{H}{|}}{\overset{\overset{H}{|}}{H-N:}} + H:Cl \rightleftharpoons \underset{\underset{H}{|}}{\overset{\overset{H}{|}}{H-N-H^+}} + Cl^- \quad (10.23)$$

aggregates that donate protons are known as *Brønsted acids*. Chemical aggregates that accept protons are known as *Brønsted bases* (Chapter 16).

Many reactions can be viewed as *proton transfer reactions*. When a strong acid such as hydrogen chloride dissolves in water, a proton transfer occurs.

$$HCl(aq) + H_2O(l) \rightleftharpoons H_3O(aq)^+ + Cl^-(aq) \quad (10.24)$$

$$\underset{\underset{H}{|}}{H-\overset{..}{O}:} + H-Cl \rightleftharpoons \underset{\underset{H}{|}}{H-\overset{..}{O}-H^+} + Cl^-$$

(Proton transferred from HCl to $H_2O$.)

What happens when ammonia is dissolved in water (see Eq. 9.16)?

$$NH_3(aq) + H_2O(l) \rightleftharpoons NH_4^+(aq) + OH^-(aq) \qquad (10.25)$$

(Proton transferred from water to ammonia.)

Neutralization is a type of displacement reaction discussed in Section 10.4. For example, Eq. 10.12 is:

$$NaOH(aq) + HCl(aq) \rightleftharpoons NaCl(aq) + H_2O(l)$$

A solution of NaOH contains Na$^+$ ions and OH$^-$. A solution of HCl contains H$_3$O$^+$ ions and Cl$^-$. The product of mixing these two solutions is a solution of NaCl (Na$^+$ ions and Cl$^-$ ions) and water. Rewriting the equation in terms of ions gives us:

$$Na^+(aq) + OH^-(aq) + H_3O^+(aq) + Cl^-(aq) \rightleftharpoons$$
$$Na^+(aq) + Cl^-(aq) + 2H_2O(l) \qquad (10.26)$$

Note that the sodium ion and chloride ion are present in the reactants and products. The actual reaction is a proton transfer between the OH$^-$ ions and the H$_3$O$^+$ ions.

$$OH^-(aq) + H_3O^+(aq) \rightleftharpoons H_2O(l) + H_2O(l) \qquad (10.27)$$

Some chemists prefer to write H$_3$O$^+$ simplified as H$^+$(aq), in which case reaction 10.27 is:

$$OH^-(aq) + H^+(aq) \rightleftharpoons H_2O(l) \qquad (10.28)$$

The displacement reactions between acids and bases have a proton transfer mechanism. The compounds listed in Table 9.3 undergo proton transfer type reactions when they are added to water. For example:

$$NH_4Cl(aq) + H_2O(l) \rightleftharpoons NH_3(aq) + H_3O^+(aq) + Cl^- \qquad (10.29)$$

$$H:\overset{H^+}{\underset{H}{\overset{..}{N}}}:(H) + Cl^- + H:\overset{..}{\underset{..}{O}}: \rightleftharpoons H:\overset{H}{\underset{H}{\overset{..}{N}}}: + Cl^- + H:\overset{..}{\underset{..}{O}}:H^+ \qquad (10.30)$$

The proton donor is NH$_4^+$. The proton acceptor is H$_2$O. ■

■ EXERCISE *The following reactions involve proton transfer. Identify the species (ion, atom, or molecule) that lose protons and the species that gain protons.*
(a) NH$_4$Br(aq) + H$_2$O(l) ⇌ NH$_3$(aq) + H$_3$O$^+$(aq) + Br$^-$(aq)
(b) CaO(s) + H$_2$O(l) ⇌ Ca$^{2+}$(aq) + 2(OH)$^-$(aq)
(c) LiOH(aq) + H$_2$S(aq) ⇌ LiHS(aq) + H$_2$O(l)

### Electron pair donation reactions

Each of the reactions in the preceding section involves the transfer of a proton to a species with an unshared electron pair. Look at Eq. 10.23 again. In each case an electron pair is donated to a proton. However, an electron pair can be donated to many other species. Consider the following example.

Stirring a silver chloride-water mixture does not cause the solid to dissolve. It is only slightly soluble in water. However, if a few drops of aqueous ammonia are added to the mixture, the silver chloride dissolves readily (Fig. 10.13). How does the ammonia cause the silver chloride to dissolve?

Chemists believe that ammonia molecules react with silver ions to form the soluble $Ag(NH_3)_2^+$ ion.

$$AgCl(s) + 2NH_3(aq) \rightleftharpoons Ag(NH_3)_2^+(aq) + Cl^-(aq) \qquad (10.31)$$

The reaction is due to the breaking of the bond between $Ag^+$ and $Cl^-$ and the formation of a bond between $Ag^+$ and $NH_3$.

$$Ag^+(aq) + 2NH_3(aq) \rightleftharpoons Ag(NH_3)_2^+(aq) \qquad (10.32)$$

Why do silver ions and ammonia molecules combine? The chemist's model is that the unshared pair of electrons on the nitrogen of the ammonia molecule is attracted to the positive silver ion. Two molecules approach the ion and are held to it by this attraction. Each nitrogen atom donates a pair of electrons to the silver ion. The silver ion is referred to as the central ion. Each $NH_3$ molecule is known as *a coordinating group*.

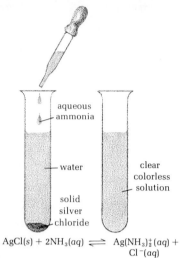

$AgCl(s) + 2NH_3(aq) \rightleftharpoons Ag(NH_3)_2^+(aq) + Cl^-(aq)$

$[H_3N:Ag:NH_3]^+$

**FIGURE 10.13**
Silver chloride is insoluble in water. When aqueous ammonia is added to the mixture, the solid dissolves readily. The chemist's model is that the ammonia molecule shares its electron pair with the positive silver ion.

$$H:\overset{..}{\underset{..}{N}}:H + Ag^+ + :\overset{..}{\underset{..}{N}}:H \rightleftharpoons \left[ H:\overset{..}{\underset{..}{N}}:Ag:\overset{..}{\underset{..}{N}}:H \right]^+ \qquad (10.33)$$
(with H's above and below each N)

Many molecules ($NH_3$) or negative ions ($OH^-$) will donate an electron pair to an element with a positive oxidation state ($Ag^+$, $Zn^{2+}$).

**Precipitation reactions**

We saw in Section 10.4 that a number of displacement reactions involve the formation of a precipitate. These precipitates form when a solution of one soluble substance is mixed with a solution of another soluble substance. What causes the precipitate to form?

When solutions of silver nitrate and hydrochloric acid are mixed, a white precipitate forms. According to Eq. 10.11:

$$AgNO_3(aq) + HCl(aq) \rightleftharpoons AgCl(s) + HNO_3(aq)$$

Aqueous silver nitrate contains $Ag^+$ ions and $NO_3^-$ ions; aqueous hydrochloric acid contains $H_3O^+$ ions and $Cl^-$. Aqueous nitric acid contains $H_3O^+$ ions and $NO_3^-$ ions. The solid precipitate may be thought of as an aggregate containing many $Ag^+$ and $Cl^-$ ions.

$$Ag^+ + NO_3^-(aq) + H_3O^+(aq) + Cl^-(aq) \rightleftharpoons$$
$$AgCl(s) + H_3O^+(aq) + NO_3^-(aq) \qquad (10.34)$$

## TABLE 10.3

The solubility of a number of substances. There is no clear-cut line between soluble and slightly soluble. If more than 0.1 g/100 ml $H_2O$ dissolve, it is usually considered to be soluble.

| Compound | Solubility at 20°C (g/100 ml $H_2O$) | Classification |
|---|---|---|
| AgBr | $8.4 \times 10^{-6}$ | Slightly soluble |
| AgCl | $1.5 \times 10^{-4}$ | Slightly soluble |
| AgI | $2.0 \times 10^{-7}$ | Slightly soluble |
| $AgNO_3$ | 222 | Soluble |
| $Ag_2S$ | $1.4 \times 10^{-5}$ | Slightly soluble |
| $Ba(NO_3)_2$ | 9.2 | Soluble |
| $BaSO_4$ | $2.4 \times 10^{-7}$ | Slightly soluble |
| $Ca(NO_3)_2$ | 56 | Soluble |
| $CaSO_4$ | $2 \times 10^{-1}$ | Slightly soluble |
| KCl | 34 | Soluble |
| $KNO_3$ | 31.6 | Soluble |
| $NH_4Cl$ | 37.2 | Soluble |
| $NH_4NO_3$ | 1922 | Soluble |
| NaCl | 36.0 | Soluble |
| $NaNO_3$ | 88 | Soluble |
| $Na_2S$ | 15.8 | Soluble |

The reaction involves the $Ag^+$ and $Cl^-$ ions.

$$Ag^+(aq) + Cl^-(aq) \rightleftharpoons AgCl(s) \qquad (10.35)$$

The affinity of the $Ag^+$ ions for the $Cl^-$ ions results in the formation of AgCl ion pairs in solution. These ions continue to stick together and a precipitate forms. Some ions such as $Ag^+$ and $Cl^-$ stick together in the presence of water and form precipitates. Other ions such as $Na^+$ and $Cl^-$ do not stick together, but remain apart in the presence of water.

Other precipitation reactions occur in much the same way as the silver chloride precipitation. Any time that the mixing of solutions results in the formation of a slightly soluble substance, a precipitation reaction will occur. Table 10.3 lists the solubilities of many substances. Factors that determine whether ions will stick together include charge and ion size.

Barium sulfate is listed in Table 10.3 as being only slightly soluble. Would a precipitate result from mixing a solution of barium hydroxide with a solution of sodium sulfate? ■

■ EXERCISE   Using Table 10.3, state whether or not a precipitate will form in the following cases:
(a) $AgNO_3(aq) + Na_2S(aq) \rightleftharpoons$
(b) $KNO_3(aq) + NaCl(aq) \rightleftharpoons$
(c) $AgNO_3(aq) + KCl(aq) \rightleftharpoons$
(d) $AgNO_3(aq) + NH_4Cl(aq) \rightleftharpoons$

## Summary

This chapter deals with reactions and equations. A balanced equation represents a reaction. The equation tells us what the reactants are and what the products are. Symbols may be used in an equation to indicate in which phases the reactants and products exist, what energy requirements are needed, and whether a catalyst is present. The coefficients in a balanced equation tell us the simplest ratios that exist between moles of reactants and moles of products. From these mole ratios we can calculate weight ratios.

Reactions can occur when molecules or atoms collide. For a collision to produce a reaction, the molecules must have sufficient energy and the proper orientation. The rate of a reaction may be affected by:

(1) A temperature change. With an increase in temperature more particles have sufficient energy to react and the reaction rate increases.

(2) A catalyst. A catalyst lowers the energy required for a reaction and as a result, the reaction speeds up.

(3) Concentration. Increasing the concentration of a reactant will increase the number of collisions between reacting particles. This will increase the rate of the reaction.

(4) Surface area. In a solid, only atoms or molecules at the surface can react. Decreasing the particle size of a sample increases its surface area and more molecules can react in any instant.

Using Dalton's model of a chemical reaction, we can classify a reaction as either: (1) a combination of atoms, (2) a disunion of atoms (decomposition), (3) a replacement of atoms in a substance, or (4) a displacement of atoms.

Reactions can also be classified according to the mechanism that occurs during the reaction: (1) electron transfer, (2) proton transfer, (3) electron pair donation, or (4) precipitation.

## Objectives

By the end of this chapter, you should be able to do the following:

1. Given a word equation, write a balanced formula equation. (Question 2)

2. Given a balanced equation, state the mole ratios of reactants and products. (Question 3)

3. Given a balanced equation and the number of moles of one substance, calculate the number of moles of any other substance needed or produced in the reaction. (Question 4)

4. Given a balanced equation and the number of molecules of one substance, calculate the number of molecules of any other substance needed or produced in the reaction. (Question 5)

5. Given a balanced equation and the number of moles of one substance, calculate the weight of any other substance needed or produced in the reaction. (Question 7)

6. Given a balanced equation (including energy terms) and the number of moles of one substance, calculate the energy needed or produced in the reaction. (Question 8)

7. In specific examples, state the effect on the rate of reaction of the following factors: temperature, concentration, catalysts, surface areas. (Questions 10–16)

8. Given a balanced equation, classify it in terms of a rearrangement of atoms as either: combination, decomposition, replacement, or displacement. (Question 17)

9. Given a balanced equation, classify it in terms of a mechanism as either: electron transfer, proton transfer, electron pair donation, or precipitation. (Questions 18, 19)

## Questions

1. Explain the meaning of the following terms (if possible, give examples): *formula equations; orientation of molecular collisions; bond breaking; surface area; catalyst; reaction rate; concentration; combination; decomposition; replacement; displacement; Dalton's model of a chemical reaction; acid; base; neutralization; unshared electron pair; proton; electron transfer; proton transfer; electron pair donation; precipitation; solubility.*

2. Qualitative analysis is a technique in chemistry used to identify unknown substances. Many metal ions in solution can be identified by the characteristic solubilities of their compounds, by the color of their solutions, and by the color of the precipitates they form. The following word equations represent reactions commonly performed in qualitative analysis. Write the balanced formula equations.
(a) aqueous barium nitrate + aqueous ammonium carbonate $\rightleftharpoons$ aqueous ammonium nitrate + solid barium carbonate
(b) aqueous copper(II) nitrate + aqueous ammonium sulfide $\rightleftharpoons$ solid copper(II) sulfide + aqueous ammonium nitrate
(c) solid lead(II) sulfate + acetic acid $\rightleftharpoons$ aqueous lead(II) acetate + sulfuric acid
(d) solid cobalt(II) sulfide + nitric acid $\rightleftharpoons$ aqueous cobalt(II) nitrate + gaseous nitrogen monoxide + solid sulfur + water
(e) solid nickel(II) sulfide + nitric acid $\rightleftharpoons$ aqueous nickel(II) nitrate + gaseous nitrogen monoxide + solid sulfur + water

3. Balance the following equations. Indicate the mole relationship of reactants to products.
(a) $KBrO_3(s) \xrightleftharpoons{\text{heating}} 2KBr(s) + 3O_2(g)$
(b) $Mn(OH)_3(s) + HNO_3(aq) \rightleftharpoons Mn(NO^2)_2(aq) + H_2O(l)$
(c) $BF_3(s) + H_2O(l) \rightleftharpoons H_3BO_3(aq) + HF(g)$
(d) $PH_3(g) \xrightleftharpoons{\text{heating}} P_4(s) + H_2(g)$
(e) $N_2H_4(g) + H_2O_2(l) \rightleftharpoons N_2(g) + H_2O(l)$

4. Silicon tetrafluoride can be prepared by heating a mixture of calcium fluoride, silicon dioxide, and sulfuric acid.

$$2CaF_2(s) + SiO_2(s) + 2H_2SO_4(l) \rightleftharpoons 2CaSO_4(s) + SiF_4(g) + 2H_2O(l)$$

(a) If 2.00 moles of $CaF_2$ reacted, how many moles of $CaSO_4$ were produced? (b) If 6.00 moles of $CaF_2$ reacted, how many moles of $SiF_4$ were produced? (c) If 1.50 moles of $CaF_2$ reacted, how many moles of $H_2O$ were produced? (d) If 0.358 mole of $CaF_2$ reacted, how many moles of $H_2O$ were produced? (e) How many moles of $SiO_2$ are needed to react with 0.772 mole of $CaF_2$? (f) If 0.0319 mole of $SiF_4$ is produced, how many moles of $SiO_2$ are reacted? (g) If 4.71 moles of $CaSO_4$ are produced, how many moles of $SiF_4$ are also produced?

5. The oxygen content of metal oxides can be determined by reacting the metal oxide with bromine trifluoride. For example:

$$3TiO_2 + 4BrF_3 \rightleftharpoons 3TiF_4 + 2Br_2 + 3O_2$$

(a) If 3 formula units of $TiO_2$ react, how many molecules of $BrF_3$ react? (b) If 12 formula units of $TiO_2$ react, how many molecules of $TiF_4$ are produced? (c) If 24 formula units of $TiO_2$ react, how many molecules of $Br_2$ are produced? (d) If 127 formula units of $TiO_2$ react, how many molecules of $O_2$ are produced? (e) If $2 \times 10^{23}$ molecules of $BrF_3$ react, how many molecules of $O_2$ are produced?

6. Why have chemists invented the mole concept? What advantages are there in using it?

7. Iodine is found in the crust of the earth as part of different compounds. It usually is found as the iodide or iodate ion. The free element can be prepared from Chilean sodium iodate as follows:

$$2NaIO_2 + 5NaHSO_3 \rightleftharpoons I_2 + 2Na_2SO_4 + 3NaHSO_4 + H_2O$$

(a) What is the weight of 1.00 mole of each reactant and of each product? (b) If 2.00 moles of $NaIO_3$ react, what weight of $I_2$ will be produced? (c) If 2.00 moles of $NaIO_3$ react, what weight of $NaHSO_3$ will react? (d) If 5.50 moles of $NaIO_3$ react, what weight of $Na_2SO_4$ will be produced? (e) If 5.00 moles of $I_2$ are to be produced, what weight of $NaIO_3$ is needed? (f) If 2.75 moles of $I_2$ are to be produced, what weight of $NaHSO_3$ is needed?

8. When diborane, $B_2H_6$, burns, it releases large amounts of energy per unit weight. For this reason it has been investigated as a rocket fuel. The reaction is:

$$B_2H_6(g) + 3O_2(g) \rightleftharpoons B_2O_3(s) + 3H_2O(g) + 482 \text{ kcal/mole}$$

(a) How much energy is released when 1.00 mole of $B_2H_6$ is burned? (b) How much energy is released when 5.00 moles of $B_2H_6$ are burned? (c) How many moles of $B_2H_6$ must burn to produce 1000 kcal of energy? (d) How many moles of $O_2$ must react to produce $1.00 \times 10^{10}$ kcal of energy?

9. (a) What is the surface area of a cube that measures 5.0 cm × 5.0 cm × 5.0 cm? (b) This cube was then broken up into 125 smaller cubes that measured 1.0 cm × 1.0 cm × 1.0 cm. What is their total surface area?

10. Would you expect ¼ teaspoon of sugar to dissolve faster in hot tea or iced tea? Why?

11. Why does whiskey have a more noticeable effect on biological systems than an equal volume of beer?

12. Would you expect grape juice to ferment faster in a cool room or warm room?

13. Place an egg in a large beaker of water at 100°C. After 3 minutes,

examine the inside of the egg. Place another egg in a large beaker of water at 90°C. After 3 minutes, examine the inside of the egg. Compare and explain your observations.

14. On a cool day in Alaska, a fly was observed to behave normally in the sunlight. However, when the sun was covered by a cloud, the fly fell to the ground motionless. Explain.

15. A motorist reported on a cold morning that his car battery was almost "dead." The next day was warm and he reported that the same battery was just like new. Explain.

16. Solid zinc reacts very slowly if at all with solid copper(II) sulfate, yet solid zinc reacts rapidly with an aqueous solution of copper(II) sulfate. Explain.

17. Classify the following reactions as either combination, decomposition, replacement, or displacement.
   (a) $2AsH_3(g) \rightleftharpoons 2As(s) + 3H_2(g)$
   (b) $2Sb(s) + 5F_2(g) \rightleftharpoons 2SbF_5(s)$
   (c) $2Al(s) + Fe_2O_3(s) \rightleftharpoons Al_2O_3(s) + 2Fe(s)$
   (d) $PCl_3(l) + AsF_3(s) \rightleftharpoons PF_3(g) + AsCl_3(s)$
   (e) $As_2O_3(s) + 6H_2(g) \rightleftharpoons 2As(s) + 3H_2O(l)$
   (f) $2NaI(s) + Br_2(l) \rightleftharpoons 2NaBr(s) + I_2(s)$
   (g) $N_2(g) \rightleftharpoons 2N(g)$
   (h) $2NO(g) + O_2(g) \rightleftharpoons 2NO_2(g)$
   (i) $Pb(NO_3)_2(aq) + Na_2SO_4(aq) \rightleftharpoons PbSO_4(s) + 2NaNO_3(aq)$
   (j) $FeCl_2(aq) + (NH_4)_2S(aq) \rightleftharpoons FeS(s) + 2NH_4Cl(aq)$

18. Which of the reactions in Question 17 involve electron transfer?

19. Classify the following reactions as either electron transfer, proton transfer, electron pair donation, or precipitation.
   (a) $N_2H_4 + H_2O \rightleftharpoons N_2H_5^+ + OH^-$
   (b) $2NH_3 \rightleftharpoons NH_4^+ + NH_2^-$
   (c) 
$$\underset{\underset{O}{|}}{\overset{\overset{O}{\|}}{O-S}}+:\underset{H}{\overset{H}{N-H}} \rightleftharpoons \underset{\underset{O}{|}}{\overset{\overset{O}{\|}}{O-S}}-\underset{H}{\overset{H}{N-H}}$$
   (d) $FeCl_2(aq) + (NH_4)_2S(aq) \rightleftharpoons FeS(s) + 2NH_4Cl(aq)$
   (e) $2SO_2(g) + O_2(g) \rightleftharpoons 2SO_3(s)$
   (f) $H_3PO_4(l) + H_2O(l) \rightleftharpoons H_2PO_4^-(aq) + H_3O^+(aq)$
   (g) $S^{2-}(aq) + H_2O(l) \rightleftharpoons HS^-(aq) + OH^-(aq)$
   (h) $Cd^{2+}(aq) + 6NH_3(aq) \rightleftharpoons Cd(NH_3)_6^{2+}(aq)$
   (i) $2Au_2O_3(s) \rightleftharpoons 4Au(s) + 3O_2(g)$

# 11

# Selected elements and compounds

## 11.1 Introduction

In this chapter we shall consider selected reactions of the metals of group IA and the nonmetals of group VIIA because they serve as good examples of the relationship between atomic structure and group properties.

Where might we encounter the group IA or VIIA elements or their compounds outside of the chemical laboratory? The human body may be considered as a complex chemical system whose reactions are controlled in part by these elements or their ions. Sodium, potassium, and chloride ions influence the metabolism of body cells. Small amounts of fluoride ion affect the ability of teeth to resist decay, and iodine helps regulate the thyroid gland.

In Chapter 10 we considered some of the factors that affect the rate of a reaction. In this chapter we shall examine the effect of two factors—state of subdivision and concentration—on the reactions between metals and acids. The final section of this chapter will involve practice in predicting the products of a reaction.

## 11.2 Group IA elements

TABLE 11.1

The atomic structure of the group IA elements (except for hydrogen). Notice that each element has one electron in an s orbital in the valence shell.

The group IA elements are called the *alkali metals*. Hydrogen is not usually considered a member of this group because its properties are quite different from those of the other elements in the group. What is similar about the electronic structure of these elements (Table 11.1)?

|  | Lithium | Sodium | Potassium | Rubidium | Cesium | Francium |
|---|---|---|---|---|---|---|
| Atomic number | 3 | 11 | 19 | 37 | 55 | 87 |
| Valence shell configuration | $2s^1$ | $3s^1$ | $4s^1$ | $5s^1$ | $6s^1$ | $7s^1$ |
| Atomic radius (A) | 1.23 | 1.57 | 2.03 | 2.16 | 2.35 | |
| Ionic radius (A) | 0.93 | 0.95 | 1.33 | 1.48 | 1.69 | |
| Melting point (°C) | 180 | 97.5 | 63.4 | 38.8 | 28.7 | |
| Boiling point (°C) | 1336 | 880 | 760 | 700 | 670 | |
| Density (g/ml) | 0.53 | 0.97 | 0.86 | 1.53 | 1.90 | |

### Physical properties

The group IA elements are solids that can be cut with a knife at room temperature. Their freshly cut surfaces have a metallic luster and the substances conduct heat and electricity well. When heated gently on an asbestos pad, a small piece of the solid melts into a shiny silvery ball before burning. The low melting points of these elements indicate that there is a weak bond between these atoms. The weak bond is due in part to the large size of the metallic atoms.

In the vapor state these elements exist as diatomic molecules, such as $Li_2$, $Na_2$, and $K_2$. Draw atomic orbital diagrams for these molecules. Why is the formula in the vapor not $Na_3$ or $Na_4$?

TABLE 11.2

Some chemical properties of the alkali metals.

### Chemical properties

The elements in group IA react with many nonmetallic elements to form compounds with ionic bonds (Table 11.2). In

|  | Lithium | Sodium | Potassium | Rubidium | Cesium | Francium |
|---|---|---|---|---|---|---|
| Ionization energy (kcal/mole) | 124 | 118 | 100 | 96 | 90 | — |
| Electronegativity | 1.0 | 0.9 | 0.8 | 0.8 | 0.7 | 0.7 |
| Common oxidation number in compounds | 1+ | 1+ | 1+ | 1+ | 1+ | 1+ |
| Reducing agent ability | Strong | Strong | Strong | Strong | Strong | Strong |
| Oxidation potential | +3.04 | +2.71 | +2.92 | +2.92 | +3.08 | |
| Formula of fluoride | LiF | NaF | KF | RbF | CsF | FrF |
| Formula of oxide | $Li_2O$ | $Na_2O$ | $K_2O$ | $Rb_2O$ | $Cs_2O$ | $Fr_2O$ |

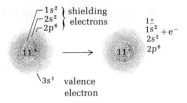

$Na(g) + 118 \text{ kcal} \rightleftharpoons Na^+(g) + e^-$

**FIGURE 11.1**
*Ionization energy of sodium. As with the other alkali metals, sodium loses its valence electron with relative ease.*

these compounds the group IA elements exhibit a 1+ oxidation number since each has one valence electron which can be removed (Fig. 11.1).

The relatively low ionization energy of the group IA elements was discussed in Chapter 5. The electronegativities of the elements are low for similar reasons.

*Reaction with water* The reaction of sodium with water was described in Chapter 9. Potassium reacts in a faster but similar manner, producing hydrogen gas and potassium hydroxide.

$$2K(s) + 2H_2O(l) \rightleftharpoons 2KOH(aq) + H_2(g) \tag{11.1}$$

What is a reasonable mechanism for this reaction? A check of oxidation numbers shows that electrons have been transferred from potassium to hydrogen. Rewriting Eq. 11.1 with oxidation numbers:

$$2K(s) + 2H_2O(l) \rightleftharpoons KOH(aq) + H_2(g) + \text{energy}$$
$$(0) \quad (1+)(2-) \quad (1+)(2-)(1+) \quad (0)$$

We may write half-reactions for the oxidation of potassium and for the reduction of hydrogen.

$$2K(s) \underset{\text{oxidation}}{\rightleftharpoons} 2K^+(aq) + 2e^- \tag{11.2}$$

$$2H^+(aq) + 2e^- \underset{\text{reduction}}{\rightleftharpoons} H_2(g) \tag{11.3}$$

Note that the number of electrons lost by the potassium equals the number of electrons gained by the hydrogen.

Why does potassium react faster with water than does sodium? Since the reaction involves the loss of an electron from atoms of the metal, which metal loses electrons more readily, sodium or potassium? Compare their ionization energies. For sodium it is 118 kcal/mole, and for potassium, 100 kcal/mole. Since potassium has a lower ionization energy, it loses electrons more readily, and it should therefore react more readily. Actually, ionization energy is only one of several factors that influence the rate of this reaction. ■

■ **EXERCISE** *Which of the alkali metals should react slowest and which fastest with water? Lithium, sodium, potassium, rubidium, cesium, or francium? Write the equations for the reactions.*

**Reactions with group VIIA elements: observations and equations**

The reaction between sodium metal and chlorine gas was described in Chapter 6. How does sodium metal react with other elements from group VIIA? How do the other group IA elements react with the group VIIA elements? Consider the reaction between solid potassium and solid iodine. When an equal number of potassium atoms and iodine atoms are heated, a violent reaction occurs. After a few seconds, only a white solid is left.

What is this white solid? When mixed with water, the solid dissolves readily, leaving a clear, colorless solution. When a drop of this solution is put in a flame, the flame turns purple. Other substances that contain potassium ions turn flames purple, so we conclude that the white solid yields *potassium ions* when it dissolves in water. When a drop of the solution of the white solid is added to a few milliliters of silver nitrate, a pale yellow solid forms. This is a test for iodide ions, since previous experiments have shown that iodide ions react with silver nitrate solutions to produce a yellow solid. It is concluded that the white solid contains potassium ions and iodide ions; therefore, potassium and iodine reacted to form potassium iodide.

$$2K(s) + I_2(s) \xrightleftharpoons{\text{heating}} 2KI(s) + \text{energy} \qquad (11.4)$$

What is the mechanism for the reaction? Check the oxidation numbers; the half-reactions are:

$$2K(s) \rightleftharpoons 2K^+(s) + 2e^- \qquad (11.5)$$
$$I_2(s) + 2e^- \rightleftharpoons 2I^-(s) \qquad (11.6)$$

Each potassium atom loses an electron and is oxidized. Each iodine atom gains an electron and is reduced. The atomic orbital diagrams for the valence shell of each atom are:

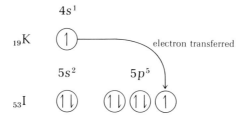

The other metals in group IA undergo similar reactions with the group VIIA elements. *Each reaction involves the loss of an electron by the less electronegative group IA element and the gaining of an electron by the more electronegative group VIIA element.*

■ EXERCISE  Write completed balanced equations for the reactions between (a) sodium and bromine, and (b) cesium and fluorine.

$$2Li(s) + I_2(s) \rightleftharpoons 2LiI(s) + \text{energy} \qquad (11.7)$$
$$2Cs(s) + Cl_2(g) \rightleftharpoons 2CsCl(s) + \text{energy} \qquad (11.8)$$

### Reducing agents

Examine the equation for the reaction between potassium and water (Eq. 11.1). Note that the hydrogen gained electrons and was reduced in oxidation number from 1+ to 0. In the reaction between potassium and iodine, the iodine gained electrons and was reduced from 0 to 1− (Eq. 11.4). Because of its ability to lose an electron easily, the *potassium atom* reduces the hydrogen in water and the iodine atom in iodine. A substance which gives up electrons and causes other substances to undergo reduction is called a *reducing agent*. Potassium and the other group IA elements are strong reducing agents (see Eqs. 11.7 and 11.8). ■

■ EXERCISE   Examine Eqs. 11.7 and 11.8. Which substance in each equation is the reducing agent? If an oxidizing agent causes oxidation, which substance in each equation is the oxidizing agent?

### Heats of formation

Equations 11.4, 11.7, and 11.8 indicate that energy is a product of the reaction. The amount of heat liberated for each mole of substance formed is listed in Table 11.3.

The amount of energy liberated depends on the *relative energy levels* of the reactants and products. A mole of the compound lithium iodide(s) is 64.8 kcal less energetic than the elements lithium metal and iodine(s) from which it is made. A mole of the compound sodium iodide is 68.8 kcal less energetic than the elements from which it is made. In each example given here, the elements (lithium and iodine or sodium and iodine) have a larger amount of chemical energy than the compounds (lithium iodide or sodium iodide). The relative energy levels of each iodide compound are shown in Fig. 11.2. Do you note any trend in going down the group?

The energy liberated or absorbed during the formation of a mole of a compound from its elements is known as the heat of formation, or enthalpy of formation, for that compound. It is symbolized by $\Delta H_F$.

**TABLE 11.3**

The heats of formation in kcal/mole of compounds formed between the group IA and group VIIA elements. The negative sign indicates that energy is given off when the compounds are formed.

|          | Lithium | Sodium | Potassium | Rubidium | Cesium |
|----------|---------|--------|-----------|----------|--------|
| Fluorine | LiF     | NaF    | KF        | RbF      | CsF    |
|          | −146    | −136   | −134      | −131     | −127   |
| Chlorine | LiCl    | NaCl   | KCl       | RbCl     | CsCl   |
|          | −97.7   | −98.2  | −104      | −103     | −104   |
| Bromine  | LiBr    | NaBr   | KBr       | RbBr     | CsBr   |
|          | −83.7   | −86.0  | −93.7     | −93.0    | −94.3  |
| Iodine   | LiI     | NaI    | KI        | RbI      | CsI    |
|          | −64.8   | −68.8  | −78.3     | −78.5    | −80.5  |

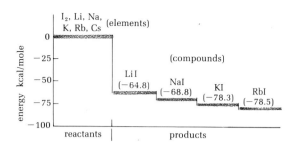

FIGURE 11.2
The relative energy levels between the group IA elements and the compounds they form with iodine.

Whenever a group IA metal reacts with a group VIIA nonmetal, there is a release of heat. Reactions which release heat are said to be *exothermic*. In an exothermic reaction, the chemical system loses energy, and the value of $\Delta H_F$ has a negative sign. Some reactions absorb energy. These reactions are endothermic and $\Delta H_F$ has a positive value since the system has more energy after the reaction than it did previously.

■ EXERCISE  Are heats of formation periodic properties? Examine Table 11.3.

## Solubility

Most compounds of the group IA elements are water soluble. The solubility is due in part to the large size of the group IA ions and their low charge of 1+. The forces between ions in a solid crystal increase as the size of the ions decreases. The forces between the ions in a crystal also increase as the charge on the ions increases. Thus, salts become less soluble as the charge on the ions increases and as their size decreases.

## Electrolysis

In Chapter 6 we saw that molten salts of ionic compounds are good conductors of electricity. What happens when electricity is passed through lithium chloride?

When solid lithium chloride is heated to 610°C two things are observed (Fig. 11.3). The lithium chloride melts and the light bulb begins to glow. We can conclude that the molten lithium chloride conducts electricity. After a few minutes a yellow-green gas forms at one electrode. Analysis of this gas shows it to have the properties of chlorine. At the other electrode a silvery metallic liquid is observed. Analysis of the substance shows it to have the properties of lithium. Evidently, lithium chloride is being decomposed to lithium and chlorine by electrical energy. *The decomposition of a compound by electricity is called electrolysis.*

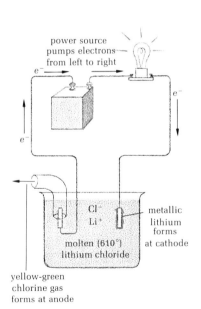

FIGURE 11.3
When solid lithium chloride melts, the light bulb glows brightly, indicating that an electrical current is flowing. After a while a green gas is noticed at one electrode and a metallic liquid at the other.

$$2\text{LiCl}(l) + \text{energy} \rightleftharpoons 2\text{Li}(l) + \text{Cl}_2(g) \tag{11.9}$$

The equation can also be written in ionic form since molten LiCl consists of lithium ions, $\text{Li}^+$, and chloride ions, $\text{Cl}^-$.

$$2\text{Li}^+(l) + 2\text{Cl}^-(l) + \text{energy} \rightleftharpoons 2\text{Li}(l) + \text{Cl}_2(g)$$

How does electrical energy flow through molten lithium chloride? The power source pumps electrons into the positive electrode. The excess of negatively charged particles causes this electrode to attract positive lithium ions. These ions flow toward the negative electrode, where they pick up an electron and form neutral lithium atoms.

$$\text{Li}^+(l) + e^- \rightleftharpoons \text{Li}^0 \tag{11.10}$$

The formation of a large number of lithium atoms accounts for the appearance of lithium at the negative electrode. This reaction also accounts for what happens to the electrons that enter the electrode on the opposite side. The lithium ions were *reduced* from 1+ to 0 at the *cathode*.

The power source removes electrons from one electrode, giving it a positive charge. The negative chloride ions in the molten salt are attracted to this electrode. At this electrode the chloride ions lose an electron and form neutral atoms which form diatomic molecules.

$$2\text{Cl}^-(l) \rightleftharpoons \text{Cl}_2(g) + 2e^- \tag{11.11}$$

The formation of a large number of chlorine molecules accounts for the observation of a yellow-green gas at one electrode. This reaction also explains where the electrons that leave the electrode come from. The chloride ions are *oxidized* from 1− to 0 at the *anode*.

The two products of the reaction, chlorine and lithium, must be kept separated. Why? Why does molten lithium chloride conduct electricity when solid lithium chloride does not? The flow of electricity is the movement of charged particles. In solid lithium chloride there are charged particles, $\text{Li}^+$ and $\text{Cl}^-$, but they are held in position and are *not* free to move. Hence the solid does not conduct. In molten lithium chloride the ions in the liquid are less strongly attracted to each other, so they are free to move about. Under the influence of electrical energy the positive lithium ions move toward the negative electrode and the negative chloride ions move toward the positive electrode (Fig. 11.4). ■

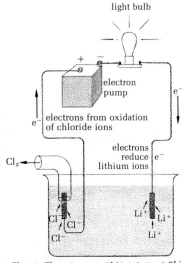

**FIGURE 11.4**

*In molten lithium chloride the ions are free to move about. Negative chloride ions migrate to the positive electrode and positive lithium ions migrate to the negative electrode.*

■ **EXERCISE** *Write ionic equations for the electrolysis of (a) sodium fluoride and (b) potassium iodide.*

| TABLE 11.4 | | Fluorine | Chlorine | Bromine | Iodine | Astatine |
|---|---|---|---|---|---|---|
| The atomic structure of the group VIIA elements. | Atomic number | 9 | 17 | 35 | 53 | 85 |
| | Valence shell configuration | $2s^2 2p^5$ | $3s^2 3p^5$ | $4s^2 4p^5$ | $5s^2 5p^5$ | $6s^2 6p^5$ |
| | Atomic radius (A) | 0.72 | 0.99 | 1.14 | 1.33 | |
| | Ionic radius (A) | 1.36 | 1.81 | 1.95 | 2.16 | |
| | Melting point (°C) | −220 | −101 | −7.2 | 114 | |
| | Boiling point (°C) | −188 | −34 | 58 | 184 | |
| | Density (g/ml) | 1.11 (liquid) | 1.56 (liquid) | 3.12 | 4.93 | |

## 11.3 Group VIIA elements

The group VIIA elements, or the *halogens*, are fluorine, chlorine, bromine, iodine, and astatine. Astatine is a rare, radioactive element, and many of its properties are unknown. The group VIIA elements and their structures are shown in Table 11.4. What is similar about their electronic structure?

**Physical properties**

The group VIIA elements are characteristic nonmetals. At room temperature, fluorine and chlorine are gases, bromine is a volatile liquid, and iodine is a soft solid. Each element exists as a diatomic molecule. Whether as solids, liquids, or gases, they are all poor conductors of heat and electricity. The solids are soft, indicating weak forces of attraction between molecules.

Chlorine, bromine, and iodine are each slightly soluble in water. A solution of chlorine in water—chlorine water—has a slight yellow-green color; bromine water is brownish; and iodine water is yellow-brown. The *nonpolar halogens* are much more soluble in *nonpolar solvents*, such as carbon tetrachloride or carbon disulfide. In carbon tetrachloride, chlorine is pale yellow, bromine is brown, and iodine is purple.

When carbon tetrachloride is added to aqueous solutions of chlorine, bromine, and iodine and the systems are vigorously mixed, the water layers become colorless. The carbon tetrachloride layers take on the particular color of each halogen (Fig. 11.5). What has happened? The halogens have been ex-

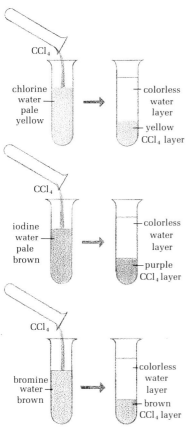

**FIGURE 11.5**

*Solubility of halogens. When carbon tetrachloride is added to aqueous solutions of chlorine, bromine, and iodine, and the systems are vigorously mixed, the water layers become colorless. The nonpolar carbon tetrachloride layers take on the particular color of each halogen.*

■ EXERCISE  *What element other than bromine is a liquid at room temperature?*

tracted from the water layer by the carbon tetrachloride. The halogens move from the polar liquid in which they are less soluble to the nonpolar liquid in which they are more soluble.

$$I_2(aq) \rightleftharpoons I_2(CCl_4)$$ ■

**Chemical properties**

The halogens react with most metals to form ionic compounds. The atomic orbital diagrams of the halogens

indicate that they have room for one more electron in their valence shell. Consequently, they readily form the ions $F^-$, $Cl^-$, $Br^-$, and $I^-$ (Table 11.5). In addition, the halogens may share electrons with electronegative elements such as oxygen or nitrogen. In these compounds the common oxidation states are 1+, 3+, 5+, and 7+.

Because halogens are both reactive and volatile, they should be handled with care. They react readily with many compounds in the body.

*Reaction with water* When chlorine is bubbled through water, the water takes on a slight pale-yellow color (see Section 9.4). This is evidence that some chlorine dissolves. Bromine and iodine are slightly soluble in water. The solubility of these elements is:

$Cl_2$: 0.091 mole in a liter of solution
$Br_2$: 0.21 mole in a liter of solution
$I_2$: 0.0013 mole in a liter of solution

**TABLE 11.5**

Some chemical properties of the halogens.

| | | Fluorine | Chlorine | Bromine | Iodine | Astatine |
|---|---|---|---|---|---|---|
| Ionization energy (kcal/mole) | | 402 | 300 | 273 | 241 | |
| Electronegativity | | 4.0 | 3.0 | 2.8 | 2.5 | |
| Common oxidation number in compounds | | 1− | 1−, 1+, 3+, 5+, 7+ | 1−, 1+, 5+ | 1−, 1+, 5+, 7+ | |
| Oxidizing agent ability | | Very strong | Strong | Strong | Strong | |
| Oxidation potential | | −2.87 | −1.36 | −1.07 | −0.53 | |
| Sodium compound | | NaF | NaCl | NaBr | NaI | |
| Aluminum compound | | $AlF_3$ | $AlCl_3$ | $AlBr_3$ | $AlI_3$ | |

When a solution of silver nitrate is added to a solution of chlorine, a precipitate of AgCl forms. Where did the Cl⁻ ion come from? Chemists believe that chlorine reacts with water as follows:

$$Cl_2(aq) + H_2O(l) \rightleftharpoons H^+(aq) + Cl^-(aq) + HClO(aq) \qquad (11.12)$$

This is an example of an *auto-oxidation reduction*, or *disproportionation*. One chlorine atom is oxidized from 0 to 1+ and one chlorine atom is reduced from 0 to 1−.

$$Cl_2(s) + H_2O(l) \rightleftharpoons H^+(aq) + Cl^-(aq) + HClO(aq) \qquad (11.13)$$
(0)   (1+)(2−)   (1+)   (1−)   (1+)(1+)(2−)

The compound HClO is hypochlorous acid and it is a good oxidizing agent. Solutions containing hypochlorite ions are sold commercially as household bleaching agents. Solutions of chlorine or bromine are often used as swimming pool disinfectants. These substances are used as bleaches and disinfectants because of their ability to gain electrons and thus oxidize other substances. For example, a colored compound (dye) may be oxidized to a colorless compound, and thereby bleached.

*Reaction with hydrogen*  When a burning hydrogen flame is thrust into a bottle containing heated bromine vapor, the hydrogen continues to burn. The bromine reacts with the hydrogen to form hydrogen bromide gas.

$$H\!:\!H + :\!\ddot{\underset{..}{Br}}\!:\!\ddot{\underset{..}{Br}}\!: \rightleftharpoons 2H\!:\!\ddot{\underset{..}{Br}}\!: \qquad \Delta H_F = -8.65 \text{ kcal/mole}$$

This reaction is similar to the one between hydrogen and chlorine and hydrogen and fluorine.

$$H\!:\!H + :\!\ddot{\underset{..}{Cl}}\!:\!\ddot{\underset{..}{Cl}}\!: \rightleftharpoons 2H\!:\!\ddot{\underset{..}{Cl}}\!: \qquad \Delta H_F = -22.0 \text{ kcal/mole}$$

$$H\!:\!H + :\!\ddot{\underset{..}{F}}\!:\!\ddot{\underset{..}{F}}\!: \rightleftharpoons 2H\!:\!\ddot{\underset{..}{F}}\!: \qquad \Delta H_F = -64.2 \text{ kcal/mole}$$

The reaction between hydrogen and solid iodine is not exothermic. Energy equivalent to 6.20 kcal is needed to produce a mole of hydrogen iodide:

$$H\!:\!H + :\!\ddot{\underset{..}{I}}\!:\!\ddot{\underset{..}{I}}\!: \longrightarrow 2H\!:\!\ddot{\underset{..}{I}}\!: \qquad \Delta H_F = +6.20 \text{ kcal/mole}$$

Heats of formation serve to indicate how stable a compound is compared to its constituent elements. *In order to decompose a compound, the amount of energy liberated (or used up) during its formation must be put back into (or taken out of) the system.* The energy liberated when a bond forms must be put back into the system before the bond can be broken. Since the

■ EXERCISE  Notice that the heats of formation of the hydrogen halides decrease in regular fashion down the group. (a) What relationship can you find between heat of formation and electronegativity? (b) Predict the trend in heat of formation for $H_2O$, $H_2S$, $H_2Se$, $H_2Te$, and $H_2Po$.

formation of HF liberates 64.2 kcal, the decomposition of a mole of hydrogen fluoride would require 64.2 kcal of energy.

$$HF(g) + 64.2 \text{ kcal} \rightleftharpoons \tfrac{1}{2}H_2(g) + \tfrac{1}{2}F_2(g) \tag{11.14}$$

Since the formation of HI requires 6.2 kcal, the decomposition of hydrogen iodide would liberate 6.2 kcal.

$$HI(g) \rightleftharpoons \tfrac{1}{2}H_2(g) + \tfrac{1}{2}I_2(s) + 6.2 \text{ kcal} \quad \blacksquare \tag{11.15}$$

In Section 9.4 we saw that hydrogen chloride dissolves readily in water to form an acidic solution. The reaction is a proton transfer type.

$$H\!:\!\ddot{\underset{..}{Cl}}\!:\!(g) + H\!:\!\ddot{\underset{..}{O}}\!:\!H(l) \rightleftharpoons H\!:\!\overset{H\;+}{\underset{..}{\ddot{O}}}\!:\!H(aq) + :\!\ddot{\underset{..}{Cl}}\!:^{-}(aq) \tag{11.16}$$

When HCl, HBr, or HI dissolves in water, the resulting solution is *strongly acidic* because of the large number of $H_3O^+$ ions produced. The solution turns litmus bright red and reacts with zinc metal rapidly. Almost *all of the hydrogen halide* molecules react with water molecules, but to varying degrees. Because solutions of HF are *weakly acidic,* only a *low percentage of HF molecules react* with water molecules. Recall that the heat of formation of HF is high. The bond between hydrogen and fluorine is strong and therefore only a few hydrogen ions can be transferred to water molecules.

$$HF(aq) + H_2O(l) \rightleftharpoons H_3O^+(aq) + F^-(aq) \tag{11.17}$$

### Alkali metal halide solutions

*Reactions with silver nitrate solutions*  Silver nitrate aqueous solution is often used to test solutions for the presence of dissolved halide ions. We saw in Section 10.4 that when HCl and $AgNO_3$ are mixed, a white precipitate forms. What happens when a solution of $AgNO_3$ is added to clear, colorless solutions of KF, KCl, KBr, and KI? When three drops of clear colorless silver nitrate solution were added to these solutions, the following observations were made (Fig. 11.6):

1. The $KF(aq) + AgNO_3(aq)$ mixture remained clear and colorless.
2. The $KCl(aq) + AgNO_3(aq)$ mixture turned a cloudy white.
3. The $KBr(aq) + AgNO_3(aq)$ mixture turned a cloudy pale yellow.
4. The $KI(aq) + AgNO_3(aq)$ mixture turned a cloudy yellow.

How can these observations be explained? The mixture of $KF(aq)$ and $AgNO_3(aq)$ remains clear and colorless; there is no apparent reaction. The mixture of $KCl(aq)$ and $AgNO_3(aq)$ forms a white precipitate. We saw in Section 10.4 that $HCl(aq)$ and $AgNO_3(aq)$ react to form solid AgCl. We also might expect

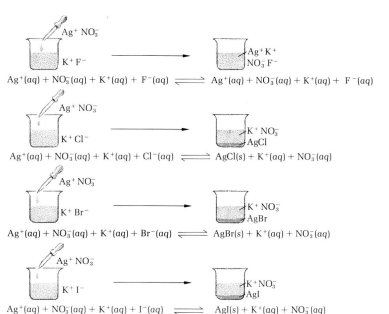

FIGURE 11.6
Aqueous silver nitrate was added to solutions of KF, KCl, KBr, and KI. In the latter three systems precipitates were formed. In each of these three cases the silver ion combined with a halide ion to form a solid.

KCl(aq) and AgNO$_3$(aq) to react and form solid AgCl since both solutions contain the ions Ag$^+$ and Cl$^-$, which react to form the solid AgCl.

$$\text{KCl(aq)} + \text{AgNO}_3\text{(aq)} \rightleftharpoons \text{AgCl(s)} + \text{KNO}_3\text{(aq)} \quad (11.18)$$

The potassium bromide and potassium iodide also react with silver nitrate to form precipitates.

$$\text{KBr(aq)} + \text{AgNO}_3\text{(aq)} \rightleftharpoons \text{AgBr(s)} + \text{KNO}_3\text{(aq)} \quad (11.19)$$
$$\text{KI(aq)} + \text{AgNO}_3\text{(aq)} \rightleftharpoons \text{AgI(s)} + \text{KNO}_3\text{(aq)} \quad (11.20)$$

Each of these reactions is a precipitation reaction between the positive silver ion and a negative halide ion.

$$\text{Ag}^+\text{(aq)} + \text{Cl}^-\text{(aq)} \rightleftharpoons \text{AgCl(s)} \quad (11.21)$$
$$\text{Ag}^+\text{(aq)} + \text{Br}^-\text{(aq)} \rightleftharpoons \text{AgBr(s)} \quad (11.22)$$
$$\text{Ag}^+\text{(aq)} + \text{I}^-\text{(aq)} \rightleftharpoons \text{AgI(s)} \quad (11.23)$$

■ EXERCISE Aqueous silver nitrate forms precipitates when added to other solutions that contain halide ions. Complete and balance the following equations:
AgNO$_3$(aq) + MgCl$_2$(aq) $\rightleftharpoons$
AgNO$_3$(aq) + AlBr$_3$(aq) $\rightleftharpoons$

The potassium ion and nitrate ion are in solution initially and remain there. They do not actively participate in the reaction. Alkali metal ions other than potassium could have been used. ■

*Reactions with halogens* Halogens in the elemental state are diatomic molecules which readily accept electrons and become negative halide ions (see Chapter 6).

$$\text{Cl}_2 + 2e^- \rightleftharpoons 2\text{Cl}^- \quad (11.24)$$

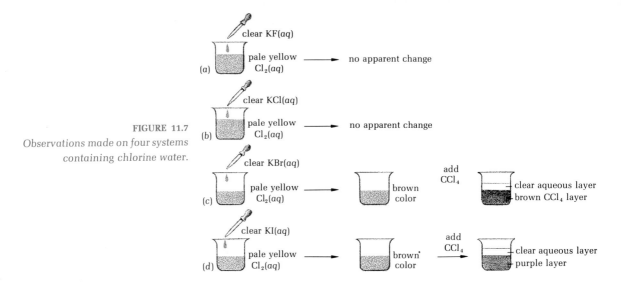

FIGURE 11.7
Observations made on four systems containing chlorine water.

The ease with which halogens gain electrons is not the same for $F_2$, $Cl_2$, $Br_2$, or $I_2$. Consider an experiment designed to measure the ease with which the halogen atoms gain electrons to become halide ions. Recall that the halide ions, $Cl^-$, $Br^-$, and $I^-$, are colorless in aqueous solution but that chlorine water is yellow, bromine water is red-brown, and iodine water is brown. In carbon tetrachloride, chlorine is yellow, bromine is brown, and iodine is purple. When aqueous solutions of KF, KCl, KBr, and KI were added to four different test tubes of chlorine water, the following observations were made (Fig. 11.7):

1. $Cl_2$(aq) + KF(aq): No apparent reaction.
2. $Cl_2$(aq) + KCl(aq): No apparent reaction.
3. $Cl_2$(aq) + KBr(aq): A brown color appears as the two solutions mix. When a few milliliters of carbon tetrachloride are mixed in, the water layer becomes colorless and the carbon tetrachloride layer turns brown. This is evidence that molecular bromine, $Br_2$, was formed.
4. $Cl_2$(aq) + KI(aq): A brown color appears as the two solutions mix. When a few milliliters of carbon tetrachloride are mixed in, the water layer becomes colorless and the carbon tetrachloride layer turns purple. This is evidence that molecular iodine, $I_2$, was formed.

How can these observations be explained? In cases 1 and 2, assume that there was no reaction since there was no visible change. In case 3, the presence of a brown color in the water and then again in the carbon tetrachloride indicates the production of bromine, $Br_2$, from KBr. Evidently, the chlorine atoms in $Cl_2$(aq) accept electrons from the bromide ions in KBr(aq), forming chloride ions and bromine.

$$2KBr(aq) + Cl_2(aq) \rightleftharpoons 2KCl(aq) + Br_2(aq) \qquad (11.25)$$

If we omit the potassium ion from the equation, we have:

$$2Br^-(aq) + Cl_2(aq) \rightleftharpoons 2Cl^-(aq) + Br_2(aq) \qquad (11.26)$$

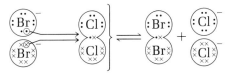

Chlorine atoms have accepted electrons from the negative bromide ions. This is an electron-transfer reaction.

Why do chlorine atoms attract electrons away from bromide ions? Which atoms have the greater attraction for electrons, chlorine or bromine? The electronegativity of chlorine is 3.0 and for bromine, 2.8. The ionization energy of chlorine is 300 kcal/mole; for bromine it is 273 kcal/mole. Chlorine atoms have a greater attraction for electrons than bromine atoms; thus one may conclude that electrons will flow from $Br^-$ to $Cl_2$ when they collide in solution.

In case 4, the presence of a brown color in the water and a purple color in the carbon tetrachloride indicates the production of iodine, $I_2$, from KI.

$$2KI(aq) + Cl_2(aq) \rightleftharpoons 2KCl(aq) + I_2(aq) \qquad (11.27)$$
$$2I^-(aq) + Cl_2(aq) \rightleftharpoons 2Cl^-(aq) + I_2(aq) \qquad (11.28)$$

Again we consider this as an electron-transfer reaction. Electrons are transferred from the iodide ions to the chlorine atoms.

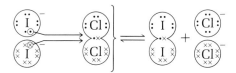

Why doesn't the mixture of aqueous potassium fluoride and chlorine water produce a reaction?

$$KF(aq) + Cl_2(aq) \rightleftharpoons \text{no reaction}$$
$$2KCl(aq) + Cl_2(aq) \rightleftharpoons 2KCl(aq) + Cl_2(aq) \qquad (11.29)$$

We see that *in a "contest" for electrons, chlorine wins over bromine and iodine, but not over fluorine.* In competition for electrons between bromine and iodine, the more electronegative bromine wins. *The ability of group VIIA atoms to gain*

■ EXERCISE  Write the potential products of the following reactions. If no reaction is expected, label "no reaction."
(a) NaCl(aq) + Br$_2$(aq) $\rightleftharpoons$
(b) LiBr(aq) + Cl$_2$(aq) $\rightleftharpoons$
(c) LiBr(aq) + Br$_2$(aq) $\rightleftharpoons$
(d) RbI(aq) + Br$_2$(aq) $\rightleftharpoons$

electrons or to be reduced follows the same trend as electronegativity and ionization energy.

Electronegativity        F > Cl > Br > I
Ionization energy        F > Cl > Br > I
Tendency to undergo reduction    F > Cl > Br > I  ■

## 11.4 Reactions of metals with acids

One characteristic property of aqueous solutions of acids is that they react with and dissolve most metals. Is this characteristic true of all acids? Do acids react only with certain metals? What factors affect the rate of these reactions?

**Changing the metal**

The effect of hydrochloric acid on four different metals can be studied by the following means. Place about 10 ml of dilute acid into each of four test tubes. To the first test tube add a small piece of sodium. *Caution:* Use a piece that is somewhat smaller than a grain of rice. Add the sodium with tongs and use extreme care. To the second test tube add magnesium metal, to the third add nickel, and to the fourth add copper (Fig. 11.8).

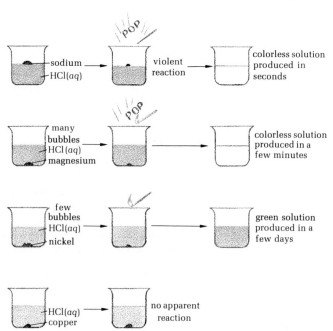

FIGURE 11.8
The reaction between dilute hydrochloric acid and four different metals. (a) The sodium reacts rapidly, producing a combustible gas. (b) The magnesium reacts rapidly, producing a combustible gas. (c) The nickel reacts slowly, producing a gas, but under these conditions it fails to ignite. (d) The copper does not appear to react.

**TABLE 11.6**

*Activity series: The ease with which metals replace hydrogen from an acid.*

| | | |
|---|---|---|
| **More active metals** | K<br>Ba<br>Sr<br>Ca<br>Na | Metal reacts violently with acid to produce salt plus hydrogen. |
| | Mg<br>Al<br>Mn<br>Zn<br>Cr | Metal reacts rapidly with acid to produce salt plus hydrogen. |
| | Fe<br>Cd<br>Co<br>Ni<br>Sn<br>Pb | Metal reacts moderately to slowly with acid to produce salt plus hydrogen. |
| | H | Reference |
| **Less active metals** | Cu<br>Hg<br>Ag<br>Pt<br>Au | Metal does not replace hydrogen in an acid. |

Test for hydrogen gas by holding a burning splint over the mouth of each test tube. You should observe that:

1. The sodium reacts *violently* with the acid. The metal moves about the top of the solution and rapidly disappears. The burning splint produces a "pop" when put near the mouth of the test tube.

2. The magnesium metal slowly sinks into the acid as colorless bubbles begin rising from the surface of the metal. The burning splint at the mouth of the test tube produces a "pop." In a short period of time the magnesium disappears.

3. The nickel metal sinks into the acid. In a short time *a few colorless bubbles* begin rising from the surface of the metal. The burning splint at the mouth of the test tube fails to produce any noticeable effect. In a few hours the solution turns green but some solid is left at the bottom of the tube.

4. The copper metal sinks in the acid. No reaction is noticeable even after several hours.

The results of the experiment indicate that metals have different tendencies to react with acid. The relative rates of reaction are similar to the activity series for the metals developed in Chapter 9. Compare Table 11.6 with Table 9.2.

When a metal such as magnesium is added to an acid, it appears to replace the hydrogen in the acid.

$$Mg(s) + 2HCl(aq) \rightleftharpoons MgCl_2(aq) + H_2(g) \qquad (11.30)$$

Writing the equation in terms of ions, we see that chloride ions are not involved in the reaction:

$$Mg(s) + 2H^+(aq) + 2Cl^-(aq) \rightleftharpoons Mg^{2+}(aq) + 2Cl^-(aq) + H_2(g)$$
$$Mg(s) + 2H^+(aq) \rightleftharpoons Mg^{2+}(aq) + H_2(g) \qquad (11.31)$$

Equation 11.31 shows that electrons are transferred from magnesium atoms to aqueous hydrogen ions.

$$Mg + \begin{Bmatrix} \rightarrow H^+(aq) \\ \rightarrow H^+(aq) \end{Bmatrix} \rightleftharpoons Mg^{2+}(aq) + \begin{pmatrix} H \\ \cdots \\ H \end{pmatrix} \qquad (11.32)$$

■ **EXERCISE** Use Table 11.6 to determine which of the following metals could theoretically serve as a construction material in a railroad tank car used to carry solutions of hydrochloric acid: tin, aluminum, silver, calcium, copper, iron, platinum. Which of these metals do you think might be too expensive?

■ **EXERCISE** Complete and balance the following equations. If no reaction, write no reaction.
(a) Ca(s) + HBr(aq) $\rightleftharpoons$
(b) Ba(s) + HBr(aq) $\rightleftharpoons$
(c) Al(s) + HBr(aq) $\rightleftharpoons$
(d) Ag(s) + HBr(aq) $\rightleftharpoons$

The other metals that react with acids do so in a similar manner.

$$2Na(s) + 2HCl(aq) \rightleftharpoons 2NaCl(aq) + H_2(g) \qquad (11.33)$$
$$Ni(s) + 2HCl(aq) \rightleftharpoons NiCl_2(aq) + H_2(g) \qquad (11.34)$$

During the reaction, the metal atom is removed from the crystal lattice (endothermic process), ionized (endothermic process), and dissolved (exothermic process) (Fig. 11.9). Evidently, the energy requirement of these processes for copper is so high that it does not react with the aqueous hydrogen ion. (Copper can be dissolved by the aqueous nitrate ion under appropriate conditions.) ■

**Changing the acid**

Do all acids react with the same metal at the same rate? Add equal amounts of identical-size pieces of zinc to each of three test tubes, one containing HCl, another $H_2SO_4$, and the third $HCH_3CO_2$(acetic acid). The reactions are as follows:

$$Zn(s) + 2HCl(aq) \underset{}{\overset{fast}{\rightleftharpoons}} ZnCl_2(aq) + H_2(g) \qquad (11.35)$$

$$Zn(s) + H_2SO_4(aq) \underset{}{\overset{fast}{\rightleftharpoons}} ZnSO_4(aq) + H_2(g) \qquad (11.36)$$

$$Zn(s) + 2HCH_3CO_2(aq) \underset{}{\overset{slow}{\rightleftharpoons}} Zn(CH_3CO_2)_2(aq) + H_2(g) \quad (11.37)$$

Judging by the different rates of reaction, acetic acid is different from hydrochloric acid and sulfuric acid. What makes acetic acid different? Remember, the metal reacts with aqueous hydrogen ions in solution.

$$Zn(s) + 2H^+(aq) \rightleftharpoons Zn^{2+}(aq) + H_2(g) \qquad (11.38)$$

Only a small percentage of the acetic acid molecules form ions in water. The acetic acid molecules are in equilibrium with the $H^+(aq)$ and $CH_3CO_2^-(aq)$ ions.

$$HCH_3CO_2(aq) \rightleftharpoons H^+(aq) + CH_3CO_2^-(aq) \qquad (11.39)$$

Since the solution contains a substantially smaller number of *aqueous hydrogen ions* than solutions of sulfuric acid or hydrochloric acid, it reacts with zinc at a slower rate. Acids, such as acetic acid, that form fewer aqueous hydrogen ions in water are called *weak acids*. Other common weak acids are $H_2SO_3$ (sulfurous acid), $HNO_2$ (nitrous acid), $HClO$ (hypochlorous acid), $H_2S$ (hydrosulfuric acid), and $H_2CO_3$ (carbonic acid). ■

■ EXERCISE  *A certain bottle of acetic acid solution contains 0.5 mole of acetic acid. If 0.3 percent of the acetic acid molecules react to form ions, how many moles of aqueous hydrogen ions are in the bottle? How many ions is this? Does this number "seem" small to you?*

**Other factors**

Usually a metal will react faster with a more concentrated acidic solution than with a less concentrated one. However, if the acid contains no water at all, then no $H_3O^+$ ions may be present and the acid may be unreactive. As with other solids, metals will usually react faster if they are divided into small pieces. Changing the pressure usually has no effect on reactions involving solids or liquids.

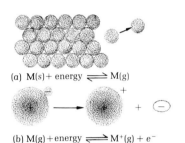

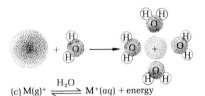

FIGURE 11.9
*Dissolving a metal. During the reaction of a metal with an acid, the metal atom may be thought of as going through the following steps: (a) atom goes from solid state to gaseous state, (b) atom loses one (or more) electrons, (c) metal ion is hydrated.*

## 11.5 Practice in determining products of a reaction

In the first eleven chapters of this book we have discussed many different reactions. Many reactions fit into general categories that are easy to learn. For example, in general, a metal will react with a nonmetal to form a salt. If you are given a periodic chart and asked what the product is when magnesium reacts with bromine, you should be able to:

1. Recognize that magnesium is a metal
2. Recognize that bromine is a nonmetal
3. Determine that a magnesium atom is likely to lose two electrons since it is in group IIA
4. Determine that a bromine atom is likely to gain one electron since it is in group VIIA
5. Write the formula of the product, $MgBr_2$
6. Write the balanced equation for the reaction:

$$Mg(s) + Br_2(l) \rightleftharpoons MgBr_2(s)$$

The statement "metals react with nonmetals to form salts" is, however, a general statement or guideline, and there are exceptions to it. Some metals do not react with nonmetals. In other cases a reaction may occur only under certain conditions of temperature, pressure, concentration, or physical states. For instance, generally, "Metals react with acids to form salts and hydrogen gas." But you have already learned in this chapter that copper does not react with hydrochloric acid and neither do silver and gold.

Chemists develop models to help explain the "exceptions" and predict the existence of other exceptions. The prediction of products of a reaction is difficult at best. Another complicating factor is that many chemical systems undergo several different reactions at the same time. *One cannot know for sure what the products of a reaction will be until the reactants are mixed and the products are analyzed.*

### Metals with nonmetals

In general, metals react with nonmetals to produce salts. The metals tend to lose electrons to nonmetals. Use the periodic chart to determine the possible oxidation number of the products. Some examples of reactions between metals and nonmetals are:

■ EXERCISE *Write the equation for the reaction between aluminum and oxygen.*

$$2Na(s) + S(s) \rightleftharpoons Na_2S(s) \tag{11.40}$$

$$6Li(s) + N_2(g) \rightleftharpoons 2Li_3N(s) \tag{11.41}$$

$$2Al(s) + 3I_2(s) \rightleftharpoons 2AlI_3(s) \tag{11.42}$$

### Elements with oxygen

Under appropriate conditions most elements combine with oxygen to form oxides. Nonmetals such as carbon burn in oxygen to form *nonmetallic oxides.*

$$C(s) + O_2(g) \rightleftharpoons CO_2(g) + energy \tag{11.43}$$

If the amount of oxygen is limited, then the product may consist largely of carbon monoxide.

$$2C(s) + O_2(g) \rightleftharpoons 2CO(g) + energy \tag{11.44}$$

Often, more than one product is possible when a nonmetal reacts with oxygen.

$$S(s) + O_2(g) \rightleftharpoons SO_2(g) + energy \tag{11.45}$$

In the presence of a catalyst, sulfur dioxide may react further with oxygen.

$$2SO_2(g) + O_2(g) \rightleftharpoons 2SO_3(g) + \text{energy} \qquad (11.46)$$

Gasoline and many other fuels contain carbon and hydrogen bonded together. The combustion of gasoline gives carbon dioxide and water.

$$2C_8H_{18}(g) + 25O_2(g) \rightleftharpoons 16CO_2(g) + 18H_2O(l) + \text{energy} \qquad (11.47)$$

The complete combustion of compounds which contain only carbon and hydrogen—hydrocarbons—produces only carbon dioxide and water as products. ■

■ EXERCISE  Write completed and balanced equations for reactions between (a) selenium and oxygen and (b) $C_3H_8(g) + O_2(g)$.

**Oxides with water**

There are two broad categories of oxides: metallic or ionic oxides, and nonmetallic or covalent oxides. When added to water, the *ionic oxides produce basic solutions*. Such solutions contain hydroxide ions. When added to water, the *covalent oxides produce acidic solutions*. Such solutions contain hydronium ions (aqueous hydrogen ions).

The common ionic oxides consist of oxygen bound to the active metals of groups IA and IIA. These substances consist of positive metal ions and negative oxide ions. Sodium oxide reacts with water as follows:

$$Na_2O(s) + H_2O(l) \rightleftharpoons 2NaOH(aq) + \text{energy} \qquad (11.48)$$

A hydrogen ion is transferred from water to the oxide ion.

$$\left[ :\ddot{\text{O}}: \right]^{2-} + \text{H} \!:\! \ddot{\text{O}} \!:\! \text{H} \longrightarrow \left[ :\ddot{\text{O}} \!:\! \text{H} \right]^{-} + \left[ :\ddot{\text{O}} \!:\! \text{H} \right]^{-}$$

Similar reactions are:

$$BaO(s) + H_2O(l) \rightleftharpoons Ba(OH)_2(aq) + \text{energy} \qquad (11.49)$$
$$CrO(s) + H_2O(l) \rightleftharpoons Cr(OH)_2(aq) + \text{energy} \qquad (11.50)$$

The common covalent oxides consist of oxygen bound to the nonmetals or to metals in high oxidation states—for example, $CrO_3$. These substances react with water to produce acidic solutions. For example, aqueous solutions of sulfur dioxide are acidic. The solution contains sulfurous acid produced as follows:

$$SO_2(g) + H_2O(l) \rightleftharpoons H_2SO_3(aq) \qquad (11.51)$$

$$\ddot{\text{O}} \!:\!:\! \ddot{\text{S}} \!:\! \overset{:\ddot{\text{O}}:}{} + \text{H} \!:\! \ddot{\text{O}} \!:\! \text{H} \rightleftharpoons :\ddot{\text{O}} \!:\! \ddot{\text{S}} \!:\! \overset{\text{H}}{\underset{}{:\ddot{\text{O}}:}} \!:\! \ddot{\text{O}} \!:\! \text{H}$$

Solid tetraphosphorus decaoxide, $P_4O_{10}$, dissolves in water to give an acidic solution.

$$P_4O_{10}(s) + 6H_2O(l) \rightleftharpoons 4H_3PO_4(aq) \qquad (11.52)$$

Tetraphosphorus hexaoxide, $P_4O_6$, dissolves in water forming a weakly acidic solution.

$$P_4O_6(s) + 6H_2O(l) \rightleftharpoons 4H_3PO_3(aq) \qquad (11.53)$$

How can you know whether $H_3PO_4$ or $H_3PO_3$ will form? Notice in the three preceding reactions that the oxidation states of sulfur and phosphorus do not change. Phosphorus is 5+ in $P_4O_{10}$ and in $H_3PO_4$. Phosphorus is 3+ in $P_4O_6$ and $H_3PO_3$.

■ EXERCISE  Complete and balance the following:
(a) $SO_3(g) + H_2O(l) \rightleftharpoons$
(b) $K_2O(s) + H_2O(l) \rightleftharpoons$

### Hydrides with water

There are two broad categories of hydrides: metallic or ionic hydrides, and nonmetallic or covalent hydrides. When added to water, the *ionic hydrides produce basic solutions*. When added to water, the *covalent hydrides of group VIA and group VIIA produce acidic solutions*.

In group IA and IIA, metals form ionic hydrides. When mixed with water, the hydride ion gains a hydrogen ion from the water molecule.

$$LiH(s) + H_2O(l) \rightleftharpoons LiOH(aq) + H_2 \qquad (11.54)$$

$$:H^- + H:\overset{..}{\underset{H}{O}}: \rightleftharpoons H:H + :\overset{..}{O}:H^-$$

The reaction of hydrogen halides with water was discussed in Section 11.3. These substances produce acidic solutions. The hydrides of sulfur, selenium, and tellurium are weakly acidic.

$$H_2S(g) + H_2O(l) \rightleftharpoons HS^-(aq) + H_3O^-(aq) \qquad (11.55)$$

Nonpolar methane, $CH_4$, is not appreciably soluble in water, and therefore it does not react with water. Ammonia, being a polar molecule, is quite soluble in water and reacts to give basic solutions.

■ EXERCISE  How many moles of water will react with: (a) 3.00 moles of $BaH_2$? (b) 3.00 moles of HBr?

$$NH_3(g) + H_2O(l) \rightleftharpoons NH_4^+(aq) + OH^-(aq) \qquad ■ \qquad (11.56)$$

### Active metals with water

As we learned in the beginning of this chapter and in Chapter 9, active metals can replace one hydrogen atom in the water molecule.

$$2\text{Na(s)} + 2\text{H}_2\text{O(l)} \rightleftharpoons 2\text{NaOH(aq)} + \text{H}_2\text{(g)} + \text{energy} \qquad (11.57)$$

**Metals with acids**

In Section 11.4 we learned that many metals react with acids and form hydrogen gas and salts. The acid consists of two parts: (1) positive hydrogens, and (2) negative nonmetals or negative groups. The metal replaces the hydrogen in each case.

$$\text{Mg(s)} + \text{H}_2\text{SO}_4\text{(aq)} \rightleftharpoons \text{MgSO}_4\text{(aq)} + \text{H}_2\text{(g)} \qquad (11.58)$$
$$\text{Zn(s)} + 2\text{HC}_2\text{H}_3\text{O}_2\text{(aq)} \rightleftharpoons \text{Zn(C}_2\text{H}_3\text{O}_2)_2\text{(aq)} + \text{H}_2\text{(g)} \qquad (11.59)$$

**Mixing solutions**

When silver nitrate and hydrochloric acid are mixed, the silver(I) ions combine with the chloride ions to form slightly soluble silver chloride. The reaction is a displacement type.

$$\text{AgNO}_3\text{(aq)} + \text{HCl(aq)} \rightleftharpoons \text{AgCl(s)} + \text{HNO}_3\text{(aq)} \qquad (11.60)$$

On the other hand, mixing silver nitrate and hydrofluoric acid produces no precipitate. *One of the conditions* under which mixed solutions react is when a precipitate forms. (Note: There are many other possible types of reactions.) How can one determine in advance whether a precipitate will form? Check the *possible products* of a double displacement reaction with the solubility listings in Table 11.7. If one or both of the poten-

TABLE 11.7 — The solubilities of common substances in water.

| | | Soluble | Very slightly soluble |
|---|---|---|---|
| Common acids | | All | |
| Nitrates | | All | |
| Chlorates | | All | |
| Chlorides | | All, except those listed as slightly soluble | copper(I) lead(II) mercury(I) silver |
| Sulfates | | All, except those listed as slightly soluble | strontium barium lead(II) |
| Hydroxides | | sodium potassium ammonium strontium barium | All, except those listed as soluble |
| Sulfides | | ammonium group IA group IIA | All, except those listed as soluble |
| Arsenates, borates, carbonates, phosphates | | ammonium group IA | All, except those listed as soluble |

tial products is only slightly *soluble*, then a reaction will occur. For instance, will a precipitate form when a solution of barium chloride mixes with a solution of nitric acid?

$$BaCl_2(aq) + HNO_3(aq) \longrightarrow ?$$

If we assume that a displacement is possible, what would be the products? If the barium and hydrogen displace each other, the equation would be:

$$BaCl(aq) + 2HNO_3(aq) \rightleftharpoons Ba(NO_3)_2 + 2HCl \quad (11.61)$$

Are the possible products soluble? A check of Table 11.7 shows that both are soluble. Therefore a precipitate does *not* form. Other than mixing, no reaction occurs.

What happens when sulfuric acid is added to barium chloride? What are the possible products?

$$BaCl_2(aq) + H_2SO_4(aq) \longrightarrow ?$$

If barium and hydrogen displace each other, the products would be:

$$BaCl_2(aq) + H_2SO_4(aq) \rightleftharpoons BaSO_4 + 2HCl \quad (11.62)$$

A check of solubilities in Table 11.7 shows that barium sulfate is only very slightly soluble. Therefore, a precipitate does form; a reaction does occur. Some other examples are:

$$FeCl_3(aq) + 3NaOH(aq) \rightleftharpoons Fe(OH)_3(s) + 3NaCl(aq) \quad (11.63)$$

$$Zn(NO_3)_2(aq) + Na_2S(aq) \rightleftharpoons ZnS(s) + 2NaNO_3(aq) \quad (11.64)$$

■ EXERCISE  Use Table 11.7 to predict whether mixing the following solutions will produce a precipitate.
(a) $ZnCl_2(aq) + Na_2CO_3(aq) \rightarrow$
(b) $Cd(NO_3)_2(aq) + KCl(aq) \rightarrow$

Summary  The alkali metals of group IA and the halogens of group VIIA are common representative elements. The alkali metals are soft, highly reactive and good conductors of electricity. Their freshly cut surfaces are bright and shiny but in the air these surfaces quickly become coated with an oxide film. The metals have low melting points and boiling points because the forces between their large atoms are weak.

Because of their low ionization energies, the alkali metals react vigorously with water and with halogens. In the reaction with water, the products are hydrogen gas and soluble basic hydroxides. In the reaction with halogens, the products are white solid salts. Most alkali metal compounds are water soluble. Molten alkali metal halides conduct electricity. The electricity decomposes the compound into its elements. The reactivity of the elements *increases down the group.*

The halogens are reactive diatomic nonmetals. Fluorine and chlorine are gases, bromine is a volatile liquid, and iodine is a solid. The melting points and boiling points of these elements are low because of the weak forces between the molecules. Like most nonmetals, the halogens are poor conductors of electricity.

Chlorine reacts with water yielding some HClO and HCl. Hydrogen reacts with halogens. The reaction is highly exothermic with fluorine, less so with chlorine and bromine, and endothermic with solid iodine. The reactivity of these elements *decreases down the group.*

Substances containing soluble chloride, bromide, or iodide precipitate silver chloride, silver bromide, or silver iodide from silver nitrate solutions. A solution of chlorine water will react readily with bromide or iodide ions. The chlorine atoms readily remove electrons from bromide ions and iodide ions.

Most metals dissolve in acidic solutions. The rate of the reaction depends upon which metal and which acid is used.

Chemists group reactions into categories in order to simplify the process of what to expect when two different substances are mixed. Classifications based on types of reactants include:

1. Metals + nonmetals → salts
2. Elements + oxygen → oxides
3. Metallic oxides + water → bases
4. Nonmetallic oxides + water → acids
5. Metallic hydrides + water → bases + hydrogen
6. Group VI and VIIA nonmetal hydrides + water → acids
7. Active metals + water → bases + hydrogen
8. Metals + acids → salts + hydrogen
9. Mixing solutions may yield a precipitate.

## Objectives

*By the end of this chapter, you should be able to do the following:*

1. Write equations for the reactions between the group IA metals and water. (Question 3)

2. Write equations for the reactions between the group IA metals and the halogens. (Questions 2, 3)

3. State what trends occur for the group IA elements in atomic size, ionization energy, and reactivity. (Question 4)

4. Diagram an electrical cell used for the decomposition of an alkali metal halide. Label the movement of electrons and ions, state what the products are, and write equations for the reactions. (Question 5)

5. State the relative size of the attractive forces (large, medium, small) that exist between halogen molecules. (Question 9)

6. Write equations for the reactions between halogens and water. (Question 3)

7. Write equations for the reactions between halogens and hydrogen. (Question 7)

8. Write equations for the reactions between aqueous silver nitrate and halide ions. (Question 8)

9. State the products of a reaction (if any) between one halide ion and another halogen atom. (Question 10)

10. State what factors affect the rate of reaction between a metal and an acid. (Question 11)

11. Write equations for the reactions between selected substances (see Section 11.5). (Questions 8, 12)

Questions

1. Define each of the following terms (where possible, give examples): alkali metals; halogens; weak acid; precipitation; halide ion; base; acid; oxidizing agent; reducing agent; heat of formation; exothermic; endothermic; electrolysis; auto-oxidation; chlorine water; metallic oxide; nonmetallic oxide; metallic hydride; nonmetallic hydride.

2. What is the formula of the compound that forms when the following reactions take place?
(a) lithium reacts with chlorine; (b) potassium reacts with bromine; (c) lithium oxide reacts with water; (d) potassium oxide reacts with water.

3. Write equations for the reactions between the following substances:
(a) Rb and $H_2O$; (b) Cs and $F_2$; (c) $Cl_2$ and K; (d) $Br_2$ and $H_2O$; (e) $Br_2$ and $I^-$; (f) Mg and $N_2$; (g) $SO_2$ and $H_2O$; (h) $Cl_2$ and $Br^-$; (i) Co and HCl; (j) $NH_3$ and $H_2O$.

4. Rearrange the following substances in order of increasing:
(a) atomic size: Li, Rb, Cs, Na, K
(b) atomic size: Cl, Br, F, I
(c) valence shell number: Rb, Li, Cs, K, Na
(d) valence shell number: F, Br, I, Cl
(e) electronegativity: Li, Na, Cs, Rb, K
(f) electronegativity: Br, F, Cl, I
(g) reactivity: K, Na, Cs, Rb, Li
(h) reactivity: Cl, I, F, Br
(i) ionization energy: Li, Na, Cs, Rb, K
(j) ionization energy: I, Br, F, Cl

5. Explain the electrolysis of molten potassium fluoride, showing the flow of electrons and ions, and the reactions that occur at each electrode.

6. The potassium content of a 150-pound adult is about 150 grams. This is combined potassium in the form of salts. (a) How many moles of potassium is this? (b) How many atoms of potassium is this?

7. Write balanced equations for the reactions between:
(a) $K(s) + H_2O(l)$
(b) $Rb(s) + I_2(s)$
(c) $Rb(s) + Br_2(l)$
(d) $H_2(g) + Br_2(l)$
(e) $Sr(s) + H_2SO_4(aq)$
(f) $Ca(s) + HBr(aq)$

8. Write balanced equations for the reaction (if any) between:
(a) $AgNO_3(aq) + NaI(aq) \rightarrow$
(b) $AgNO_3(aq) + RbI(aq) \rightarrow$
(c) $AgNO_3(aq) + K_2CO_3(aq) \rightarrow$
(d) $FeCl_3(aq) + (NH_4)_2S(aq) \rightarrow$
(e) $Hg_2(NO_3)_2(aq) + HCl(aq) \rightarrow$
(f) $Pb(NO_3)_2(aq) + Na_2S(aq) \rightarrow$

9. In a reference text look up the boiling points of NaF, NaCl, NaBr, and NaI. Compare these values with the boiling points of the group IA and group VIIA elements. Explain, in terms of the forces between particles, why each group of substances has a different boiling point.

10. State whether mixing the following substances will produce a reaction. If a reaction does occur, describe what you would expect to observe.
(a) $MgCl_2(aq) + Br_2(aq)$
(b) $AlCl_3(aq) + Br_2(aq)$
(c) $FeBr_2(aq) + Br_2(aq)$
(d) $FeBr_2(aq) + Cl_2(aq)$
(e) $HCl(aq) + I_2(aq)$
(f) $HI(aq) + Br_2(aq)$
(g) $HI(aq) + Cl_2(aq)$
(h) $Cl_2(aq) + Br_2(aq)$

11. A student was given two pieces of 6 inch iron wire and told to dissolve them in acid. He was asked to dissolve one piece slowly and the other piece rapidly. Name four factors that he could manipulate in order to speed up or slow down the rate of reaction. Explain how each of these factors works.

12. Complete and balance the following equations:
(a) $CaH_2(s) + H_2O(l)$
(b) $N_2O_3(g) + H_2O(l)$
(c) $N_2O_5(g) + H_2O(l)$
(d) $Cl_2O_7(g) + H_2O(l)$
(e) $CaO(s) + H_2O(l)$
(f) $Ni(s) + Cl_2(g)$
(g) $Zn(s) + O_2(g)$
(h) $O_2(g) + F_2(g)$

13. The average human body contains about 25 milligrams of combined iodine. How many moles of *iodine atoms* is this?

14. All vertebrates have thyroid glands. A lack of iodide ion in the diet may result in underactivity of the thyroid; lack of iodide from birth may result in mental retardation. One biologically active compound in the thyroid is thyroxine, $C_{14}H_{11}O_4NI_4$. (a) What is the percentage by weight of iodine in thyroxine? (b) How many moles of combined *iodine atoms* are there in 0.0020 mole of thyroxine?

15. A liter of blood plasma contains about 140 millimoles of sodium ions, 5 millimoles of potassium ions, and 103 millimoles of chloride ions. (a) What is the weight of sodium, potassium, and chlorine in a liter of blood? (b) What is the ratio of sodium ions to potassium ions in blood? (This same ratio occurs in sea water.)

16. Two common minerals that contain halogens are halite, NaCl, and fluorite, $CaF_2$. (a) What is the percent by weight of chlorine in halite? (b) What is the percent by weight of fluorine in fluorite?

17. The chloride ion can be oxidized to elemental chlorine by strong oxidizing agents. One such agent is potassium permanganate.

$2KMnO_4(aq) + 16HCl(aq) \longrightarrow$
$\qquad 2MnCl_2(aq) + 5Cl_2(g) + 2KCl(aq) + 8H_2O(l)$

(a) How many moles of potassium permanganate are needed to react with 8.00 moles of HCl? (b) How many moles of molecular chlorine can be produced from 8.00 moles of HCl?

18. Many nonmetal chlorides react vigorously with water. For example, phosphorus trichloride reacts with water to give phosphorous acid and hydrochloric acid.

$$PCl_3(l) + 3H_2O(l) \rightleftharpoons H_3PO_3(aq) + 3HCl(aq)$$

(a) How many moles of HCl can be produced from $5.0 \times 10^3$ moles of $PCl_3$? (b) What is the weight of the HCl obtained in part (a)?

19. Concentrated sulfuric acid reacts with sodium chloride. The products are hydrogen chloride and sodium hydrogen sulfate. Write the balanced equation for this reaction.

20. When gently heated, potassium chlorate decomposes into potassium perchlorate and potassium chloride.

$$4KClO_3(s) \rightleftharpoons 3KClO_4(s) + KCl(s)$$

What are the oxidation numbers of each element in the products and in the reactants? What element has undergone oxidation? What element has undergone reduction?

# 12

## The gas laws

### 12.1 Introduction

Why study gases? The earth on which we live is surrounded by a mixture of gases which is called the atmosphere. This relatively thin atmosphere is the only region of known space in which man can survive unassisted. The gases in the atmosphere are primarily oxygen and nitrogen.

Gases undergo many similar physical changes. Gases undergo similar volume changes under the same changes of temperature or pressure. For example, a gas contained in a cylinder with a movable piston will expand to a larger volume when it is heated. Expansion upon heating is a behavior common to all gases not confined in sealed containers of constant volume. The pressure exerted by a gas in a sealed container of constant volume increases as the temperature of the gas is increased. Again, this behavior is exhibited by all gases. Mathematical models will be developed in this chapter which will enable us to predict and to explain the behavior of various samples of gases under various conditions of temperature and pressure.

The kinetic molecular theory model, KMT, was introduced in Chapter 8 to help explain the properties of gases. The KMT model is based on the following assumptions about gases:

1. Gases consist of individual particles called molecules.
2. The particles of gases are in constant, random motion.
3. The particles move in straight lines.
4. The kinetic energy of a gas particle is given by the relationship, K.E. = $\frac{1}{2}mv^2$.

5. Energy may be transferred from one particle to another during a collision but there is no net change in the energy of the system. These are called elastic collisions.

6. The gas particles are small compared to the volume occupied by the gas; therefore the space occupied by the gas particles may be neglected.

7. The particles of a gas are so far apart that the attractive and repulsive forces acting between them are negligible.

These assumptions make up an idealized model which helps us understand the behavior of gases. Although the assumptions of the "ideal" model do not hold exactly for real gases, at atmospheric temperatures and pressures, most real gases behave so closely to the "ideal" that the deviations are often neglected. Unless a statement is made to the contrary, the ideal model will be assumed to hold true.

## 12.2 Gas densities

The scientific community has agreed to compare densities of gases at a given or *standard* set of conditions of *temperature and pressure* (Table 12.1). The conditions chosen were 0°C (273°K) and 1.00 atmosphere (760 mm Hg) pressure. The abbreviation, STP, is often used to designate the standard temperature and pressure. The units of density most often used for gases are grams/liter instead of grams/milliliter as is the case for solids and liquids. Why?

**TABLE 12.1**

*The weights of 1.00-liter samples of gases. Compare the mass of 1.00 liter of the gases with their molecular weights.*

| Gas | Formula | Molecular weight | Volume of 1.00 mole at 0°C, 1.00 atm (liters) | Mass of 1.00 liter at 0°, 1.00 atm (g) | Density at 0°C, 1.00 atm (g/liter) |
|---|---|---|---|---|---|
| Hydrogen | $H_2$ | 2.02 | 22.4 | 0.0898 | 0.090 |
| Helium | He | 4.00 | 22.4 | 0.178 | 0.178 |
| Methane | $CH_4$ | 16.0 | 22.4 | 0.717 | 0.717 |
| Ammonia | $NH_3$ | 17.0 | 22.4 | 0.771 | 0.771 |
| Neon | Ne | 20.2 | 22.4 | 0.900 | 0.900 |
| Nitrogen | $N_2$ | 28.0 | 22.4 | 1.25 | 1.250 |
| Carbon monoxide | CO | 28.0 | 22.4 | 1.25 | 1.250 |
| Oxygen | $O_2$ | 32.0 | 22.4 | 1.43 | 1.429 |
| Hydrogen chloride | HCl | 36.5 | 22.4 | 1.64 | 1.639 |
| Carbon dioxide | $CO_2$ | 44.0 | 22.4 | 1.98 | 1.977 |
| Chlorine | $Cl_2$ | 71.0 | 22.4 | 3.21 | 3.214 |
| Xenon | Xe | 131.3 | 22.4 | 5.85 | 5.851 |

Referring to Table 12.1, why does 1.00 liter of neon weigh more than 1.00 liter of helium? Why does 1.00 liter of oxygen weigh more than 1.00 liter of neon? Notice that the densities of the gases are in the same order as their molecular weights. Why is this the case? The volume of 1.00 mole of each of the gases is also included in the table.

The weight of 1.00 mole of helium is 4.00 g. The volume occupied by 1.00 mole of He at STP can be calculated using the density of He at STP, 0.178 g/liter as a unity term.

$$\text{Volume}_{He} = \frac{4.00 \text{ g}}{1.00 \text{ mole}} \times \frac{1.00 \text{ liter}}{0.178 \text{ g}} = 22.4 \text{ liters/mole}$$

The volume occupied by 1.00 mole of Ne and by 1.00 mole of $O_2$ at STP can be calculated in the same manner.

$$V_{Ne} = \frac{20.2 \text{ g}}{1.00 \text{ mole}} \times \frac{1.00 \text{ liter}}{0.900 \text{ g}} = 22.4 \text{ liters/mole}$$

$$V_{O_2} = \frac{32.0 \text{ g}}{1.00 \text{ mole}} \times \frac{1.00 \text{ liter}}{1.43 \text{ g}} = 22.4 \text{ liters/mole}$$

Notice that the volume occupied by 1.00 mole of each of these gases at STP is 22.4 liters. *The volume which 1.00 mole of a gas occupies is independent of the kind of gas.* How can this relationship help explain the relative densities of He, Ne, and $O_2$? Remember that 1.00 mole of He contains $6.02 \times 10^{23}$ molecules of He. How many molecules are there in 1.00 mole of oxygen? 1.00 mole of neon? Since the volume occupied by 1.00 mole of He at STP is 22.4 liters, there must be $6.02 \times 10^{23}$ molecules in 22.4 liters at STP. At STP there would be $6.02 \times 10^{23}$ oxygen molecules contained in 22.4 liters and under the same conditions 22.4 liters of neon contain $6.02 \times 10^{23}$ neon molecules. *When $6.02 \times 10^{23}$ molecules of a gas occupy a volume of 22.4 liters at STP, the gas is said to behave ideally.* Gases which behave in this way are called ideal gases.

How many molecules are there in 1.00 liter of a gas at STP? Since 22.4 liters of an ideal gas contain $6.02 \times 10^{23}$ molecules, the number of molecules contained in 1.00 liter of an ideal gas can be calculated using this relationship.

$$\text{Number of moles} = 1.00 \text{ liter} \times \frac{1.00 \text{ mole}}{22.4 \text{ liters}} = \frac{1.00}{22.4} \text{ mole}$$

Number of molecules in 1.00 liter of ideal gas

$$= \frac{1.00}{22.4} \text{ mole} \left( 6.02 \times 10^{23} \frac{\text{molecules}}{\text{moles}} \right)$$

$$= 2.68 \times 10^{22} \text{ molecules}$$

12.2 Gas densities

The number of molecules in 1.00 liter of any gas which behaves ideally at STP is $2.68 \times 10^{22}$ regardless of which gas is considered. *Equal volumes of ideal gases contain the same number of molecules under the same conditions of temperature and pressure so long as the gases considered behave ideally.*

Since we know the volume occupied by 1.00 mole of gas at STP, we can determine the number of moles or grams in a given volume of gas. This is valuable when you desire to calculate the number of moles and grams in a sample of gas that has been collected from a certain reaction.

EXAMPLE 1  A student heated some limestone (solid $CaCO_3$) to form solid CaO and $CO_2$. He collected 1.75 liters of dry $CO_2$ at STP. How many grams and how many moles of $CO_2$ did he collect?

Find: (a) moles $CO_2$ and (b) g $CO_2$ in the sample

(a)

Given: $V_{CO_2}$ at STP = 1.75 liters

*Relationship:* 1.00 mole occupies 22.4 liters at STP.

1.75 liters × ? = ? mole $CO_2$

$$1.75 \text{ liters} \times \frac{1.00 \text{ mole}}{22.4 \text{ liters}} = 0.0882 \text{ mole } CO_2$$

(b)

*Relationship:* 1.00 mole of $CO_2$ weighs 44.0 g.

0.0882 mole × ? = ? grams of $CO_2$

$$0.0882 \text{ mole} \times \frac{44.0 \text{ grams}}{1.00 \text{ mole}} = 3.88 \text{ g } CO_2$$

In 1812 Avogadro proposed the concept that equal volumes of gases under the same conditions of temperature and pressure contain the same number of molecules. Avogadro did not know about or use the number $6.02 \times 10^{23}$ but the number of molecules in 1.00 mole of gas or in 22.4 liters at STP now bears his name. Avogadro has been bestowed this honor for advancing such a useful concept.

## 12.3 Pressure, temperature, and volume changes

**The relationship between pressure and volume**

What is the relationship between the *pressure* exerted by a sample of a gas and the *volume* occupied by the sample at constant temperature? To investigate this relationship, use a syringe with a needle sealed so that no gas can pass through. If the syringe is filled with nitrogen and placed in a closed tube connected to a vacuum pump and a sensitive pressure gauge, as shown in Fig. 12.1, the syringe serves as a cylinder with a movable piston. The tube and vacuum pump serve as a means of changing the external pressure exerted on the syringe. The pressure gauge allows this external pressure to be accurately read. Since the plunger of the syringe is free to move, the external pressure can be assumed to be the same as the internal pressure exerted by the nitrogen. During an experiment with such an apparatus, when the pressure gauge was at 1.00 atmosphere, the volume of nitrogen in the syringe at 25°C was 4.14 ml. After the volume of nitrogen at 1.00 atmosphere was read, the vacuum pump was used to adjust the pressure to 2.00 atmosphere but the temperature was held constant at 25°C. The volume of $N_2$ in the syringe at this pressure was determined to be 2.07 ml.

Why does the volume of the nitrogen in the syringe decrease when the pressure outside the syringe is increased? The nitrogen in the syringe consists of particles, nitrogen molecules, which are in constant random motion. These molecules exert pressure on the walls of the syringe and on the plunger because of their collisions with these surfaces. The molecules of $N_2$ in the syringe move with a certain average velocity at a given temperature. At 25°C, for instance, the $N_2$ molecules move at the same average velocity regardless of the pressure and volume. At a temperature of 25°C and an external pressure of 1.00 atm, the $N_2$ sample occupies a volume of 4.14 ml. When the external pressure is doubled and the temperature is held at 25°C, the $N_2$ sample occupies a volume of 2.07 ml. The molecules move with the same velocity as before since the temperature is the same in both cases, but the molecules now have only half as much room in which to move. With only half as much room the molecules strike the walls twice as often as before, and thus they exert twice as much pressure as before.

Can a mathematical model be derived which will describe the relationship between the pressure and volume of an ideal gas at constant temperature? Remember that in the nitrogen experiment (Fig. 12.1), when the pressure was increased, the volume of $N_2$ was observed to decrease. These observations can be summarized by saying that *at constant temperature the volume occupied by a sample of an ideal gas is inversely proportional to the pressure exerted by the gas.* This can be stated mathematically as:

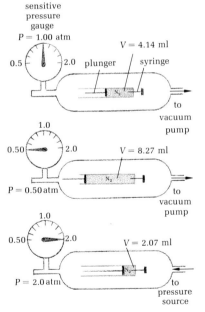

FIGURE 12.1

An apparatus used to study the relationship between pressure and volume.

$$V \propto \frac{1}{P} \quad (T \text{ and } n \text{ are constants}) \tag{12.1}$$

where V = volume, P = pressure, n = number of moles, and $\propto$ means "is proportional to." This equation states that V is proportional to 1/P or V is inversely proportional to P. The equation could also have been written as:

$$P \propto \frac{1}{V}$$

This equation states the same proportionality.

An equation which contains the proportionality sign, $\propto$, can be stated more exactly in slightly different terms. The proportionality between two quantities must involve a constant factor, and the proportionality can be replaced by an equals sign, =, if a constant term is inserted into the equation. The equation V $\propto$ 1/P becomes V = k (1/P), where k is the constant of proportionality.

$$V = k\left(\frac{1}{P}\right) \quad \text{or} \quad V = \frac{k}{P} \tag{12.2}$$

Equation 12.2 can be simplified further by multiplying both sides of the equation by P.

$$P(V) = P\left(\frac{k}{P}\right) = k$$
$$PV = k \tag{12.3}$$

Equation 12.3 states that for a sample of an ideal gas, the product of its volume and its pressure is a constant at constant temperature. Is the volume occupied by a gas multiplied by the pressure it exerts really a constant? What is the proportionality constant for this gas? The product, PV, for these data is included in Table 12.2. For the pressure-volume data taken from the experiment with a volume of nitrogen in a syringe, the product, PV, is found to be constant over the range of pressures from 0.25 atm to 4.00 atm. The constant, k, is found to be 4.14 ml-atm in this case.

The equation PV = k has the same form as the equation for a hyperbola. When the pressure and volume data of Table 12.2 are plotted, the resulting curve is a hyperbola as seen in Fig. 12.2.

The equation PV = k is a mathematical model which describes the relationship between the pressure and volume of a gas. This equation states that the volume of a gas decreases with an increase in pressure, and a decrease in pressure causes an

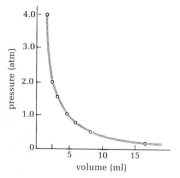

FIGURE 12.2
Pressure-volume relationship for a sample of gas. This graph presents the data from Table 12.2.

**TABLE 12.2**

*The product, PV, for a quantity of gas is found to be a constant (within experimental error, of course).*

| Volume, V (ml) | Pressure, P (atm) | Product, PV = K (ml-atm) |
|---|---|---|
| 16.5 | 0.25 | 4.15 |
| 8.28 | 0.50 | 4.14 |
| 5.51 | 0.75 | 4.14 |
| 4.14 | 1.00 | 4.14 |
| 2.76 | 1.50 | 4.14 |
| 2.07 | 2.00 | 4.14 |
| 1.04 | 4.00 | 4.16 |
| | Average PV = 4.14 ml-atm | |

increase in volume. The mathematical model is known as Boyle's law, since Robert Boyle was the first to describe the pressure-volume relationship for gases in mathematical terms.

**The relationship between volume and the number of moles of gas in the sample**

The syringe which was considered earlier contained a certain quantity of $N_2$. How does the volume change when the number of moles is doubled at the same temperature? Experiments show that the volume occupied by 0.100 mole of $N_2$ at 1.00 atm pressure and 0°C is 2.24 liters but the volume occupied by 0.200 mole of $N_2$ at this temperature and pressure is 4.48 liters. The dependence of the volume upon the quantity of gas present at a constant temperature and pressure is indicated in Table 12.3.

Doubling the quantity of gas at constant temperature and pressure doubles the volume. The volume occupied by a gas at constant temperature and pressure is proportionally increased as the quantity of gas is increased. The volume, at constant T and P, is directly proportional to the quantity of gas present:

$$V \propto n \qquad (12.4)$$

**TABLE 12.3**

*The volume occupied by a gas is increased if the number of moles of gas is increased at a constant temperature of 0°C and a constant pressure of 1.00 atm.*

| Pressure (atm) | Temperature (°C) | Mass of $N_2$ (g) | $N_2$ (moles) | Volume (liters) |
|---|---|---|---|---|
| 1.00 | 0 | 2.80 | 0.100 | 2.24 |
| 1.00 | 0 | 5.60 | 0.200 | 4.48 |
| 1.00 | 0 | 11.20 | 0.400 | 8.96 |
| 1.00 | 0 | 22.40 | 0.800 | 17.9 |
| 1.00 | 0 | 28.00 | 1.00 | 22.4 |
| 1.00 | 0 | 44.8 | 1.60 | 35.8 |

where $V$ = volume and $n$ = number of moles of gas. As before, the equation $V \propto n$ can be replaced by the more exact equation $V = k'n$, where $k'$ is a constant and $T$ and $P$ are constant.

$$V = k'n \qquad (12.5)$$

Equation 12.5 can now be solved for $k'$:

$$k' = \frac{V}{n} \qquad (12.6)$$

Substitution of the first set of data from Table 12.3 in this equation allows $k'$ to be calculated:

$$k' = \frac{V}{n} = \frac{2.24 \text{ liters}}{0.100 \text{ mole}} = \frac{22.4 \text{ liters}}{1.00 \text{ mole}} = 22.4 \text{ liters/mole}$$

The constant, $k'$, is equal to 22.4 liters/1.00 mole. Check the remainder of the data in Table 12.3 to be sure that $k'$ is really a constant. Equation 12.5 is the algebraic model of the observation that at constant temperature and pressure the volume of a gas increases proportionally as the number of moles of gas is increased.

### The relationship between the volume of a gas and the temperature in degrees Kelvin

The effect of heating a quantity of gas upon the volume occupied by the gas at constant pressure is shown in Table 12.4. The volume of a gas expands with an increase in temperature and contracts with a decrease in temperature, if the pressure and quantity are held constant. According to the data in Table 12.4, the volume doubles with a doubling of the temperature in °K. The volume is directly proportional to the temperature in °K.

$$V \propto T \qquad (12.7)$$

TABLE 12.4

An increase in the temperature of a gas causes an increase in the volume occupied by the gas if the pressure and quantity of gas are constant.

| Quantity of gas (moles) | Pressure (atm) | Temperature (°K) | Volume (ml) | $C = \dfrac{\text{volume}}{\text{temperature}}$ (ml/°K) |
|---|---|---|---|---|
| 0.00406 | 1.00 | 150 | 50.0 | 0.333 |
| 0.00406 | 1.00 | 200 | 66.6 | 0.333 |
| 0.00406 | 1.00 | 300 | 100.0 | 0.333 |
| 0.00406 | 1.00 | 400 | 133.3 | 0.333 |
| 0.00406 | 1.00 | 500 | 166.6 | 0.333 |
| 0.00406 | 1.00 | 600 | 200.0 | 0.333 |

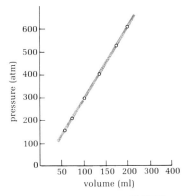

**FIGURE 12.3**
Temperature-volume relationship for a sample of gas. This graph presents the data from Table 12.4.

This proportionality is expressed more exactly with the use of a constant.

$$V = cT \tag{12.8}$$

where $c$ is a constant. Equation 12.8 is the mathematical model which indicates the relationship between the temperature in °K and the volume occupied by a given quantity of gas at constant pressure.

A graphic model of the data in Table 12.4 is presented in Fig. 12.3. These data give a straight line as might be expected from the direct proportionality $V = cT$. This straight line can be used to determine the volume occupied by this quantity of gas at any temperature between 150°K and 600°K, but what volume will the same quantity of gas occupy at 700°K? The straight line can be extended so that an extrapolation to the volume at 700°K can be made. The volume occupied by the gas at 700°K is found from the extrapolation to be 233.3 ml. This is in close agreement with the experimentally determined value.

The relationship between the volume and temperature (°K) of a gas were first studied by Jacques Charles in 1803. For this reason, the mathematical model which was described earlier, $V = cT$, is known as Charles' law.

If the volume is proportional to the absolute temperature and the volume occupied by a gas decreases with a decrease in temperature, what volume would a sample of gas occupy at 0.0°K? What happens to all gases before they reach this temperature?

## 12.4 Gas law calculations

If a quantity of gas occupies a volume of 54.3 ml at 27.0°C and 745 mm Hg, what volume will this quantity of gas occupy at STP? Both chemists and chemistry students are faced with problems similar to this. Several of the more common calculations involving gases will be investigated in this section. In addition to Boyle's law and Charles' law, the use of a more general relationship for gases will be introduced in the following section.

**Pressure-volume changes**

When the pressure exerted by a quantity of gas was increased from 2.01 atm to 4.02 atm with the temperature held constant, the volume decreased from 408 ml to a new volume. What was the final volume? The final volume can be conveniently calculated with the use of Boyle's law.

$$PV = k$$

If we designate the initial pressure and volume as $P_1$ and $V_1$, we can likewise designate the final pressure and volume as $P_2$ and $V_2$. The product of the initial pressure and volume, $P_1V_1$, is equal to the constant k.

$$P_1V_1 = k \tag{12.9}$$

The product of the final pressure and volume, $P_2V_2$, is also equal to the constant k.

$$P_2V_2 = k \tag{12.10}$$

Since the two different products, $P_1V_1$ and $P_2V_2$, are equal to the same thing, they must be equal to each other.

$$P_1V_1 = k = P_2V_2$$
$$P_1V_1 = P_2V_2 \tag{12.11}$$

The expression (12.11) can be solved for the final volume, $V_2$.

$$V_2 = \frac{P_1}{P_2} V_1 \tag{12.12}$$

Initial volume = $V_1$ = 408 ml
Initial pressure = $P_1$ = 2.01 atm
Final pressure = $P_2$ = 4.02 atm

$$V_2 = \frac{2.01 \text{ atm}}{4.02 \text{ atm}} (408 \text{ ml}) = 204 \text{ ml}$$

When the pressure exerted on a quantity of gas is increased from 2.01 atm to 4.02 atm, the original volume of 408 ml must decrease to a final volume of 204 ml if the temperature is held constant.

We have already noted that the volume of a gas is halved when the pressure is doubled at constant temperature. How can this observation be used in a calculation such as the one noted previously?

EXAMPLE 1

Given: $V_1$ = initial volume = 408 ml
$P_1$ = initial pressure = 2.01 atm
$P_2$ = final pressure = 4.02 atm

Find: $V_2$ = final volume = ? What relationship is necessary to convert the initial volume to the final volume?

$$V_2 = V_1 \times ? = ? \text{ ml}$$

Solution: Since a pressure change causes the volume change, a term which can be used to convert the initial volume, $V_1$, to the final volume, $V_2$, should contain the pressure terms, $P_1$ and $P_2$. Since an increase in pressure caused a decrease in volume, the term which we seek in this case must be less than 1.00. This term is obtained for this case by dividing the smaller pressure by the larger one 2.01 atm/4.02 atm. This expression is exactly the same as was obtained previously using Boyle's law.

$$V_2 = 408 \text{ ml} \, \frac{2.01 \text{ atm}}{4.02 \text{ atm}} = 204 \text{ ml}$$

Boyle's law was used to solve the problem in one case; logic was used in the other case, but the end result was the same. In order to solve such problems using logic, you should note whether the pressure is being increased or decreased. If the pressure increases, the initial volume is multiplied by a fraction, involving $P_1$, and $P_2$, which is less than 1.00. For a pressure decrease, the volume increases so the initial volume, $V_1$, is multiplied by a number, involving $P_1$ and $P_2$, which must now be greater than 1.00.

EXAMPLE 2  The pressure which was exerted on a gas was changed from 761 mm Hg to 652 mm Hg at constant temperature. The final volume occupied by the gas at 652 mm Hg was 2.30 liters. What was the volume occupied by the gas at the initial pressure of 761 mm Hg?

Given: $P_1$ = initial pressure = 761 mm Hg
$P_2$ = final pressure = 652 mm Hg
$V_2$ = final volume = 2.30 liters

Find: $V_1$ = initial volume = ?

Solution: The volume of the gas was changed from an unknown initial volume to a final volume of 2.30 liters when the pressure decreased from 761 mm Hg to a final pressure of 652 mm Hg. Which volume is smaller, $V_1$ or $V_2$? Since the pressure was greater initially, the initial volume must have been smaller than the final volume. The final volume must be multiplied by a factor involving $P_1$ and $P_2$ which is less than 1.00.

$$V_1 = 2.30 \text{ liters} \times \frac{652 \text{ mm Hg}}{761 \text{ mm Hg}} = 1.97 \text{ liters}$$

EXAMPLE 3  A gas was compressed from an initial volume of 5.07 liters to a final volume of 0.780 liter. The initial pressure was 1.00 atm. What pressure, $P_2$, did the gas exert at a volume of 0.780 liter?

Given: $V_1$ = initial volume = 5.07 liters
$V_2$ = final volume = 0.780 liter
$P_1$ = initial pressure = 1.00 atm

Find: $P_2$ = final pressure = ?

■ EXERCISE  The volume occupied by 1.00 mole of oxygen at 1.00 atm pressure and 0°C is 22.4 liters. What volume does 1.00 mole of oxygen occupy at 5.00 atmospheres and 0°C?

**EXERCISE** *A gas occupies a volume of 7.52 ml at a pressure of 751 mm Hg. What pressure would be required in order to reduce the volume of the gas to 5.01 ml?*

*Solution:* The volume occupied by the gas was compressed from an initial volume of 5.07 liters to a final volume of 0.780 liter. Would the pressure have to be increased or decreased to cause a decrease in volume? The initial pressure must be multiplied by a fraction larger than 1.00 in order to obtain the final pressure.

$$P_2 = 1.00 \text{ atm} \times \frac{5.07 \text{ liters}}{0.780 \text{ liter}} = 6.52 \text{ atm} \quad \blacksquare$$

## Temperature-volume changes

A student increased the temperature of a 125 ml sample of helium from 236°K to 472°K while holding the pressure constant at 1.00 atmosphere. The total quantity of gas in the sample remained constant throughout the experiment. What was the volume occupied by this sample of helium at 472°K? According to Charles' law, the volume of a gas increases in proportion to the absolute temperature (°K) if the pressure and quantity of gas remain constant.

Can logic be used to solve problems which involve temperature-volume changes, as was the case for pressure-volume changes? The final volume of a sample of gas which undergoes a temperature-volume change can be calculated using the initial temperature and volume and the final temperature. The initial volume, $V_1$, must be multiplied by some factor involving the initial and final temperatures, $T_1$ and $T_2$.

What factor can be used to calculate the final volume from the initial volume?

### EXAMPLE 4

Given: $V_1$ = initial volume = 125 ml
$T_1$ = initial temperature = 236°K
$T_2$ = final temperature = 472°K

Find: $V_2$ = final volume = ?

Solution: $V_1 \times ? = V_2$
$V_2$ = 125 ml × ? = ? ml

Since the temperature is the only other property—aside from volume—of the system which changes, the factor being sought must contain the temperature terms, $T_1$ and $T_2$. The increase in temperature in this case causes the volume to increase, therefore the factor which is sought must be greater than 1.00. The term sought is 472°K/236°K.

$$V_2 = 125 \text{ ml} \times \frac{472°K}{236°K} = 250 \text{ ml}$$

Problems of this type can also be solved using Charles' law, which states that the volume occupied by a gas is directly proportional to the absolute temperature if the pressure and quantity of gas are maintained constant. The mathematical expression of Charles' law for a quantity of gas at an initial state of temperature and volume at constant pressure is:

$$V_1 = cT_1, \qquad c = \frac{V_1}{T_1} \tag{12.13}$$

At a state of final temperature and volume for the same quantity of gas at the same pressure, the expression becomes:

$$V_2 = cT_2, \qquad c = \frac{V_2}{T_2} \tag{12.14}$$

The constant term, $c$, is the same in the two cases; thus we can set $V_1/T_1$ equal to $V_2/T_2$ since quantities which are equal to the same term are equal to each other.

$$\frac{V_1}{T_1} = c = \frac{V_2}{T_2} \quad \text{or} \quad \frac{V_1}{T_1} = \frac{V_2}{T_2} \tag{12.15}$$

Equation 12.15 is a statement of Charles' law. To calculate the final volume for the previous problem, Eq. 12.15 is solved for $V_2$ and the appropriate values substituted into the resulting equation.

$$V_2 = \frac{T_2}{T_1} V_1 = \frac{472°K}{236°K} (125 \text{ ml}) = 250 \text{ ml}$$

This is exactly the same result as was obtained by a "common sense" or logical approach.

EXAMPLE 5  A chemist measured the volume of oxygen in a cylinder and found that it contained 85.6 ml at 100°C and 1.00 atm pressure. What volume would this same quantity of oxygen occupy if the temperature were lowered to −10.0°C at a constant pressure of 1.00 atm?

Given: $T_1$ = initial temperature = 100°C = 373°K
$T_2$ = final temperature = −10.0°C = 263°K
$V_1$ = initial volume = 85.6 ml

Find: $V_2$ = final volume = ?

Relationship: The temperature decreases so the volume must decrease.

Solution:

$$V_2 = 85.6 \text{ ml} \times \frac{263°K}{373°K} = 60.4 \text{ ml}$$

**EXAMPLE 6** A quantity of gas reached a final volume of 0.632 liter when the temperature of the gas was increased from 27.0°C to 200°C. What was the initial volume of the gas at 27.0°C?

Given: $T_1 = 27.0°C = 300°K$
$T_2 = 200°C = 473°K$
$V_2 = 0.632$ liter

Find: $V_1 =$ initial volume $= ?$

Solution:

$$V_1 = 0.632 \text{ liter} \times \frac{300°K}{473°K} = 0.401 \text{ liter}$$

**EXAMPLE 7** A certain quantity of argon occupied a volume of 22.4 liters at 273°K and 1.00 atm. (What quantity of Ar was this?) The volume occupied by this quantity of Ar was reduced from 22.4 liters to 19.7 liters by a temperature change at constant pressure. What was the final temperature?

Given: $V_1 = 22.4$ liters
$V_2 = 19.7$ liters
$T_1 = 273°K$

Find: $T_2 =$ final temperature $= ?$

Solution: $T_2 = T_1 \times ? = ? °K$

$$T_2 = 273°K \times \frac{19.7 \text{ liters}}{22.4 \text{ liters}} = 240°K$$
$$°C = 240°K - 273°K = -33°C \quad \blacksquare$$

■ **EXERCISE** The volume occupied by 0.500 mole of $H_2$ at STP is 11.2 liters. What volume would this quantity of $H_2$ occupy if the temperature were raised to 100°C while the pressure was held constant?

■ **EXERCISE** A quantity of nitrogen occupies a volume of 23.1 ml at 35°C and 1.00 atm. What will be the temperature at which the same quantity of nitrogen will occupy 46.2 ml at 1.00 atm?

The pressure exerted by a gas in a sealed container is directly proportional to the absolute temperature of the gas (Fig. 12.4).

$$P \propto T$$

The pressure exerted by a gas is equal to the product of the absolute temperature, $T$, and a constant, $k''$.

$$P = k''T \tag{12.16}$$

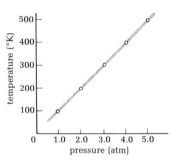

FIGURE 12.4
Temperature-pressure relationship for a sample of gas.

At 273°K, 1.00 mole of helium in a 22.4 liter sealed container exerts a pressure of 1.00 atm. What will be the pressure exerted by 1.00 mole of helium if it occupies 22.4 liters at 546°K? The absolute temperature has been increased by a factor of 546°K/273°K or a factor of 2/1. The absolute temperature has doubled; therefore the pressure should also double. The final pressure can be calculated by multiplying the initial pressure by the factor, 546°K/273°K.

$$P_2 = P_1 \frac{546°K}{273°K} = 1.00 \text{ atm} \times \frac{546°K}{273°K} = 1.00 \text{ atm} \times 2.00 = 2.00 \text{ atm}$$

Equation 12.16 can be solved to obtain the same result. For the initial state, this equation becomes:

$$P_1 = k''T_1, \text{ or } k'' = \frac{P_1}{T_1}$$

For the final state:

$$P_2 = k''T_2, \text{ or } k'' = \frac{P_2}{T_2}$$

The constants are the same in the two cases, so $P_1/T_1$ can be set equal to $P_2/T_2$, since terms which are equal to the same thing are equal to each other.

$$\frac{P_1}{T_1} = \frac{P_2}{T_2} \qquad (12.17)$$

$$P_2 = \frac{T_2}{T_1} P_1$$

For the previous case:

$$P_2 = \frac{546°K}{273°K}(1.00 \text{ atm}) = 2.00 \text{ atm}$$

The following example involves a sample of gas whose volume is held constant in a sealed container while the temperature and pressure are varied.

EXAMPLE 8  A sample of gas in a sealed container exerts a pressure of 1.00 atm at a temperature of $-21°C$. What pressure will the gas in this container exert at $27°C$?

Given: $T_1$ = initial temperature = $-21°C + 273°C = 252°K$
$T_2$ = final temperature = $27°C + 273°C = 300°K$
$P_1$ = initial pressure = 1.00 atm

Find: $P_2$ = final pressure = ?

Solution: $P_2 = \dfrac{300°K}{252°K}(1.00 \text{ atm}) = 1.19 \text{ atm}$  ■

■ EXERCISE  *A sample of $N_2$ in a sealed container was cooled from $673°K$ to $473°K$. The pressure exerted by the gas at $673°K$ was 12.1 atm. What is the pressure of the gas at $473°K$?*

## 12.5  General gas laws

**The combined gas laws**

Carefully examine the mathematical models of gaseous behavior which have been presented thus far.

These four models are:

1. $PV = k$ (at constant $T$ and constant $n$)
2. $V = k'n$ (at constant $P$ and constant $T$)
3. $V = cT$ (at constant $P$ and constant $n$)
4. $P = K''T$ (at constant $V$ and constant $n$)

The symbols $k$, $k'$, $c$, and $k''$ represent constants. These models are quite useful for the solution of some problems dealing with gases. If one of these models is used, two of the four possible variables must remain constant. Can a model be applied to those cases in which only $n$ is held constant? For example, the volume occupied by a sample of carbon dioxide at standard temperature and pressure is 22.4 liters. What volume will this same quantity of $CO_2$ occupy at 100°C and 620 mm pressure? A mathematical model, which is more general than the previous ones, is required for solving this problem.

For a given number of moles of gas, $n$, an equation for a more general mathematical model is:

$$\frac{P_1 V_1}{T_1} = C \tag{12.18}$$

where $C$ is a constant (See Table 12.5). Likewise for a second set of conditions:

$$\frac{P_2 V_2}{T_2} = C \tag{12.19}$$

Quantities which are equal to the same thing are equal to each other; thus,

$$\frac{P_1 V_1}{T_1} = \frac{P_2 V_2}{T_2} \tag{12.20}$$

Equation 12.20 is known as the Combined Gas law.

TABLE 12.5

The data in this table show that PV/T is equal to a constant for 1.00 mole of He under the experimental conditions examined. The average value of PV/T in this case is 0.0821 liter-atm/°K.

| n (mole) | P (atm) | V (liters) | T (°K) | $\frac{PV}{T}\left(\frac{\text{liters-atm}}{°K}\right)$ |
|---|---|---|---|---|
| 1.00 | 1.00 | 22.4 | 273 | 0.0821 |
| 1.00 | 2.12 | 35.1 | 906 | 0.0821 |
| 1.00 | 0.532 | 62.3 | 404 | 0.0820 |
| 1.00 | 0.643 | 50.1 | 392 | 0.0822 |
| 1.00 | 2.82 | 15.3 | 525 | 0.0822 |
| 1.00 | 1.53 | 46.2 | 861 | 0.0821 |

EXAMPLE 9  A quantity of nitrogen occupies a volume of 7.23 ml in a gas syringe at 273°K and 1.00 atm pressure. What volume would this same quantity of $N_2$ occupy at 546°K and 2.00 atm?

Given: $P_1 = 1.00$ atm, $V_1 = 7.23$ ml, $T_1 = 273$°K
$P_2 = 2.00$ atm, $V_2 = ?$, $T_2 = 546$°K

Find: $V_2 = ?$

Relationship: $\dfrac{P_1 V_1}{T_1} = \dfrac{P_2 V_2}{T_2}$

Solution: $V_2 = \dfrac{V_1 P_1 T_2}{P_2 T_1}$

$$V_2 = (7.23 \text{ ml}) \left(\dfrac{1.00 \text{ atm}}{2.00 \text{ atm}}\right) \left(\dfrac{546°K}{273°K}\right) = 7.23 \text{ ml}$$

The final volume is the same as the original volume.

Can Example 9 be solved using logic, as was the case for the previous gas problems? The temperature change and the pressure change can be treated separately. The pressure increased from 1.00 atm to 2.00 atm—that is, it doubled; thus the volume should have been halved since an increase in pressure causes a decrease in volume.

$$V_2 = 7.23 \text{ ml} \times \dfrac{1.00 \text{ atm}}{2.00 \text{ atm}}$$

The temperature increased from 273°K to 546°K, or doubled. Doubling the temperature should have resulted in doubling the volume.

$$V_2 = 7.23 \text{ ml} \left(\dfrac{1.00 \text{ atm}}{2.00 \text{ atm}}\right) \left(\dfrac{546°K}{273°K}\right) = 7.23 \text{ ml}$$

This is the same result as in Example 9.

**The ideal gas law**

From the Combined Gas law, the following equation can be derived.

$$\dfrac{PV}{T} = nR \qquad (12.21)$$

where $n =$ number of moles of gas and $R$ is a constant.

Solving for $R$, Eq. 12.21 becomes:

$$\dfrac{PV}{nT} = R \qquad (12.22)$$

TABLE 12.6

The data in this table show that $PV/nT$ is constant for He over a range of experimental conditions. The average value of $PV/nT$ was found to be 0.0821 liter-atm/°K-mole.

| n (moles) | P (atm) | V (liters) | T (°C) | $\frac{PV}{nT}\left(\frac{\text{liters-atm}}{\text{°K-mole}}\right)$ |
|---|---|---|---|---|
| 1.00 | 1.00 | 22.4 | 273 | 0.0821 |
| 3.12 | 2.00 | 30.1 | 235 | 0.0821 |
| 0.822 | 0.716 | 43.3 | 459 | 0.0822 |
| 1.25 | 0.517 | 60.5 | 305 | 0.0820 |
| 2.37 | 0.823 | 92.3 | 390 | 0.0822 |
| 2.00 | 2.00 | 44.8 | 546 | 0.0821 |

What happens if the number of moles of gas, $n$, is not held constant? Is Eq. 12.21 still valid? If $PV/nT$ is constant when $P$, $V$, $T$, and $n$ are all made to vary, then Eq. 12.21 is still valid. Experimental data for helium are listed in Table 12.6. Note that even when $P$, $V$, $n$, and $T$ vary, $PV/nT$ remains constant.

Equation 12.21 is one statement of the general Ideal Gas equation. The equation can be rearranged to give:

$$PV = nRT \qquad (12.23)$$

The constant, $R$, is called the universal gas constant, $R$ is equal to 0.0821 liter-atm/°K mole when the unit of $P$ is atm, the unit of $V$ is liters, the unit of $T$ is °K and the unit of $n$ is moles. $R$ can have other values when different units of $P$, $V$, $T$, and $n$ are used. For instance, the value of $R$ is 62.4 liters mm Hg/°K-mole when the pressure is given in mm Hg instead of atm.

EXAMPLE 10   What volume will 1.52 moles of He occupy at 720 mm Hg and 325°K?

Given: $n = 1.52$ moles
$P = 720$ mm Hg
$T = 325$°K

Find: $V = ?$

Relationship: $PV = nRT$, $R = 62.4 \dfrac{\text{liters-mm Hg}}{\text{°K-mole}}$

Solution: $V = \dfrac{nRT}{P}$

$$V = \frac{(1.52 \text{ moles})\left(62.4 \dfrac{\text{liters-mm Hg}}{\text{°K-mole}}\right)(325\text{°K})}{720 \text{ mm Hg}} = 42.8 \text{ liters}$$

EXAMPLE 11   At what temperature will 1.25 moles of helium occupy a volume of 22.4 liters and exert a pressure of 1.00 atm?

Given: n = 1.25 moles He
V = 22.4 liters
P = 1.00 atm

Find: T = ?

Relationship: $PV = nRT$, $R = 0.0821 \dfrac{\text{liter-atm}}{\text{°K-mole}}$

Solution: $T = \dfrac{PV}{nR}$

$$T = \dfrac{(1.00 \text{ atm})(22.4 \text{ liters})}{(1.25 \text{ moles})\left(0.0821 \dfrac{\text{liter-atm}}{\text{°K-mole}}\right)} = 218\text{°K}$$

EXAMPLE 12  What is the weight of a mole of a gas if $5.32 \times 10^{-2}$ gram of the gas occupies a volume of 20.0 ml at 293°K and 750 mm Hg pressure?

Given: Weight of gas = $5.32 \times 10^{-2}$ gram
V = 20.0 ml = 0.0200 liter
T = 293°K
P = 750 mm Hg

Find: Weight of 1.00 mole of the gas

Relationship: $PV = nRT$, $R = \dfrac{62.4 \text{ liters-mm Hg}}{\text{°K-mole}}$

Solution: The number of moles, n, of gas which is equivalent to $5.32 \times 10^{-2}$ g must first be calculated.

$$n = \dfrac{PV}{RT} = \dfrac{(750 \text{ mm Hg})(0.0200 \text{ liter})}{\left(62.4 \dfrac{\text{liters-mm Hg}}{\text{°K-mole}}\right)(293\text{°K})}$$

$n = 8.07 \times 10^{-4}$ mole

The number of grams per mole is:

$$\dfrac{\text{grams}}{\text{moles}} = \dfrac{5.32 \times 10^2 \text{ g}}{8.07 \times 10^{-4} \text{ mole}} = 65.8 \text{ g/mole}$$

The molecular weight of the gas is 65.8 amu.

## 12.6 Partial pressures

An increase in the number of collisions of gas molecules per unit time causes the pressure exerted by a gas to be increased. A decrease in the number of collisions per unit time will result in a decrease in the pressure exerted by a gas. The number of collisions per unit time is influenced by two factors: (1) The "concentration" of the gas, or the number of gas molecules contained in a unit volume of gas and (2) the average velocity of gas molecules.

An increase in the average velocity of gas molecules causes an increase in the number of collisions per unit time and an increase in the force of the collisions. If there is no temperature change in the system, the average velocity of the gas molecules remains constant.

The number of collisions per second in a container of gas can also be increased by adding more gas to the sample at constant temperature and constant volume. An increase in the number of gas molecules in a given volume will result in an increase in the number of collisions and an increase in the pressure exerted by the gas.

EXAMPLE 13  The pressure exerted by 1.00 mole of carbon dioxide contained in a 22.4-liter flask at 273°K is 1.00 atm. What pressure would be exerted if an additional 1.00 mole of $CO_2$ were added to this 22.4-liter container at the same temperature?

Given: $T = 273°K$
$V = 22.4$ liters
$n = $ total $= 1.00$ mole $+ 1.00$ mole $= 2.00$ moles of $CO_2$

Find: $P = ?$

Solution: $P = \dfrac{nRT}{V}$

$$P = \dfrac{(2.00 \text{ moles})\left(0.0821 \dfrac{\text{liter-atm}}{°\text{K-mole}}\right)(273°\text{K})}{22.4 \text{ liters}} = 2.00 \text{ atm}$$

Doubling the number of moles of $CO_2$ at constant temperature and volume doubled the number of collisions; thus the pressure was doubled.

EXAMPLE 14  What is the pressure exerted by 1.00 mole of $CO_2$ and 1.00 mole of $N_2$ if both are in the 22.4-liter container at 273°K (Fig. 12.5)? Since 1.00 mole of $CO_2$ contains $6.023 \times 10^{23}$ molecules and 1.00 mole of $N_2$ also contains $6.023 \times 10^{23}$ molecules, there will be $12.046 \times 10^{23}$ molecules or 2.00 moles of molecules in the container. The pressure depends only upon the number of collisions if the temperature and volume are held constant; thus the pressure is independent of the type of molecule that is doing the colliding. In this case there are 2.00 moles of molecules, 1.00 mole of $N_2$, and 1.00 mole of $CO_2$.

Given: $V = 22.4$ liters
$T = 273°K$
$n = n_{CO_2} + n_{N_2} = 1.00$ mole $N_2 + 1.00$ mole $CO_2 = 2.00$ moles

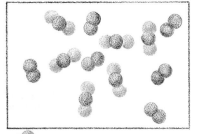

$CO_2 = 1.00$ mole

$N_2 = 1.00$ mole

$V = 22.4$ l
$T = 273°K$
$P_{CO_2} = 1.00$ atm
$P_{N_2} = 1.00$ atm
$P_T = P_{CO_2} + P_{N_2} = 2.00$ atm

FIGURE 12.5

*Partial pressure of two gases. One mole of $CO_2$ in a 22.4-liter container at standard temperature exerts a pressure of 1.00 atm. One mole of $N_2$ in the same container also exerts a pressure of 1.00 atm. The partial pressure of $CO_2$ is 1.00 atm, the partial pressure of $N_2$ is 1.00 atm, and the total pressure of both gases is the sum of the partial pressures, or 2.00 atm.*

Find: $P = ?$

$$P = \frac{nRT}{V} = \frac{(2.00 \text{ moles})\left(0.0821 \frac{\text{liter-atm}}{°\text{K-mole}}\right)(273°\text{K})}{22.4 \text{ liters}}$$

$P = 2.00$ atm

When the total pressure exerted by a mixture of gases is calculated using the Ideal Gas law, the *total* number of moles of *all* gases in the mixture must be used.

For a mixture of three gases the total number of moles is:

$n_t = n$ (total) $= n_1 + n_2 + n_3$
$n_1 =$ number of moles of gas 1
$n_2 =$ number of moles of gas 2
$n_3 =$ number of moles of gas 3

For this mixture of three gases, the Ideal Gas law becomes:

$$PV = n_t RT = (n_1 + n_2 + n_3) RT \tag{12.24}$$

or

$$P_t = \frac{n_1 RT}{V} + \frac{n_2 RT}{V} + \frac{n_3 RT}{V}$$

We can see that the term $n_1 RT/V$ is equal to a pressure term, $P_1$, and likewise for the other terms.

$$P_t = \frac{n_t RT}{V}, \quad P_1 = \frac{n_1 RT}{V}, \quad P_2 = \frac{n_2 RT}{V}, \quad P_3 = \frac{n_3 RT}{V}$$

The total pressure must be the sum of the three pressure terms, $P_1$, $P_2$, and $P_3$.

$$P \text{ total} = P_1 + P_2 + P_3 \tag{12.25}$$

The terms, $P_1$, $P_2$, and $P_3$, are called *partial pressures*. The partial pressures are the pressures each gas would exert if it alone were contained in the given volume at the same temperature. Equation 12.25 is a mathematical statement of Dalton's law of partial pressures. Dalton's law states that the total pressure exerted by a mixture of gases is the sum of the partial pressures exerted by the individual gases.

EXAMPLE 15  What will be the total pressure exerted when 2.00 moles of He, 1.00 mole of $N_2$, and 1.00 mole of $O_2$ are placed in a 35.0 liter-container at 273°K? What will be the partial pressure exerted by each gas?

Given: $V = 35.0$ liters
$T = 273\,°K$
$n_{He} = 2.00$ moles
$n_{N_2} = 1.00$ mole
$n_{O_2} = 1.00$ mole

Find: $P_{total} = ?\ P_{He} = ?\ P_{N_2} = ?\ P_{O_2} = ?$

Relationship: $P_t = \dfrac{n_t RT}{V}$

Solution: What is the *total* number of moles of *all* gases in the container?

$n_t = n_{He} + n_{N_2} + n_{O_2}$
$= 2.00 \text{ moles} + 1.00 \text{ mole} + 1.00 \text{ mole} = 4.00 \text{ moles}$

$P_t = \dfrac{n_t RT}{V}$

$P_t = \dfrac{(4.00 \text{ moles})\left(0.0821\,\dfrac{\text{liter-atm}}{°K\text{-mole}}\right)(273\,°K)}{35.0 \text{ liters}} = 2.56 \text{ atm}$

Now, each of the partial pressures can be calculated individually.

$P_{He} = \dfrac{n_{He} RT}{V} = \dfrac{(2.00 \text{ moles})\left(0.0821\,\dfrac{\text{liter-atm}}{°K\text{-mole}}\right)(273\,°K)}{35.0 \text{ liters}} = 1.28 \text{ atm}$

$P_{N_2} = \dfrac{n_{N_2} RT}{V} = \dfrac{(1.00 \text{ mole})\left(0.0821\,\dfrac{\text{liter-atm}}{°K\text{-mole}}\right)(273\,°K)}{35.0 \text{ liters}} = 0.640 \text{ atm}$

$P_{O_2} = \dfrac{n_{O_2} RT}{V} = \dfrac{(1.00 \text{ mole})\left(0.0821\,\dfrac{\text{liter-atm}}{°K\text{-mole}}\right)(273\,°K)}{35.0 \text{ liters}} = 0.640 \text{ atm}$

The total pressure is the sum of the partial pressures.

$P_t = P_{He} + P_{N_2} + P_{O_2}$
$= 1.28 \text{ atm} + 0.640 \text{ atm} + 0.640 \text{ atm} = 2.56 \text{ atm}$

This is the same result as was obtained for P total previously.

Gases that are only slightly soluble in water can be collected in bottles by displacing water. Figure 12.6 illustrates the collection of a sample of hydrogen gas. In addition to the hydrogen gas being collected, the sample is saturated with water vapor. The total pressure of the system is due to both the partial pressure of the hydrogen gas and the partial pressure of the water vapor.

$P_t = P_{H_2} + P_{H_2O}$

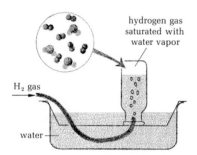

FIGURE 12.6
Collecting a sample of hydrogen gas. When hydrogen gas is collected by the displacement of water, the sample is saturated with water vapor. In this example, the temperature is 25°C and the atmospheric pressure is 755 mm Hg.

**TABLE 12.7**

The vapor pressure of water at various temperatures. These values of the vapor pressure can be used to obtain the pressure exerted by a sample gas when the sample gas is mixed with $H_2O$ vapor. A plot of these data is shown in Fig. 12.7.

| Temperature (°C) | Vapor pressure of $H_2O$ (mm Hg) |
|---|---|
| 5 | 6.5 |
| 10 | 9.2 |
| 15 | 12.8 |
| 20 | 17.5 |
| 25 | 23.8 |
| 30 | 31.8 |
| 40 | 55.3 |
| 50 | 92.5 |
| 60 | 149.4 |
| 70 | 233.7 |
| 80 | 355.1 |
| 90 | 525.8 |
| 100 | 760.0 |

■ EXERCISE   A 525 ml sample of nitrogen gas was collected over water at 30.0°C and 760 mm Hg. (a) What is the partial pressure of the nitrogen gas? and (b) What would be the volume of dry nitrogen gas at STP?

If the pressure of the system shown in Fig. 12.6 is 755 mm Hg and the temperature is 25°C, what is the partial pressure of the hydrogen? The total pressure is known ($P_t = 755$ mm Hg) and the partial pressure of the water vapor can be obtained from a table of vapor pressures (Table 12.7 and Fig. 12.7). At 25°C the vapor pressure is 23.8 mm Hg.

$P_t = 755$ mm Hg
$P_{H_2O} = 23.8$ mm Hg
$P_t = P_{H_2} + P_{H_2O}$
$P_{H_2} = P_t - P_{H_2O} = (755 - 23.8)$ mm Hg $= 731$ mm Hg

**EXAMPLE 16**   A 271 ml sample of oxygen gas was collected over water at 20.0°C and 763 mm Hg. (a) What is the partial pressure of the oxygen? and (b) What volume would the dry oxygen occupy at STP?

(a)

Given: $P_t = 763$ mm Hg
$T_1 = 20.0$°C
$V_1 = 271$ ml
$P_{H_2O} = 17.5$ mm Hg (from Table 12.7)

Find: $P_{O_2} = ?$

Relationship: $P_t = P_{O_2} + P_{H_2O}$

Solution: $P_{O_2} = P_t - P_{H_2O} = (763 - 17.5)$ mm Hg $= 745$ mm Hg

(b)

Given: $P_{O_2} = 745$ mm Hg (This represents the oxygen pressure of the system if it were dry.)

$T_1 = 20$°C $= (20 + 273)$ °K $= 293$°K
$V_1 = 271$ ml

Find: $V_2 = ?$ at STP

Relationship: $\dfrac{P_1 V_1}{T_1} = \dfrac{P_2 V_2}{T_2}$

Solution: In the above relationship $P_{O_2} = P_1$. Solving for $V_2$ gives:

$V_2 = V_1 \times \dfrac{P_1}{P_2} \times \dfrac{T_2}{T_1}$

$V_2 = 271$ ml $\times \dfrac{745 \text{ mm Hg}}{760 \text{ mm Hg}} \times \dfrac{273\text{°K}}{293\text{°K}} = 248$ ml ■

## 12.7 Relative rates of diffusion

Which molecules move more rapidly on the average under the same conditions, $H_2$ molecules or $O_2$ molecules? The average

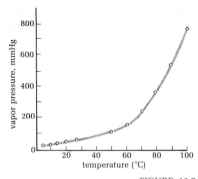

**FIGURE 12.7**
A plot of the vapor pressure of water versus temperature. This graph presents the data from Table 12.7.

kinetic energy of molecules is directly proportional to their absolute temperature.

$$\text{K.E.} \propto T \tag{12.26}$$

Recall that the kinetic energy, K.E., of any moving particle, including gas molecules, is given by:

$$\text{K.E.} = \tfrac{1}{2}mv^2$$

Since gas molecules at the same temperature have the same average kinetic energy regardless of mass, the lighter molecules move more rapidly than heavier ones. Therefore, at a given temperature hydrogen molecules have a greater average velocity than oxygen molecules. Can their relative velocities be compared mathematically?

At a given temperature, the average kinetic energy of any two gases is equal, thus $\text{K.E.}_{H_2}$ is equal to $\text{K.E.}_{O_2}$.

$$\text{K.E.}_{H_2} = \text{K.E.}_{O_2}$$

Since $\text{K.E.} = \tfrac{1}{2}mv^2$,

$$\tfrac{1}{2}m_{H_2} \times (v_{H_2})^2 = \tfrac{1}{2}m_{O_2} \times (v_{O_2})^2$$

or

$$m_{H_2} \times (v_{H_2})^2 = m_{O_2} \times (v_{O_2})^2 \tag{12.27}$$

Equation 12.27 can be rearranged by dividing both sides by $m_{H_2}(v_{O_2})^2$

$$\frac{m_{H_2}(v_{H_2})^2}{m_{H_2}(v_{O_2})^2} = \frac{m_{O_2}(v_{O_2})^2}{m_{H_2}(v_{O_2})^2}$$

$$\frac{v_{H_2}^2}{v_{O_2}^2} = \frac{m_{O_2}}{m_{H_2}} \tag{12.28}$$

The relative velocity, $v_{H_2}/v_{O_2}$, of the two gases can be obtained by extracting the square root of both sides of Eq. 12.28.

$$\frac{v_{H_2}}{v_{O_2}} = \left(\frac{m_{O_2}}{m_{H_2}}\right)^{1/2} \tag{12.29}$$

Equation 12.29 can be used to calculate the relation between the velocities of hydrogen and oxygen. The mass of an oxygen molecule is 32.0 amu; the mass of a hydrogen molecule is 2.0 amu.

$$\frac{v_{H_2}}{v_{O_2}} = \left(\frac{m_{O_2}}{m_{H_2}}\right)^{1/2} = \left(\frac{32.0 \text{ amu}}{2.0 \text{ amu}}\right)^{1/2} = (16.0)^{1/2} = 4.0$$

$$v_{H_2} = 4.0\, v_{O_2}$$

Calculation shows that a hydrogen molecule, on the average, moves 4.0 times faster than an oxygen molecule.

When a gas is placed in an empty container it moves very rapidly throughout the container, filling it completely. Because of the rapid motion of its molecules, a gas has the tendency to move out and fill its container. The molecular motion also causes the molecules of gases to intermingle rapidly when two gases are mixed.

The effect of the molecular mass of gas particles on the average velocity of the particles is illustrated by the laws of diffusion and effusion. In 1831 Thomas Graham discovered the *Law of Diffusion*. According to Graham's observations, *at uniform pressure the rate at which two gases will interdiffuse is inversely proportional to the square roots of their molecular masses.*

Consider an experiment in which the volume of neon that diffuses out is 48.4 ml and the volume of air that diffuses in is 40.0 ml. The ratio of volumes are:

$$\frac{V_{Ne}}{V_{air}} = \frac{48.4 \text{ ml}}{40.0 \text{ ml}} = 1.21$$

Is this ratio inversely proportional to the square root of the molecular masses? Air is about 80 percent nitrogen and 20 percent oxygen and their average mass is 29.0. The mass of neon is 20.2.

$$\left(\frac{m_{air}}{m_{Ne}}\right)^{1/2} = \left(\frac{29.0}{20.2}\right)^{1/2} = (1.43)^{1/2} = 1.20 = \frac{V_{Ne}}{V_{air}}$$

Graham's *Law of Effusion* was published in 1846. This law states that *the rate of effusion of two gases at the same pressure is inversely proportional to the square roots of their molecular masses.* This sounds similar to Graham's Law of Diffusion and in fact, both laws give the same mathematical result. During effusion a gas passes through *a small hole in a thin plate into a vacuum* (Fig. 12.8). It is important to note as Graham did, that " . . . the phenomena of effusion and diffusion are distinct and essentially different in their nature." Both laws can be explained using more rigorous mathematical KMT models than we will cover here.

The relative rate of effusion of two gases can be determined from their relative molecular masses. Consider the following example.

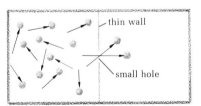

FIGURE 12.8

*Effusion apparatus. The rate at which a gas will effuse into a vacuum is inversely proportional to the square root of its molecular mass. For effusion to occur, the hole must be small and the wall thin.*

EXAMPLE 17 Which gas would effuse faster from an apparatus under

■ EXERCISE  What will be the relative rate of diffusion of nitrogen in carbon dioxide?

■ EXERCISE  A sample of carbon monoxide gas effuses from an apparatus in 25 minutes. A gas of unknown mass effuses from the same apparatus in 45 minutes. (a) Which gas effuses at the greater rate? (b) Which gas has the lower molecular weight? (c) What is the molecular weight of the unknown gas?

the same set of conditions, helium or carbon dioxide? What are their relative rates of effusion?

Given: Carbon dioxide, $m_{CO_2}$ = 44.0 amu
Helium, $m_{He}$ = 4.00 amu

Find: (a) Which gas effuses faster; and (b) $\dfrac{\text{Rate}_{He}}{\text{Rate}_{CO_2}}$ = ?

Relationship: $\dfrac{\text{Rate}_{He}}{\text{Rate}_{CO_2}} = \left(\dfrac{m_{CO_2}}{m_{He}}\right)^{1/2}$

Solution: (a) The lighter gas will effuse faster; in this case helium effuses faster. For (b),

$$\dfrac{R_{He}}{R_{CO_2}} = \left(\dfrac{m_{CO_2}}{m_{He}}\right)^{1/2} = \left(\dfrac{44.0}{4.00}\right)^{1/2} = (11.0)^{1/2} = 3.32 \quad ■$$

## 12.8  Real and ideal gases

What is an ideal gas? Review the assumptions of the kinetic molecular theory included in the introduction to this chapter and note the last two assumptions:

> The gas particles are small compared to the volume occupied by the gas.
>
> The particles of gas are so far apart that the attractive and repulsive forces acting between them are negligible.

An ideal gas is a gas that obeys the gas laws such as Boyle's law, Charles' law, and the Ideal Gas law. The assumptions of the kinetic molecular theory are part of the ideal gas model. There are no completely ideal gases because the gases such as $O_2$, $N_2$, He, Ar, $CO_2$, and others are all *real gases*, that is, their molecules do occupy some space and some forces exist between them.

The assumptions of the kinetic molecular theory are approximately true for real gases. The molecules in a sample of $O_2$ at STP are so far apart that the attraction between the oxygen molecules is extremely small, so small that it can be ignored for practical purposes. The volume occupied by the oxygen molecules at STP is also quite small compared to the volume occupied by the gas.

What happens if a sample of oxygen is cooled while at the same time the external pressure is increased? As the temperature is lowered, the molecules will move more and more slowly. As the pressure is increased, the molecules will be forced closer and closer together. The process is schematically

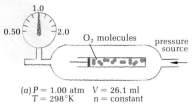

(a) P = 1.00 atm   V = 26.1 ml
T = 298°K   n = constant

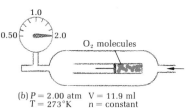

(b) P = 2.00 atm   V = 11.9 ml
T = 273°K   n = constant

FIGURE 12.9
(a) A sample of oxygen occupies a volume of 26.1 ml at 1.00 atm and 298°K. (b) When the temperature of this sample is lowered to 273°K and the external pressure increased to 2.0 atm, the volume occupied is 11.9 ml.

demonstrated in Fig. 12.9. As the molecules move closer together, the attractive forces increase; the volume occupied by the gas becomes smaller and smaller. As we can see from Fig. 12.9, the molecules themselves begin to occupy a larger and larger percentage of the volume as the total volume decreases. As the gas is cooled and the external pressure is increased, the assumptions of the kinetic molecular theory become less true. The gas laws are no longer valid. Under conditions of high pressures and low temperatures, the ideal gas model must be modified.

## Summary

The volume occupied by a sample of gas depends upon the number of moles of gas in the sample, the pressure, and the temperature of the sample. If the temperature and number of moles of gas are held constant, the volume occupied by a gas is inversely proportional to the pressure which it exerts:

*Boyle's law*     $PV = k$

If the pressure and the number of moles of gas are held constant, the volume occupied by a gas is directly proportional to its absolute temperature:

*Charles' law*     $V = cT$

The Combined Gas law puts the observations of both Boyle's law and Charles' law into a single equation. The number of moles, n, is the only quantity which must be held constant for this law to apply:

*Combined Gas law*     $\dfrac{PV}{T} = C$

A gas will behave ideally if:

1. The space occupied by the gas molecules themselves is small compared to the total volume occupied by the gas.
2. The gas molecules are far enough apart so that the attractive and repulsive forces between them are negligible.

Gases deviate from this behavior at high pressures and low temperatures. At all temperatures and pressures considered in this text, the common gases will be assumed to exhibit ideal gas behavior.

Gases which exhibit ideal gas behavior not only obey Boyle's law, Charles' law, and the Combined Gas law, but they also obey the Ideal Gas law:

*Ideal Gas law*     $PV = nRT$

In fact, Boyle's law, Charles' law, and the Combined Gas law can be considered as special cases of the Ideal Gas law.

In a mixture of gases, each gas exerts its own pressure, as if the other gases were not even present. The total pressure exerted by all of the gases present in a mixture of gases is the sum of the partial pressures exerted by the individual gases. This statement is known as Dalton's Law of Partial Pressures:

$$P_{total} = P_1 + P_2 + P_3$$

The tendency of gases to spread through another gas is known as diffusion. The rates of diffusion of gases have been observed experimentally to be inversely proportional to the square roots of their molecular weights. This proportionality is known as Graham's Law of Diffusion.

$$\text{Graham's Law of Diffusion} \quad \frac{V_1}{V_2} = \left(\frac{M_2}{M_1}\right)^{1/2}$$

## Objectives

By the end of this chapter, you should be able to do the following:

1. List the differences between *real gases* and *ideal gases*. (Questions 2, 4)
2. State the assumptions of the kinetic molecular theory. (Question 2)
3. Calculate the density of a gas at STP. (Question 10)
4. Given the pressure and volume of a gas in an initial state, calculate the final volume (or pressure) when the pressure (or volume) has changed. (Question 7)
5. Given the temperature and volume of a gas in an initial state, calculate the final volume (or temperature) when the temperature (or volume) has changed. (Questions 15, 20)
6. Given the temperature, pressure, and volume of a gas in an initial state, calculate one of these variables in a final state if the other two are known. (Questions 8, 14)
7. Use the Ideal Gas law, $PV = nRT$, for calculations. (Questions 9, 12, 13)
8. Use Dalton's Law of Partial Pressures for calculations. (Questions 17, 19, 22)
9. Calculate the relative rates of diffusion of two gases. (Questions 11, 18, 21)

## Questions

1. Define each of the following terms: (Give examples where possible.) *real gas; ideal gas; kinetic molecular theory; kinetic energy; Boyle's law; Charles' law; Combined Gas law; Ideal Gas law; Graham's Law of Diffusion; of Effusion; Dalton's Law of Partial Pressures; gas constant, R; partial pressure; diffusion; effusion.*

2. What part of the ideal gas model does not apply to gases at high pressures and low temperatures?

3. Use the kinetic molecular theory to explain why the temperature of a real gas rises when it is compressed.

4. List the differences between real gases and ideal gases.

5. When bubbles of gas are released below the surface of water, they get larger as they approach the surface. Explain this phenomenon.

6. What is the effect of an increase in temperature upon the kinetic energy of molecules? Upon their velocities?

7. The pressure exerted by a 0.51 liter sample of gas at 25°C is 2.01 atm. What pressure will this same sample of gas exert at 25°C and a volume of 1.02 liters?

8. A sample of hydrogen occupies a volume of 312 ml at 27.0°C and 793 mm Hg pressure. What volume will this same quantity of $H_2$ occupy at STP?

9. The volume of an automobile tire is about 100 liters. How many moles of gas (or gases) would be required to fill a 100 liter tire so that a pressure of 3.00 atm is reached at 25°C? How many molecules would this be?

10. Natural gas consists largely of methane, $CH_4$. Calculate the density of methane at STP.

11. Which will diffuse faster, $CH_4$ or $CO_2$? Calculate their relative rates of diffusion.

12. At what temperature will 2.15 moles of helium exert a pressure of 3.50 atm while occupying a volume of 32.2 liters?

13. A sample of gas whose mass is 0.532 gram occupies a volume of 243 ml at 373°K and 1.05 atm pressure. What is the molecular weight of this substance?

14. A sample of acetylene $C_2H_2$, occupies a volume of 12.2 liters at 298°K and 720 mm Hg. What volume will the same sample of acetylene occupy at 350°K and 760 mm Hg pressure?

15. A sample of nitrogen occupies 12.2 ml in a gas syringe at 273°K and 1.00 atm pressure. What volume will this same sample of $N_2$ occupy at 373°K and 1.00 atm pressure?

16. What is the mass of 1.00 liter of methane at STP?

17. A mixture of gases is made by adding the following weights of gases to a 10.0 liter flask at 273°K: 6.4 g $O_2$, 2.8 g $N_2$, and 4.0 g He. What is the total pressure exerted by the mixture? What is the partial pressure of each gas?

18. An unknown gas effuses at a rate $1/8$ as fast as hydrogen. What is the molecular weight of the unknown gas?

19. A mixture of helium and argon is formed by adding 0.14 mole of argon and 0.35 mole of He to a 4.53 liter container at 373°K. What is the partial pressure of He? Ar?

20. Assuming that the pressure is held constant at 1.00 atm, calculate the final volume of each of the following:

(a) 5.13 liters of Ne at 27.0°C are cooled to −123°C
(b) 4.02 ml of $H_2$ at 273°K are heated to 373°K
(c) 1.02 liters of $N_2$ at 1000°K is cooled to 200°K
(d) 22.4 liters of $O_2$ at 25.0°C are heated to 323°C.

21. Which will diffuse faster, neon gas or mercury vapor? Calculate their relative rates of diffusion.

22. What will be the total pressure exerted by a mixture of gases if each of the gases in the mixture exerts the following: Gas A: 0.15 atm; Gas B: 0.35 atm; Gas C: 0.65 atm; Gas D: 1.05 atm.

# 13

# Calculations involving chemical equations

## 13.1 Introduction

Many electrical power plants burn sulfur-containing coal. When the coal is burned, the sulfur is oxidized to sulfur dioxide, a common pollutant. Sulfur dioxide, which is the oxide of a nonmetal, is an acidic oxide, and therefore it can be removed by a basic oxide such as CaO. Calcium oxide is used for this purpose because it is quite inexpensive to prepare by decomposing $CaCO_3$. What weight of calcium oxide is needed to remove $7.2 \times 10^5$ grams of sulfur dioxide each day from the exhaust gas of an electrical power station? The answer to this problem involves the use of a chemical equation since calcium oxide and sulfur dioxide react to form calcium sulfite. The process of using *chemical equations* in making chemical calculations is known as *stoichiometry*.

In this chapter we will apply the concept of unity terms from balanced equations to solve stoichiometry problems. The main ideas and techniques used in solving these problems have already been introduced in Chapters 1–12. This chapter organizes these techniques into a systematic approach to solving such problems.

## 13.2 Chemical unity terms—a review

A *unity term* is a fraction whose numerator and denominator are equivalent (see Chapter 2). The numerator and denominator must be different expressions of the same thing. For example:

(1) $\dfrac{1000 \text{ mg}}{1.00 \text{ g}} = 1$

(2) $\dfrac{454 \text{ g}}{1.00 \text{ lb}} = 1$

(3) $\dfrac{1.00 \text{ mole } O_2}{22.4 \text{ liters } O_2} = 1$

(4) $\dfrac{1.00 \text{ mole } O_2}{32.0 \text{ g } O_2} = 1$

A balanced chemical equation indicates the number of moles of one substance that are chemically equivalent to a given number of moles of another substance in the reaction which the equation represents (see Chapters 9 and 11). Unity terms derived from balanced equations relate the amount of one substance (in the numerator) that is *chemically equivalent* to the amount of another substance (in the denominator) in this reaction. For example:

$$4KClO_3(s) \xrightleftharpoons{\text{gentle heating}} 3KClO_4(s) + KCl(s) \qquad (13.1)$$

According to the coefficients in this equation, 4 moles of $KClO_3$ are equivalent to (will produce) 3 moles of $KClO_4$.

$$4 \text{ moles } KClO_3 \approx 3 \text{ moles } KClO_4 \qquad (13.2)$$

The same equation also indicates that 4 moles of $KClO_3$ will produce 1 mole of KCl.

$$4 \text{ moles } KClO_3 \approx 1 \text{ mole } KCl \qquad (13.3)$$

When 3 moles of $KClO_4$ are produced, 1 mole of KCl will be produced.

$$3 \text{ moles } KClO_4 \approx 1 \text{ mole } KCl \qquad (13.4)$$

Equations 13.2, 13.3, and 13.4 may be expressed as the following unity terms:

$$\frac{4 \text{ moles KClO}_3}{3 \text{ moles KClO}_4} = 1 \quad \text{or} \quad \frac{3 \text{ moles KClO}_4}{4 \text{ moles KClO}_3} = 1$$

$$\frac{4 \text{ moles KClO}_3}{1 \text{ mole KCl}} = 1 \quad \text{or} \quad \frac{1 \text{ mole KCl}}{4 \text{ moles KClO}_3} = 1$$

$$\frac{3 \text{ moles KClO}_4}{1 \text{ mole KCl}} = 1 \quad \text{or} \quad \frac{1 \text{ mole KCl}}{3 \text{ moles KClO}_4} = 1$$

**Grams per mole**

The weight of a mole of a substance is equal to its atomic weight, molecular weight, or formula weight expressed in grams. It is often necessary to use a grams per mole unity term to solve chemical problems. For example, one mole of *chlorine atoms* weighs 35.5 grams:

1.00 mole Cl atoms ≈ 35.5 g Cl atoms

One mole of *chlorine molecules* weighs 71.0 grams:

1.00 mole $Cl_2$ ≈ 71.0 g $Cl_2$

One mole of *nitric acid molecules* weighs 63.0 grams:

1.00 mole $HNO_3$ ≈ 63.0 g $HNO_3$

The relationship between moles and grams in these examples may be expressed as the following unity terms:

(1) $\dfrac{1.00 \text{ mole Cl atoms}}{35.5 \text{ g Cl atoms}} = 1 \quad \text{or} \quad \dfrac{35.5 \text{ g Cl atoms}}{1.00 \text{ mole Cl atoms}} = 1$

(2) $\dfrac{1.00 \text{ mole Cl}_2 \text{ molecules}}{71.0 \text{ g Cl}_2 \text{ molecules}} = 1 = \dfrac{71.0 \text{ g Cl}_2 \text{ molecules}}{1.00 \text{ mole Cl}_2 \text{ molecules}}$

(3) $\dfrac{1.00 \text{ mole HNO}_3}{63.0 \text{ g HNO}_3} = 1 \quad \text{or} \quad \dfrac{63.0 \text{ g HNO}_3}{1.00 \text{ mole HNO}_3} = 1$

■ EXERCISE  Write unity terms expressing the relationship between grams and moles for the following substances: (a) fluorine molecules, (b) xenon tetrafluoride.

**Liters of gas per mole**

At STP one mole of an ideal gas occupies 22.4 liters. Thus 22.4 liters of gas is physically equivalent to 1.00 mole of gas at STP:

1.00 mole gas ≈ 22.4 liters

Specific examples are:

1.00 mole $CO_2$ ≈ 22.4 liters $CO_2$ at STP
1.00 mole He ≈ 22.4 liters He at STP

■ EXERCISE  Write unity terms expressing the relationship between the number of moles and the volume in liters for samples of the following gases at STP: (a) xenon; (b) $NH_3$.

The relationship between moles and volume for the given examples results in the following unity terms:

(1) $\dfrac{1.00 \text{ mole } CO_2}{22.4 \text{ liters } CO_2} = 1$   or   $\dfrac{22.4 \text{ liters } CO_2}{1.00 \text{ mole } CO_2} = 1$

(2) $\dfrac{1.00 \text{ mole He}}{22.4 \text{ liters He}} = 1$   or   $\dfrac{22.4 \text{ liters He}}{1.00 \text{ mole He}} = 1$ ■

**Particles per mole**

One mole of a substance contains $6.02 \times 10^{23}$ particles of that substance. Thus $6.02 \times 10^{23}$ particles are equivalent to 1.00 mole of a substance. For example, one mole of manganese contains $6.02 \times 10^{23}$ magnanese atoms:

1.00 mole Mn ≈ $6.02 \times 10^{23}$ Mn atoms

One mole of bromine *molecules* contains $6.02 \times 10^{23}$ bromine *molecules*:

1.00 mole $Br_2$ molecules ≈ $6.02 \times 10^{23}$ $Br_2$ molecules

One mole of calcium fluoride contains $6.02 \times 10^{23}$ calcium fluoride formula units, $CaF_2$.

1.00 mole $CaF_2$ ≈ $6.02 \times 10^{23}$ $CaF_2$ formula units

The relationship between moles and particles in the previous examples gives the following unity terms:

(1) $\dfrac{1.00 \text{ mole Mn}}{6.02 \times 10^{23} \text{ Mn atoms}} = 1$   or   $\dfrac{6.02 \times 10^{23} \text{ Mn atoms}}{1.00 \text{ mole Mn}} = 1$

(2) $\dfrac{1.00 \text{ mole } Br_2 \text{ molecules}}{6.02 \times 10^{23} \text{ } Br_2 \text{ molecules}} = 1$   or   $\dfrac{6.02 \times 10^{23} \text{ } Br_2 \text{ molecules}}{1.00 \text{ mole } Br_2 \text{ molecules}} = 1$

■ EXERCISE  Write unity terms expressing the relationship between moles and particles for the following substances: (a) gold atoms; (b) $NH_3$.

(3) $\dfrac{1.00 \text{ mole } CaF_2}{6.02 \times 10^{23} \text{ } CaF_2 \text{ formula units}}$   or   $\dfrac{6.02 \times 10^{23} \text{ } CaF_2 \text{ formula units}}{1.00 \text{ mole } CaF_2} = 1$ ■

**Using unity terms**

The relationship between moles, grams, and number of particles is illustrated in Fig. 13.1. Multiplying a quantity by a frac-

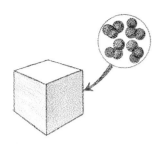

substance: nitrogen dioxide, NO$_2$
conditions: STP
state: gas
volume: 22.4 liters
amount: 1.00 mole
weight: 46.0 grams
particles: 6.02 × 10$^{23}$ NO$_2$ molecules

FIGURE 13.1

*The box contains one mole of nitrogen dioxide at STP. One mole of NO$_2$ weighs 46.0 grams, has a volume of 22.4 liters, and contains 6.02 × 10$^{23}$ molecules.*

weight of mole ⟶ grams

number of particles ⟶ 6.02 × 10$^{23}$

volume of gas at STP ⟶ 22.4 liters

FIGURE 13.2

*One mole of a substance is equivalent to a certain weight, to a definite number of particles, and to a definite volume (of a gas at STP).*

tion equal to unity does not change the value of that quantity. In the same manner, multiplying a chemical quantity by a unity term does not change that quantity but rather gives an *equivalent* expression. In the following examples, we will demonstrate the use of unity terms as just discussed in solving chemical problems. The method of approach used in these problems is shown in Fig. 13.2.

*Mole-gram conversions*

EXAMPLE 1   A flask of argon gas contains 0.137 gram of argon. How many moles of argon is this?

*Given:* 0.137 g argon

*Find:* The number of moles of argon equivalent to 0.137 g argon

*Relationship:* From the periodic chart: 1.00 mole argon ≈ 39.9 g argon

*Solution:* $0.137 \text{ g Ar} \times \dfrac{1.00 \text{ mole Ar}}{39.9 \text{ g Ar}} = 0.00344$ mole Ar

$= 3.44 \times 10^{-3}$ mole Ar

Note that the unity term has "grams Ar" in the denominator so the two terms "grams Ar" will mathematically "cancel" one another.

EXAMPLE 2   A flask of nitrogen gas contains 0.984 gram of nitrogen. How many moles of nitrogen is this?

*Given:* 0.984 g nitrogen

*Find:* The number of moles of nitrogen equivalent to 0.984 g nitrogen. *Note:* For a diatomic gas like nitrogen the statement "moles of nitrogen" is assumed to mean "moles of nitrogen molecules."

*Relationship:* 1 mole of nitrogen, N$_2$ ≈ 28.0 g nitrogen

*Solution:* $0.984 \text{ g N}_2 \times \dfrac{1.00 \text{ mole N}_2}{28.0 \text{ g N}_2} = 0.0351$ mole N$_2$

$= 3.51 \times 10^{-2}$ mole N$_2$

EXAMPLE 3   A chemistry laboratory manual calls for the use of 0.0240 mole of sulfuric acid. What weight of hydrogen sulfate must be in the sample of sulfuric acid solution which is used for this experiment?

*Given:* 0.0240 mole of sulfuric acid

*Find:* The number of grams of sulfuric acid equivalent to 0.0240 mole of sulfuric acid.

*Relationship:* 1.00 mole H$_2$SO$_4$ ≈ 98.1 g H$_2$SO$_4$

*Solution:* $0.0240 \text{ mole H}_2\text{SO}_4 \times \dfrac{98.1 \text{ g H}_2\text{SO}_4}{1.00 \text{ mole H}_2\text{SO}_4} = 2.35 \text{ g H}_2\text{SO}_4$

*Mole-particle conversions*

**EXAMPLE 4** How many atoms are present in a 4.57 mole sample of gold?

Given: 4.57 moles Au

Find: The number of atoms of gold equivalent to 4.57 moles of gold

Relationship: 1 mole Au $\approx 6.02 \times 10^{23}$ Au atoms

Solution: 4.57 moles Au $\times \dfrac{6.02 \times 10^{23} \text{ atoms Au}}{1.00 \text{ mole Au}}$

$= 2.75 \times 10^{24}$ atoms Au

**EXAMPLE 5** A chemistry professor told his class he could carry 4,600,000,000,000,000,000,000 water molecules on one finger. How many moles of water is this? How many grams?

Given: 4,600,000,000,000,000,000,000 or $4.6 \times 10^{21}$ molecules of water

(a)

Find: The number of moles of water equivalent to $4.6 \times 10^{21}$ molecules of water

Relationship: 1.00 mole $H_2O \approx 6.02 \times 10^{23}$ Molecules $H_2O$

Solution: $4.6 \times 10^{21}$ molecules $H_2O \times \dfrac{1.00 \text{ mole } H_2O}{6.02 + 10^{23} \text{ molecules } H_2O}$

$= 7.6 \times 10^{-3}$ mole $H_2O$

(b)

Find: The number of grams of $H_2O$ in $7.6 \times 10^{-3}$ mole of $H_2O$

Relationship: 18.0 g $H_2O \approx 1.00$ mole $H_2O$

Solution: $7.6 \times 10^{-3}$ mole $H_2O \times \dfrac{18.0 \text{ g } H_2O}{1.00 \text{ mole } H_2O} = 0.14$ g $H_2O$

Since 1.00 drop of water weighs 0.050 g, this is almost 3 drops.

*Mole-volume conversions*

**EXAMPLE 6** An "empty" test tube has a capacity of 0.100 liter. How many moles of "air" molecules are there in the "empty" test tube at STP. Note: Even though air is a mixture, we can consider it a substance for the purpose of solving this problem.

Given: Volume of a gas is 0.100 liter

Find: The number of moles of gas equivalent to 0.100 liter

Relationship: 1 mole of gas $\approx 22.4$ liters (at STP)

Solution: 0.100 liter gas $\times \dfrac{1.00 \text{ mole gas}}{22.4 \text{ liters gas}} = .00446$ mole gas

$= 4.46 \times 10^{-3}$ mole gas

■ EXERCISE  A sample of carbon tetrachloride weighs 0.814 gram. (a) How many moles of carbon tetrachloride are there in the sample? (b) Caluculate the number of molecules in the sample.

■ EXERCISE  A certain telephone booth has a volume of $3.0 \times 10^3$ liters. How many moles of gas can the telephone booth hold at STP?

EXAMPLE 7  What will be the volume of 0.0649 mole of fluorine gas at STP?

Given: 0.0649 mole $F_2$

Find: The number of liters of fluorine that are equivalent to 0.0649 mole of $F_2$

Relationship: 1.00 mole $F_2 \approx$ 22.4 liters $F_2$

Solution: 0.0649 mole $F_2 \times \dfrac{22.4 \text{ liters } F_2}{1.00 \text{ mole } F_2} = 1.45$ liters $F_2$   ■

## Using the mole concept

What volume will 11.03 grams of carbon dioxide gas occupy at STP? We could find the volume of 11.03 grams carbon dioxide if we first found how many moles 11.03 grams are.

$$\text{grams} \longrightarrow \text{moles} \longrightarrow \text{volume}$$

*Problems such as this are readily solved by converting whatever is given to moles, and then changing moles to what is asked for.*

$$\text{given} \longrightarrow \text{moles} \longrightarrow \text{find}$$

EXAMPLE 8  A small piece of dry ice (solid carbon dioxide) weighs 11.03 grams. What volume will the carbon dioxide occupy after it sublimes to a gas at STP?

Given: 11.03 grams of carbon dioxide

Find: Volume of 11.03 grams carbon dioxide at STP

Relationship: 1.00 mole $CO_2 \approx$ 44.0 g $CO_2$
             1.00 mole $CO_2 \approx$ 22.4 liters $CO_2$

Solution: 11.03 g $CO_2 \times \dfrac{1.00 \text{ mole } CO_2}{44.0 \text{ g } CO_2} \times \dfrac{22.4 \text{ liters } CO_2}{1.00 \text{ mole } CO_2} = 5.75$ liters $CO_2$

— changes moles to liters
— changes grams to moles

The general outline to the solution of this problem was:

given ⟶ moles ⟶ find
(g $CO_2$)  (moles $CO_2$)  (liters $CO_2$)

EXAMPLE 9  How many molecules of $CO_2$ are present in 11.03 grams of carbon dioxide?

Given: 11.03 grams $CO_2$

Find: The number of molecules equivalent to 11.03 grams $CO_2$
Relationship: 1.00 mole $CO_2 \approx 44.0$ g $CO_2$
1.00 mole $CO_2 \approx 6.02 \times 10^{23}$ molecules $CO_2$

Solution: given ⟶ moles ⟶ find
(g $CO_2$)　(moles $CO_2$)　(molecules $CO_2$)

$$11.03 \text{ g } CO_2 \times \frac{1.00 \text{ mole } CO_2}{44.0 \text{ g } CO_2} \times \frac{6.02 \times 10^{23} \text{ molecules } CO_2}{1.00 \text{ mole } CO_2}$$
$$= 1.51 \times 10^{23} \text{ molecules } CO_2$$

**EXAMPLE 10** What is the weight in grams of one atom of lead?

Given: 1 atom of lead

Find: The number of grams equivalent to one atom of lead

Relationship: 1.00 mole Pb $\approx 207$ g Pb
1.00 mole Pb $\approx 6.02 \times 10^{23}$ Pb atoms

Solution: given ⟶ moles ⟶ find
(Pb atom)　(moles Pb atoms)　(g Pb)

$$1 \text{ Pb atom} \times \frac{1.00 \text{ mole Pb}}{6.02 \times 10^{23} \text{ Pb atoms}} \times \frac{207 \text{ g Pb}}{1.00 \text{ mole Pb}} = 3.4 \times 10^{-22} \text{ g Pb}$$

changes moles to grams

changes atoms to moles

**EXAMPLE 11** Silicon tetrafluoride is a gas at STP. What is the weight of 15.0 ml $SiF_4$ at STP?

Given: 15.0 ml or 0.0150 liter $SiF_4$

Find: The number of grams of $SiF_4$ equivalent to 0.0150 liter $SiF_4$

Relationship: 1.00 mole $SiF_4 \approx 104$ g $SiF_4 \approx 22.4$ liters $SiF_4$

Solution: given ⟶ moles ⟶ find
(liters $SiF_4$)　(moles $SiF_4$)　(g $SiF_4$)

$$0.0150 \text{ liter } SiF_4 \times \frac{1.00 \text{ mole } SiF_4}{22.4 \text{ liters } SiF_4} \times \frac{104 \text{ g } SiF_4}{1.00 \text{ mole } SiF_4}$$
$$= 0.0697 \text{ g } SiF_4 \quad \blacksquare$$

■ EXERCISE  A balloon has a volume of 3.79 liters and contains helium at STP. (a) How many moles of helium are in the balloon? (b) How many atoms of helium are in the balloon? (c) What weight of helium is in the balloon?

■ EXERCISE  A bottle of a certain beverage contains 200 grams of ethyl alcohol, $CH_3CH_2OH$. How many molecules of alcohol does it contain?

## 13.3 The mole relationship in a chemical equation

A chemical equation gives us the *mole ratios* of the substances involved in a reaction. For example, ethane gas burns in air to form carbon dioxide and water vapor.

$$2C_2H_6(g) + 7O_2(g) \rightleftharpoons 4CO_2(g) + 6H_2O(g)$$

How many moles of oxygen are needed to burn 0.600 mole of ethane? According to the coefficients in the equation, the mole ratio of ethane to oxygen is 2 to 7. In this particular reaction, two moles of ethane are chemically equivalent to seven moles of oxygen. The unity terms that relate ethane and oxygen are:

$$\frac{2 \text{ moles } C_2H_6}{7 \text{ moles } O_2} = 1 \quad \text{or} \quad \frac{7 \text{ moles } O_2}{2 \text{ moles } C_2H_6} = 1$$

Since we want to convert moles of ethane to the chemically equivalent number of moles of oxygen, we should arrange the unity term so that moles of ethane cancel out.

$$0.600 \text{ mole } C_2H_6 \times \frac{7 \text{ moles } O_2}{2 \text{ moles } C_2H_6} = 2.10 \text{ moles } O_2$$

Therefore 0.600 mole of ethane required 2.10 moles of oxygen for reaction. Figure 13.3 illustrates a technique used for solving problems where you are given moles of substance A and asked to find moles of substance B.

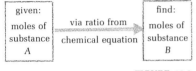

FIGURE 13.3
*Use the coefficients from the chemical equation to calculate the moles of substance B that are equivalent to moles of substance A.*

EXAMPLE 12   Calcium oxide and tetraphosphorus decaoxide react to form calcium phosphate.

$$6CaO(s) + P_4O_{10}(s) \rightleftharpoons 2Ca_3(PO_4)_2(s)$$

(a) How many moles of $P_4O_{10}$ are needed to react with 5.72 moles of CaO? (b) How many moles of $Ca_3(PO_4)_2$ will be produced when 5.72 moles of CaO react?

(a)

Given: 5.72 moles CaO

Find: The number of moles of $P_4O_{10}$ equivalent to 5.72 moles of CaO

Relationship: From the equation, 6 moles CaO ≈ 1 mole $P_4O_{10}$

Solution: $5.72 \text{ moles CaO} \times \dfrac{1.00 \text{ mole } P_4O_{10}}{6.00 \text{ moles CaO}} = 0.953 \text{ mole } P_4O_{10}$

(b)

Given: 5.72 moles CaO

Find: The number of moles of $Ca_3(PO_4)_2$ equivalent to 5.72 moles of CaO

Relationship: From the equation, 6 moles CaO ≈ 2 moles $Ca_3(PO_4)_2$

Solution: $5.72 \text{ moles CaO} \times \dfrac{2.00 \text{ moles } Ca_3(PO_4)_2}{6.00 \text{ moles CaO}}$

$= 1.91 \text{ moles } Ca_3(PO_4)_2$

13.3 The mole relationship in a chemical equation

■ **EXERCISE** Proteins are large molecules necessary for life. Some proteins act as enzymes and hormones; they are necessary for muscle contraction, vision, and breathing. Proteins are made up of smaller molecules called amino acids. One amino acid called glycine can be synthesized as follows:

$$CH_2ClCO_2H(aq) + NH_3(aq) \rightleftharpoons$$
chloroacetic acid
$$CH_2NH_2CO_2H(aq) + HCl(aq)$$
glycine

How many moles of ammonia are needed to make 0.173 mole of glycine?

■ **EXERCISE** Concentrated nitric acid reacts with nonmetals such as sulfur.

$$S(s) + 6HNO_3(aq) \rightleftharpoons$$
$$H_2SO_4(aq) + 6NO_2(g) + 2H_2O(l)$$

(a) If 0.100 mole of sulfur is dissolved, how many moles of nitric acid react?
(b) How many moles of each product result from the reaction of 0.100 mole of sulfur?

**EXAMPLE 13** Calcium oxide, like other ionic oxides, reacts with sulfuric acid. The products are calcium sulfate and water.

$$CaO(s) + H_2SO_4(aq) \rightleftharpoons CaSO_4(s) + H_2O(l)$$

How many moles of calcium oxide are needed to neutralize 0.0375 mole of sulfuric acid?

Given: 0.0375 mole $H_2SO_4$

Find: The number of moles calcium oxide equivalent to 0.0375 mole of $H_2SO_4$

Relationship: From the equation, 1 mole $H_2SO_4 \approx$ 1 mole CaO

Solution: 0.0375 mole $H_2SO_4 \times \dfrac{1.00 \text{ mole CaO}}{1.00 \text{ mole } H_2SO_4}$

$$= 0.0375 \text{ mole CaO}$$

**EXAMPLE 14** Ammonia gas will react with phosphoric acid and form ammonium phosphate.

$$3NH_3(g) + H_3PO_4(l) \rightleftharpoons (NH_4)_3PO_4(s)$$

How many moles of ammonia are needed to neutralize 4.13 moles of phosphoric acid? How would you solve this problem?

Find: The number of moles of ammonia equivalent to 4.13 moles of phosphoric acid

Relationship: From the equation, 3.00 moles $NH_3 \approx$ 1.00 mole $H_3PO_4$

Solution: 4.13 moles $H_3PO_4 \times \dfrac{3.00 \text{ moles } NH_3}{1.00 \text{ mole } H_3PO_4}$

$$= 12.4 \text{ moles } NH_3 \quad ■$$

## 13.4 Calculations involving chemical equations

In Section 13.1 we asked, "What weight of calcium oxide is needed to remove $7.2 \times 10^5$ grams of sulfur dioxide from an exhaust gas?" To answer this question, we must know the products of the reaction and what the mole ratios are of calcium oxide to sulfur dioxide. The product of the reaction between calcium oxide and sulfur dioxide is calcium sulfite:

$$CaO(s) + SO_2(g) \rightleftharpoons CaSO_3(s)$$

The coefficients in the equation show us that one mole of calcium oxide is equivalent to one mole of sulfur trioxide.

1.00 mole CaO $\approx$ 1.00 mole $SO_3$

We were given *grams* of sulfur dioxide, though, *not moles* of sulfur dioxide. If we can convert *grams* of sulfur dioxide to moles of sulfur dioxide, then we can use the *mole ratio* to calculate the number of *moles* of calcium oxide needed. Since the question asks for grams of calcium oxide, we must convert *moles* of calcium oxide to *grams* of calcium oxide. An outline of these steps is:

$$\underset{given}{\text{grams SO}_2} \longrightarrow \text{moles SO}_2 \longrightarrow \text{moles CaO} \longrightarrow \underset{find}{\text{grams CaO}}$$

$$7.2 \times 10^5 \text{ g SO}_2 \times \underset{\substack{\uparrow \\ \text{converts grams SO}_2 \\ \text{to moles SO}_2}}{\frac{1.00 \text{ mole SO}_2}{64.1 \text{ g SO}_2}} \times \underset{\substack{\uparrow \\ \text{converts moles of} \\ \text{SO}_2 \text{ to moles of CaO}}}{\frac{1.00 \text{ mole CaO}}{1.00 \text{ mole SO}_2}} \times \underset{\substack{\uparrow \\ \text{converts moles CaO} \\ \text{to grams CaO}}}{\frac{56.1 \text{ g CaO}}{1.00 \text{ mole CaO}}} = 6.3 \times 10^5 \text{ CaO}$$

Notice that in solving this problem we converted the units given (grams) to moles. We then used the mole ratio from the equation to convert from moles of one substance to moles of the other substance. Finally, we converted from moles to the units desired (grams). In problems such as this we will follow the same technique (Fig. 13.4). Remember, the units that are initially *given* may commonly be in grams, particles, liters of gas, or moles. The units of the quantity that you are to *find* may also commonly be in grams, particles, liters of gas, or moles. You will find the following examples more beneficial if you attempt to solve them by yourself before reading the solution.

**EXAMPLE 15** When heated in oxygen at 500°C, barium oxide reacts to form barium peroxide.

$$2\text{BaO}(s) + \text{O}_2(g) \rightleftharpoons 2\text{BaO}_2(s)$$

(a) How many liters of oxygen at STP are needed to react with 13.7 grams of barium oxide? (b) How many grams of barium peroxide can be produced from 13.7 grams of barium oxide?

(a)

Given: 13.7 g BaO

Find: The number of liters of oxygen that are equivalent to 13.7 g BaO

Outline of Solution: grams BaO → moles BaO → moles $O_2$ → liters $O_2$

Relationships: 1.00 mole BaO ≈ 153 g BaO
2.00 moles BaO ≈ 1.00 mole $O_2$
1.00 mole $O_2$ ≈ 22.4 liters $O_2$

Solution: $13.7 \text{ g BaO} \times \dfrac{1.00 \text{ mole BaO}}{153 \text{ g BaO}} \times \dfrac{1.00 \text{ mole O}_2}{2.00 \text{ moles BaO}}$
$\times \dfrac{22.4 \text{ liters O}_2}{1.00 \text{ mole O}_2} = 1.00 \text{ liter O}_2$

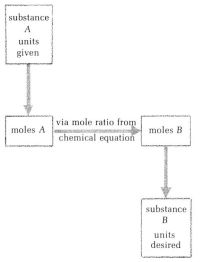

**FIGURE 13.4**
An outline of a technique for solving problems involving chemical equations. The units given may be grams, particles, or liters of substance A. These are converted to moles of substance A. Moles of substance A are converted to moles of substance B via the mole ratio from the chemical equation. Moles of substance B are then converted to the units desired.

(b)

Given: 13.7 g BaO

Find: The number of grams of barium peroxide equivalent to 13.7 g barium oxide

Outline of Solution: grams BaO → moles BaO → moles $BaO_2$ → grams $BaO_2$

Relationships: 1.00 mole BaO ≈ 153 g BaO
2.00 moles BaO ≈ 2.00 moles $BaO_2$
169 g $BaO_2$ ≈ 1.00 mole $BaO_2$

Solution: $13.7 \text{ g BaO} \times \dfrac{1.00 \text{ mole BaO}}{153 \text{ g BaO}} \times \dfrac{2.00 \text{ moles BaO}_2}{2.00 \text{ moles BaO}}$
$\times \dfrac{169 \text{ g BaO}_2}{1.00 \text{ mole BaO}_2} = 15.1 \text{ g BaO}_2$

**EXAMPLE 16** Nitric acid dissolves copper (II) sulfide as follows:

$3CuS(s) + 8HNO_3(aq) \rightleftharpoons 3Cu(NO_3)_2(aq) + 3S(s) + 2NO(g) + 4H_2O(l)$

What *weight of each product* is formed, if 0.601 mole of nitric acid react?

Given: 0.601 mole $HNO_3$

Find: The number of grams of each product equivalent to 0.601 mole $HNO_3$

Outline of Solution: moles $HNO_3$ → moles of product → grams of product

Relationships: 8 moles $HNO_3$ ≈ 3 moles $Cu(NO_3)_2$
8 moles $HNO_3$ ≈ 3 moles S
8 moles $HNO_3$ ≈ 2 moles NO
8 moles $HNO_3$ ≈ 4 moles $H_2O$
1.00 mole $Cu(NO_3)_2$ ≈ 187 g $Cu(NO_3)_2$
1.00 mole S ≈ 32.1 g S
1.00 mole NO ≈ 30.0 g NO
1.00 mole $H_2O$ ≈ 18.0 g $H_2O$

Solutions:

For $Cu(NO_3)_2$,

$0.601 \text{ mole HNO}_3 \times \dfrac{3 \text{ moles Cu(NO}_3)_2}{8 \text{ moles HNO}_3} \times \dfrac{187 \text{ g Cu(NO}_3)_2}{1.00 \text{ mole Cu(NO}_3)_2}$
$= 42.2 \text{ g Cu(NO}_3)_2$

For S,

$0.601 \text{ mole HNO}_3 \times \dfrac{3 \text{ moles S}}{8 \text{ moles HNO}_3} \times \dfrac{32.1 \text{ g S}}{1.00 \text{ mole S}} = 7.23 \text{ g S}$

For NO,

$0.601 \text{ mole HNO}_3 \times \dfrac{2 \text{ moles NO}}{8 \text{ moles HNO}_3} \times \dfrac{30.0 \text{ g NO}}{1.00 \text{ mole NO}} = 4.52 \text{ g NO}$

For $H_2O$,

$$0.601 \text{ mole } HNO_3 \times \frac{4 \text{ moles } H_2O}{8 \text{ moles } HNO_3} \times \frac{18.0 \text{ g } H_2O}{1.00 \text{ mole } H_2O} = 5.41 \text{ g } H_2O$$

EXAMPLE 17  At elevated temperatures, methane reacts with steam forming carbon monoxide and hydrogen.

$$CH_4(g) + H_2O(g) \rightleftharpoons CO(g) + 3H_2(g)$$

(a) If 1000 liters of methane (STP) react, how many liters of carbon monoxide will be formed? (b) If 1000 liters of methane (STP) react, how many liters of hydrogen will be formed?

Given: 1000 liters methane

Find: The number of liters of (a) carbon monoxide and (b) hydrogen, equivalent to 1000 liters of methane

Outline of Solution: liters methane → moles methane → moles product → liters product

Relationships:
(a) 22.4 liters $CH_4$ ≈ 1.00 mole $CH_4$
    1.00 mole $CH_4$ ≈ 1.00 mole CO (ratio from equation)
    1.00 mole CO ≈ 22.4 liters CO
(b) 22.4 liters $CH_4$ ≈ 1.00 mole $CH_4$
    1.00 mole $CH_4$ ≈ 3.00 moles $H_2$ (ratio from equation)
    1.00 mole $H_2$ ≈ 22.4 liters $H_2$

Solutions:

(a) $1000 \text{ liters } CH_4 \times \dfrac{1.00 \text{ mole } CH_4}{22.4 \text{ liters } CH_4} \times \dfrac{1.00 \text{ mole } CO}{1.00 \text{ mole } CH_4}$

$\times \dfrac{22.4 \text{ liters } CO}{1.00 \text{ mole } CO} = 1000 \text{ liters } CO$

(b) $1000 \text{ liters } CH_4 \times \dfrac{1.00 \text{ mole } CH_4}{22.4 \text{ liters } CH_4} \times \dfrac{3.00 \text{ moles } H_2}{1.00 \text{ mole } CH_4}$

$\times \dfrac{22.4 \text{ liters } H_2}{1.00 \text{ mole } H_2} = 3000 \text{ liters } H_2 = 3.00 \times 10^3 \text{ liters } H_2$

EXAMPLE 18  A component of smelling salts, ammonium carbonate, decomposes into ammonia, water, and carbon dioxide. How many moles of each product form when 10.0 grams of ammonium·carbonate decompose?

$$(NH_4)_2CO_3(s) \rightleftharpoons 2NH_3(g) + H_2O(g) + CO_2(g)$$

Given: 10.0 g $(NH_4)_2CO_3$

Find: The number of moles of each product equivalent to 10.0 g $(NH_4)_2CO_3$

Outline of Solution: grams $(NH_4)_2CO_3$ → moles $(NH_4)_2CO_3$ → moles of product

Relationships: 1.00 mole $(NH_4)_2CO_3 \approx 96.0$ g $(NH_4)_2CO_3$
1.00 mole $(NH_4)_2CO_3 \approx 2.00$ moles $NH_3$
1.00 mole $(NH_4)_2CO_3 \approx 1.00$ mole $H_2O$
1.00 mole $(NH_4)_2CO_3 \approx 1.00$ mole $CO_2$

Solutions:
For $NH_3$,

$$10.0 \text{ g } (NH_4)_2CO_3 \times \frac{1.00 \text{ mole } (NH_4)_2CO_3}{96.0 \text{ g } (NH_4)_2CO_3} \times \frac{2 \text{ moles } NH_3}{1 \text{ mole } (NH_4)_2CO_3}$$
$$= 0.208 \text{ mole } NH_3$$

For $H_2O$,

$$10.0 \text{ g } (NH_4)_2CO_3 \times \frac{1.00 \text{ mole } (NH_4)_2CO_3}{96.0 \text{ g } (NH_4)_2CO_3} \times \frac{1 \text{ mole } H_2O}{1 \text{ mole } (NH_4)_2CO_3}$$
$$= 0.104 \text{ mole } H_2O$$

For $CO_2$,

$$10.0 \text{ g } (NH_4)_2CO_3 \times \frac{1.00 \text{ mole } (NH_4)_2CO_3}{96.0 \text{ g } (NH_4)_2CO_3} \times \frac{1 \text{ mole } CO_2}{1 \text{ mole } (NH_4)_2CO_3}$$
$$= 0.104 \text{ mole } CO_2$$

EXAMPLE 19  In many cases the amount of product formed during a reaction is *less* than the theoretical amount. This may be because not all of the initial substances react or because several reactions may occur at once producing products other than those expected. How many grams of ethylene can be produced from 2.00 moles ethyl alcohol if the actual yield is 65 percent of the theoretical yield?

Given: 2.00 moles ethyl alcohol; yield = 65%

$$CH_3CH_2OH(aq) \underset{\text{in acid}}{\overset{\text{heating}}{\rightleftarrows}} CH_2CH_2(g) + H_2O(l)$$
ethyl alcohol        ethylene       water

Find: The number of grams of $CH_2CH_2$ equivalent to 2.00 moles of $CH_3CH_2OH$

Outline of Solution: moles alcohol → moles ethylene → theoretical grams ethylene → actual grams ethylene

Relationships: 1.00 mole $CH_3CH_2OH \approx 1.00$ mole $CH_2CH_2$
1.00 mole $CH_2CH_2 \approx 2.80$ g $CH_2CH_2$

Solution: After obtaining the theoretical amount of $CH_2CH_2$ produced, we will multiply by 65/100 to obtain the actual amount produced, since only 65 g of $CH_2CH_2$ will be formed for every 100 g of $CH_2CH_2$ which are theoretically predicted.

$$2.00 \text{ moles } CH_3CH_2OH \times \frac{1 \text{ mole } CH_2CH_2}{1.00 \text{ mole } CH_3CH_2OH} \times \frac{28.0 \text{ g } CH_2CH_2}{1.00 \text{ mole } CH_2CH_2}$$
$$= 54.0 \text{ g } CH_2CH_2$$

$$54.0 \text{ g } CH_2CH_2 \times \frac{65}{100} = 35.1 \text{ g } CH_2CH_2$$

EXAMPLE 20  A chemical equation gives the mole ratio of reactants to products. Often, however, a chemist will use an *excess* of one reactant for a certain reason. Consider the following case. A chemist mixed a solution containing 0.0200 mole of lead(II) nitrate with a solution containing 0.0600 mole of hydrochloric acid. How many moles of lead(II) chloride formed?

$$Pb(NO_3)_2(aq) + 2HCl(aq) \rightleftharpoons PbCl_2(s) + 2HNO_3(aq)$$

The balanced equation shows that 1.00 mole $Pb(NO_3)_2$ is equivalent to 2.00 moles HCl.

$$1.00 \text{ mole } Pb(NO_3)_2 \approx 2.00 \text{ moles HCl}$$

The ratio in which the reactants were mixed, however, is

$$\frac{0.0600 \text{ mole HCl}}{0.0200 \text{ mole } Pb(NO_3)_2} = \frac{3.00 \text{ moles HCl}}{1.00 \text{ mole } Pb(NO_3)_2}$$

There is an excess of HCl. The amount of product that can form in this case depends on the amount of $Pb(NO_3)_2$. This substance is called the *limiting reagent*. If all of the limiting reagent reacts, there will still be some HCl left over. We will base our calculations on the amount of limiting reagent present, since all of it reacts. Another way of finding which reagent is the limiting reagent follows:

*Given:* 0.0200 mole $Pb(NO_3)_2$; 0.0600 moles HCl

*Find:* The number of moles of $PbCl_2$ equivalent to the given moles of reactants

*Relationships:* 1.00 mole $Pb(NO_3)_2 \approx$ 2.00 moles HCl
   1.00 mole $Pb(NO_3)_2 \approx$ 1.00 mole $PbCl_2$

*Solution:* To determine which is the limiting reagent, we determine the amount of HCl needed to react with 0.0200 mole $Pb(NO_3)_2$.

$$0.0200 \text{ mole } Pb(NO_3)_2 \times \frac{2 \text{ moles HCl}}{1 \text{ mole } Pb(NO_3)_2} = 0.0400 \text{ mole HCl}$$

We need 0.0400 mole HCl but were given 0.0600 mole. Therefore, as noted above, HCl is in excess and the $Pb(NO_3)_2$ is the limiting reagent.

$$0.0200 \text{ mole } Pb(NO_3)_2 \times \frac{1 \text{ mole } PbCl_2}{1 \text{ mole } Pb(NO_3)_2} = 0.0200 \text{ mole } PbCl_2$$

EXAMPLE 21  A chemist mixed 50.0 grams of zinc with 50.0 grams of sulfur and ignited the mixture. Which element was in excess? How many grams of zinc sulfide were formed?

$$Zn(s) + S(s) \rightleftharpoons ZnS(s)$$

*Given:* 50.0 g Zn
   50.0 g S

*Find:* (a) Which reagent is in excess? (b) How many grams of zinc sulfide formed?

*Outline of Solution:*

$$\left.\begin{array}{l}\text{grams zinc} \longrightarrow \text{moles zinc} \\ \text{grams sulfur} \longrightarrow \text{moles sulfur}\end{array}\right\} \text{which is in excess?}$$

moles of limiting reagent $\longrightarrow$ moles ZnS $\longrightarrow$ grams ZnS

*Relationships:* 1.00 mole Zn $\approx$ 65.4 g Zn
1.00 mole S $\approx$ 32.1 g S
1.00 mole Zn $\approx$ 1.00 mole ZnS
1.00 mole S $\approx$ 1.00 mole ZnS

*Solution:* The number of moles of each reactant are:

$$50.0 \text{ g Zn} \times \frac{1.00 \text{ mole Zn}}{65.4 \text{ g Zn}} = 0.764 \text{ mole Zn}$$

$$50.0 \text{ g S} \times \frac{1.00 \text{ mole S}}{32.1 \text{ g S}} = 1.56 \text{ moles S}$$

The amount of sulfur required to react with 0.764 mole of zinc is:

$$0.764 \text{ mole Zn} \times \frac{1 \text{ mole S}}{1 \text{ mole Zn}} = 0.764 \text{ mole S}$$

Sulfur is in excess because we have 1.56 moles but need only 0.764 mole. Using the amount of limiting reagent to calculate the weight of zinc sulfide formed, we have:

$$0.764 \text{ mole Zn} \times \frac{1.00 \text{ mole ZnS}}{1.00 \text{ mole Zn}} \times \frac{97.5 \text{ g ZnS}}{1.00 \text{ mole ZnS}} = 74.5 \text{ g ZnS}$$

## Summary

A chemical equation is a source of information about a given chemical reaction. The equation indicates what the reactants are and what the products are. The coefficients in the equation indicate the *mole ratios* for the substances involved in the reaction. If we are given a certain number of moles of one substance, this mole ratio allows us to calculate the number of moles of another substance needed or produced in the reaction.

If in a problem we are given an amount of a substance in units other than moles, we simply need to convert these units to moles and then again use the mole ratio to calculate the number of moles of another substance. If the units asked for are other than moles, then convert moles to units desired.

Up to now we have discussed the relationships between moles and grams, moles and particles, and moles and liters (for gases at STP). In the next three chapters we will consider solutions and reactions which occur in solutions. In these chapters a new idea involving moles will be introduced. This concept is called *molarity*; this term means the number of moles of a substance dissolved in one liter of solution.

*Objectives* By the end of this chapter, you should be able to do the following:
1. Given either (a) the weight of a substance, (b) the number of particles in a substance, (c) the volume (for a gas) of a substance or, (d) the number of moles of a substance, calculate any of the other terms. (Questions 2, 4)

2. Given the moles of one substance taking part in a chemical reaction, calculate the number of moles of another substance needed or produced in the reaction. (Questions 8–10)

3. Given the amount of one substance taking part in a chemical reaction, calculate the amount of another substance needed or produced in a reaction. Units of amounts given and amount to find may be in either (a) moles, (b) grams, (c) particles, (d) liters (of a gas). (Questions 11–27)

*Questions* 1. Define the following terms: *stoichiometry; molar volume; mole; unity term; chemically equivalent; limiting reagent; mole ratio; excess reagent.*

2. Write as unity terms the conversion factors necessary to make each of the following conversions: (a) Grams of silver to moles of silver; (b) grams of lead(II) sulfide to moles of lead(II) sulfide; (c) moles of oxygen to grams of oxygen; (d) moles of nitrogen gas to liters of nitrogen gas at STP; (e) liters of hydrogen at STP to grams of hydrogen; (f) liters of carbon dioxide at STP to moles of carbon dioxide; (g) grams of sugar, $C_{12}H_{22}O_{11}$, to moles of sugar.

3. Write all of the mole ratios as unity terms which can be obtained from the following balanced equation.

$$MnO_2(s) + 4HCl(aq) \rightleftharpoons MnCl_2(aq) + Cl_2(g) + 2H_2O(l)$$

4. How many atoms of hydrogen are there in two grams of hydrogen?

5. Why is the chemist interested in stoichiometric calculations?

6. Balance the following equations:

(a) $Fe(s) + HCl(l) \rightleftharpoons FeCl_2(aq) + H_2(g)$
(b) $Al(s) + H_2SO_4(aq) \rightleftharpoons Al_2(SO_4)_3(aq) + H_2(g)$
(c) $C_2H_6(g) + O_2(g) \rightleftharpoons CO_2(g) + H_2O(g)$
(d) $C_4H_{10}(g) + O_2(g) \rightleftharpoons CO_2(g) + H_2O(g)$
(e) $P_4O_{10}(s) + H_2O(l) \rightleftharpoons H_3PO_4(aq)$

7. Write balanced chemical equations for the following word equations:

(a) iron(s) + chlorine(g) yields iron(III) chloride(s)
(b) aluminum(s) + oxygen(g) yields aluminum oxide(s)
(c) phosphorus (V) oxide(s) + water(l) yields phosphoric acid(aq)
(d) copper(II) oxide(s) + hydrogen(g) yields copper(s) + water(l)
(e) magnesium hydroxide(s) + sulfuric acid(aq) yields magnesium sulfate(aq) + water(l).

8. When mercury(II) oxide is heated, mercury and oxygen are formed:

$$HgO(s) \rightleftharpoons Hg(l) + O_2(g)$$

(a) Balance the equation; (b) Use the balanced equation to calculate the number of moles of oxygen which could be prepared from 5.15 moles of the oxide.

9. When ethane, $C_2H_6$, is burned in oxygen, carbon dioxide and water are the products.

$$2C_2H_6(g) + 7O_2(g) \rightleftharpoons 4CO_2(g) + 6H_2O(g)$$

(a) How many moles of ethane would be necessary to prepare 176 moles of carbon dioxide? (b) How many moles of ethane would be necessary to prepare 176 grams carbon dioxide?

10. Use the following balanced equation to fill in the blanks on the basis of the information supplied.

$$2AgNO_3(aq) + H_2S(g) \rightleftharpoons Ag_2S(s) + 2HNO_3(aq)$$

(a) 5.0 moles of $AgNO_3$ would produce _____ moles of $HNO_3$.
(b) 14 moles of $H_2S$ would produce _____ moles of $HNO_3$.
(c) 7.0 moles of $H_2S$ would produce _____ moles of $Ag_2S$.
(d) 3.0 moles of $AgNO_3$ would produce _____ moles of $Ag_2S$.

11. Hydrogen peroxide, $H_2O_2$, is often used to "bleach" hair. Many times a "pfft" is heard when a bottle of $H_2O_2$ is opened, indicating that a gas has been released. Hydrogen peroxide decomposes according to the following equation:

$$2H_2O_2(l) \rightleftharpoons 2H_2O(l) + O_2(g)$$

Use this balanced equation to fill in the following blanks.
(a) 4.0 moles of $O_2$ require _____ moles of $H_2O_2$.
(b) 12 moles of $H_2O$ require _____ moles of $H_2O_2$.
(c) 22.4 liters of $O_2$ require _____ moles of $H_2O_2$.
(d) 224 liters of $O_2$ require _____ moles of $H_2O_2$.
(e) 12 molecules of $H_2O$ require _____ molecules of $H_2O_2$.

12. One of the chemical reactions of importance to the beer industry is the fermenting of glucose, $C_6H_{12}O_6$, to produce ethyl alcohol, $C_2H_5OH$, and carbon dioxide. Using the following balanced equation, calculate the number of moles of glucose necessary to produce 454 grams of alcohol.

$$C_6H_{12}O_6(s) \rightleftharpoons 2C_2H_5OH(l) + 2CO_2(g)$$

13. Common household bleach is the solution resulting when chlorine gas reacts with a solution of sodium hydroxide. Using the following equation determine the number of moles of NaOH necessary to react with 200 grams of chlorine gas.

$$Cl_2(g) + 2NaOH(aq) \rightleftharpoons NaClO(aq) + NaCl(aq) + H_2O(l)$$

14. The air purification system in Apollo spacecraft included the use of lithium hydroxide to remove excess carbon dioxide from the environment. Using the following balanced equation, calculate how many grams of lithium carbonate would result from the removal of 30.5 moles of $CO_2$.

$2\text{LiOH}(aq) + \text{CO}_2(g) \rightleftharpoons \text{Li}_2\text{CO}_3(s) + \text{H}_2\text{O}(l)$

Use this balanced equation to fill in the blanks:
(a) 5.0 moles of $\text{CO}_2$ would produce _____ moles of $\text{H}_2\text{O}$.
(b) 4.0 moles of LiOH would produce _____ moles of $\text{Li}_2\text{CO}_3$.
(c) 22.4 liters of $\text{CO}_2$ would produce _____ moles of $\text{H}_2\text{O}$.
(d) 15 molecules of $\text{CO}_2$ would produce _____ molecules of $\text{Li}_2\text{CO}_3$.

15. One battery system prepared for use in an electric car would harness the reaction between sodium and sulfur to supply energy. Using the following equation (unbalanced), calculate the number of grams of sodium required to react with 4.54 grams of sulfur.

$\text{Na}(s) + \text{S}(s) \rightleftharpoons \text{Na}_2\text{S}(s)$

16. Green plants use water and carbon dioxide to produce glucose, $\text{C}_6\text{H}_{12}\text{O}_6$, and oxygen. Using the following equation (unbalanced) calculate the number of moles of glucose produced from 10.0 liters of $\text{CO}_2$ at STP.

$\text{CO}_2(g) + \text{H}_2\text{O}(l) \rightleftharpoons \text{C}_6\text{H}_{12}\text{O}_6(aq) + \text{O}_2(g)$

17. Some cigarette lighters supply heat by burning butane gas, $\text{C}_4\text{H}_{10}$. Using the following balanced equation calculate the number of liters of $\text{CO}_2$ gas produced from the burning of 116 grams of butane.

$2\text{C}_4\text{H}_{10}(g) + 13\text{O}_2(g) \rightleftharpoons 8\text{CO}_2(g) + 10\text{H}_2\text{O}(g)$

Use this balanced equation to fill in the blanks:
(a) 2.0 moles of $\text{C}_4\text{H}_{10}$ would produce _____ moles of $\text{H}_2\text{O}$.
(b) 5.0 moles of $\text{C}_4\text{H}_{10}$ would produce _____ moles of $\text{CO}_2$.
(c) 13 moles of $\text{O}_2$ would produce _____ moles of $\text{CO}_2$.
(d) 26 moles of $\text{O}_2$ would produce _____ moles of $\text{H}_2\text{O}$.
(e) 20 moles of $\text{C}_4\text{H}_{10}$ would produce _____ moles of $\text{CO}_2$.

18. Smelling salts contain ammonium carbonate which readily decomposes into ammonia, a heart stimulant. Using the following equation, calculate the weight of $(\text{NH}_4)_2\text{CO}_3$ necessary to produce 1.00 liter of $\text{NH}_3$ at STP.

$(\text{NH}_4)_2\text{CO}_3(s) \rightleftharpoons 2\text{NH}_3(g) + \text{CO}_2(g) + \text{H}_2\text{O}(g)$

19. Laughing gas, $\text{N}_2\text{O}$, is a mild anesthetic. It can be made by gently heating ammonium nitrate. Using the following balanced equation, calculate the weight of $\text{N}_2\text{O}$ available from 5.00 grams of $\text{NH}_4\text{NO}_3$.

$\text{NH}_4\text{NO}_3(s) \rightleftharpoons \text{N}_2\text{O}(g) + 2\text{H}_2\text{O}(g)$

20. Under certain conditions, arsenic reacts with oxygen to form arsenic(III) oxide, a common insecticide and weed killer. The fatal human dose is about 0.10 gram. Using the following equation (unbalanced), calculate the number of moles of arsenic necessary to make 0.10 gram of $\text{As}_2\text{O}_3$.

$\text{As}(s) + \text{O}_2(g) \rightleftharpoons \text{As}_2\text{O}_3(s)$

21. A minor product of the combustion of gasoline in a car engine is the poisonous gas carbon monoxide. This gas is a good fuel and will easily burn to form the less dangerous carbon dioxide. Using the fol-

lowing equation (unbalanced), determine the volume of $CO_2$ gas at STP that would result from burning 3.00 liters of CO gas.

$$CO(g) + O_2(g) \rightleftharpoons CO_2(g)$$

22. The noble gas xenon will react with fluorine to produce xenon hexafluoride. Using the following equation (unbalanced) calculate: (a) The volume of fluorine gas at STP that will react with 2.00 liters of xenon gas; (b) The weight of $XeF_6$ that can be produced from 2.00 liters of xenon gas.

$$Xe(g) + F_2(g) \rightleftharpoons XeF_6(s)$$

23. Superphosphate fertilizer is a mixture resulting when sulfuric acid is added to phosphate rock, $Ca_3(PO_4)_2$. Using the following balanced equation, calculate the weight of sulfuric acid necessary to produce 454 grams of $Ca(H_2PO_4)_2$.

$$Ca_3(PO_4)_2(s) + 2H_2SO_4(aq) \rightleftharpoons Ca(H_2PO_4)_2(aq) + 2CaSO_4(aq)$$

24. Many emergency lamps contain calcium carbide. This chemical reacts with water to produce the combustible gas acetylene, $C_2H_2$. Using the following equation (unbalanced), calculate the weight of $CaC_2$ necessary to produce 11.2 liters of $C_2H_2$ gas at STP.

$$CaC_2(s) + H_2O(l) \rightleftharpoons Ca(OH)_2(aq) + C_2H_2(g)$$

25. A mixture of baking soda, $NaHCO_3$, and vinegar, $HC_2H_3O_2$, produces carbon dioxide. Using the following equation calculate the weight of baking soda needed to produce 0.010 liters of $CO_2$.

$$NaHCO_3(s) + HC_2H_3O_2(aq) \rightleftharpoons NaC_2H_3O_2(aq) + H_2O(l) + CO_2(g)$$

26. Much of the scale that accumulates in boilers, pipes, and tea kettles is iron(III) carbonate. This can be removed easily with dilute hydrochloric acid. Two moles of HCl were added to 0.150 mole of $Fe_2(CO_3)_3$. Which reactant is in excess? How many moles of $FeCl_3$ were produced?

$$Fe_2(CO_3)_3(s) + 6HCl(aq) \rightleftharpoons 2FeCl_3(aq) + 3H_2O(l) + 3CO_2(g)$$

27. Propane, $C_3H_8$, a gas obtained from petroleum, is used for heating and cooking, and as a fuel for trucks and tractors. A chemist burned an 8.8 gram sample of propane in the laboratory. The volume of oxygen used was 2.24 liters at STP. Which reactant was in excess? How many moles of carbon dioxide would be produced in the process? What weight of water would be produced? The balanced equation for the burning of propane is:

$$C_3H_8(g) + 5O_2(g) \rightleftharpoons 3CO_2(g) + 4H_2O(g)$$

# 14

# Solutions

## 14.1 Introduction

A *solution* is a homogenous mixture of two or more substances. In Chapter 1, we learned that a solution consists of two or more substances mixed together at the particulate or molecular level. Solutions may have a variable composition of their component substances.

Why do we study solutions? Since many reactions occur readily only in solution, a knowledge of solutions gives the chemist more insight into certain chemical systems. Other scientists also have an interest in solutions. Biologists believe that life on earth began in a solution, namely the ocean. Bodily reactions that are vital to life take place in solutions within and outside of cells. The nitrogenous waste materials which are the end products of body metabolism are excreted from the body in a solution, urine.

Solution chemistry is important in the administration and use of medicines. Many drugs are absorbed into the body in solution form. Water solutions of glucose are given intravenously to provide energy for patients who are unable to eat food or who have excess loss of fluids.

Solutions are also important in industry and in the home. Aqueous sulfuric acid is used in automobile batteries to conduct electricity. Aqueous hydrochloric acid solutions are used to clean industrial metal parts. Aqueous ethylene glycol solutions are used in car radiators to prevent freezing during winter months. Carbonated beverages are water solutions of flavorings, sugar, and carbon dioxide.

## 14.2 Examples of solutions

Consider the solutions prepared by mixing the ten substances listed in Table 14.1 with water. When small amounts of these substances are added to water and mixed, eight of them form clear, colorless, homogenous, liquid systems called solutions. Each of these eight substances is called a *solute*; a solute is the substance that is dissolved. The water is called the *solvent*; a solvent is the substance that does the dissolving.

Neither silver chloride nor carbon tetrachloride dissolve appreciably in water. They are said to be very slightly soluble in water. If two *liquids* mix, they are often said to be miscible. If they do not mix to any appreciable extent they are said to be *immiscible*. Water and ethyl alcohol are miscible; water and carbon tetrachloride are immiscible.

To learn more about these solutions, one could make several physical property measurements on them. Four such measure-

**TABLE 14.1**

*The influence of some common substances on the properties of water.*

| Substance | Initial phase | Observation on aqueous mixture | Electrical conductivity of mixture | Effect on freezing point compared to water | Effect on boiling point compared to water | Amount dissolved (g/100 ml $H_2O$) |
|---|---|---|---|---|---|---|
| Sugar, $C_{12}H_{22}O_{11}$ | solid | clear homogeneous liquid | none | lower | higher | 204 |
| Sodium chloride, NaCl | solid | clear homogeneous liquid | strong | lower | higher | 36 |
| Hydrogen chloride, HCl | gas | clear homogeneous liquid | strong | lower | higher | 82 |
| Acetic acid, $HC_2H_3O_2$ | liquid | clear homogeneous liquid | weak | lower | higher | unlimited |
| Ethyl alcohol, $CH_3CH_2OH$ | liquid | clear homogeneous liquid | none | lower | — | unlimited |
| Ammonia, $NH_3$ | gas | clear homogeneous liquid | weak | lower | — | 90 |
| Sodium hydroxide, NaOH | solid | clear homogeneous liquid | strong | lower | higher | — |
| Calcium chloride, $CaCl_2$ | solid | clear homogeneous liquid | strong | lower | — | — |
| Silver chloride, AgCl | solid | white, solid and clear liquid | none | none | none | very low |
| Carbon tetrachloride, $CCl_4$ | liquid | two clear liquids | none | none | — | very low |

ments are (1) the solubility or amount of solute dissolved in 100 grams of water, (2) the electrical conductivity of the solution, (3) the freezing point of the solution, and (4) the boiling point of the solution (Table 14.1).

**Solubility in water**

The amount of solute that will dissolve in 100 grams of water at a given temperature varies widely with the solute as Table 14.1 shows. Only about $10^{-4}$ gram of silver chloride dissolves, whereas about 36 grams of sodium chloride and 204 grams of sugar dissolve in water. Ethyl alcohol and acetic acid are soluble in water in *all* proportions. These values are the maximum amounts that will dissolve at a given temperature.

**Electrical conductivity**

The tendency of these solutions to conduct an electrical current varies widely (Table 14.1). Substances or solutions which conduct electricity are called *electrolytes*. Aqueous solutions of sodium chloride, hydrogen chloride, sodium hydroxide, and calcium chloride are good conductors of electricity and are called *strong electrolytes*. Aqueous solutions of acetic acid and ammonia are weak conductors of electricity and are called *weak electrolytes*. Sugar and alcohol solutions do not conduct any appreciable electrical current; they are called *nonelectrolytes*. The ability of strong and weak electrolytes to conduct electricity is due to the presence of ions in solution. The properties of electrolytes are discussed in Chapter 15.

**Freezing points and boiling points**

Pure water freezes at 0.00°C and boils at 100.00°C at 1.00 atm pressure. Table 14.1 shows the freezing points of solutions to be below 0°C at 1.00 atm pressure. The particles of solute prevent the water molecules from freezing together at 0°C. This phenomenon is apparent when bodies of salt water remain liquid even below 0°C.

Note also in Table 14.1 that the boiling point of many solutions is above 100°C at 1.00 atm pressure. Thus particles of solute dissolved in water prevent the water molecules from evaporating as rapidly as they do in pure water.

The number of degrees by which the freezing point is lowered and the boiling point is raised depends upon the number of particles dissolved in a given amount of water. The more par-

■ EXERCISE  To a large pot of rapidly boiling water, add some salt. Describe what happens. Repeat, using sugar instead of salt. Place some salt on an ice cube. Describe what happens. Repeat, using sugar.

ticles of solute in one kilogram of water, the lower the freezing point and the higher the boiling point will be. Properties such as freezing points and boiling points, which depend upon the *number of dissolved particles*, are called *colligative properties*. Two other colligative properties of solutions are vapor pressure and osmotic pressure. ■

## 14.3  Types of solutions

**Phases**

In the previous section, we saw that a solid (sugar, salt), a liquid (alcohol, acetic acid), or a gas (ammonia, hydrogen chloride) may dissolve in water to form an aqueous solution. Other liquids besides water may serve as the solvent. For example, many medicines given intramuscularly are dissolved in alcohol.

Solids may also serve as solvents. Many types of steel contain small percentages of carbon dissolved in iron. Graphite and ice may contain small amounts of gas dissolved in them.

■ EXERCISE  Identify the solvent and solute in each of the following beverages whose contents are: (a) 70% water, 30% ethyl alcohol; (b) 70% ethyl alcohol, 30% water; (c) 50% water, 50% ethyl alcohol.

Gases also serve as solvents. Water vapor dissolves in air and makes the atmosphere humid. The odor of moth balls (naphthalene) is readily detected in a closet in which the solid moth balls have been placed, indicating that vapor from the moth balls has dissolved in air, a mixture of gases (Table 14.2). ■

**Dilute—concentrated**

A reagent bottle in the chemical laboratory is labled "Concentrated Hydrochloric Acid." A liter of this solution normally contains *about* twelve moles of hydrogen chloride dissolved in water. A *concentrated* solution is one that contains a *relatively large amount* of solute dissolved in a given volume of solution. The word "concentrated" does not indicate exactly how much solute is present.

| TABLE 14.2 | | Solid | Liquid | Gas |
|---|---|---|---|---|
| Example of the nine possible simple types of solutions classified by original phases of solute and solvent. The solute may be a solid, liquid, or gas. The solvent may be a solid, liquid, or gas. | Solid | solid in solid:<br>**nickel in copper** | solid in liquid:<br>**salt in water** | solid in gas:<br>**naphthalene in air** |
| | Liquid | liquid in solid:<br>**mercury in silver** | liquid in liquid:<br>**alcohol in water** | liquid in gas:<br>**water in air** |
| | Gas | gas in solid:<br>**hydrogen in palladium** | gas in liquid:<br>$CO_2$ **in water** | gas in gas:<br>**oxygen in air** |

**TABLE 14.3**

*The solubilities of various substances in water.*

| Substance | Grams dissolved/100 g $H_2O$ |
|---|---|
| NaCl | 36 |
| KCl | 28 |
| AgCl | $8.9 \times 10^{-5}$ |
| NaF | 4.0 |
| AgF | 182 |
| Sugar, $C_{12}H_{22}O_{11}$ | 204 |
| Xe | 0.062 |

A student pours 10.0 ml of concentrated sodium chloride solution into a flask and adds enough water to make 1.00 liter of solution. The new solution is said to be dilute. A *dilute* solution is one that contains a relatively small amount of solute dissolved in a given volume of solution. The word "dilute" does not indicate exactly how much solute is dissolved.

**Saturated**

When small amounts of sodium chloride are added to a beaker of water and stirred, the sodium chloride dissolves. If the process is continued, the salt will continue to dissolve until a specific amount of salt has been added. After this point is reached, any additional salt added will sink to the bottom without dissolving. The solution holds as much salt as it can at that temperature. Such a solution is said to be *saturated*. A saturated solution of sodium chloride contains 36 grams of salt per 100 grams of water at 20°C. If a solution holds less solute than it can at saturation, the solution is said to be *unsaturated*.

Table 14.3 lists the amounts of solute contained in various saturated solutions. Note that the word "saturated" does not tell us whether a given solution contains a large amount of solute, as in the case of AgF, or a small amount of solute, as in the case of AgCl.

A warm solution of saturated aqueous sodium acetate contains more solute than a cold solution of saturated aqueous sodium acetate. When a warm saturated solution of sodium acetate is carefully cooled, nothing appears to happen. If, however, the solution is disturbed, by the addition of a small crystal or by stirring, a large amount of solid quickly forms. This is an example of *supersaturation* (Fig. 14.1). The carefully cooled solution holds more solute than it normally can and is said to be supersaturated. The added crystal acts as a nucleus for the formation of crystals which remove the excess solute from the solution phase.

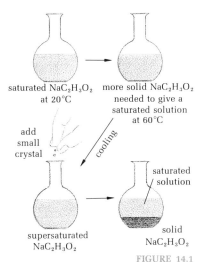

**FIGURE 14.1**
*A saturated solution of sodium acetate contains more solute at 60°C than at 20°C. If the saturated warm solution is slowly cooled, an unstable supersaturated solution forms. The addition of a small crystal to the supersaturated solution causes a large precipitate to form.*

## 14.4 Factors affecting solubility

Sodium chloride is soluble in water, slightly soluble in ethyl alcohol, and only very slightly soluble in carbon tetrachloride. Lead(II) chloride is slightly soluble in cold water, yet soluble in hot water. Carbon dioxide remains dissolved in soda when the cap is on the bottle, but it leaves the solution when the cap is removed. What factors affect the solubility of substances?

TABLE 14.4

The relative solubility of substances in a polar and nonpolar solvent.

| Substance | Water: polar | Carbon tetrachloride: nonpolar |
|---|---|---|
| KCl: ionic | soluble | slightly soluble |
| NaCl: ionic | soluble | slightly soluble |
| HCl: polar | soluble | slightly soluble |
| $C_{12}H_{22}O_{11}$, Sugar: polar | soluble | slightly soluble |
| $Cl_2$: nonpolar | slightly soluble | soluble |
| $I_2$: nonpolar | slightly soluble | soluble |
| $CS_2$: nonpolar | slightly soluble | soluble |
| $CCl_4$: nonpolar | slightly soluble | soluble |

**Nature of solute and solvent**

Note in Table 14.4 that the substances which dissolve to any extent in water are either ionic or polar. Water is also polar, so we may generalize and say that *polar solvents dissolve polar solutes*. The substances that dissolve in nonpolar carbon tetrachloride are nonpolar, so we may generalize and say that *nonpolar solvents dissolve nonpolar solutes*.

Why does water dissolve polar substances? Consider a molecule of ethyl alcohol, $C_2H_5OH$. Its molecular structure is

$$\begin{array}{c} \phantom{H-}H\phantom{-C-}H \\ \phantom{H-}|\phantom{-C-}| \\ H-C-C-O-H \\ \phantom{H-}|\phantom{-C-}| \\ \phantom{H-}H\phantom{-C-}H \end{array}$$

The vicinity of the molecule about the oxygen atom is negative because the oxygen has a high electronegativity. The other parts of the molecule are slightly positive. The molecule is polar. There are attractive forces between water molecules and ethyl alcohol molecules. The positive ends of the water molecule are attracted to the negative ends of the alcohol molecule. The negative ends of the water molecule are attracted to the positive ends of the alcohol molecule. Due to these attractive forces, the polar alcohol molecule is surrounded by water molecules. In effect it dissolves or mixes with the water (Fig. 14.2). These same forces occur between water molecules and other polar and ionic substances.

Why don't large amounts of iodine dissolve in water? The iodine molecule is nonpolar $:\!\overset{..}{\underset{..}{I}}\!-\!\overset{..}{\underset{..}{I}}\!:$. It has no positive or negative end. Since water molecules are polar, each water molecule is attracted to other water molecules. The iodine molecule cannot attract water molecules away from one another and

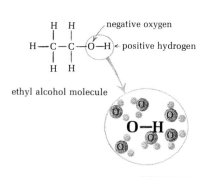

FIGURE 14.2
Dissolution of ethyl alcohol in water.

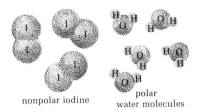

nonpolar iodine    polar water molecules

**FIGURE 14.3**
*Iodine is only slightly soluble in water.*

■ **EXERCISE** Name five substances used around the house or in cars that are not soluble in water.

■ **EXERCISE** Soaps and detergents are long molecules. One end of the molecule is nonpolar and the other end is polar. Can you use this model to explain the ability of soaps and detergents to dissolve nonpolar greases in polar water?

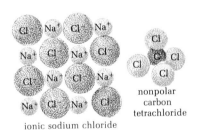

ionic sodium chloride    nonpolar carbon tetrachloride

**FIGURE 14.4**
*Sodium chloride does not dissolve appreciably in carbon tetrachloride.*

mix with them. Consequently, it does not dissolve appreciably (Fig. 14.3). For the same reason, other nonpolar molecules do not dissolve in water.

Why doesn't sodium chloride dissolve in carbon tetrachloride? The structure of carbon tetrachloride is

$$\text{Cl}-\underset{\underset{\text{Cl}}{|}}{\overset{\overset{\text{Cl}}{|}}{\text{C}}}-\text{Cl}$$

It is a nonpolar molecule. The positive sodium ions in sodium chloride are strongly attracted to the negative chloride ions. The nonpolar carbon tetrachloride molecules exert no attractive forces on these ions and hence do not pull them away from one another. The sodium chloride therefore does not dissolve in the carbon tetrachloride (Fig. 14.4). For the same reason, other polar or ionic substances do not dissolve in nonpolar solvents.

Why does nonpolar iodine dissolve in nonpolar carbon tetrachloride (Fig. 14.5)? ■

**Temperature**

What effect does temperature have on the solubility of different substances? The water solubilities of samples of sodium nitrate, sodium chloride, and sodium sulfate may be measured at various temperatures. Samples of one of the salts are weighed out and slowly mixed in small portions with 100 grams of water at various temperatures. When it is apparent that no more solid will dissolve, the solution is filtered, the water evaporated, and the deposited salt weighed. The amount of solid that dissolved at that temperature is the same as the amount deposited when the water is evaporated.

Note in Table 14.5 that the solubility of sodium nitrate increases with an increase in temperature. For sodium chloride the solubility remains approximately constant with only a slight increase as the temperature increases. For sodium sul-

| TABLE 14.5 | Temperature (°C) | $NaNO_3$ (g) | $NaCl$ (g) | $Na_2SO_4$ (g) |
|---|---|---|---|---|
| The solubility of three salts at four different temperatures. | 25 | 86 | 34 | 58 |
| | 50 | 107 | 36 | 49 |
| | 75 | 137 | 37 | 42 |
| | 100 | 180 | 37 | 40 |

14.4 Factors affecting solubility

■ EXERCISE  Design and execute an experiment at home to see if sugar is more soluble in iced tea or in hot tea.

■ EXERCISE  A student was given a beaker of newly distilled water. If he left the beaker on his desk for several days, what substances probably dissolved in the water?

fate, the solubility decreases as the temperature increases. The data are plotted in Fig. 14.6. ■

**Pressure**

Carbonated beverages are bottled under pressure. Carbon dioxide is added to the soda. With the bottle top on, a clear soda appears to be a homogeneous substance. When the top is removed, bubbles can be seen rising up and out of the soda. These bubbles consist of carbon dioxide that is no longer soluble in the soda. The solubility of gases is proportional to the pressure of that gas above the solution. As the pressure increases, the solubility increases. Pressure has little effect on the solubility of solids or liquids. ■

## 14.5  Factors affecting the rate of dissolution

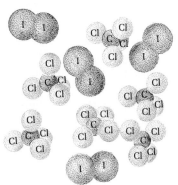

FIGURE 14.5
Both iodine and carbon tetrachloride are nonpolar; therefore the attractive forces between molecules are weak. When placed in contact with one another, there are no strong forces to overcome and the two mix.

**Agitation**

After adding a teaspoon of sugar to a cup of coffee, most people stir it. This causes the sugar to dissolve more rapidly. Stirring or agitation speeds up the rate at which substances dissolve. It helps the solute particles mix with solvent molecules.

**Temperature**

Salt is approximately as soluble in hot water as in cold water. Yet a teaspoon of salt dissolves faster in a glass of hot water than in a glass of cold water. Why?

The average kinetic energy of the water molecules is greater in hot water than it is in cold water. Consequently sodium ions and chloride ions are struck by water molecules more frequently in hot water. The rapid motion of the hot water molecules pulls the ions away from the solid and into the solution more quickly. In general, heating speeds the dissolution of substances.

**State of subdivision**

Increasing the state of subdivision of a substance increases the surface area of the substance. Thus, when placed in water, powdered copper(II) sulfate pentahydrate dissolves much faster than large crystals of the same salt, since the

312  14: Solutions

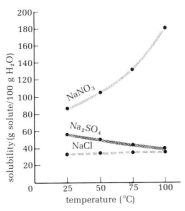

FIGURE 14.6

The solubility of three substances as a function of temperature. This graph represents the data from Table 14.5.

larger surface area is struck by many more water molecules in a given time. The greater the subdivision, the more rapid the rate of dissolution of the substance.

## 14.6 Concentrations of solutions

The concentration of a solution is a measure of the quantity of solute contained in a given quantity of the solution. If we are to prepare a solution, we can measure the quantity of solute and solvent used and then calculate the concentration of the solution. The rate of a chemical reaction in solution is influenced by the concentration of the solution. In medicine, the difference between life and death may depend upon the administration of the proper concentration of a particular medication. We shall consider three common methods of expressing concentration: weight percent, molality, and molarity.

**Weight percent**

A concentration unit widely used in industry is the weight percent of solute in solution. The weight percent of a solute is calculated by dividing the weight of the solute by the total weight of the solution and then multiplying by one hundred.

$$\text{wt \%} = \frac{\text{wt of solute}}{\text{wt of solute} + \text{wt of solvent}} \times 100$$

What is the weight percent of solute in a solution made by dissolving 15.0 grams of sugar in 85.0 grams of water? The weight of the solute plus the weight of the solvent is 15.0 g + 85.0 g = 100.0 g.

$$\text{wt \%} = \frac{15.0 \text{ g}}{100 \text{ g}} \times 100 = 15.0\% \text{ sugar by weight}$$

EXAMPLE 1  How would you prepare 200 grams of a 4.50 percent by weight solution of salt?

Given: 200 g solution; 4.50% salt by weight

Find: Weight of salt needed and weight of water needed

Relationships: 100 g solution ≈ 4.50 g salt

Solution: In 200 g of solution we have:

$$200 \text{ g solution} \times \frac{4.50 \text{ g salt}}{100 \text{ g solution}} = 9.00 \text{ g salt}$$

The amount of water needed is 200 g − 9.00 g = 191 g.

EXAMPLE 2  If 1.00 liter of a solution has a density of 1.10 g/ml and contains 14.6 percent by weight of solute, what weight of solute does the solution contain?

Given: Volume of solution = 1.00 liter
Density of solution = 1.10 g/ml
Weight percent of solute = 14.6%

Find: Weight of solute

Outline of solution:

volume of solution ⟶ weight of solution ⟶ weight of solute

Relationships: 1.10 g ≈ 1.00 ml
100 g solution ≈ 14.6 g solute

Solution: In order to find the weight of the solute in solution, we must first find the weight of the solution.

$$1.00 \text{ liter} \times \frac{1000 \text{ ml}}{\text{liters}} \times \frac{1.10 \text{ g}}{\text{ml}} = 1100 \text{ g of solution}$$

$$1100 \text{ g solution} \times \frac{14.6 \text{ g solute}}{100 \text{ g solution}} = 160 \text{ g solute}$$

We could have set up all the terms at once:

$$1.00 \text{ liter solution} \times \frac{1000 \text{ ml}}{\text{liter}} \times \frac{1.10 \text{ g}}{\text{ml}} \times \frac{14.6 \text{ g solute}}{100 \text{ g solution}} = 160 \text{ g solute}$$

— converts g solution to g solute
— converts ml solution to g solution
— converts liters solution to ml solution

■ EXERCISE  What is the percent weight of ethyl alcohol in a solution made by mixing 25.3 grams alcohol with 150 grams water?

■ EXERCISE  How would you make 150 grams of a 12.5 percent weight solution of potassium nitrate?

■ EXERCISE  How many grams of solute are there in 1.00 liter of a solution that has a density of 1.40 g/ml. The weight percent of solute is 70.0 percent.

**Molality**

It is frequently useful to know the number of moles of solute dissolved in a given mass of solvent. *Molality* is the number of moles of solute dissolved in 1000 grams or 1.00 kilogram of solvent. It is symbolized by m.

$$m = \text{molality} = \frac{\text{moles of solute}}{1.00 \text{ kg solvent}}$$

The molality of solutions is the basis for calculating changes in freezing points, which we shall study in the next section. Some examples of typical molality problems follow.

EXAMPLE 3  What is the molality of a solution that contains 0.685 mole HCl dissolved in 500 grams of water?

Given: Moles of solute = 0.685 mole HCl
Weight of solvent = 500 g $H_2O$

*Find:* Molality of solution

*Relationship:* Molality = $\dfrac{\text{moles solute}}{1.00 \text{ kg solvent}}$

*Solution:* To calculate the molality, we will convert grams of solvent to kilograms of solvent first.

$$500 \text{ g H}_2\text{O} \times \dfrac{1.00 \text{ kg}}{1000 \text{ g}} = 0.500 \text{ kg H}_2\text{O}$$

Using the definition of molality we have:

$$m = \dfrac{\text{moles solute}}{\text{kg solvent}} = \dfrac{0.685 \text{ mole HCl}}{0.500 \text{ kg H}_2\text{O}} = \dfrac{1.37 \text{ moles HCl}}{1.00 \text{ kg H}_2\text{O}} = 1.37 \ m$$

■ **EXERCISE** *What is the molality of a solution prepared by dissolving 0.274 gram of sodium phosphate in 34.7 grams of water?*

We could have set this up all at once as:

$$m = \dfrac{0.685 \text{ mole HCl}}{500 \text{ g H}_2\text{O}} \times \dfrac{100 \text{ g}}{1.00 \text{ kg}} = \dfrac{1.37 \text{ moles HCl}}{1.00 \text{ kg H}_2\text{O}} = 1.37 \ m$$

**Molarity**

When solutions are to be used in reactions, a convenient concentration unit is molarity. *Molarity is the number of moles of solute per liter of solution.* The symbol for molarity is *M*.

$$M = \text{molarity} = \dfrac{\text{moles solute}}{\text{liter solution}}$$

This concentration term is useful for two reasons. First, by expressing the amount of solute present in *units of moles*, this concentration unit may easily be employed in calculations involving chemical reactions. Second, by expressing the amount of solution present in volume units, this concentration unit lends itself to volume calculations. Volumes of liquids are easily and accurately measured with simple and inexpensive equipment, such as graduated cylinders, pipettes, burettes, and volumetric flasks.

How is 1.00 liter of a 1.00 *M* solution of sodium chloride prepared? A 1.00 *M* NaCl solution contains 1.00 mole of NaCl per liter of solution:

$$1.00 \ M \text{ NaCl} = \dfrac{1.00 \text{ mole NaCl}}{1.00 \text{ liter solution}}$$

Weigh out 1.00 mole or 58.5 grams of NaCl on a balance and place the NaCl in a flask. Then add to it enough water so that the final solution has a volume of 1.00 liter. This is usually done in a volumetric flask similar to the one shown in Fig. 14.7.

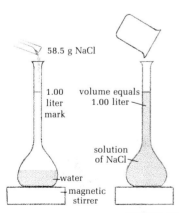

FIGURE 14.7
*Preparation of a 1.00 M solution of sodium chloride.*

EXAMPLE 4   A student dissolved 0.200 mole of barium nitrate in enough water to make 2.00 liters of solution. What is the molarity of the solution?

Given: Moles of solute = 0.200 mole $Ba(NO_3)_2$
       Volume of solution = 2.00 liters

Find: Molarity or $\dfrac{\text{moles solute}}{\text{liters solution}}$

Solution: Molarity = $\dfrac{\text{moles solute}}{\text{liter solution}} = \dfrac{0.200 \text{ mole } Ba(NO_3)_2}{2.00 \text{ liter solution}}$

$= \dfrac{0.100 \text{ mole } Ba(NO_3)_2}{1.00 \text{ liter solution}} = 0.100 \text{ M}$

EXAMPLE 5   A chemist dissolved 0.519 gram of iron(III) chloride in enough water to make 500 ml of solution. What is the molarity of the solution?

Given: Weight of solute = 0.519 g $FeCl_3$
       Volume of solution = 500 ml

Find: The molarity of the solution

Relationships: 1.00 mole $FeCl_3$ = 162 g $FeCl_3$
               1000 ml = 1.00 liter

Solution: To calculate the molarity, we need to know how many moles of $FeCl_3$ and how many liters of solution we have.

$0.519 \text{ g } FeCl_3 \times \dfrac{1.00 \text{ mole } FeCl_3}{162 \text{ g } FeCl_3} = 0.00320 \text{ mole } FeCl_3$

$500 \text{ ml solution} \times \dfrac{1.00 \text{ liter}}{1000 \text{ ml}} = 0.500 \text{ liter solution}$

Molarity = $\dfrac{0.00320 \text{ mole } FeCl_3}{0.500 \text{ liter}} = \dfrac{0.00640 \text{ mole } FeCl_3}{\text{liter}}$

We could have set these calculations up all at once.

$\dfrac{0.519 \text{ g } FeCl_3}{500 \text{ ml solution}} \times \dfrac{1000 \text{ ml}}{1.00 \text{ liter}} \times \dfrac{1.00 \text{ mole } FeCl_3}{162 \text{ g } FeCl_3} = 0.00640 \text{ M } FeCl_3$

EXAMPLE 6   Which contains a greater number of moles of sulfuric acid, (a) 150 ml of 0.376 M $H_2SO_4$, or (b) 168 ml of 0.346 M $H_2SO_4$?

Given: (a) 1.00 liter ≈ 0.376 mole
       (b) 1.00 liter ≈ 0.346 mole

Find: Number of moles of $H_2SO_4$ in solution (a) and (b).

Relationships: (a) 1.00 liter ≈ 0.376 mole
               (b) 1.00 liter ≈ 0.346 mole

Solutions: (a) $150 \text{ ml} \times \dfrac{1.00 \text{ liter}}{1000 \text{ ml}} \times \dfrac{0.376 \text{ mole}}{1.00 \text{ liter}} = 0.0564 \text{ mole}$

(b) $168 \text{ ml} \times \dfrac{1.00 \text{ liter}}{1000 \text{ ml}} \times \dfrac{0.346 \text{ mole}}{1.00 \text{ liter}} = 0.0582 \text{ mole}$

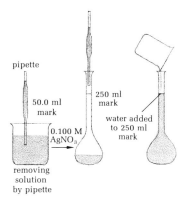

FIGURE 14.8

A given solution of known concentration can be used to make a more dilute solution whose concentration may be calculated. Place 50 ml of a 0.100 M $AgNO_3$ solution in a 250 ml volumetric flask. Add enough water to make the final volume 250 ml.

■ EXERCISE  What is the molarity of a solution that contains 3.97 moles of acetic acid in 4.50 liters of solution?

■ EXERCISE  How many moles of nitric acid are there in 100 ml of 12.0 M solution?

■ EXERCISE  A technician took a test tube containing 0.25 ml of 12 M HCl and added distilled water to the test tube until the final volume was 12.0 ml. What is the molarity of the HCl solution in the test tube?

The calculations show that 168 ml of 0.346 M solution contains more moles of $H_2SO_4$.

**EXAMPLE 7**  A student was given a bottle of 0.100 M $AgNO_3$ solution and asked to make up a more dilute solution. He withdrew 0.0500 liter (50.0 ml) of the solution and added it to a 0.250 liter (250 ml) volumetric flask (Fig. 14.8). He then added enough water to dilute the solution to 0.250 liter. What is the new concentration?

*Given:* Initial solution: 0.0500 liter of 0.100 M $AgNO_3$
Final solution: 0.250 liter of unknown molarity

*Find:* Molarity of final solution

*Solution:* Which solution contains more silver nitrate, the 0.0500 liter of 0.100 M initial solution or the 0.250 liter of the final solution? The key to the answer to our problem lies in the fact that both solutions contain the *same number of moles* of silver nitrate. We have simply added distilled water to the initial solution in order to obtain the final solution. This observation allows us to write the following relationship.

initial moles of $AgNO_3$ = final moles $AgNO_3$
(initial solution)           (final solution)

The number of moles of solute in a solution can be obtained by multiplying the molarity of the solution by the volume in liters.

$$\frac{\text{moles of solute}}{\text{liters of solution}} \times \text{liters of solution} = \text{number of moles of solute}$$

molarity × liters = number of moles of solute

For this problem we have:

initial moles $AgNO_3$ = final moles $AgNO_3$
initial molarity × initial liters = final molarity × final liters

$$\frac{0.100 \text{ mole } AgNO_3}{1.00 \text{ liter}} \times 0.0500 \text{ liter} = \text{final molarity} \times 0.250 \text{ liter}$$

$$\frac{0.100 \text{ mole } AgNO_3}{1.00 \text{ liter}} \times \frac{0.0500 \text{ liter}}{0.250 \text{ liter}} = \text{final molarity } AgNO_3$$

$$\frac{0.0200 \text{ mole } AgNO_3}{1.00 \text{ liter}} = \text{final molarity}$$

The dilute solution contains 0.0200 M $AgNO_3$.  ■

## 14.7  Properties of solutions

We saw in Section 14.2 that the presence of a solute in a solvent affects the boiling point, the freezing point, and in some cases the electrical conductivity of the solution. In this section we shall look at the quantitative effect of a solute on the vapor pressure and boiling point of a solution. In order to simplify

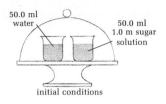

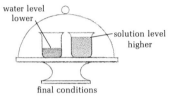

**FIGURE 14.9**

*Two beakers were placed inside a sealed glass vessel. One beaker contained distilled water, the other beaker contained a 1.0 molal sugar solution. After a period of time, the level in the beaker of water fell and the level in the beaker of sugar solution rose.*

our discussion, we will consider only ideal, nonvolatile nonelectrolytes. An ideal substance is one whose behavior fits our simple model. A nonvolatile substance is one that has no appreciable vapor pressure at room temperature. Sugar and urea are examples of nonvolatile nonelectrolytes.

**Vapor pressure**

Vapor pressure is a measure of the rate at which particles of liquid or solid substance leave the surface. We saw in Chapter 8 that the vapor pressure of water varies with the temperature. Does the presence of a solute affect the vapor pressure? Consider the simple experiment shown in Fig. 14.9. Two beakers are sealed inside of a glass vessel. One beaker contains 50 ml of 1.0 molal sugar solution while the other contains distilled water. The beakers in the glass vessel were set aside for a long period of time. When they were observed again, their volumes had changed. The level in the distilled water beaker had dropped but the level in the sugar solution rose. How is this possible?

When a beaker of distilled water is placed in a closed vessel, water molecules continually evaporate from the surface of the liquid. Once a sufficient amount of water vapor forms, molecules in the vapor state begin to collide with the liquid surface. These vapor molecules may be recaptured by the liquid. Soon the rate at which vapor molecules enter the liquid is equal to the rate at which water molecules leave the liquid.

rate of leaving liquid = rate of entering liquid

Why did the level in the pure water beaker fall and the level in the solution beaker rise? Evidently, water has left the water beaker and entered the solution beaker. We could explain this observation if we assume that water molecules leave the *solution* at a *slower rate* than they leave the *pure water,* but that water vapor molecules re-enter both beakers at the same rate. Suppose that from a given surface area of pure water 1000 $H_2O$ molecules vaporize per second, but only 996 $H_2O$ molecules leave the solution per second. In that same time period 998 $H_2O$ molecules re-enter the pure water beaker and 998 $H_2O$ molecules re-enter the solution beaker. What change is there in each beaker?

| | Number $H_2O$ entered | − | Number $H_2O$ left | = | Net change |
|---|---|---|---|---|---|
| For pure water | 998 | − | 1000 | = | −2 |
| For solution | 998 | − | 996 | = | +2 |

TABLE 14.6 The vapor pressure of pure water, a 0.10 m sugar solution, and a 1.0 m sugar solution at various temperatures.

| Temperature (°C) | Vapor pressure of water (mm Hg) | Vapor pressure of 0.10 m solution (mm Hg) | Vapor pressure of 1.0 m solution (mm Hg) |
|---|---|---|---|
| 0 | 4.58 | 4.57 | 4.50 |
| 10 | 9.21 | 9.19 | 9.05 |
| 20 | 17.54 | 17.51 | 17.23 |
| 30 | 31.82 | 31.76 | 31.25 |
| 40 | 55.32 | 55.22 | 54.32 |
| 50 | 92.51 | 92.34 | 90.86 |
| 60 | 149.38 | 149.31 | 146.70 |
| 70 | 233.7 | 233.3 | 229.5 |
| 80 | 355.1 | 354.5 | 348.7 |
| 90 | 525.8 | 524.8 | 516.3 |
| 100 | 760.0 | 758.7 | 747.0 |
| 110 | 1075.0 | 1073.0 | 1058.0 |

In that given surface area, the water beaker loses two $H_2O$ molecules per second while the solution beaker gains two $H_2O$ molecules. Over a large surface area and a long period of time we observed that the level in the water beaker falls and the level in the solution beaker rises.

A nonvolatile solute such as sugar causes the vapor pressures of solutions to be less than those of pure solvents. This drop in vapor pressure is proportional to the number of solute particles per given number of solvent molecules. In *dilute solutions* the drop in vapor pressure is proportional to the molality. Table 14.6 compares the vapor pressure of pure water at various temperatures with the theoretical vapor pressure of a 1.0 molal solution of sugar and a 0.10 molal solution of sugar.

What is the vapor pressure of pure water, a 0.100 m sugar solution, and a 1.00 m sugar solution at 50.0°C?

| Vapor pressure | | Change (difference between solution and pure water) |
|---|---|---|
| water | = 92.51 mm Hg | — |
| 0.010 m | = 92.34 mm Hg | 0.17 |
| 1.00 m | = 90.86 mm Hg | 1.65 |

■ EXERCISE What is the vapor pressure of pure water, a 0.100 m sugar solution, and a 1.00 m sugar solution at 40°C? By what amount is the vapor pressure of each solution lowered?

Note that the drop in vapor pressure for the 1.00 m solution is about ten times that of the 0.100 m solution. This indicates that the change in vapor pressure is a function of the molality. ■

### Boiling point

Compare the boiling points of the three different liquids in Table 14.7. Note that a 1.0 m sugar solution raises the boiling point of water by 0.52°C; a 0.10 molal sugar solution raises the boiling point of water by 0.05°C. We might expect a 10 m sugar solution to raise the boiling point by 5.2°C. Evidently each mole of solute per 1000 gram of solvent is capable of raising the boiling point of water by 0.52°C. This is true for other nonvolatile nonelectrolytes besides sugar. The number 0.52°C is called the molal-boiling point constant of water and is given the symbol $K_b$.

TABLE 14.7

The boiling points of three liquids.

| Liquid | Temperature (°C) |
|---|---|
| Water | 100.00 |
| 0.10 m sugar soln | 100.05 |
| 1.0 m sugar soln | 100.52 |

$$K_b = \frac{0.52°C}{\text{molal solution}}$$

The change in boiling point of a solution ($\Delta T_b$) is equal to the $K_b$ times the molality:

$$\Delta T_b = K_b \times \text{molality}$$

EXAMPLE 8  What is the change in the boiling point of a 0.376 molal aqueous solution?

Given: 0.376 m solution

Find: Change in boiling point

Relationship: $\Delta T_b = K_b \times m$
$\Delta 0.52°C \approx 1.0\ m$

Solution: $\Delta T_b = \dfrac{0.52°C}{m} \times 0.371\ m = 0.19°C$

EXAMPLE 9  What is the molarity of a solution whose boiling point has changed 0.47°C?

Given: $\Delta T_b = 0.47°C$

Find: The molality of the solution

Relationship: $0.52°C \approx 1.0\ m$

Solution: The $K_b$ is a relationship between the change in boiling point of a solution and the molality of the solution. It is a ratio of change in temperature to molality and can be used as a unity term.

$$0.47°C \times \frac{1.0\ m}{0.52°C} = 0.90\ m$$

EXAMPLE 10  How many moles of solute are there in a solution that contains 0.300 kilogram of solvent and raises the boiling point 0.192°C?

Given: Solution raises boiling point by 0.192°C and contains 0.300 kg solvent

■ EXERCISE   Examine Examples 10 and 11 and see if you can suggest a method for determining the molecular weight of an unknown, nonvolatile, soluble nonelectrolyte using a balance, a thermometer, glass apparatus, and heating equipment as instruments.

■ EXERCISE   What is the change in boiling point of a 0.638 m solution of sugar in water?

■ EXERCISE   What is the molality of a solution of sugar in water whose boiling point is 100.73°C?

Relationship: $\Delta 0.52°C \approx 1.0$ m

$$1.0 \text{ m} \approx \frac{1.0 \text{ mole solute}}{1.0 \text{ kg solvent}}$$

Solution: First, we calculate the molality of the solution from the change in boiling point. Then we calculate the number of moles of solute from the molality and weight of solvent.

$$0.192°C \times \frac{1.0 \text{ m}}{0.52°C} = 0.37 \text{ m} = \frac{0.37 \text{ mole solute}}{1 \text{ kg solvent}}$$

$$\frac{0.37 \text{ mole solute}}{1 \text{ kg solvent}} \times 0.300 \text{ kg solvent} = 0.11 \text{ mole solute}$$

EXAMPLE 11   Check Example 10 again. Suppose the weight of the solute had been 8.52 grams; what would be the weight of one mole of the solute?

Given: Weight of solute = 8.52 g
Moles of solute = 0.11 mole

Find: Weight per mole

Solution: $\dfrac{8.52 \text{ g}}{0.11 \text{ mole}} = 76$ g/mole   ■

Why is the boiling point of a solution higher than that of pure solvent? The boiling point of a substance is the temperature at which the vapor pressure of the substance is equal to 760 mm Hg. The presence of a nonvolatile solute in a solvent lowers the vapor pressure of the substance. The presence of a nonvolatile particle on the surface of the solution slows down the rate at which the solvent vaporizes. A higher temperature is needed to raise the vapor pressure to 760 mm Hg and cause it to boil.

**Freezing point depression**

The freezing point of a substance is the temperature at which the solid and liquid are in equilibrium. This equilibrium occurs *when the solid and liquid have the same vapor pressure.* The vapor pressure of ice and pure water are equal at 0°C. The vapor pressure of a 1.0 m sugar solution and ice are equal at $-1.86°C$. Therefore a 1.0 m sugar solution does not freeze until the temperature reaches $-1.86°C$. This quantity, $-1.86°C$, is the molal freezing point constant for water, symbolized by the expression $K_f$.

$$K_f = \frac{-1.86°C}{1 \text{ molal}}$$

Just as we can calculate the change in boiling point of a given solution, we can also calculate the change in freezing point.

Both changes are caused by the effect that the solute has on the vapor pressure of the system.

The effect of a solute on the vapor pressure and thus on the boiling and freezing points of substances varies from solvent to solvent. For example, in benzene $K_f = -2.53$, $K_b = 5.12$.

## Summary

Solutions are homogenous mixtures of two or more substances and may vary in composition. The substance dissolved is called the solute; the substance doing the dissolving is called the solvent. The solute or solvent may be a solid, liquid, or gas.

A concentrated solution contains a large amount of solute in a given amount of solution. A dilute solution contains less solute. A saturated solution holds as much solute as it can at a given temperature. An unsaturated solution contains less solute than it could hold at a given temperature. Supersaturated solutions are unstable. They contain more solute than they normally should. Some combinations of liquids mix in all proportions; they are said to be miscible. Other combinations of liquids do not seem to mix at all; they are said to be immiscible.

Ionic and polar substances will usually dissolve only in polar solvents such as water. The polar water molecules attract the positive and negative ends of the ionic or polar solutes. Nonpolar substances will usually dissolve only in nonpolar solvents.

The concentration of a solution is the amount of solute dissolved in a certain amount of solvent or solution. Three common concentration units are percent by weight, molality, and molarity. Molality is the number of moles of solute in 1.00 kg of solvent. Molarity is the number of moles of solute in 1.00 liter of solution. Colligative properties of solutions depend upon the number of solute particles per number of solvent particles. Vapor pressure, boiling point, freezing point, and osmotic pressure are colligative properties.

## Objectives

By the end of this chapter, you should be able to do the following:

1. Point out examples of common solutions, and identify the solute and the solvent. (Questions 2, 3)

2. Given a solute and a solvent, make a saturated and an unsaturated solution if possible. (Question 4)

3. Given the polar or nonpolar nature of various solutes and solvents, predict their expected solubility. (Question 5)

4. Predict the effect of pressure on the solubility of selected substances. (Question 6)

5. Given the weights or number of moles of the solute and solvent, calculate the percent composition of a solution. (Questions 9, 13)

6. Given the percent composition of substances in solution, calculate the weight of each substance in a given amount of solution. (Questions 7, 8, 10, 11, 12, 14)

7. Given the weight or number of moles of solute and solvent used in solution, calculate the molality of the solution. (Questions 9, 13, 26)

8. Given the weight or number of moles of solute and the volume of solution, calculate the molarity of a solution. (Questions 9, 10, 11)

9. Given the volume and molarity of a solution, calculate the amount of solute present. (Questions 15, 16, 17)

10. Given the molality of a solution, calculate the change in freezing point or boiling point. (Question 18)

11. Given the change in freezing point or boiling point of a solution, calculate its molality. (Questions 19, 20, 22)

12. From freezing point or boiling point data (weight of solute and solvent and temperature change), calculate the molecular weight of the solute. (Question 21)

13. Given the initial volume and concentration of a solution, calculate the concentration after solvent has been added or removed. (Questions 23, 24)

## Questions

1. In your own words, define each of the following terms. Where possible, give examples. How is each term related to the area of solutions? How is each related to the area of (1) bonding, or (2) chemical reactions? *solution; solvent; solute; miscible; immiscible; weak electrolyte; strong electrolyte; nonelectrolyte; nonvolatile; freezing point; saturated; unsaturated; supersaturated; soluble; slightly soluble; weight percent; molality; molarity; boiling point; colligative properties; dilute; concentrated; vapor pressure; $K_b$; $K_f$.*

2. Select five liquids that may be purchased in drug stores or supermarkets. Classify them as pure substances, solutions, or mixtures. For each solution, identify the solute and solvent.

3. Identify the solute and solvent in the following household solutions: vinegar, iodine, peroxide, ammonia, chlorine bleach.

4. Make a saturated solution and an unsaturated solution using one of the following substances: sugar, washing soda, copper(II) sulfate, alum.

5. Water and acetic acid are polar solvents. Carbon tetrachloride and benzene are nonpolar solvents. Attempt to predict the solubility of the following solutes in each of these solvents. Check your prediction in a reference text such as *The Handbook of Chemistry and Physics.* (a) octane (nonpolar); (b) HBr(g); (c) HI(g); (d) Br$_2$(l); (e) O$_2$(g); (f) S=C=S(l); (g) NaBr(s); (h) H—CH$_3$(g); (i) H—CH$_2$—O—H(l); (j) AgF(s); (k) phenol (polar) (l); (l) sulfur(s); (m) helium(g).

6. What effect will an increase of pressure have on the solubility of the following substances in benzene? (a) helium(g); (b) oxygen(g); (c) carbon tetrachloride(l); (d) sugar(s); (e) hexane(l); (f) carbon dioxide(g).

7. A bottle of aqueous hydrogen peroxide bought at a drug store was 3.0 percent peroxide. The solution weighed 227 grams. How many grams of hydrogen peroxide were bought? How many grams of water were bought? How many moles of hydrogen peroxide were bought?

8. An aqueous solution of sodium thiosulfate is often used to remove chlorine from water for aquariums. How many grams of sodium thiosulfate does a 13.0 gram bottle of 0.01 percent solution contain? If the bottle of solution costs 29 cents, how much would a pound of sodium thiosulfate cost?

9. A student prepared the following solutions:
(a) 1.00 gram $C_{12}H_{22}O_{11}$ in 100 ml of water
(b) 1.00 gram NaCl in 100 ml of water
(c) 1.00 gram $CuSO_4 \cdot 5H_2O$ in 100 ml of water
(d) 1.00 gram $KAl(SO_4)_2 \cdot 12H_2O$ in 100 ml of water

What is the percent composition of each solute in solution? What is the molality of each solution? Assume that the final volume of each solution is approximately 100 ml. What is the molarity of each solution?

10. A certain liter of beer is 5.0 percent ethyl alcohol by weight. If the density of the solution is 1.0 g/ml, what is the weight of the alcohol? What is the molarity of the solution? The formula of ethyl alcohol is $CH_3CH_2OH$.

11. A certain liter of wine is 14.0 percent ethyl alcohol by weight. If the density of the wine is 1.0 g/ml, what is the weight of the alcohol? What is the molarity of the solution?

12. A liter of vinegar contains 4.5 percent acetic acid by weight. (a) If the density of the vinegar is 1.0 g/ml, what is the weight of the acetic acid? (b) How many moles of acetic acid are there? (c) If 4.0 percent of the acetic acid molecules react with water to form hydronium ions, how many moles of hydronium ions are in the solution?

13. If 0.478 mole of urea, $CO(NH_2)_2$, were dissolved in 100 g of water: (a) What is the percent composition? (b) What is the molality?

14. There are many kinds of sugar. Sucrose or cane sugar, $C_{12}H_{22}O_{11}$, is the most common. Glucose, $C_6H_{12}O_6$, also called dextrose, makes up 0.10 percent by weight of human blood. Because it can be used by the body immediately, solutions of glucose are often injected directly into the blood of persons needing nourishment. How would you make one quart of 0.10 percent glucose solution?

15. If a solution of stomach acid is $10^{-3}$ M HCl, how many moles of HCl will 5.0 ml contain?

16. How many moles of solute are there in the following amounts of 3.5 M KBr solution: (a) 1.00 liter; (b) 100 ml; (c) 10 ml; (d) 1.0 ml.

17. What weight of KBr is there in each solution given in Question 16?

18. What are the freezing points and boiling points of the following solutions? (a) 0.01 m sugar; (b) 0.01 m urea; (c) 0.001 m sugar; (d) 0.001 m urea.

19. A solution of a nonvolatile nonelectrolyte has a boiling point of 100.73°C. What is the molality of the solution?

20. A solution of a nonelectrolyte has a freezing point of −0.97°C. What is the molality of the solution?

21. A student dissolved 35.2 grams of solute in 641 grams of water. The freezing point of the solution was −1.38°C. (a) From the temperature reading, calculate the molality of the solution. (b) How many moles of solute are there per 1000 g of water? (c) How many moles of solute are there per 641 g of water? (d) What is the mole weight of the solute? (How many grams per 1.00 mole?)

22. The solution in a car radiator froze at 12°F. (a) What was the temperature in °C? (b) What was the molality of the solution?

23. A 50.0 ml sample of 18.0 M $H_2SO_4$ was diluted to 1.00 liter. What is the molarity of the diluted solution?

24. One liter of 0.00100 M $CaCl_2$ solution was evaporated down to 25.0 ml. What is the molarity of the concentrated solution?

25. A 35.0 ml sample of 0.100 M NaOH was mixed with a 65.0 ml sample of 0.700 M NaOH. What is the molarity of the mixture? *Hint:* How many moles of NaOH are in each original solution? in the final solution?

26. A jeweler dissolved 10.0 grams of silver in 90 grams of gold. What is the molality of the solution?

# Theory of ionization and ionic reactions

## 15.1 Introduction

Biologists believe that "messages" from the brain are transmitted to different parts of the body by electrical currents. Metal wires and molten salts are good conductors of electrical currents. Are electrical messages conducted in the human body in the same way as they are through a copper wire or through a molten salt? Body fluids and cellular fluids contain solutes, such as sodium chloride, whose aqueous solutions are good conductors of electrical currents. Although scientists do not yet fully understand the role that sodium chloride plays in bodily functions, its behavior and the behavior of other electrolytes is being carefully studied. In this chapter we will investigate some of the properties of electrolytes and some typical reactions involving aqueous solutions of electrolytes.

## 15.2 Conductivity of electrolytic solutions

In the previous chapter we divided substances into those whose aqueous solutions do not conduct electricity under certain experimental conditions (nonelectrolytes) and those whose solutions do conduct electricity under these conditions (electrolytes). Aqueous solutions of substances such as HCl and

**TABLE 15.1**

*Examples of common strong and weak electrolytes.*

| Strong electrolytes | Species in soln in large amounts | Species in soln in small amounts |
|---|---|---|
| NaCl | $Na^+(aq)$ $Cl^-(aq)$ | |
| $Na_2SO_4$ | $Na^+(aq)$ $SO_4^{2-}(aq)$ | |
| $CaCl_2$ | $Ca^{2+}(aq)$ $Cl^-(aq)$ | |
| $H_2SO_4$ | $H_3O^+(aq)$ $HSO_4^-(aq)$ | $SO_4^{2-}$ |
| HCl | $H_3O^+(aq)$ $Cl^-(aq)$ | |
| NaOH | $Na^+(aq)$ $OH^-(aq)$ | |
| $Ba(OH)_2$ | $Ba^{2+}(aq)$ $OH^-(aq)$ | |
| Weak electrolytes | | |
| $NH_3$ | $NH_3$ | $NH_4^+(aq)$, $OH^-(aq)$ |
| $HC_2H_3O_2$ | $HC_2H_3O_2$ | $H_3O^+(aq)$, $C_2H_3O_2^-$ |
| $HNO_2$ | $HNO_2$ | $H_3O^+(aq)$, $NO_2^-(aq)$ |
| $H_2CO_3$ | $H_2CO_3$ | $H_3O^+(aq)$, $HCO_3^-(aq)$, $CO_3^{2-}(aq)$ |

NaCl are good conductors of electricity, or *strong electrolytes*. Aqueous solutions of $NH_3$ and $HC_2H_3O_2$ are poorer conductors of electricity, or *weak electrolytes* (Table 15.1).

The ability of a solution to conduct electricity depends roughly on the concentration of ions present in solution since the current is conducted through the solution by ions. Solutions of strong electrolytes contain significantly larger concentrations of ions than solutions of weak electrolytes. An aqueous solution of the strong electrolyte NaCl consists primarily of $Na^+(aq)$ ions and $Cl^-(aq)$ ions in addition to the solvent, $H_2O$. A solution of $HNO_3$ contains primarily $H_3O^+(aq)$ ions and $NO_3^-(aq)$ ions together with solvent molecules, $H_2O$.

A solution of ammonia, $NH_3$, in water contains some $NH_4^+(aq)$ ions and some $OH^-(aq)$ ions from the reaction between the solute, $NH_3$, and solvent $H_2O$.

$$NH_3(aq) + H_2O(l) \rightleftharpoons NH_4^+(aq) + OH^-(aq) \qquad (15.1)$$

Since $NH_3(aq)$ is a poor conductor, or weak electrolyte, there must be relatively few ions in solution. Most of the dissolved ammonia remains as $NH_3(aq)$ molecules. Other weak electrolytes such as $HC_2H_3O_2(aq)$ and $H_2S(aq)$ exist in solution primarily as the molecules $HC_2H_3O_2$ and $H_2S$.

How does an aqueous solution of a strong electrolyte conduct electricity? Consider what happens when a direct electrical current is passed through an aqueous solution of dilute copper (II) chloride (Fig. 15.1). After the current has run for a period of time, one electrode becomes coated with a copper-colored, me-

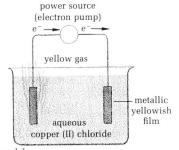

model:
oxidation at anode $2Cl^-(aq) \rightleftharpoons Cl_2(g) + 2e^-$
reduction at cathode $Cu^{2+}(aq) + 2e^- \rightleftharpoons Cu°(s)$

**FIGURE 15.1**

*Aqueous copper(II) chloride is a good conductor of electricity.*

15.2 Conductivity of electrolytic solutions

tallic film. At the other electrode a pale yellow gas is seen. What reactions are occurring?

The electrical generator or pump removes electrons from one electrode and puts them on the other electrode. The electrode that has lost electrons becomes positively charged with respect to the solution; thus, it attracts the negatively charged particles in the solution. The chloride ions in the solution are attracted to this electrode. When they reach it, each ion loses an electron and forms a neutral atom.

$$2Cl^-(aq) \rightleftharpoons Cl_2(g) + 2e^- \tag{15.2}$$

This explains the appearance of the yellow gas at one electrode.

■ EXERCISE What chemical species (that is, ions and molecules) besides water molecules are present in aqueous solutions of the following electrolytes? Indicate the species present in large amounts and those in small amounts. Strong electrolytes: NaBr and HClO$_4$. Weak electrolytes: H$_2$S and AgCl(s).

The electrode that gains electrons is negatively charged and attracts the positive copper (II) ions in the solution. When these ions reach the electrode, they gain two electrons and form neutral copper atoms.

$$Cu^{2+}(aq) + 2e^- \rightleftharpoons Cu^0(s) \tag{15.3}$$

This explains the appearance of the metal at one electrode. The equation for the electrolysis of CuCl$_2$ is obtained by summing the two half-reactions.

$$2Cl^-(aq) \rightleftharpoons Cl_2(g) + 2e^-$$
$$Cu^{2+}(aq) + 2e^- \rightleftharpoons Cu^0(s)$$
$$Cu^{2+}(aq) + 2Cl^-(aq) \rightleftharpoons Cl_2(g) + Cu^0(s)$$

■ EXERCISE When electricity is passed through an aqueous solution of iron(III) bromide, a dark, metallic solid forms at the negative electrode and a brown, volatile liquid forms at the positive electrode. What substances are produced at each electrode? Trace the flow of charge through this circuit. Write equations for any reactions that occur.

Often reactions such as this are complicated by the presence of the solvent water in the chemical system. The water may react at the anode to produce oxygen, or at the cathode to produce hydrogen. ■

## 15.3 Freezing points and boiling points of electrolytes

A 1.0 m solution of sugar or other nonelectrolyte will lower the freezing point of water by 1.86°C and raise the boiling point by 0.52°C. How does an electrolyte dissolved in water affect the freezing point and boiling point of water? Consider 0.00100 m solutions of sugar, NaCl, Na$_2$SO$_4$, and AlBr$_3$ (Table 15.2).

Notice that sodium chloride lowers the freezing point of water nearly twice as much as sugar, and sodium sulfate lowers the freezing point almost three times as much as sugar. Why? We said in Chapter 14 that the freezing point change of a solu-

**TABLE 15.2**

The calculated change in freezing point of 0.00100 m solutions of several different substances.

| Substance | Calculated change in freezing point (°C) | Particles in soln per molecule |
|---|---|---|
| Sugar | −0.00186 | 1 (sugar) |
| NaCl | −0.00372<br>−(2 × 0.00186) | 2 (Na$^+$, Cl$^-$) |
| Na$_2$SO$_4$ | −0.00558<br>−(3 × 0.00186) | 3 (2Na$^+$, SO$_4^{2-}$) |
| AlBr$_3$ | −0.00744<br>−(4 × 0.00186) | 4 (Al$^{3+}$, 3Br$^-$) |

tion is a *colligative property*—that is, it depends on the number of particles present in a solution. A 0.001 m solution of sugar contains 0.001 mole of solute particles per kilogram of water. But a 0.001 m solution of sodium chloride contains 0.001 mole of sodium ions, plus 0.001 mole of chloride ions or 0.002 mole of particles per kilogram of water. Since the sodium chloride solution contains twice as many particles as does the sugar solution, it lowers the freezing point almost twice as much. Because positive and negative ions in solution still attract one another, the freezing point is not lowered as much as might be expected.

Electrolytes such as NaCl also lower the vapor pressure of water and raise its boiling point by a factor almost twice that of sugar solutions. In more concentrated solutions, the effects of electrolytes on colligative properties are somewhat less than expected. In concentrated solutions, the *chemical effectiveness* or *activity* of an ion is lowered by the closeness of oppositely charged ions. This effect can be ignored for our purposes.

EXAMPLE 1  What is the theoretical freezing point of the following solutions? (a) 0.00300 m ethanol, a nonelectrolyte and (b) 0.00300 m Na$_2$SO$_4$, a strong electrolyte.

Relationship: 1.00 m ≈ −1.86 °C change

Solution: (a) Since ethanol is a nonelectrolyte, 1.00 mole produces 1.00 mole of particles.

$$0.00300 \text{ m} \times \frac{1.86\,°C}{1.00 \text{ m}} = -0.0055\,°C$$

Solution: (b) Sodium sulfate is a strong electrolyte. It dissociates in water to produce Na$^+$ ions and SO$_4^{2-}$ ions. 1.00 mole of Na$_2$SO$_4$ produces 3.00 moles of particles.

$$(0.00300 \text{ m}) \times \left(\frac{3 \text{ particles}}{1 \text{ Na}_2\text{SO}_4 \text{ particle}}\right) \times \left(-\frac{1.86\,°C}{1.00 \text{ m}}\right) = -0.0165\,°C$$

■ EXERCISE  Calculate the freezing point changes of the following solutions. (a) 0.00170 m urea, a nonelectrolyte and (b) 0.00170 m K$_2$SO$_4$, a strong electrolyte.

## 15.4 Solubility

Many ionic compounds such as lead(II) chloride and barium sulfate are only slightly soluble (see Chapters 10 and 11). When a sample of a slightly soluble compound such as silver chloride is placed in water, most of the solid sinks to the bottom of the container and only about $1.1 \times 10^{-5}$ mole of AgCl dissolves per liter of water. Apparently the forces between silver ions and chloride ions are great enough so that the polar water molecules cannot pull them apart. Are there any periodic trends in the solubilities of common ionic and polar compounds? The areas of the periodic chart shown in Figs. 15.2–15.7 indicate positive ions that form compounds of low solubility with negative ions such as $Cl^-$, $Br^-$, $I^-$, $S^{2-}$, $OH^-$, $SO_4^{2-}$, $CO_3^{2-}$, and $PO_4^{3-}$.

Which ions in Fig. 15.2 form slightly soluble compounds with the halide group $Cl^-$, $Br^-$, and $I^-$? The group IB monovalent ions, $Cu^+$, $Ag^+$, and $Au^+$ each form slightly soluble compounds with these ions. In addition the sixth period elements from osmium to lead form slightly soluble halide compounds. Not only do elements in the same group have similar properties, but often elements near one another in the periodic chart have similar properties. Mendeleev pointed this out when he first proposed his periodic table.

Nickel(II) sulfide, mercury(II) sulfide, and gold(II) sulfide are all "insoluble." Are they "insoluble" to the same extent? The word "insoluble" is used to mean low solubility, since all substances are somewhat soluble. For NiS, HgS, and AuS the following amounts of solute will dissolve per liter of solution.

NiS $4 \times 10^{-5}$ mole/liter
HgS $1 \times 10^{-25}$ mole/liter
AuS $7 \times 10^{-40}$ mole/liter

■ EXERCISE  The electronegativity of nickel is 1.5; mercury, 1.9; and gold, 2.3. What relationship is there between these electronegativities and the solubilities of NiS, HgS, and AuS?

■ EXERCISE  Using Figs. 15.2 to 15.7, state which of the following compounds are soluble and which are only slightly soluble. (a) AgI, (b) $Na_2S$, (c) FeS, (d) $CuSO_4$, (e) $CaCO_3$, (f) KOH.

■ EXERCISE  What aqueous reagent would cause only one of the metal ions in each solution to precipitate?

The data show that even though NiS has a very low solubility, HgS is about $10^{20}$ times less soluble than NiS, and AuS is about $10^{14}$ times less soluble than HgS.

EXAMPLE 2  A solution in a test tube contains $AgNO_3$ and $Fe(NO_3)_3$. By the addition of what reagent might you separate the silver from the iron?

*Solution:* By referring to Figs. 15.2 to 15.7, can you find any case where there is a difference between the solubility of the $Ag^+$ and $Fe^{3+}$ compounds? If we added $Cl^-$ ions, $Br^-$ ions, or $I^-$ ions to the test tube, the slightly soluble silver compounds should form. Then we could pour off the solution containing the $Fe(NO_3)_3$(aq). Suppose we added

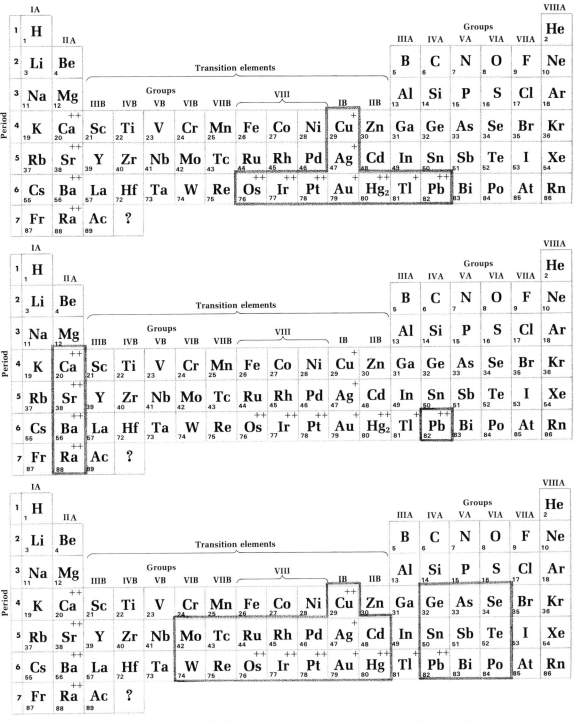

**FIGURE 15-2 (top)**
*Metal ions that form slightly soluble compounds with $Cl^-$, $Br^-$, and $I^-$.*

**FIGURE 15.3 (center)**
*Metal ions that form slightly soluble compounds with $SO_4^{2-}$.*

**FIGURE 15.4 (bottom)**
*Metal ions that form slightly soluble compounds with $S^{2-}$ ions in slightly acidic solutions.*

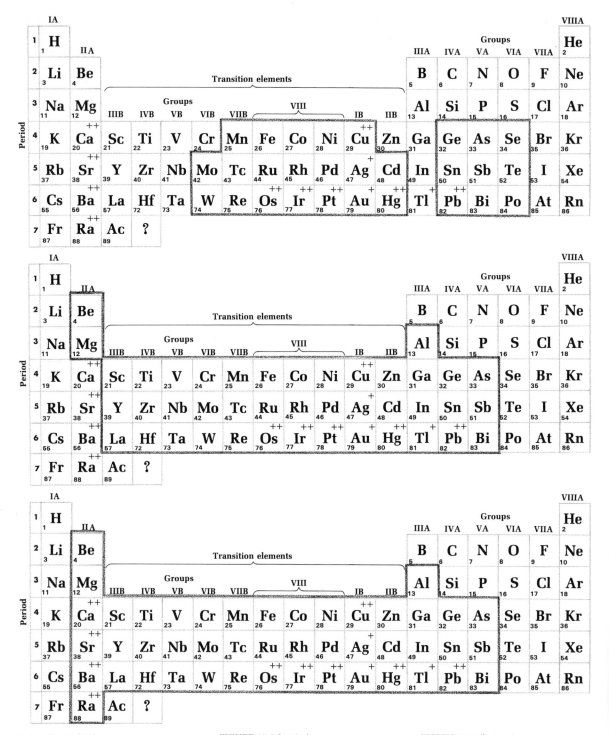

**FIGURE 15.5 (top)**
*Metal ions that form slightly soluble compounds with $S^{2-}$ ions in neutral solutions.*

**FIGURE 15.6 (center)**
*Metal ions that form slightly soluble compounds with $OH^-$ ions.*

**FIGURE 15.7 (bottom)**
*Metal ions that form slightly soluble compounds with $CO_3^{2-}$ and $PO_4^{3-}$.*

What compound precipitates? What ions are left in solution?
(a) KOH(aq) and Ba(OH)$_2$(aq)
(b) Ba(NO$_3$)$_2$(aq) and Pb(NO$_3$)$_2$(aq)
(c) CuSO$_4$(aq) and Fe(NO$_3$)$_3$(aq)
(d) NiCl$_2$(aq) and CaCl$_2$(aq).

NaBr(aq) to the original test tube. Then the aqueous bromide ions would react with the aqueous silver ions to give solid silver bromide.

$$Ag^+(aq) + Br^-(aq) \rightleftharpoons AgBr(s)$$

Left in the solution would be Fe(NO$_3$)$_3$(aq) and NaNO$_3$(aq); that is Fe$^{3+}$(aq), Na$^+$(aq), and NO$_3^-$(aq). ∎

## 15.5 Conductivity and the solvent

Pure water is a poor conductor of electricity, but molten salts are good conductors. The ability of a number of other systems to conduct an electrical current is shown in Table 15.3.

Note that pure acetic acid does not conduct a current and neither does pure water. Yet a solution of acetic acid in water does conduct a current. Pure benzene does not conduct electricity and neither does a solution of hydrogen chloride in benzene, but when immiscible water is added to the solution, the water layer conducts electricity (Fig. 15.8). How can we explain these observations?

The fact that neither pure water nor pure acetic acid conduct electricity indicates that neither substance contains a significant number of ions. When acetic acid and water are mixed, the solution conducts electricity weakly. Since the solution also has acidic properties, the acetic acid molecules must have reacted with the water molecules to form some ions.

$$HC_2H_3O_2(aq) + H_2O(l) \rightleftharpoons C_2H_3O_2^-(aq) + H_3O^+(aq) \quad (15.4)$$

or

$$\begin{array}{c}H\phantom{xx}O\\|\phantom{xx}\|\\H-C-C-O-H\\|\\H\end{array} + :\ddot{O}-H \rightleftharpoons \begin{array}{c}H\phantom{xx}O\\|\phantom{xx}\|\\H-C-C-O:^-\\|\\H\end{array} + H-\ddot{O}-H^+ \\ \phantom{xxxxxxxxxxxxxxxxxxxxxxxxxxxxxxxxxxxxxxxxxxxx}|\\ \phantom{xxxxxxxxxxxxxxxxxxxxxxxxxxxxxxxxxxxxxxxxxxxx}H \quad (15.5)$$

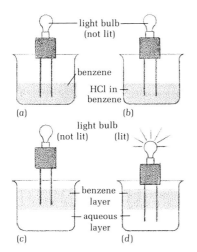

FIGURE 15.8
Neither (a) benzene, a nonpolar compound, nor (b) HCl(g) dissolved in benzene will conduct electricity. When water is added to the system, it forms a separated layer on the bottom of the beaker. The top benzene layer (c) still does not conduct electricity but the bottom water layer (d) does.

The highly polar water molecule has the ability to pull a hydrogen ion away from about 1 percent of the acetic acid molecules and form hydronium ions and acetate ions.

The fact that neither the nonpolar benzene compound nor the solution of HCl in benzene conducts electricity indicates that neither system contains a significant number of ions. When water is added to the system, only the water layer conducts a current. What ions could exist in the water? Perhaps the polar

| TABLE 15.3 | System | Conducts Electricity |
| --- | --- | --- |
| The ability of several chemical systems to conduct electricity. | Pure water | No |
| | Pure acetic acid | No |
| | Water + acetic acid | Yes (weakly) |
| | Benzene | No |
| | Benzene + HCl | No |
| | Benzene + HCl + $H_2O$ | Yes (water layer only) |
| | Molten salts, $KClO_3$ | Yes |
| | Water + soluble salts, NaCl | Yes |
| | Water + slightly soluble salts, AgCl | No |

HCl molecules left the benzene layer and entered the polar water layer. It seems reasonable that HCl would be more soluble in a polar substance than in a nonpolar one. Upon coming in contact with the water, the HCl could react to produce hydronium ions and chloride ions.

$$HCl(benzene) + H_2O(l) \rightleftharpoons H_3O^+(aq) + Cl^-(aq) \qquad (15.6)$$

The concepts developed in Sections 15.2 to 15.5 concerning ionization are summarized in Section 15.6

## 15.6 Theory of ionization

The fact that molten salts readily conduct electricity indicates that these substances already exist as ions before they are added to water. Water simply causes solid salts to *dissociate* into separate ions.

$$Na^+,Cl^-(s) \xrightleftharpoons{H_2O} Na^+(aq) + Cl^-(aq) \qquad (15.7)$$

Many substances exist as molecules. Examples are hydrogen chloride, acetic acid, and ammonia. In the pure state or in nonpolar solvents like benzene, these substances are nonconductors of electricity. Water causes these substances to *ionize*. It reacts with them to form ions. Hydrogen chloride is a strong electrolyte because virtually 100 percent of the HCl molecules in solution react to form ions.

$$HCl(g) + H_2O(l) \longrightarrow H_3O^+(aq) + Cl^-(aq) \qquad (15.8)$$

Acetic acid and ammonia are weak electrolytes because only a small percentage of their molecules react at any one instant with water molecules to form ions.

$$HC_2H_3O_2(aq) + H_2O(l) \rightleftharpoons H_3O^+(aq) + C_2H_3O_2^-(aq) \quad (15.9)$$
$$NH_3(aq) + H_2O(l) \rightleftharpoons NH_4^+(aq) + OH^-(aq) \quad (15.10)$$

*The number of ions of an electrolyte that will form depends on several factors: (1) The nature of the solvent (compare HCl in benzene and HCl in water); (2) the nature of the electrolyte (compare HCl in water and $NH_3$ in water); and (3) the temperature of the system.*

## 15.7 Mixing solutions—reactions

**Formation of slightly soluble substances**

*Observations* We saw in Section 15.4 that when two solutions are mixed, a reaction will occur if a precipitate can form. Under what other circumstances will a reaction occur? Consider the following examples.

When nitric acid is added to sodium carbonate solution, many bubbles appear in the mixture and rise to the top of the liquid. The equation for the displacement reaction is:

$$Na_2CO_3(aq) + 2HNO_3(aq) \rightleftharpoons$$
$$2NaNO_3(aq) + H_2O(l) + CO_2(g) \quad (15.11)$$

The reaction proceeds because of the formation of a *slightly soluble* substance. Instead of being a solid, however, the slightly soluble substance is a gas.

A solution of hydrochloric acid also causes bubbles to appear in the sodium carbonate solution. The reaction is similar to the previous one.

$$Na_2CO_3(aq) + 2HCl(aq) \rightleftharpoons$$
$$2NaCl(aq) + H_2O(l) + CO_2(g) \quad (15.12)$$

When solutions are mixed, a reaction will occur if one or both of the potential products of a displacement reaction are slightly soluble. The slightly soluble substance may be a solid, a liquid, or a gas.

*Net ionic equations* The addition of nitric acid or hydrochloric acid to a solution of sodium carbonate produces carbon dioxide gas in both cases. Why do both acids give the same product? Let us examine more closely the equations for these reactions. The *molecular equations* (15.11 and 15.12) for each reaction again are:

$$Na_2CO_3(aq) + 2HNO_3(aq) \rightleftharpoons 2NaNO_3(aq) + H_2O(l) + CO_2(g)$$
$$Na_2CO_3(aq) + 2HCl(aq) \rightleftharpoons 2NaCl(aq) + H_2O(l) + CO_2(g)$$

We could write the ionic compounds and strong acids in ionic form since that is how they exist in solution. We will use $H^+(aq)$ instead of $H_3O^+(aq)$. Sodium carbonate will be written as $2Na^+(aq) + CO_3^{2-}(aq)$. Sodium chloride is $Na^+(aq) + Cl^-(aq)$. Sodium nitrate is $Na^+(aq) + NO_3^-(aq)$. Since water and carbon dioxide exist primarily as molecules, we will write them as $H_2O$ and $CO_2$. Rewriting Eqs. 15.11 and 15.12 as *ionic equations* gives us:

$$2Na^+(aq) + CO_3^{2-}(aq) + 2H^+(aq) + 2NO_3^-(aq) \rightleftharpoons$$
$$2Na^+(aq) + 2NO_3^-(aq) + H_2O(l) + CO_2(g) \quad (15.13)$$

$$2Na^+(aq) + CO_3^{2-}(aq) + 2H^+(aq) + 2Cl^-(aq) \rightleftharpoons$$
$$2Na^+(aq) + 2Cl^-(aq) + H_2O(l) + CO_2(g) \quad (15.14)$$

Examine Eq. 15.14 closely. Notice that $2Na^+(aq)$ and $2Cl^-(aq)$ appear on both sides of the equation. In other words, the initial reactants and the final products contain $Na^+(aq)$ ions and $Cl^-(aq)$. Since these ions are part of the reactants and the products, they do not undergo any chemical change; they are called *spectator ions*. Writing the ionic equation without the spectator ions gives us the *net ionic equation*.

$$\cancel{2Na^+(aq)} + CO_3^{2-}(aq) + 2H^+(aq) + \cancel{2Cl^-(aq)} \rightleftharpoons$$
$$\cancel{2Na^+(aq)} + \cancel{2Cl^-(aq)} + H_2O(l) + CO_2(g) \quad (15.15)$$

or

$$CO_3^{2-}(aq) + 2H^+(aq) \rightleftharpoons H_2O(l) + CO_2(g) \quad (15.16)$$

What is the net ionic equation for the reaction between sodium carbonate and nitric acid?

$$\cancel{2Na^+(aq)} + CO_3^{2-}(aq) + 2H^+(aq) + \cancel{2NO_3^-(aq)} \rightleftharpoons$$
$$\cancel{2Na^+(aq)} + \cancel{2NO_3^-(aq)} + H_2O(l) + CO_2(g) \quad (15.17)$$

or

$$CO_3^{2-}(aq) + 2H^+(aq) \rightleftharpoons H_2O(l) + CO_2(g) \quad (15.18)$$

■ EXERCISE  Write the molecular equation, the ionic equation, and the net ionic equation for the reaction that occurs when a solution of potassium carbonate mixes with a solution of hydrobromic acid.

Compare Eqs. 15.16 and 15.18. Both net ionic equations are identical. The reactions that occur between sodium carbonate and the two strong acids, hydrochloric and nitric, are identical. Other strong acids react with soluble carbonates in the same manner. ■

What are the net ionic equations for the reaction of sodium carbonate with calcium nitrate? These two compounds are soluble salts and are written in ionic form. Calcium carbonate is only slightly soluble, and thus it is indicated as a solid, $CaCO_3(s)$.

$$Ca^{2+}(aq) + 2NO_3^-(aq) + 2Na^+(aq) + CO_3^{2-}(aq) \rightleftharpoons$$
$$CaCO_3(s) + 2Na^+(aq) + 2NO_3^-(aq) \quad (15.19)$$
$$Ca^{2+}(aq) + CO_3^{2-}(aq) \rightleftharpoons CaCO_3(s) \quad (15.20)$$
(net ionic equation)

Equation 15.20 shows that in the solution, only $Ca^{2+}$ ions and $CO_3^{2-}$ ions react. The $Na^+$ ions and $NO_3^-$ ions are spectator ions.

Is there a reaction between $Na_2CO_3(aq)$ and $KNO_3(aq)$?

$$2Na^+(aq) + CO_3^{2-}(aq) + 2K^+(aq) + 2NO_3^-(aq) \rightleftharpoons$$
$$2K^+(aq) + CO_3^{2-}(aq) + 2Na^+(aq) + 2NO_3^-(aq) \quad (15.21)$$
(no net ionic equation)

When $Na_2CO_3(aq)$ and $KNO_3(aq)$ are mixed, the potential products are all soluble. Thus, no observable reaction occurs.

**Formation of slightly ionized molecules**

When a solution of a strong acid such as HBr is mixed with a solution of a strong base such as KOH, a reaction takes place as evidenced by the heat given off. However, there is no observable color change or change of state. The products are water and a soluble salt, KBr. This reaction is an example of one that produces a *slightly ionized molecule* as a product. In this case the slightly ionized molecule is water. Compare the molecular equation, the ionic equation, and the net ionic equation for this reaction.

$$HBr(aq) + KOH(aq) \rightleftharpoons H_2O(l) + KBr(aq) \quad (15.22)$$
$$H_3O^+(aq) + Br^-(aq) + K^+(aq) + OH^-(aq) \rightleftharpoons$$
$$2H_2O + K^+(aq) + Br^-(aq) \quad (15.23)$$
$$H_3O^+(aq) + OH^-(aq) \rightleftharpoons 2H_2O \quad (15.24)$$
(net ionic equation)

*When solutions are mixed, a reaction will occur if one of the possible products is slightly ionized. Slightly ionized molecules include water, weak acids, and weak bases.* Some examples are as follows:

| | |
|---|---|
| Water, $H_2O$ | Sulfurous acid, $H_2SO_3$ |
| Ammonia, $NH_3$ | Phosphorous acid, $H_3PO_3$ |
| Acetic acid, $HC_2H_3O_2$ | Hypochlorous acid, $HClO$ |
| Hydrogen cyanide, $HCN$ | Hydrogen sulfide, $H_2S$ |
| Nitrous acid, $HNO_2$ | Phenol, $C_6H_5OH$ |

The concept of forming a slightly ionized molecule by mixing two solutions together is often used to prepare solutions of weak acids and weak bases. For example, acetic acid can be made by mixing a strong acid such as HCl(aq) with a salt containing the acetate ion.

$$HCl(aq) + NaC_2H_3O_2(aq) \rightleftharpoons NaCl(aq) + HC_2H_3O_2(aq) \quad (15.25)$$

The net ionic equation is

$$H_3O^+(aq) + C_2H_3O_2^-(aq) \rightleftharpoons HC_2H_3O_2(aq) + H_2O(l) \quad (15.26)$$

In this reaction the proton from the aqueous HCl ($H_3O^+$) is transferred to the acetate ion. To see what happens when HBr(aq) is mixed with aqueous $KC_2H_3O_2$, refer to Fig. 15.9.

The ammonium ion in a solution may be detected by adding a strong base to the solution and then warming gently. The odor of ammonia indicates that ammonium ions are present in solution.

■ EXERCISE  *Write the net ionic equation for the reaction between a solution of the salt $NH_4NO_3$ and the strong base KOH.*

$$NH_4Cl(aq) + NaOH(aq) \rightleftharpoons NaCl(aq) + NH_3(aq) + H_2O(l) \quad (15.27)$$

The net ionic equation is

■ EXERCISE  *Write the net ionic equation for the following ionic equation.*

$$Ba^{2+}(aq) + 2OH^-(aq) + 2H_3O^+(aq) + 2Cl^-(aq) \rightleftharpoons Ba^{2+}(aq) + 2Cl^-(aq) + 2H_2O(l)$$

$$NH_4^+(aq) + OH^-(aq) \rightleftharpoons NH_3(aq) + H_2O(l) \quad (15.28)$$

Warming the solution drives off some of the volatile ammonia and water. In this example a proton transfer from the $NH_4^+(aq)$ ion to the $OH^-(aq)$ ion results in the formation of two slightly ionized molecules, $NH_3$ and $H_2O$. ■

## 15.8 Dissolving a precipitate

Once two solutions have been mixed and a precipitate has formed, is there any way to get the ions of the precipitate back into solution—that is, dissolve the precipitate? Two common methods are considered. A substance may be added that will form either a slightly ionized molecule or a complex ion.

**Formation of slightly ionized molecules**

Many metal ions form precipitates in the presence of sufficient concentrations of sulfide ions (Fig. 15.4). Many of these precipitates can be dissolved by adding a solution of a strong acid to

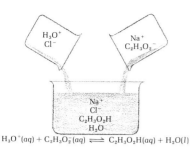

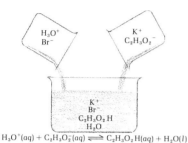

**FIGURE 15.9**
*Mixing aqueous solutions of HCl and NaC$_2$H$_3$O$_2$ gives the same net reaction as mixing aqueous solutions of HBr and KC$_2$H$_3$O$_2$.*

■ **EXERCISE** One item in a geologist's field kit is a small bottle of hydrochloric acid. The geologist uses the acid to help identify rocks that contain carbonates. A drop of acid placed on a carbonate sample causes the sample to bubble. Write the molecular equation, the ionic equation, and the net ionic equation for the reaction between solid calcium carbonate and hydrochloric acid. What causes the bubbling?

them. For example, zinc sulfide dissolves in a solution of hydrochloric acid to produce aqueous zinc chloride and hydrogen sulfide.

$$ZnS(s) + 2HCl(aq) \rightleftharpoons ZnCl_2(aq) + H_2S(aq) \qquad (15.29)$$

Examine the ionic equation and the net ionic equation.

$$ZnS(s) + 2H_3O^+(aq) + 2Cl^-(aq) \rightleftharpoons$$
$$Zn^{2+}(aq) + 2Cl^-(aq) + H_2S(aq) + 2H_2O(l) \qquad (15.30)$$

Removing the spectator ions gives us

$$ZnS(s) + 2H_3O^+(aq) \rightleftharpoons$$
$$Zn^{2+}(aq) + H_2S(aq) + 2H_2O(l) \qquad (15.31)$$

Note that in this reaction a sulfide ion in the precipitate picks up two hydrogen ions to form the *slightly ionized* H$_2$S molecule.

Many insoluble hydroxides may be dissolved by adding a strong acid to them. Consider the reaction of hydrochloric acid on copper(II) hydroxide.

$$Cu(OH)_2(s) + 2HCl(aq) \rightleftharpoons CuCl_2(aq) + 2H_2O(l) \qquad (15.32)$$

The ionic equation is

$$Cu(OH)_2(s) + 2H_3O^+(aq) + 2Cl^-(aq) \rightleftharpoons$$
$$Cu^{2+}(aq) + 2Cl^-(aq) + 4H_2O(l) \qquad (15.33)$$

Removing the spectator ions gives

$$Cu(OH)_2(s) + 2H_3O^+(aq) \rightleftharpoons Cu^{2+}(aq) + 4H_2O(l) \qquad (15.34)$$

In this reaction, aqueous protons combine with hydroxide ions to form slightly ionized water molecules.

Some precipitates dissolve to form slightly ionized, slightly soluble gas molecules. Carbonates bubble and dissolve when mixed with solutions of strong acids.

$$BaCO_3(s) + 2HBr(aq) \rightleftharpoons$$
$$BaBr_2(aq) + H_2O(l) + CO_2(g) \qquad (15.35)$$

The net ionic equation is

$$BaCO_3(s) + H_3O^+(aq) \rightleftharpoons$$
$$Ba^{2+}(aq) + 3H_2O(l) + CO_2(g) \qquad (15.36)$$

In this reaction two protons are transferred from 2H$_3$O$^+$(aq) to the carbonate ion which then decomposes into the slightly ionized water and carbon dioxide molecules. ■

15.8 Dissolving a precipitate

### Formation of a complex ion

Copper(II) hydroxide dissolves in acid solutions as Eq. 15.34 shows. This precipitate will also dissolve in ammonia solutions. What reaction is occurring?

$$Cu(OH)_2(s) + NH_3(aq) \rightleftharpoons \;?$$

It is believed that four ammonia molecules react with the solid hydroxide to produce $Cu(NH_3)_4^{2+}$ ions.

$$Cu(OH)_2(s) + 4NH_3(aq) \rightleftharpoons \\ Cu(NH_3)_4^{2+}(aq) + 2OH^-(aq) \quad (15.37)$$

The $Cu(NH_3)_4^{2+}$ ion is called the tetraamminecopper(II) ion. A *complex ion* is a central metal atom or ion that is bonded to one or more atoms, molecules, or ions. In this example $Cu^{2+}$ is the central metal ion bonded to four ammonia molecules (Fig. 15.10).

Formation of complex ions makes it possible to dissolve a large number of precipitates, as indicated by the two following examples.

Silver chloride may be dissolved by the addition of aqueous ammonia. The complex ion formed is the diamminesilver(I) ion, $Ag(NH_3)_2^+$.

$$AgCl(s) + 2NH_3(aq) \rightleftharpoons Ag(NH_3)_2^+(aq) + Cl^-(aq) \quad (15.38)$$

Zinc hydroxide may be dissolved by the addition of concentrated aqueous potassium hydroxide.

$$Zn(OH)_2(s) + 2OH^-(aq) \rightleftharpoons Zn(OH)_4^{2-}(aq) \quad (15.39)$$

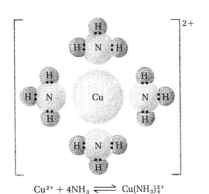

$Cu^{2+} + 4NH_3 \rightleftharpoons Cu(NH_3)_4^{2+}$

FIGURE 15.10
A typical complex ion is tetraamminecopper(II): $Cu(NH_3)_4^{2+}$.

### Summary

Substances whose solutions conduct electricity are called electrolytes. The ability of these solutions to conduct electricity is due to the formation of ions in the solution. Good conductors of electricity are called strong electrolytes. Most of the solute of a strong electrolyte exists in ionic form in solution. Poor conductors of electricity are called weak electrolytes. Only a small percent of a weak electrolyte exists in ionic form in solution.

A mole of a strong electrolyte such as sodium chloride dissolves in water and forms two moles of ions. For this reason a 1.00 molal solution of sodium chloride will have nearly twice the effect of a 1.00 molal solution of a nonelectrolyte on such colligative properties as vapor pressure, boiling point elevation, and freezing point lowering.

Substances produce ions in aqueous solutions by two common methods. Ionic substances, such as sodium chloride, conduct electricity in the molten state. They must consist of individual ions before water is added to them. Water serves to pull these ions apart and dissolve the substance. Covalent substances such as HCl and $NH_3$ exist as molecules in the pure state. Water reacts with these molecules, usually via proton transfer, to form ions.

When solutions of electrolytes are mixed, a reaction will occur if one of the products is slightly soluble or slightly ionized. The slightly soluble substance may be a solid, liquid, or gas. The slightly ionized substance is usually a weak acid, weak base, or water.

A precipitate may be dissolved via the formation of a slightly ionized molecule or by the formation of a complex ion.

## Objectives

By the end of this chapter, you should be able to do the following:

1. Identify the ions present in a solution of a strong electrolyte or a weak electrolyte. (Questions 2, 4, 5)

2. Given a description of the electrolysis of an aqueous solution of a simple salt, identify the expected products, write equations for any reactions, and trace the flow of electrical charge through the system. (Question 6)

3. Given the molality of a solution of an electrolyte, calculate the expected change in freezing point or boiling point. (Questions 7, 8, 9)

4. Given a solubility chart, identify substances in a list that are slightly soluble. (Question 10)

5. Using solubility charts, write the formula of a regaent that will cause one metal ion in solution to precipitate. (Question 14)

6. Given the formulas of substances in a simple chemical system that conducts electricity, write equations for the formation of any ions present. (Question 15)

7. Given two aqueous solutions, write the molecular equation and net ionic equation for a displacement reaction that may occur when the solutions are mixed. (Question 16)

8. Given the word equation or molecular equation for the dissolution of a precipitate, write the ionic equation and the net ionic equation. (Question 17)

## Questions

1. Define the following terms. Give operational and theoretical definitions where possible. *nonelectrolyte; dissociation; ionization; molecular equations; ionic equations; net ionic equations; spectator ions; slightly ionized; complex ion.*

2. Solutions of the following compounds were tested in the laboratory and found to be strong electrolytes. What ions are present in large amounts in these solutions? (a) KOH; (b) $Sr(OH)_2$; (c) $NaClO_3$; (d) $NiCl_2$; (e) $Mn(NO_3)_2$; (f) $Na_2CO_3$; (g) $Na_2SO_4$; (h) $K_2NaPO_4$; (i) $K_3Al(SO_4)_3$.

3. Which of the compounds listed in Question 2 would you expect to be conductors in the molten or liquid state?

4. Solutions of the following compounds were tested in the laboratory and found to be weak electrolytes. What ions are present in these solutions? What molecules are in solution? Which are present in large amounts? (a) HIO; (b) $H_2Te$; (c) $NH_3$; (d) $CH_3NH_2$ (similar to $NH_3$); (e) $H_2SeO_3$ (similar to $H_2SO_3$); (f) $HClO_2$.

5. The following compounds exist in solution in the human body. Which compounds are strong electrolytes? Which are weak electrolytes? List the ions and molecules, if any, that these substances exist as in solution. (a) NaCl; (b) KCl; (c) $H_2CO_3$; (d) $NaH_2PO_4$.

6. Solutions of the following substances were electrolyzed. From the data given, write equations for the formation of the products, and trace the flow of electrical current through the circuit. (a) Nickel(II) chloride: At one electrode a dark, metallic powder forms. At the other electrode a yellow-green gas forms. (b) Hydrochloric acid: At one electrode a colorless, combustible gas forms. At the other electrode a yellow-green gas forms.

7. What would be the expected (theoretical) freezing point depression of the following solutions of strong electrolytes? (a) $1.0 \times 10^{-4}$ m KI; (b) $1.0 \times 10^{-3}$ m KI; (c) $1.0 \times 10^{-2}$ m $K_3PO_4$; (d) $1.0 \times 10^{-2}$ m $K_3Al(PO_4)_2$.

8. Which of the following solutions has the lowest freezing point? (a) $1.00 \times 10^{-3}$ m urea (nonelectrolyte) or (b) $7.00 \times 10^{-4}$ m HCl.

9. Which of the following solutions has the highest boiling point? (a) $1.00 \times 10^{-3}$ m sugar or (b) $3.00 \times 10^{-4}$ m AgCl.

10. Using Figs. 15.2 to 15.7 as a reference, state which of the following compounds is slightly soluble. (a) $Al_2(CO_3)_3$; (b) $K_2S$; (c) CaS; (d) $Rb_2CO_3$; (e) $NiSO_4$; (f) $Ag_2SO_4$; (g) $BaSO_4$; (h) TiS; (i) $SrCO_3$; (j) CoS; (k) $As_2S_3$; (l) $MgSO_4$.

11. If a pond of water contained a high concentration of chloride ion, which of the following ions could not be in solution in appreciable amounts? (a) $NH_4^+$; (b) $Sn^{2+}$; (c) $Pb^{2+}$; (d) $Cu^+$; (e) $Hg^{2+}$; (f) $Ba^{2+}$; (g) $Al^{3+}$; (h) $Hg_2^{2+}$; (i) $Au^+$.

12. If a beaker contained a solution with a high concentration of phosphate ions, which of the following ions could not be in solution in appreciable amounts? (a) $Hg^{2+}$; (b) $Co^{2+}$; (c) $Ni^{2+}$; (d) $Mg^{2+}$; (e) $Pb^{2+}$; (f) $Sn^{2+}$; (g) $Ag^+$; (h) $NH_4^+$; (i) $Na^+$.

13. Certain areas in the United States contain soil with large amounts of carbonate. If appreciable amounts of $CO_3^{2-}$ are in solution, which of the following ions would not be soluble in such soils? (a) $Na^+$; (b) $Ca^{2+}$; (c) $Mg^{2+}$; (d) $Fe^{3+}$; (e) $Cu^{2+}$; (f) $K^+$.

14. Name a reagent that will cause one of the following ions to precipitate, but not the other. (a) Precipitate $Ag^+$ but not $Na^+$; (b) Precipitate

Ni$^{2+}$ but not Ba$^{2+}$; (c) Precipitate Pb$^{2+}$ but not Ca$^{2+}$; (d) Precipitate Pb$^{2+}$ but not Zn$^{2+}$; (e) Precipitate Cu$^{2+}$ but not Ba$^{2+}$.

15. The aqueous solutions of the following contain ions. Write equations for the formation of these ions. (a) KCl(s); (b) H$_2$S(g); (c) NH$_3$(g); (d) HI; (e) H$_2$SO$_4$; (f) CoCl$_2$; (g) Zn(NO$_3$)$_2$; (h) Na$_3$PO$_4$.

16. Write molecular equations, ionic equations, and net ionic equations for reactions between the following solutions.
(a) CdCl$_2$(aq) + Ba(OH)$_2$(aq)
(b) CdSO$_4$(aq) + LiOH(aq)
(c) Zn(NO$_3$)$_2$(aq) + K$_2$S(aq)
(d) NaBr(aq) + K$_2$SO$_4$(aq)
(e) KNO$_2$(aq) + HCl(aq)
(f) K$_2$SO$_3$(aq) + HI(aq)
(g) (NH$_4$)$_2$SO$_4$(aq) + Ba(OH)$_2$(aq)
(h) HNO$_2$(aq) + Ba(OH)$_2$(aq)

17. Write net ionic equations for the following.
(a) SnS(s) + Na$_2$S(aq) $\rightleftharpoons$ SnS$_2^{2-}$(aq) + Na$^+$(aq)
(b) Ti(OH)$_2$(s) + 2HBr(aq) $\rightleftharpoons$ TiBr$_2$(aq) + 2H$_2$O(l)
(c) BaSO$_3$(s) + 2HCl(aq) $\rightleftharpoons$ BaCl(aq) + H$_2$SO$_3$(aq)
(d) Cd(OH)$_2$(s) + 2HNO$_3$(aq) $\rightleftharpoons$ Cd(NO$_3$)$_2$(aq) + 2H$_2$O(l)
(e) Zn(OH)$_2$(s) + 2KOH(aq) $\rightleftharpoons$ Zn(OH)$_4^{2-}$(aq) + 2K$^+$(aq)

18. Metabolism of proteins in humans produces sulfuric acid. This acid is neutralized by sodium hydrogen carbonate. The carbon dioxide produced is excreted by the lungs. (a) Write the net ionic equation for the neutralization of this acid.

H$_2$SO$_4$(aq) + 2NaHCO$_3$(aq) $\rightleftharpoons$ Na$_2$SO$_4$(aq) + H$_2$O(l) + CO$_2$(g)

(b) How many moles of NaHCO$_3$ are needed to neutralize 0.0761 mole of H$_2$SO$_4$?

19. The average amount of urine voided in a 24-hour period is about 1300 ml. (a) If the density is 1.03 g/ml, what is the weight of this amount of urine? (b) If 60 grams of urine are dissolved solids, what is the weight of the water? (c) If 15 grams of the solids in urine are sodium chloride, what is the molality of NaCl in urine? What is the molarity of NaCl in urine? (d) If 30 grams of solids in urine are urea, CH$_2$(NH$_2$)$_2$, what is the molality of urea in urine? What is the molarity of urea in urine?

20. Tooth enamel contains, among other things, 35.8 percent by weight calcium ions, and 17.40 percent by weight phosphate ions. Calculate the number of moles of calcium ions and the number of moles of phosphate ions present in 10.00 grams of tooth enamel.

21. Solutions of aqueous acids can be made by adding another acid to a soluble salt and producing a precipitate. From the following word equations write molecular equations and net ionic equations.
(a) Aqueous silver nitrate + hydrochloric acid $\rightleftharpoons$
solid silver chloride + nitric acid
(b) Aqueous silver nitrate + hydrobromic acid $\rightleftharpoons$
solid silver bromide + nitric acid.
(c) Aqueous barium chloride + sulfuric acid $\rightleftharpoons$
solid barium sulfate + hydrochloric acid.

(d) Aqueous barium chlorate + sulfuric acid $\rightleftharpoons$
solid barium sulfate + chloric acid.

(e) Aqueous lead(II) nitrate + hydrochloric acid $\rightleftharpoons$
solid lead(II) chloride + nitric acid.

22. Solutions of hydrogen peroxide can be made by adding dilute sulfuric acid to barium peroxide. Write the molecular equation.

Barium peroxide + aqueous sulfuric acid $\rightleftharpoons$
solid barium sulfate + aqueous hydrogen peroxide.

23. The white paint pigment, lithopone, is made by mixing aqueous barium sulfide with aqueous zinc sulfate. Using solubility diagrams, determine the products. Write a molecular and net ionic equation for the reaction.

# Acid-base reactions

## 16.1 Introduction

Chemists have attempted to classify substances into meaningful categories. A substance may be classified by how it behaves when added to other substances. We have already classified substances as (1) elements or compounds, and (2) solids, liquids, or gases. In the seventeenth century, Robert Boyle classified some substances as either acids or bases.

Acids and bases play a major role in the biology of living organisms. The human body can function properly only when the proper amounts of acids and bases are present in each part of the body. Blood contains a slight excess of base, whereas the stomach contains an excess of acid. Too much acid in the stomach may cause ulcers. Antacids, which are bases, are sometimes taken to control excess stomach acid. A base neutralizes the effect of acids by a chemical reaction.

In this chapter we shall discuss the reactions that acids and bases undergo and the models which chemists use to explain these reactions. We have already studied proton transfer reactions and electron donor-acceptor reactions. Both types of reactions will be discussed here because they are acid-base reactions. Knowledge of the concentration of an acid or base in solution is necessary in order to calculate the amount of reactants needed or products formed in a reaction. We will expand the concentration term molarity to introduce a new method, pH, for describing solutions of acids or bases.

# 16.2 Aqueous solutions of acids and bases

**Properties of acids**

What taste is common to vinegar, unsweetened grapefruit juice, and lemon juice? Each of these substances has a sour taste because each contains acids. (Remember that laboratory chemicals should never be tasted or touched.) Acids have a number of other properties in common. We can use these properties to write the following operational definition of aqueous solutions of acids. They (a) taste sour; (b) react with many metals including zinc to form hydrogen gas; (c) react with carbonates to form carbon dioxide gas; (d) conduct electricity; (e) turn the dye litmus from blue to red; and (f) react with aqueous basic solutions to form water; that is, they neutralize the properties of bases. This last property of acids links them to bases.

**Properties of bases**

The following properties constitute an *operational definition of aqueous solutions of bases*. They (a) taste bitter; (b) feel soapy; (c) conduct electricity; (d) turn the dye litmus from red to blue; and (e) neutralize the properties of acids. Basic solutions feel soapy because they react with fats in skin tissue and change them to soaps. Remember that these chemicals are very dangerous; *all* spills should be washed with large amounts of water.

**The model of Arrhenius**

*Acids* We might expect that since aqueous solutions of acids have many macroscopic properties in common, they must have some submicroscopic properties in common as well. Lavoisier proposed in 1790 that all acids contain oxygen. About twenty years later, Davy showed that hydrochloric acid did not contain oxygen. He proposed instead that all acids contain hydrogen.

In an attempt to explain the ability of acids to conduct electricity, Arrhenius proposed in 1880 that solutions of acids contain ions. He said that the ion responsible for the properties of acids is the *hydrogen ion*, $H^+$, a proton. According to Arrhenius, when a substance such as hydrogen chloride is added to water it ionizes according to the following equation:

$$HCl(g) \xrightleftharpoons{H_2O} H^+(aq) + Cl^-(aq) \qquad (16.1)$$

Other acids react the same way when added to water.

$$H_2SO_4(l) \xrightleftharpoons{H_2O} H^+(aq) + HSO_4^-(aq) \qquad (16.2)$$

A bare proton cannot exist by itself in an aqueous medium; rather, it attaches to a water molecule to form the hydronium ion, $H_3O^+(aq)$. The sour taste of acids and their ability to turn blue litmus to red is due to the $H_3O^+(aq)$ ion. The reaction of zinc metal with acids to produce hydrogen gas is a reaction between $H_3O^+(aq)$ ions and zinc atoms.

$$2H_3O^+(aq) + Zn(s) \rightleftharpoons Zn^{2+}(aq) + H_2(g) + 2H_2O(l) \qquad (16.3)$$

*Bases* Common bases include NaOH, KOH, $Ba(OH)_2$, and $NH_3$. According to the Arrhenius model, the properties of aqueous bases were due to the aqueous *hydroxide ion*, $OH^-(aq)$. When solid sodium hydroxide is placed in water, it dissociates into aqueous sodium ions and aqueous hydroxide ions.

$$NaOH(s) \xrightleftharpoons{H_2O} Na^+(aq) + OH^-(aq) \qquad (16.4)$$

When ammonia gas dissolves in water, it reacts with the water to produce hydroxide ions.

$$NH_3(g) + H_2O(l) \rightleftharpoons NH_4^+(aq) + OH^-(aq) \qquad (16.5)$$

The bitter taste of bases, their soapy feel, their ability to turn red litmus to blue, and their ability to neutralize acids is due to the aqueous hydroxide ion.

### Neutralization

What is neutralization? Consider what happens when 100.0 ml of 0.100 M KOH is mixed with 100.0 ml of 0.100 M HCl. The resulting solution is clear and colorless like the original solutions. It tastes neither sour nor bitter, but salty. It does not feel soapy. The new solution conducts electricity, yet has no noticeable effects on samples of zinc or calcium carbonate. The new solution does not change the color of red litmus or blue litmus. The appearance of new properties (Table 16.1) when acids and bases are mixed is called *neutralization*.

How did Arrhenius explain neutralization? According to his model, the $H^+(aq)$ ion from the acid combines with the $OH^-(aq)$ ion from the base to form water.

$$H^+(aq) + OH^-(aq) \rightleftharpoons H_2O(l) + heat \qquad (16.6)$$

**TABLE 16.1**

Comparison of the properties of 0.100 M HCl, 0.100 M KOH, and a mixture of 100.0 ml of each.

| Properties and reactions | Acid HCl(aq) | Base KOH(aq) | Equivalent amounts of acid and base(aq) |
|---|---|---|---|
| Taste | sour | bitter | salty |
| Feel | — | slippery | — |
| Zinc | dissolves; produces $H_2(g)$ | dissolves; produces $H_2(g)$ | no effect |
| Calcium carbonate | dissolves; produces $CO_2(g)$ | no effect | no effect |
| Litmus | turns blue to red | turns red to blue | no effect |
| Electricity | conducts | conducts | conducts |
| Evaporation | no residue | KOH residue | KCl residue |

Since all aqueous acids contain $H^+$(aq) ions and all aqueous bases contain $OH^-$(aq) ions, Eq. 16.6 will occur whenever aqueous acids are mixed with aqueous bases.

A handy rule is "acid plus base yields salt plus water." This rule holds for Arrhenius acids and bases. The water forms as shown in Eq. 16.6. Where does the salt come from? Consider the reaction between HCl(aq) and KOH(aq). The products are $H_2O(l)$ and KCl(aq), a salt.

$$HCl(aq) + KOH(aq) \rightleftharpoons H_2O(l) + KCl(aq) \qquad (16.7)$$

The reaction involves the transfer of a proton to the $OH^-$ ion in KOH. The salt is formed from the positive ion in the base ($K^+$) and the negative ion in the acid ($Cl^-$). Since all aqueous bases contain a positive ion and all aqueous acids contain a negative ion, salts are always produced when acids and bases neutralize each other. For example:

Molecular equation:
$$HNO_3(aq) + NaOH(aq) \rightleftharpoons NaNO_3(aq) + H_2O(l) \qquad (16.8)$$
Ionic equation:
$$H_3O^+(aq) + NO_3^-(aq) + Na^+(aq) + OH^-(aq) \rightleftharpoons$$
$$Na^+(aq) + NO_3^-(aq) + 2H_2O(l) \qquad (16.9)$$
Net ionic equation:
$$H_3O^+(aq) + OH^-(aq) \rightleftharpoons 2H_2O(l) \qquad ■ \qquad (16.10)$$

■ **EXERCISE** Complete the following molecular equation. Write the ionic equation and the net ionic equation.

HBr(aq) + NaOH(aq) $\rightleftharpoons$

■ **EXERCISE** When aqueous barium hydroxide is neutralized by aqueous sulfuric acid, one product is the insoluble salt, barium sulfate. Would you expect the neutralized mixture to conduct electricity in a light bulb conductivity apparatus? Use a net ionic equation to explain why or why not.

## 16.3 Proton transfer reactions the Brønsted-Lowry model

According to Arrhenius, the hydroxide ions in a base have the power to react with an acid. What gives the hydroxide ion this property? Evidently it is due to its ability to combine with or accept a hydrogen ion.

When a liter of solution containing 1.00 mole of hydrochloric acid was added to 0.500 mole of solid sodium carbonate, vigorous bubbling occurred and the sodium carbonate dissolved. The molecular equation for the reaction is:

$$2HCl(aq) + Na_2CO_3(s) \rightleftharpoons 2NaCl(aq) + H_2CO_3(aq) \rightleftharpoons$$
$$2NaCl(aq) + H_2O(l) + CO_2(g) \quad (16.11)$$

The sodium carbonate neutralized the properties of the hydrochloric acid. Should it be called a base? In 1923 Brønsted and Lowry proposed that substances such as sodium carbonate should be classified as bases even though they do not contain hydroxide ions. The carbonate ion reacts with acids by *accepting hydrogen ions* and forming carbonic acid which decomposes into $H_2O$ and $CO_2$. Both hydroxide ions and carbonate ions neutralize acids by accepting protons. According to the Brønsted-Lowry model:

*A base is a species that can accept a proton.*

*An acid is a species that can donate a proton to another species.*

In the reaction between aqueous hydrogen chloride and sodium carbonate, the $H_3O^+$ donates a hydrogen ion to the carbonate ion (Fig. 16.1).

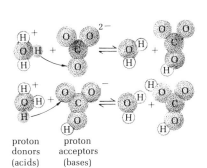

proton donors (acids)　　proton acceptors (bases)

**FIGURE 16.1**
*According to the Brønsted-Lowry model, a base is a proton acceptor.*

$$2HCl(aq) + Na_2CO_3(s) \rightleftharpoons 2NaCl(aq) + H_2CO_3(aq) \quad (16.12)$$

↑　　　　　　↑
proton donor　proton acceptor
(acid)　　　　(base)
↓　　　　　　↓

$$H_3O^+ + CO_3^{2-} \rightleftharpoons HCO_3^- + H_2O \quad (16.13)$$
$$H_3O^+ + HCO_3^- \rightleftharpoons H_2CO_3 + H_2O \quad (16.14)$$

The Brønsted-Lowry model includes the Arrhenius model and more. In the Brønsted-Lowry model, *the hydroxide ion is just one of many possible bases.* In addition the Arrhenius Model is limited to aqueous solutions, and therefore cannot be used with the many reactions that occur only in nonaqueous solvents. The Brønsted-Lowry model can be extended to other solvent systems. The following sections further illustrate the Brønsted-Lowry concept.

**Ionization of strong acids**

When added to water, HCl(g) dissolves and reacts with it:
$$HCl(g) + H_2O(l) \rightleftharpoons H_3O^+(aq) + Cl^-(aq) \quad (16.15)$$

$$H:\ddot{\underset{..}{Cl}}: + H:\ddot{\underset{H}{O}}: \rightleftharpoons H:\overset{H^+}{\underset{H}{\ddot{O}}}: + :\ddot{\underset{..}{Cl}}: \quad (16.16)$$

proton    proton<br>donor    acceptor<br>(acid)    (base)

In this reaction the HCl is the proton donor, or Brønsted-Lowry acid. The $H_2O$ is the proton acceptor, or Brønsted-Lowry base. Virtually all of the HCl reacts with water. A strong acid such as HCl is one that reacts extensively with water to produce $H_3O^+$ ions. ∎

■ EXERCISE *Complete the following equations. Identify the Brønsted-Lowry acid and base.*
(a) $HI(g) + H_2O(l) \rightleftharpoons$
(b) $HClO_4(l) + H_2O(l) \rightleftharpoons$

### Ionization of weak acids

Weak acids are operationally defined as those whose aqueous solutions are weak conductors of electricity. Theoretically, in solutions of weak acids only a small percent of the acid molecules react with water molecules. In the equation for the reaction between acetic acid and water, the long arrow pointing to the left indicates that less than 50 percent of the acid molecules have reacted.

$$CH_3COOH(l) + H_2O(l) \rightleftharpoons CH_3COO^-(aq) + H_3O^+(aq) \quad (16.17)$$

In this reaction the acetic acid molecule donates a proton to the water molecule.

structural formula showing acetic acid + water ⇌ acetate ion + hydronium ion<br>proton donor   proton acceptor<br>(acid)   (base)   (16.18)

The water molecule acts as the base in this reaction. Other weak acids are similar to acetic acid.

### Ionization of weak bases

Ammonia is a *weak* conductor of electricity. When ammonia gas reacts with water, only a small number of ammonia molecules react at any one time, as indicated by the long arrow pointing to the left.

$$NH_3(aq) + H_2O(l) \rightleftharpoons NH_4^+(aq) + OH^-(aq) \quad (16.19)$$

Which species is the base in this reaction? Which species is the acid? Examine the structural formulas.

$$H_3N: + H_2O: \rightleftharpoons H_3NH^+ + :OH^-  \qquad (16.20)$$

proton acceptor (base) — proton donor (acid)

■ EXERCISE  When rubbed between the fingers, dilute solutions of strong bases such as sodium hydroxide feel very soapy. (Caution.) Use a model to explain why dilute solutions of ammonia feel only slightly slippery.

The proton acceptor, or base, is ammonia. The proton donor, or acid, is water. Compare the function of water in this equation with its function when mixed with HCl or $CH_3COOH$. Previously, water behaved as a base; with ammonia, water behaves as an acid. Substances that can act as either acids or bases are said to be *amphiprotic*. Amphiprotic substances can either accept protons or donate protons. ■

**Conjugate acids and bases**

Let us examine the ammonia-water system further. When ammonia is added to water, some $NH_4^+$ ions and $OH^-$ ions immediately form. These ions can react with each other to form the original substances, $NH_3$ and $H_2O$. In an aqueous ammonia solution both the forward and reverse reaction continuously occur. When two opposing reactions are continuously occurring at the same time and the overall number of ions and molecules remains constant, the system is in a state of *dynamic equilibrium*. (Dynamic means motion; equilibrium means no macroscopic change.)

$$NH_3(aq) + H_2O(l) \underset{\text{reverse reaction}}{\overset{\text{forward reaction}}{\rightleftharpoons}} NH_4^+(aq) + OH^-(aq) \qquad (16.21)$$

Consider the reverse reaction for a moment. The $NH_4^+$ ion donates a proton to the $OH^-$ ion. Therefore, the $NH_4^+$ ion is a Brønsted acid and the $OH^-$ is a Brønsted base.

$$NH_3(aq) + H_2O(l) \rightleftharpoons NH_4^+(aq) + OH^-(aq) \qquad (16.22)$$
base + acid ⇌ acid + base

Compare the $NH_3$ and $NH_4^+$ particles. One acts as a base, the other as an acid. They are related to one another via a proton.

$$NH_3 \underset{-H^+}{\overset{+H^+}{\rightleftharpoons}} NH_4^+ \qquad (16.23)$$
base₁     acid₁

The ammonia molecule and the ammonium ion are *conjugate*, *acid-base pairs*. The *conjugate base* of an acid has lost a proton; $NH_3$ is the conjugate base of $NH_4^+$. The *conjugate acid* of a base has gained a proton; $NH_4^+$ is the conjugate acid of $NH_3$.

It is often useful to keep track of conjugate acid-base pairs in a reaction. In the reaction between ammonia and water, we will call $NH_3$ base$_1$ and its conjugate, $NH_4^+$ acid$_1$. Since $H_2O$ and $OH^-$ are also conjugate acid-base pairs, we will label them acid$_2$ and base$_2$.

$$NH_3(aq) + H_2O(l) \rightleftharpoons NH_4^+(aq) + OH^-(aq) \qquad (16.24)$$

$$\underset{base_1}{} + \underset{acid_2}{} \rightleftharpoons \underset{acid_1}{} + \underset{base_2}{}$$

(with $-H^+$ from base$_1$/acid$_1$ pair and $+H^+$ to acid$_2$/base$_2$ pair)

■ **EXERCISE** Label the conjugate acid-base pairs in the following reactions.
(a) $HNO_2(aq) + H_2O(l) \rightleftharpoons NO_2^-(aq) + H_3O^+(aq)$
(b) $CH_3NH_2(aq) + H_2O(l) \rightleftharpoons CH_3NH_3^+(aq) + OH^-(aq)$

All Brønsted-Lowry acid-base reactions involve conjugate acid-base pairs. For example:

$$HCl(g) + H_2O(l) \rightleftharpoons H_3O^+ + Cl^-(aq) \qquad (16.25)$$

$$\underset{acid_1}{} + \underset{base_2}{} \rightleftharpoons \underset{acid_2}{} + \underset{base_1}{}$$

### Neutralization

According to the Arrhenius model, a neutralization reaction occurs when an aqueous acid reacts with an aqueous base to form a salt and water. *According to the Brønsted-Lowry model a neutralization reaction occurs whenever the solvent is a product of the reaction.* When KOH(aq) and HCl(aq) are mixed, more solvent is produced.

$$HCl(aq) + KOH(aq) \rightleftharpoons KCl(aq) + H_2O(l) \qquad (16.26)$$

The net ionic equation is:

$$H_3O^+(aq) + OH^-(aq) \rightleftharpoons \underset{\text{Solvent} + \text{Solvent}}{H_2O + H_2O} \qquad (16.27)$$

$$\underset{acid_1}{} + \underset{base_2}{} \rightleftharpoons \underset{base_1}{} + \underset{acid_2}{}$$

The Brønsted-Lowry model includes the Arrhenius model of neutralization and extends it to other solvents.

### Prediction of products

Whenever a Brønsted acid is mixed with a Brønsted base, a reaction can occur via a proton transfer. All *Brønsted acids* must contain a transferable hydrogen ion or proton. All *Brønsted bases* must contain an unshared pair of electrons that is capable of accepting a proton. A number of these acids and bases are listed in Table 16.2 as conjugate pairs.

Compare the two acids, HCl and $H_3O^+$. Which is the stronger acid? The *stronger acid* is the one that succeeds in giving away its proton. Since HCl gives away its proton and forms $Cl^-$, it is the stronger acid. Which is the stronger base, $H_2O$ or $Cl^-$? The

**TABLE 16.2**

The relative strengths of acids and their conjugate bases.

| | Conjugates | |
|---|---|---|
| | Acid | Base |
| | $HClO_4$ | $ClO_4^-$ |
| | $HCl$ | $Cl^-$ |
| | $HNO_3$ | $NO_3^-$ |
| | $H_3O^+$ | $H_2O$ |
| | $H_2SO_3$ | $HSO_3^-$ |
| | $HSO_4^-$ | $SO_4^{2-}$ |
| Stronger acids | $H_3PO_4$ | $H_2PO_4^-$ Weaker bases |
| | $CH_3CO_2H$ | $CH_3CO_2^-$ |
| | $H_2CO_3$ | $HCO_3^-$ |
| Weaker acids | $H_2S$ | $HS^-$ Stronger bases |
| | $HCN$ | $CN^-$ |
| | $NH_4^+$ | $NH_3$ |
| | $HS^-$ | $S^{2-}$ |
| | $H_2O$ | $OH^-$ |
| | $OH^-$ | $O^{2-}$ |
| | $NH_3$ | $NH_2^-$ |

stronger base is the one that succeeds in attracting and holding the proton. Since the $H_2O$ gains a proton to form $H_3O^+$, it is the stronger base.

$$HCl(aq) + H_2O(l) \rightleftharpoons H_3O^+(aq) + Cl^-(aq) \qquad (16.28)$$

stronger acid + stronger base $\rightleftharpoons$ weaker acid + weaker base

A useful rule of the Brønsted-Lowry model is that the *stronger acid and stronger base will react more than 50 percent to form the weaker acid and weaker base.*

Notice the acids stronger than $H_3O^+$ listed in Table 16.2. When placed in an equal number of moles of water, HCl, $H_2SO_4$, $HNO_3$, and $HClO_4$ always react more than 50 percent to form the weaker acid $H_3O^+$. Weak acids such as $H_2S$ react less than 50 percent with an equal number of moles of water.

$$H_2S + H_2O \rightleftharpoons HS^- + H_3O^+ \qquad (16.29)$$

weaker acid + weaker base $\rightleftharpoons$ stronger base + stronger acid

■ **EXERCISE** *Identify the Brønsted acid and base and complete the equations. Use Table 16.2 to determine which reactions will proceed more than 50 percent.*
(a) $HClO_4(aq) + H_2O(l) \rightleftharpoons$
(b) $NH_4^+(aq) + H_2O(l) \rightleftharpoons$

## 16.4 Electron donor reactions— the Lewis model

When sodium oxide is added to water, the following reaction takes place:

$$Na_2O(s) + H_2O(l) \rightleftharpoons 2NaOH(aq) \qquad (16.30)$$

This is a Brønsted acid-base reaction because a hydrogen ion is transferred from the water molecule to the oxide ion.

$$:\!\ddot{\underset{..}{O}}\!:^{2-} + H\!:\!\ddot{\underset{..}{O}}\!:\underset{H}{\overset{}{}} \rightleftharpoons :\!\ddot{\underset{..}{O}}\!:H^- + :\!\ddot{\underset{..}{O}}\!:H^- \qquad (16.31)$$

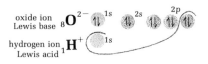

FIGURE 16.2
A Lewis acid is an electron-pair acceptor. A Lewis base is an electron-pair donor.

How does the hydrogen ion attach itself to the oxide ion? It makes use of an unshared electron pair on the oxide ion to form a covalent bond. In 1923, G. N. Lewis proposed that the formation of a covalent bond was an acid-base reaction.

Lewis called the oxide ion a base. According to the Lewis model, *a base is any chemical species that donates a share in an electron pair*. Lewis called the hydrogen ion (from water) an acid. According to the Lewis model, *an acid is any chemical species that accepts a share in an electron pair* (Fig. 16.2).

The Lewis concept of a base is similar to the Brønsted-Lowry concept. In the Brønsted-Lowry Model, a base donated an electron pair *to the hydrogen ion*. In the Lewis model, a base donates an electron pair *to any species*. The Lewis model greatly expands the idea of acids. The Brønsted and Arrhenius acids had to have *hydrogen*. A Lewis acid can be *any species* that accepts a share in an electron pair. Some examples of Lewis acid-base reactions are:

$$\underset{\substack{\text{electron pair acceptor}\\\text{(acid)}}}{\overset{H^+}{H\!:\!\ddot{\underset{H}{O}}\!:}} + \underset{\substack{\text{electron pair donor}\\\text{(base)}}}{:\!\ddot{\underset{..}{O}}\!:H^-} \rightleftharpoons H\!:\!\overset{H}{\underset{..}{\ddot{O}}}\!: + H\!:\!\overset{H}{\underset{..}{\ddot{O}}}\!: \qquad (16.32)$$

This reaction is also an acid-base reaction according to Arrhenius and Brønsted.

$$Cu^{2+} + 4\underset{\substack{\text{electron pair}\\\text{acceptor}\\\text{(acid)}}}{\begin{bmatrix}H\\|\\:N\!\!-\!\!H\\|\\H\end{bmatrix}}\underset{\substack{\text{electron pair}\\\text{donor}\\\text{(base)}}}{} \rightleftharpoons \begin{bmatrix}&&H&&\\&&|&&\\&H&H\!\!-\!\!N\!\!-\!\!H&H&\\&|&\ddot{}&|&\\H\!\!-\!\!N&:&Cu&:&N\!\!-\!\!H\\&|&\ddot{}&|&\\&H&H\!\!-\!\!N\!\!-\!\!H&H&\\&&|&&\\&&H&&\end{bmatrix}^{2+} \qquad (16.33)$$

This reaction is an acid-base reaction according to the Lewis model only. Given two reactants, how can you recognize whether they are Lewis acids or bases? Potential Lewis bases have an unshared electron pair available for sharing. Notice

■ EXERCISE  *The following equations represent Lewis acid-base reactions. Identify the acid and the base. Draw the structure of the expected product.*

(a)  $\underset{\substack{..\\F}}{\overset{\substack{..\\F}}{F\!:\!\ddot{B}}} + :\!\underset{H}{\overset{H}{\ddot{N}}}\!:H \rightleftharpoons$

(b) 
$:\ddot{\underset{..}{C}}l:\ddot{\underset{..}{A}}l + :\ddot{\underset{..}{C}}l:\rightleftharpoons$
$\phantom{(b) :}:\ddot{\underset{..}{C}}l:$
$\phantom{(b) ::\ddot{C}l:}:\ddot{\underset{..}{C}}l:$

NH$_3$ and OH$^-$ in the previous examples. The more available this electron pair is, the stronger the base will be.

Potential Lewis acids have less than an octet of electrons about one atom (BF$_3$ or Cu$^{2+}$). Other Lewis acids have an octet of electrons via double bonding (SO$_2$). ■

## 16.5 Preparation of Arrhenius acids and bases

Acids and bases are made in the laboratory in small quantities and by industry in large quantities. Often the industrial manufacturing process of chemicals is performed at high temperatures and pressures using specially constructed equipment and materials. These methods are often not suitable for laboratory use. The following reactions may be used to prepare Arrhenius acids in the laboratory.

*Reacting a covalent oxide with water*

$$P_4O_{10}(s) + 6H_2O(l) \rightleftharpoons 4H_3PO_4(aq) \quad (16.34)$$

*Direct combination of the elements*

$$H_2(g) + Cl_2(g) \rightleftharpoons 2HCl(g) \quad (16.35)$$
$$H_2(g) + I_2(g) \rightleftharpoons 2HI(g) \quad (16.36)$$

*Formation of volatile acids from nonvolatile acids*

$$H_3PO_4(l) + NaBr(s) \rightleftharpoons NaH_2PO_4(s) + HBr(g) \quad (16.37)$$
$$H_2SO_4(l) + NaCl(s) \rightleftharpoons NaHSO_4(s) + HCl(g) \quad (16.38)$$

These are proton transfer reactions. The slightly soluble HBr(g) and HCl(g) are driven off from the reaction mixture by heating.

The following reactions may be used to prepare Arrhenius bases in the laboratory

*Ionic oxides with water*

$$K_2O(s) + H_2O(l) \rightleftharpoons 2KOH(aq) \quad (16.39)$$
$$CaO(s) + H_2O(l) \rightleftharpoons Ca(OH)_2(aq) \quad (16.40)$$

*Active metals with water*

$$2Na(s) + 2H_2O(l) \rightleftharpoons 2NaOH(aq) + H_2(g) \quad (16.41)$$
$$Ba(s) + 2H_2O(l) \rightleftharpoons Ba(OH)_2(aq) + H_2(g) \quad (16.42)$$

*Ammonia with water*

$$NH_3(g) + H_2O(l) \rightleftharpoons NH_4^+(aq) + OH^-(aq) \quad (16.43)$$

*Active metal hydrides with water*

$$LiH(s) + H_2O(l) \rightleftharpoons LiOH(aq) + H_2(g) \qquad (16.44)$$

## 16.6 Normality of acids and bases

The rates of many chemical reactions in solution depend upon the number of moles of hydronium ions or hydroxide ions that are present per liter of solution.

**Concentration of $H_3O^+$ ions**

How many moles of $H_3O^+$ ions are there in a 0.300 M solution of HCl(aq)? Assume that strong acids ionize 100 percent in water solutions. If we started out with 0.300 mole of HCl, then according to the balanced equation, we will end up with 0.300 mole of $H_3O^+$.

$$HCl(aq) + H_2O \rightleftharpoons H_3O^+(aq) + Cl^-(aq) \qquad (16.45)$$

$$\underbrace{\frac{0.300 \text{ mole HCl}}{\text{liter solution}}}_{\text{original concentration of acid}} \times \underbrace{\frac{1 \text{ mole } H_3O^+}{1 \text{ mole HCl}}}_{\text{mole ratio from equation}} = \frac{0.300 \text{ mole } H_3O^+}{\text{liter solution}}$$

How many moles per liter are there in a $3.0 \times 10^{-4}$ M solution of $H_2SO_4$? If we assume for very dilute solution that the $H_2SO_4$ ionizes 100 percent, then there will be 2 moles of $H_3O^+$ ions for each original mole of acid.

$$H_2SO_4(aq) + 2H_2O(l) \rightleftharpoons 2H_3O^+(aq) + SO_4^{2-}(aq) \qquad (16.46)$$

$$\underbrace{\frac{3.0 \times 10^{-4} \text{ mole } H_2SO_4}{\text{liter solution}}}_{\text{original concentration of acid}} \times \underbrace{\frac{2 \text{ moles } H_3O^+}{1 \text{ mole } H_2SO_4}}_{\text{mole ratio from equation}} = \frac{6.0 \times 10^{-4} \text{ mole } H_3O^+}{\text{liter solution}}$$

■ EXERCISE  *Calculate the concentration of hydronium ions for the following strong acids: (a) 0.095 M $HClO_4$ and (b) $1.71 \times 10^{-5}$ M $H_2SO_4$.* ■

**Concentration $OH^-$ ions**

How many moles per liter of $OH^-$ ions are there in a 0.300 M solution of KOH? Assume that strong bases dissociate 100 percent.

$$KOH(s) \xrightarrow{H_2O} K^+(aq) + OH^-(aq) \qquad (16.47)$$

$$\frac{0.300 \text{ mole KOH}}{\text{liter solution}} \times \frac{1.00 \text{ mole OH}^-}{1.00 \text{ mole KOH}} = \frac{0.300 \text{ mole OH}^-}{\text{liter solution}}$$

How many moles per liter of OH⁻ ions are there in a 0.300 M solution of Ba(OH)₂?

$$Ba(OH)_2(s) \xrightleftharpoons{H_2O} Ba^{2+}(aq) + 2OH^-(aq) \tag{16.48}$$

■ EXERCISE  *Calculate the hydroxide ion concentration for the following solutions of strong bases: (a) 1.00 M LiOH and (b) 0.0731 M Sr(OH)₂.*

$$\frac{0.300 \text{ mole Ba(OH)}_2}{\text{liter solution}} \times \frac{2.00 \text{ moles OH}^-}{1.00 \text{ mole Ba(OH)}_2}$$
$$= \frac{0.600 \text{ mole OH}^-}{\text{liter solution}} \quad ■$$

**Normality**

The potential concentration of hydronium ions that a strong acid can produce is called the *normality* of that acid. The symbol N stands for normality. A 0.300 M HCl solution could provide 0.300 M $H_3O^+$ for reaction. A 0.300 M $H_2SO_4$ solution could provide 0.600 M $H_3O^+$ for reaction.

0.300 M HCl ⟶ 0.300 M $H_3O^+$ ⟶ 0.300 N HCl
0.300 M $H_2SO_4$ ⟶ 0.600 M $H_3O^+$ ⟶ 0.600 N $H_2SO_4$

For bases, the term "normality" applies to the potential concentration of hydroxide ions that a base can produce.

■ EXERCISE  *Calculate the normality of the solutions in the previous two exercises.*

0.300 M KOH ⟶ 0.300 M OH⁻ ⟶ 0.300 N KOH
0.300 M Ba(OH)₂ ⟶ 0.600 M OH⁻ ⟶ 0.600 N Ba(OH)₂  ■

## 16.7 The pH concept

**Water equilibrium**

Sensitive electrical measuring devices indicate that pure water does conduct a small amount of electrical current, thus showing that pure water contains some ions. Water molecules react with one another as follows:

$$H_2O(l) + H_2O(l) \rightleftharpoons H_3O^+(aq) + OH^-(aq) \tag{16.49}$$

At 25°C there is 0.0000001 ($1 \times 10^{-7}$) mole of $H_3O^+$(aq) ions per liter of water and an equal number of OH⁻(aq) ions. These ions are formed by the transfer of a hydrogen ion from one water molecule to another. Note that this is the reverse of neutralization.

Molecules of water are continually reacting with one another to form $H_3O^+$ ions and OH⁻ ions. At the same time these ions

**TABLE 16.3**

The concentration of $H_3O^+$ and $OH^-$ in water solutions of HCl.

| Molarity of HCl | Molarity of $H_3O^+(aq)$ | Molarity of $OH^-(aq)$ |
|---|---|---|
| $1 \times 10^{-2}$ | $1 \times 10^{-2}$ | $1 \times 10^{-12}$ |
| $1 \times 10^{-3}$ | $1 \times 10^{-3}$ | $1 \times 10^{-11}$ |
| $1 \times 10^{-4}$ | $1 \times 10^{-4}$ | $1 \times 10^{-10}$ |

undergo the reverse reaction: hydrogen ions are transferred from $H_3O^+$ ions to $OH^-$ ions to form water molecules. The forward and reverse reactions occur at the same rate and the concentration of each ion remains constant at $1 \times 10^{-7}$ M. The system is in a state of dynamic equilibrium.

### Adding an acid to water

What effect does the addition of a strong acid to water have on the concentration of $H_3O^+$ ions and $OH^-$ ions? If 0.0010 mole of HCl(g) is added to 1.00 liter of water, it will react with the water to form 0.0010 mole of $H_3O^+$ ions and $Cl^-$. The small number of $H_3O^+$ ions originally present is negligible compared to the amount formed from the acid.

original moles of $H_3O^+$ in pure water = 0.0000001 ($1 \times 10^{-7}$)
moles of $H_3O^+$ formed by acid = 0.0010 ($1.0 \times 10^{-3}$)

total amount of $H_3O^+$ approximately = 0.0010 mole

When strong acids such as HCl are added to pure water, we can usually ignore the $H_3O^+$ ions formed by the pure water.

Are there still $1 \times 10^{-7}$ moles of $OH^-$ ions left in the 0.0010 M HCl solution? Some of the $H_3O^+$ ions react with $OH^-$ ions to form water. The concentration of $OH^-$ ions drops from $1 \times 10^{-7}$ M to $1 \times 10^{-11}$ M. Whenever the concentration of $H_3O^+$ ions increases in aqueous solutions, the concentration of $OH^-$ decreases. Data for $1 \times 10^{-2}$ M, $1 \times 10^{-3}$ M, and $1 \times 10^{-4}$ M HCl solutions are shown in Table 16.3.

### Adding a base to water

What effect does the addition of a strong base to water have on the concentration of $H_3O^+$ and $OH^-$ ions? If 0.0010 mole of NaOH(s) is added to 1.00 liter of water, it will dissociate into 0.0010 mole of $Na^+(aq)$ ions and $OH^-(aq)$ ions. The small number of $OH^-$ ions originally present is negligible compared to the amount formed from the base.

original moles of OH⁻ in pure water = 0.0000001 (1 × 10⁻⁷)
moles of OH⁻ formed by base = 0.0010 (1.0 × 10⁻³)
_____
total amount of OH⁻ approximately = 0.0010 mole

When strong bases such as NaOH are added to water we can usually ignore the OH⁻ ions formed by pure water.

Are there still $1 \times 10^{-7}$ moles of $H_3O^+$ ions left in the 0.0010 M NaOH solution? Some of the OH⁻ ions react with $H_3O^+$ ions to form water. The concentration of $H_3O^+$ ions drops from $1 \times 10^{-7}$ M to $1 \times 10^{-11}$ M. Whenever the concentration of OH⁻ ions increases in aqueous solution, the concentration of $H_3O^+$ ions decreases. Data for $1 \times 10^{-2}$ M, $1 \times 10^{-3}$ M and $1 \times 10^{-4}$ M NaOH solutions are shown in Table 16.4.

**The equilibrium constant**

We saw in the preceding three sections that in pure water the concentration of $H_3O^+$ ions equals OH⁻ ions; in acidic solutions $H_3O^+$ ions exceed OH⁻ ions; in basic solutions OH⁻ ions exceed $H_3O^+$ ions.

Note in Tables 16.3 and 16.4 that when the concentration of $H_3O^+$ is multiplied by the concentration of OH⁻, the product in each case is $1.0 \times 10^{-14}$. Experiments and theory indicate that the product of the molar concentrations of $H_3O^+$ and OH⁻ is $1.0 \times 10^{-14}$ at 20°C. This number is called the *ion product* or *ionization constant* of water, $K_w$.

$$K_w = [H_3O^+][OH^-] = 1.0 + 10^{-14} \tag{16.50}$$

The square brackets [ ] mean moles per liter (molarity).

Equation 16.50 allows us to calculate either $[H_3O^+]$ or $[OH^-]$ if we know one or the other in an aqueous solution.

■ EXERCISE  *The concentration of OH⁻ ions in a swimming pool was $5.0 \times 10^{-7}$ M. What is the concentration of $H_3O^+$ ions?*

EXAMPLE 1  The concentration of $H_3O^+$ ions in stomach acid is $1.0 \times 10^{-3}$ M. What is the concentration of OH⁻ ions?

Given: $[H_3O^+] = 1.0 \times 10^{-3}$ M

| TABLE 16.4 | Molarity of NaOH | Molarity of OH⁻(aq) | Molarity of $H_3O^+$(aq) |
|---|---|---|---|
| The concentration of $H_3O^+$ and OH⁻ in water solutions of NaOH. | $1 \times 10^{-2}$ | $1 \times 10^{-2}$ | $1 \times 10^{-12}$ |
| | $1 \times 10^{-3}$ | $1 \times 10^{-3}$ | $1 \times 10^{-11}$ |
| | $1 \times 10^{-4}$ | $1 \times 10^{-4}$ | $1 \times 10^{-10}$ |

16.7 The pH concept

■ **EXERCISE** The concentration of $OH^-$ ions in a sample of blood was $3.4 \times 10^{-7}$. What was the $H_3O^+$?

Find: $[OH^-]$

Relationship: $K_w = [H_3O^+][OH^-]\ 1.0 \times 10^{-14}$

Solution: $[OH^-] = \dfrac{1.0 \times 10^{-14}}{[H_3O^+]} = \dfrac{1.0 \times 10^{-14}}{1.0 \times 10^{-3}} = 1.0 \times 10^{-11}$  ■

### The pH scale

The concentrations of $H_3O^+$ ions are frequently given in pH units. The pH of a solution is defined as:

$$\text{pH} = \log \dfrac{1.0}{[H_3O^+]} = -\log [H_3O^+] \tag{16.51}$$

In pure water the $H_3O^+$ concentration is $1.0 \times 10^{-7}$. The pH is equal to $\log 1.0/1.0 \times 10^{-7} = \log 1.0 \times 10^7 = 0.0 + 7.0 = 7.0$.
What is the pH of an acidic solution whose $[H_3O^+]$ ion concentration is $1.0 \times 10^{-3}$?

$$\begin{aligned}\text{pH} &= \log \dfrac{1.0}{1.0 \times 10^{-3}} = \log 1.0 \times 10^3 \\ &= \log 1.0 + \log 10^3 \\ &= 0.0 + 3.0 = 3.0\end{aligned}$$

**TABLE 16.5**

*The relation of pH to concentration of hydronium ion and hydroxide ion in water solution.*

| Classification | pH | $[H_3O^+]$ moles/liter | $[OH^-]$ moles/liter | Examples |
|---|---|---|---|---|
| **Strongly acidic** | 0 | $1.0 \times 10^0$ | $1.0 \times 10^{-14}$ | |
| | 1 | $1.0 \times 10^{-1}$ | $1.0 \times 10^{-13}$ | 0.1 M HCl |
| | 2 | $1.0 \times 10^{-2}$ | $1.0 \times 10^{-12}$ | lemons |
| | 3 | $1.0 \times 10^{-3}$ | $1.0 \times 10^{-11}$ | vinegar |
| **Weakly acidic** | 4 | $1.0 \times 10^{-4}$ | $1.0 \times 10^{-10}$ | |
| | 5 | $1.0 \times 10^{-5}$ | $1.0 \times 10^{-9}$ | boric acid |
| | 6 | $1.0 \times 10^{-6}$ | $1.0 \times 10^{-8}$ | |
| **Neutral** | 7 | $1.0 \times 10^{-7}$ | $1.0 \times 10^{-7}$ | pure water |
| | 8 | $1.0 \times 10^{-8}$ | $1.0 \times 10^{-6}$ | blood |
| **Weakly basic** | 9 | $1.0 \times 10^{-9}$ | $1.0 \times 10^{-5}$ | baking soda |
| | 10 | $1.0 \times 10^{-10}$ | $1.0 \times 10^{-4}$ | household ammonia |
| | 11 | $1.0 \times 10^{-11}$ | $1.0 \times 10^{-3}$ | |
| | 12 | $1.0 \times 10^{-12}$ | $1.0 \times 10^{-2}$ | washing soda |
| **Strongly basic** | 13 | $1.0 \times 10^{-13}$ | $1.0 \times 10^{-1}$ | 0.1 M NaOH |
| | 14 | $1.0 \times 10^{-14}$ | $1.0 \times 10^0$ | |

■ **EXERCISE** A solution of acetic acid contains $1.00 \times 10^{-4}$ mole per liter of $H_3O^+$ ions. What is the pH?

The pH scale runs from about 0.0 to 14.0. The lower range of 0.0 to 3.0 represents fairly strong acid solutions. The range of 3.0 to 7.0 represents weaker acid to neutral solutions. From 7.0 to 11.0 are the weaker basic solutions. Concentrated solutions of strong acids may have a pH less than 0.0 and concentrated

■ EXERCISE *If the pH of a sample of human blood is 7.3, what is the approximate concentration of $H_3O^+(aq)$ ions?*

solutions of strong bases may have a pH of greater than 14.0. The pH scale and concentrations of $H_3O^+$ ions and $OH^-$ ions are shown in Table 16.5. ■

## 16.8 Titrations

**Stoichiometry of neutralization**

How many milliliters of 0.432 M NaOH solution are needed to neutralize 50.0 ml of 0.391 M HCl solution? The equation for the reaction is:

$$NaOH(aq) + HCl(aq) \rightleftharpoons NaCl(aq) + H_2O(l) \qquad (16.52)$$

According to the equation, 1.00 mole of NaOH is needed to neutralize 1.00 mole HCl. To answer the original question, we must first find out how many moles of HCl are in 50.0 ml of 0.391 M HCl.

$$50.0 \text{ ml} \times \frac{1.00 \text{ liter}}{1000 \text{ ml}} \times \frac{0.391 \text{ mole HCl}}{\text{liter solution}} = 0.0195 \text{ mole HCl}$$

changes ml to liter

changes liters to moles

Since the ratio of NaOH to HCl in the balanced equation is *one to one*, the moles of NaOH required to neutralize 0.0195 mole of HCl is:

$$0.0195 \text{ mole HCl} \times \frac{1.00 \text{ mole NaOH}}{1.00 \text{ mole HCl}} = 0.0195 \text{ mole NaOH}$$

The concentration term, 0.432 mole/1.00 liter, relates moles to liters. This term can be used to convert moles of NaOH to liters of NaOH.

$$0.0195 \text{ mole NaOH} \times \frac{1.00 \text{ liter solution}}{0.432 \text{ mole NaOH}}$$

$$= 0.0452 \text{ liter solution}$$

$$0.0452 \text{ liter} \times \frac{1000 \text{ ml}}{1.00 \text{ liter}} = 45.2 \text{ ml NaOH solution}$$

It takes 45.2 ml of 0.432 M NaOH to neutralize 50.0 ml of 0.391 M HCl. This procedure is outlined in Fig. 16.3. Try to solve the following sample problem before looking at the solution.

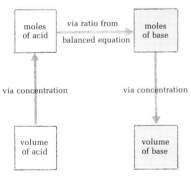

FIGURE 16.3
*The amount of base required to neutralize a given solution of acid depends upon (1) the concentration of the acid; (2) the concentration of the base; (3) the initial volume of acid; (4) the mole ratio of acid to base in the balanced equation.*

**EXAMPLE 2** How many milliliters of 0.836 M HCl are needed to neutralize 30.0 ml of 0.498 M Ba(OH)$_2$?

$$2HCl(aq) + Ba(OH)_2(aq) \rightleftharpoons BaCl_2(aq) + 2H_2O(l)$$

*Given:* Acid concentration: 0.836 mole HCl/liter
Base concentration: 0.498 mole Ba(OH)$_2$/liter
Volume of base: 30.0 ml

*Find:* Volume of HCl needed for neutralization

*Relationship:* 2.00 moles of HCl ≈ 1.00 mole Ba(OH)$_2$

*Solution:* Change volume of base to moles

$$30.0 \text{ ml} \times \frac{1.00 \text{ liter}}{1000 \text{ ml}} \times \frac{0.498 \text{ mole Ba(OH)}_2}{\text{liter}} = 0.0149 \text{ mole Ba(OH)}_2$$

Change moles of base to moles of acid

$$0.0149 \text{ mole Ba(OH)}_2 \times \frac{2.00 \text{ moles HCl}}{1.00 \text{ mole Ba(OH)}_2} = 0.0298 \text{ mole HCl}$$

Change moles of acid to volume of acid

$$0.0298 \text{ mole HCl} \times \frac{1.00 \text{ liter}}{0.836 \text{ mole}} \times \frac{1000 \text{ ml}}{1.00 \text{ liter}} = 35.7 \text{ ml HCl}$$

*Answer:* 35.7 ml of HCl solution   ■

■ **EXERCISE** *A solution of sulfuric acid had a concentration of 3.38 M. How many milliliters of 6.76 M potassium hydroxide are needed to neutralize 30.2 ml of acid?*

### Technique of a titration

How would you determine experimentally the molarity of a solution of HCl? You could add enough NaOH solution of known concentration until the base was neutralized, but how would you recognize the moment when the base was neutralized? An indicator that has been found to be most useful is phenolphthalein. This dye appears colorless in acid solution but is red in base solution.

The process of adding just enough base to an acid solution to cause neutralization is called an acid-base *titration*. (Acid could be added to base also.) During a titration, the volumes of acid and base used are carefully measured. The concentration of either the acid or the base is known and the concentration of the other can be calculated. An apparatus commonly used for a titration is shown in Fig. 16.4. The burets are carefully calibrated glass tubes that can deliver controlled amounts of liquid. The valve or stopcock at the base of the buret can be rotated to let as little as one drop of liquid out of the buret.

In a laboratory class one of the students wanted to determine the molarity of an unknown concentration of hydrobromic acid, HBr. The measurements he made were:

volume of base = final reading − initial reading:
$$43.35 - 2.15 = 41.20 \text{ ml}$$

volume of acid = final reading − initial reading:
$$35.22 - 0.00 = 35.22 \text{ ml}$$

The concentration of the base is 1.08 M. The equation for the reaction is:

$$\text{KOH(aq)} + \text{HBr(aq)} \rightleftharpoons \text{KBr(aq)} + \text{H}_2\text{O(l)} \qquad (16.53)$$

To find the molarity of the acid, we will first determine how many moles of KOH were used. Using the mole ratio we can calculate the moles of acid used. Knowing the moles of acid and the volume of acid we can calculate molarity.

(a) $41.2 \text{ ml} \times \dfrac{1.00 \text{ liter}}{1000 \text{ ml}} \times \dfrac{1.08 \text{ moles KOH}}{\text{liter}} = 0.0446 \text{ mole KOH}$

(b) $0.0446 \text{ mole KOH} \times \dfrac{1.00 \text{ mole HBr}}{1.00 \text{ mole KOH}} = 0.0446 \text{ mole HBr}$

(c) $\dfrac{0.0446 \text{ mole HBr}}{35.2 \text{ ml}} \times \dfrac{1000 \text{ ml}}{1.00 \text{ liter}} = \dfrac{1.27 \text{ moles HBr}}{\text{liter}}$

$$= 1.27 \text{ M HBr}$$

The molarity of the HBr acid solution is 1.27 M. ■

## 16.9 Salts—Brønsted acids and bases

### Salts of strong acids and strong bases

When mixed with pure water, sodium chloride dissolves but has no effect on the pH. Sodium ions and chloride ions are present, but these ions do not undergo proton transfer reactions with water molecules. Sodium chloride is an example of a *salt of a strong base and a strong acid*.

$$\text{Na}^+ \text{OH}^- + \text{H}^+ \text{Cl}^-$$
$$\downarrow \qquad \text{NaCl} \qquad \downarrow$$

The ions in a salt of a strong acid and a strong base do not accept hydrogen ions readily from water molecules; therefore, they do not function as bases. These ions do not donate hydrogen ions to water molecules; therefore, they do not function as acids (Table 16.6)

### Salts of weak bases and strong acids

What salts will change the pH of water? When either $\text{NH}_4\text{Cl}$ or $(\text{NH}_4)_2\text{SO}_4$ is placed in water, the pH drops below 7.0. What causes the change? Consider ammonium chloride. When

■ EXERCISE A solution of phosphoric acid is 1.61 M. If 41.3 ml of acid are required to neutralize 19.0 ml of sodium hydroxide, what is the molarity of the base?

$3\text{NaOH(aq)} + \text{H}_3\text{PO}_4\text{(aq)} \rightleftharpoons \text{Na}_3\text{PO}_4\text{(aq)} + 3\text{H}_2\text{O(l)}$

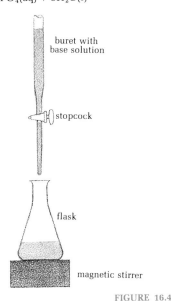

FIGURE 16.4
A titration. First a measured volume of unknown acid is added to a flask with two or three drops of phenolphthalein. A base of known concentration is then slowly added with stirring, until a red color in the solution appears for 20 seconds. Measurements are made, and the calculations are performed.

| TABLE 16.6 | Strong acids | Strong bases | Salts of strong acids and strong bases |
|---|---|---|---|
| Salts of strong acids and strong bases such as these do not affect the pH of water. | HCl | LiOH | LiCl |
| | HBr | NaOH | NaBr |
| | $HClO_3$ | RbOH | $RbClO_3$ |
| | $H_2SO_4$ | $Ba(OH)_2$ | $BaSO_4$ |
| | $HNO_3$ | $Ca(OH)_2$ | $Ca(NO_3)_2$ |

placed in water, it dissociates into $NH_4^+$ ions and $Cl^-$ ions.

$$NH_4Cl(s) \xrightleftharpoons{H_2O} NH_4^+(aq) + Cl^-(aq) \qquad (16.54)$$

The $Cl^-(aq)$ does not react with water, but the $NH_4^+$ ion donates a hydrogen ion to water molecules.

$$NH_4^+(aq) + H_2O(l) \rightleftharpoons NH_3(aq) + H_3O^+(aq) \qquad (16.55)$$

Some of the ammonium ions in the solution react with water to form $H_3O^+(aq)$. The presence of $H_3O^+$ makes the solution partially acidic and lowers the pH.

Ammonium chloride is an example of a *salt of a weak base and a strong acid*. The salt could have been made by mixing $NH_3$ and HCl together (Table 16.7).

### Salts of strong bases and weak acids

When placed in water, sodium acetate dissociates into sodium ions and acetate ions.

$$NaC_2H_3O_2(s) \xrightleftharpoons{H_2O} Na^+(aq) + C_2H_3O_2^-(aq) \qquad (16.56)$$

The sodium ions do not react with water, but the acetate ions do.

$$C_2H_3O_2^-(aq) + H_2O(l) \rightleftharpoons HC_2H_3O_2(aq) + OH^-(aq) \qquad (16.57)$$

| TABLE 16.7 | Weak bases | Strong acids | Salts of weak bases and strong acids |
|---|---|---|---|
| Salts of weak bases and strong acids such as these lower the pH of water. | $NH_3$ | HCl | $NH_4Cl$ |
| | | HBr | $NH_4Br$ |
| | | $HClO_3$ | $NH_4ClO_3$ |
| | | $H_2SO_4$ | $(NH_4)_2SO_4$ |
| | | $HNO_3$ | $NH_4NO_3$ |

**TABLE 16.8**

*Salts of strong bases and weak acids such as these raise the pH of water.*

| Strong bases | Weak acids | Salts of strong bases and weak acids |
|---|---|---|
| LiOH | $H_2S$ | $Li_2S$ |
| NaOH | $H_2SO_3$ | $Na_2SO_3$ |
| KOH | $H_2CO_3$ | $K_2CO_3$ |
| LiOH | $H_3PO_4$ | $Li_3PO_4$ |
| NaOH | $HC_2H_3O_2$ | $NaC_2H_3O_2$ |

Hydrogen ions are transferred from water molecules to acetate ions. The acetate ion is a Brønsted base. The formation of $OH^-$ ions makes the solution basic. Other negative ions from weak acids behave the same way (Table 16.8).

**Salts of weak bases and weak acids.**

When a *salt of a weak base and a weak acid* are added to water does the pH change? It depends how weak the acid is and how weak the base is. For example, when ammonium acetate is added to water, little or no pH change occurs.

$$NH_4C_2H_3O_2(s) \xrightleftharpoons{H_2O} NH_4^+(aq) + C_2H_3O_2^-(aq) \qquad (16.58)$$

■ EXERCISE *The following substances are salts. Write the equations for any reaction that each ion may undergo with water. Which salts will: (1) raise the pH, (2) lower the pH, (3) leave the pH unchanged? (a) $NaNO_3$; (b) $NH_4NO_3$; (c) $K_3PO_4$.*

The $NH_4^+$ ion reaction with water tends to increase the concentration of $H_3O^+$ ions. At the same time the $C_2H_3O_2^-$ ion reaction with water tends to increase the concentration of $OH^-$ ions. Some salts such as $(NH_4)_2S$ make aqueous solutions basic. Others make their aqueous solutions acidic. ■

## Summary

Aqueous acids turn blue litmus to red, taste sour, react with many metals releasing hydrogen, dissolve carbonates releasing carbon dioxide, and neutralize bases. Aqueous bases turn red litmus to blue, taste bitter, feel soapy, and neutralize acids. Arrhenius proposed that all aqueous acid solutions contain $H^+(aq)$ ions and all aqueous base solutions contain $OH^-(aq)$ ions. During neutralization, $H^+(aq)$ ions combine with $OH^-(aq)$ ions to form $H_2O$.

Brønsted and Lowry extended the Arrhenius model. According to them, an acid was a proton donor and a base was a proton acceptor. When a strong acid such as HCl is added to water, $H_3O^+$ ions are produced. This species is a strong proton donor, but there are many other proton donors also. The hydroxide ion is a strong proton acceptor, but there are also many other proton acceptors.

The Brønsted-Lowry model applies to many solvent systems in addition to water. Neutralization is the process whereby a proton transfer reaction produces more solvent. Using the Brønsted-Lowry table of acid and base strengths, one can predict the extent of a possible reaction. The stronger acid and stronger base will produce the weaker acid and weaker base.

G. N. Lewis extended the concept of acid-base reactions. According to his model, an acid is any electron pair acceptor. This concept includes the proton and many other chemical species. A Lewis base is any electron pair donor. This includes the hydroxide ion and many other chemical species.

Methods for making Arrhenius acids and bases include the use of oxidation–reduction reactions, $H_2(g) + Cl_2(g) \rightleftharpoons 2HCl(g)$; proton transfer reactions, $NaCl(s) + H_2SO_4(l) \rightleftharpoons NaHSO_4(s) + HCl(g)$; and precipitation reactions, $Ba(NO_3)_2(aq) + H_2SO_4(aq) \rightleftharpoons BaSO_4(s) + 2HNO_3(aq)$.

The normality of an aqueous acid or base is the potential concentration of $H_3O^+$(aq) ions or $OH^-$(aq) ions. A solution of 1.0 M $Ba(OH)_2$ contains 2.0 M $OH^-$ ions and is 2.0 normal.

The pH scale is frequently used to indicate the relative acidity or basicity of a solution. A low pH reading (0–3) indicates a strongly acidic solution. A pH of 7.0 indicates an equal number of $OH^-$ ions and $H_3O^+$ ions in solution. Such solutions are said to be neutral. A high pH (11–14) indicates a strongly basic solution.

The concentration of an unknown acid may be determined by neutralizing a known volume of the acid with a measured volume of base. The point of neutralization is recognized by use of an indicator. This process is called a titration.

Salts of strong bases and strong acids do not change the concentrations of $H_3O^+$ ions and $OH^-$ ions. Salts of weak bases and strong acids raise the concentrations of $H_3O^+$ ions and lower the concentrations of $OH^-$ ions. Salts of strong bases and weak acids lower the concentrations of $H_3O^+$ ions and raise the concentrations of $OH^-$ ions.

## Objectives

By the end of the chapter, you should be able to do the following:

1. Give an operational definition of an acid. (Question 2)

2. Give an operational definition of a base. (Question 2)

3. Give examples and use equations to illustrate and define the Arrhenius model of acids and bases. (Question 3)

4. Given the reactants, write a balanced equation for a reaction between an aqueous acid and (a) an aqueous base, (b) zinc, (c) a carbonate. (Question 4)

5. Define and illustrate by means of equations, the following Brønsted-Lowry concepts: (a) acid, (b) base, (c) neutralization. (Questions 5, 6)

6. Given a Brønsted-Lowry acid-base reaction, label the conjugate acid-base systems. (Questions 7)

7. Given a Brønsted-Lowry acid and a Brønsted-Lowry base, write the formulas of the products formed when they are mixed. (Question 8)

8. Define and illustrate by means of an equation and electron-dot formulas the following: (a) Lewis acid, (b) Lewis base. (Questions 9, 10)

9. Given the molarity of a strong acid or base, calculate the normality. (Question 11)

10. Given the concentration of $H_3O^+$ ions (or $OH^-$ ions), calculate the concentrations of $OH^-$ ions (or $H_3O^+$ ions). (Questions 12, 13)

11. Given the pH of a solution, state whether the solution is strongly acidic, weakly acidic, neutral, weakly basic, or strongly basic. (Questions 14, 15)

12. Given the concentration of $H_3O^+$ ions in powers of 10, calculate the pH. (Question 16)

13. Given the concentrations of both acid and base and the volume of one of them used in a neutralization, calculate the volume of the other. (Question 17)

14. Given the volumes of both acid and base and the concentration of one of them used in a neutralization, calculate the concentration of the other. (Question 18)

## Questions

1. Define the following terms. How do these terms and your definitions depend upon concepts developed earlier such as solubility, ionization, balanced equations, proton transfer reactions, and concentrations? *acid; base; neutralization; litmus; Arrhenius model; proton transfer; Brønsted acids; Brønsted bases; Brønsted neutralization; conjugate acids-base pairs; strong acids; weak acids; strong bases; weak bases; anphiprotic; dynamic equilibrium; Lewis base; Lewis acid; indicator; buret; normality; titration; $K_w$; ionization constant of water; pH; phenolphthalein.*

2. You are handed two unlabelled bottles of solution and told to determine if either of them is an acid or a base. List four properties of acids and four properties of bases that you could test.

3. Give an example of an aqueous acid and an aqueous base. State the Arrhenius model for the acid and the base. Write an equation to show what reaction occurs when these two solutions are mixed.

4. Complete the following molecular equations:
(a) $H_2SeO_4(aq) + LiOH(aq) \rightleftharpoons$
(b) $HBr(aq) + Zn(s) \rightleftharpoons$
(c) $HCl(aq) + Fe(s) \rightleftharpoons$
(acids oxidize iron to the 2+ oxidation state)
(d) $HNO_3(aq) + CaCO_3(s) \rightleftharpoons$

5. Give an example of and define a Brønsted-Lowry acid and base.

6. Give an example of and define neutralization using the Brønsted-Lowry model.

7. Label the conjugate acid-base pairs in the following reactions. Use the table of acid-base strengths to predict which reactions proceed over 50 percent to completion.
   (a) $NH_3(aq) + H_2S(aq) \rightleftharpoons NH_4^+(aq) + HS^-(aq)$
   (b) $O^{2-}(aq) + HSO_4^-(aq) \rightleftharpoons OH^-(aq) + SO_4^{2-}(aq)$

8. The following are reactants in Brønsted acid-base reactions. Complete the equations.
   (a) $HIO_3(aq) + Na_2S(aq) \rightleftharpoons$
   (b) $NaNO_2(aq) + NH_4Cl(aq) \rightleftharpoons$
   (c) $NO_2^-(aq) + NH_4^+(aq) \rightleftharpoons$
   (d) $NH_4^+(aq) + OH^-(aq) \rightleftharpoons$
   (e) $H_3O^+(aq) + SO_2^{2-}(aq) \rightleftharpoons$

9. Use electron-dot formulas to define and illustrate a Lewis acid and a Lewis base.

10. Complete the following Lewis acid-base reactions:
    (a) $Fe^{3+} + 6 : C \equiv N :^- \rightleftharpoons$
    (b) $: \ddot{O} :^{2-} + H^+ \rightleftharpoons$
    (c) $: \ddot{S} :^{2-} + \ddot{S} :^0 \rightleftharpoons$

11. What is the normality of the following solutions? (a) 10.0 M HCl; (b) 2.0 M $H_2SeO_4$; (c) 0.700 M $H_3PO_4$; (d) 0.00531 M $Ca(OH)_2$.

12. In a certain area of the country the ground water is $3.90 \times 10^{-8}$ M $H_3O^+$. What is the concentration of $OH^-$ ions?

13. The concentration of $OH^-$ ions in a fruit juice is $2.5 \times 10^{-11}$ M. What is the concentration of $H_3O^+$?

14. Classify the following solutions as either strongly acidic, weakly acidic, neutral, weakly basic, or strongly basic:

| Solution from: | Average pH: |
|---|---|
| apples, | 3.0 |
| eggs, | 7.8 |
| grapes, | 4.0 |
| milk of magnesia | 10.5 |
| oranges | 3.5 |
| milk | 6.4 |
| lye | 14.0 |
| muriatic acid | 0.0 |

15. Enzymes are chemicals occurring in the body that catalyze many bodily reactions. The enzyme urease decomposes urea into $CO_2$ and $NH_3$. Urease is most efficient at a pH of 6.0 to 7.0. (a) Is urease most efficient in a weakly acidic or weakly basic medium? (b) What is the $[H_3O^+]$ concentration at pH 6.0?

16. What is the pH of each of the following solutions?
    (a) $[H_3O^+] = 1.0 \times 10^0$ M
    (b) $[H_3O^+] = 1.0 \times 10^{-4}$ M
    (c) $[H_3O^+] = 1.0 \times 10^{-13}$ M
    (d) $[H_3O^+] = 1.0 \times 10^{+1}$ M

17. How many milliliters of phosphoric acid of 0.351 M are necessary to neutralize 35.4 liters of 12.0 M NaOH?

18. What is the molarity of a solution of perchloric acid if 23.9 ml of the acid are neutralized by 39.0 ml of 0.0132 M strontium hydroxide?

# Chemical kinetics and chemical equilibrium

## 17.1 Introduction

How do we control the rate and the extent of a chemical reaction? When the gases hydrogen and oxygen are mixed and ignited, they explode violently; yet the same two elements when burned under controlled conditions make a powerful rocket fuel. The speed at which substances are formed or used up in a chemical reaction is known as the *rate of reaction*. The study of the factors that affect the rate of a chemical reaction is called *chemical kinetics*. We shall review and study some of these factors in this chapter.

When a mixture of nitrogen and hydrogen is heated to 100°C at 1.00 atmosphere pressure, little formation of ammonia, $NH_3$, takes place; yet each year industry produces millions of pounds of ammonia by reacting nitrogen and hydrogen at 300 atm and 500°C. *Chemical equilibrium* is the study of those factors that influence the extent to which a reaction takes place. The quantity of products formed in relation to the quantity of reactants used is one such factor.

Equilibrium calculations are used to determine the extent to which a slightly soluble substance dissolves or the extent to which weak electrolytes ionize. We shall introduce these calculations in the last section of this chapter.

## 17.2 Reaction rates

Review Section 10.3 on the collision theory and consider that in order for gaseous hydrogen to react with gaseous iodine, the following conditions must be met:

1. Collision: The two molecules must collide.
2. Orientation: At the moment of impact, the two molecules must be oriented in such a manner that bonds may break and new bonds form; only then is a reaction possible.
3. Energy: At the moment of impact, the two molecules must possess sufficient energy to break the original bonds holding each molecule together.

If these three conditions are met, a reaction occurs:

$$\text{H—H(g)} + \text{I—I(g)} \rightleftharpoons 2\text{H—I(g)} \qquad (17.1)$$

The rate of a reaction is a function of the concentration of reacting molecules, orientation of molecules, and the energy the molecules possess:

$$\text{rate} = f(\text{concentration, orientation, energy}) \qquad (17.2)$$

What factors affect these conditions?

### Collisions

In a homogeneous system, the number of collisions that occur between two or more reactants at a given temperature depends upon the *concentration*. In solutions the concentration is usually given in molarity. When all the reactants are gases, the molarity at a given temperature depends on the pressure.

In heterogeneous systems where one or more of the reactants is a solid, the number of collisions that will occur depends upon the subdivision of the solid; that is, the *particle size*. The smaller the solid particles, the more *surface area*, and the faster the reaction occurs.

Some reactions occur as soon as the reactants come in contact with each other. This is especially true in aqueous solutions. The OH⁻(aq) from the base NaOH(aq) reacts with the $H_3O^+$(aq) from the acid HCl(aq) almost as soon as they collide. In such cases the rate of reaction and of collision depends upon the *rate of mixing*.

### Orientation

When molecules collide, they usually do so randomly. It is only by chance that they collide with the proper orientation to

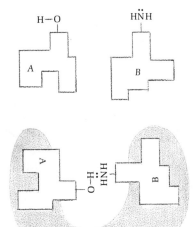

**FIGURE 17.1**
A model for the role of an enzyme in a chemical reaction. A and B represent complex molecules. The desired reaction is a proton transfer from A to B. Such a reaction is unlikely to occur when A and B collide. The enzyme serves to hold the molecules in positions so that a proton is easily transferred.

cause a reaction. With simple molecules or simple ions such as $H_3O^+$(aq) or $OH^-$(aq) the structure is uncomplicated enough to permit a high percentage of collisions to result in a reaction.

Many molecules are more complex, however. When two complex molecules collide, there is only a small chance that a given reaction may occur. Many reactions in living organisms involve such molecules. Enzymes are large protein molecules that catalyze many reactions. The role of some enzymes is to bring particular parts of complex molecules together at the proper orientation for a reaction to occur (Fig. 17.1).

**Energy**

Most reactions involve two separate steps: breaking bonds and forming new bonds.

The forming of a bond is an exothermic process and occurs readily. Reactions can take place only when bonds are broken in the reactants. For instance, before one molecule of $H_2$(g) and one molecule of $I_2$(g) can react to form two molecules of HI(g), the H—H bond and the I—I bond must be broken. There are two ways to increase the bond-breaking rate: increasing the average kinetic energy of the reactants or using a catalyst.

*Increasing the energy* The energy required to start a reaction is called the *activation energy* (Fig. 17.2). The activation energy can be visualized as a barrier which must be surmounted before a reaction takes place. In other words, the molecules cannot react unless they possess some minimum energy.

How is the activation energy barrier overcome? The rate of many reactions is increased greatly by increasing the temperature. An increase in temperature increases the average kinetic energy of the molecules in a system. Compare the kinetic energy distribution in Figs. 17.3(a) and 17.3(b). The shaded areas represent the relative number of molecules whose energy exceeds the activation energy. At the lower temperature the

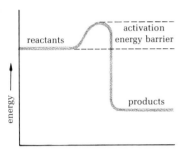

FIGURE 17.2
Before reactants can react to form products, they must overcome the activation energy barrier.

FIGURE 17.3
(a) The relative number of molecules possessing enough energy to overcome the activation energy are shown by the size of the shaded area under the curve. This area is small at low temperature. (b) When the temperature is increased, the average kinetic energy of the molecules increases. More molecules possess the activation energy as shown by the larger shaded area.

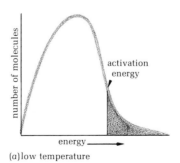
(a) low temperature

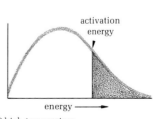
(b) high temperature

FIGURE 17.4
(a) The reaction shown has a high activation energy. At a given temperature, only a small number of molecules (shaded area) have sufficient kinetic energy to react. (b) A catalyst may speed up a reaction by lowering the activation energy. It may do this by providing an alternate path for the reaction.

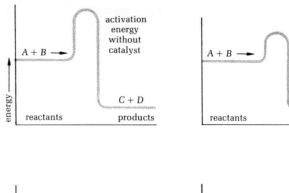

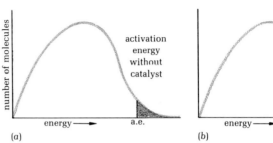

■ EXERCISE At room temperature, a mixture of hydrogen gas and chlorine gas remains unreactive in a dark container. When exposed to ultraviolet light, the two gases react violently to form hydrogen chloride. (a) Propose why hydrogen and chlorine do not react when mixed. (b) What role does ultraviolet light have in starting the reaction? (c) How else might you increase the rate of reaction?

shaded area is small; only a few molecules can react. At the higher temperature the shaded area is larger; many more molecules can react.

Using a catalyst  Catalysts, such as some enzymes, speed up the rate of reactions by lowering the *activation energy* (Fig. 17.4). Remember, however, that the catalyst itself is not consumed in the reaction. Consider the following hypothetical reaction:

$$A\text{—}A(g) + B\text{—}B(g) \rightleftharpoons 2A\text{—}B(g) \qquad (17.3)$$

Suppose the rate of the reaction is slow because the A—A bond is difficult to break. A catalyst, X, may combine with A—A is such a way as to weaken the bond:

$$A\text{—}A + X \rightleftharpoons \underbrace{A\text{----}A\text{----}X}_{\text{weakened bond}} \qquad (17.4)$$

When B—B collides with the weakened bond between atoms A—A, a reaction is more likely to occur:

$$B\text{—}B + A\text{----}A\text{----}X \rightleftharpoons 2A\text{—}B + X \qquad (17.5)$$

Note here that a *catalyst* is both a reactant and a product of a chemical reaction. ■

## 17.3 The concept of dynamic equilibrium

**Radioactive tracers**

Before studying chemical equilibrium, we will examine a chemical tool useful in equilibrium studies—radioactive tracers.

Many isotopes of different elements are *radioactive*. These isotopes emit radiation of various types that can be detected by a Geiger counter.

Radioactive isotopes have essentially the same chemical properties as nonradioactive isotopes. This makes them useful in following or tracing the whereabouts of certain atoms during and after a reaction. We simply find out which substance or substances are radioactive. We will designate radioactive elements with an asterisk (*). The radiation emitted and the methods used to detect radiation will not concern us.

**Phase equilibrium**

Consider a sealed bottle half-filled with water. The space above the water contains air and, as we saw in Chapter 8, water vapor. If the temperature of this heterogeneous system is kept constant, the amount of water in the liquid phase and the amount of $H_2O$ in the vapor phase both remain constant. A system in which the macroscopic amounts of material remain constant is said to be in a state of *equilibrium*. When two different phases are involved, it is called a *phase equilibrium*.

Is there any exchange of molecules between the liquid and vapor phases? Let us find out. Withdraw a drop of liquid $H_2O$ from the bottom of the container and replace it with a drop of water that contains some radioactive oxygen isotopes. In a short time, drops of liquid water from throughout the sample of liquid $H_2O$ show radioactivity. The molecules in the radioactive drop of water have been distributed throughout the liquid.

A little later the vapor above the liquid becomes radioactive, indicating that water molecules are leaving the liquid and entering the vapor.

$$H_2O^*(l) \longrightarrow H_2O^*(v) \tag{17.6}$$

The amount of liquid does not change, however, which means that some molecules in the vapor must be returning to the liquid at the same rate as they are vaporizing, thus maintaining this steady state.

$$H_2O(l) \longleftarrow H_2O(v) \tag{17.7}$$

Combining Eqs. 17.6 and 17.7 gives us

$$H_2O(l) \rightleftharpoons H_2O(v) \tag{17.8}$$

If we assume that both reactions are going on at the same time, we can explain both how the vapor became radioactive and why the amount of liquid and vapor remains constant.

Since motion is taking place on a molecular level in the system we call this a *dynamic equilibrium*. Note that there are no macroscopic changes. (Remember that macroscopic quantities are those we can observe.) We *explain* the dynamic equilibrium by assuming that the forward and reverse reactions are occurring at the same rate. This is known as the principle of *microscopic reversibility*. ■

■ **EXERCISE** Make a saturated solution of copper(II) sulfate. Decant the saturated solution into a clean bottle. Make a note of the shape of a large crystal of solid copper(II) sulfate pentahydrate. Place the crystal in the solution and close the system. After one week re-examine the shape of the crystal. Explain any changes, using the principle of microscopic reversibility.

### Chemical equilibrium

Radioactive tracers can be used to show that an equilibrium exists between reactants and products in a chemical reaction. Consider the following example.

When $H_2S(g)$ at a pressure of 1.00 atm is mixed with water, the gas will dissolve to the extent of 0.1 mole per liter. Some of the dissolved gas molecules undergo a proton transfer reaction with water molecules to form hydrogen sulfide ions and hydronium ions. Some of the hydrogen sulfide ions react further with water molecules to form more hydronium ions and sulfide ions:

$$H_2S(aq) + H_2O(l) \rightleftharpoons HS^-(aq) + H_3O^+(aq) \tag{17.9}$$
$$HS^-(aq) + H_2O(l) \rightleftharpoons S^{2-}(aq) + H_3O^+(aq) \tag{17.10}$$

Since $H_2S$ and $HS^-$ are both weak acids, only a very small amount of sulfide ion is formed. The sulfide ion can be detected by adding a drop of lead(II) nitrate solution. Lead(II) ions combine with sulfide ions to form a black precipitate of lead(II) sulfide:

$$Pb^{2+}(aq) + S^{2-}(aq) \rightleftharpoons PbS(s) \tag{17.11}$$

Does the system of $H_2S(g)$ and its aqueous solution of $H_2S$, $HS^-$, and $S^{2-}$ as shown in Fig. 17.5(a) constitute a dynamic equilibrium? At constant temperature and pressure the macroscopic properties of the system remain constant. To test for macroscopic reversibility some of the gaseous $H_2S$ was replaced by $H_2S^*(g)$ that contained some radioactive sulfur. If the forward and reverse reactions shown in Eqs. 17.9–17.11 are oc-

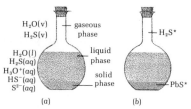

FIGURE 17.5
(a) The closed system contains three phases. The gaseous phase contains $H_2S(g)$ and $H_2O(v)$. The liquid phase contains $H_2O(l)$, $H_2S(aq)$, $H_3O^+(aq)$, $HS^-(aq)$, $S^{2-}(aq)$, and $Pb^{2+}(aq)$. The solid phase contains the precipitate $PbS(s)$. (b) Some $H_2S(g)$ is replaced with $H_2S^*(g)$. In a short time, the $PbS(s)$ contains some $PbS^*$.

■ **EXERCISE** Use the concepts of constant macroscopic properties and *microscopic reversibility to explain what is going on inside an unopened bottle of soda. Assume the bottle contains water saturated with $CO_2$ plus some sugar and flavoring.*

curring at the same rate, then eventually some of the lead(II) sulfide should become radioactive. When tested, the precipitate was radioactive, indicating that some of the radioactive sulfur from the $H_2S$ gas did indeed become part of the precipitate. The system is in a state of dynamic equilibrium. ∎

## 17.4 Shifting the equilibrium

**Le Chatlier's principle**

Ice and water in a closed system exist in equilibrium at 0.0°C *if* the pressure is 1.00 atm. If the pressure is increased, say to 400 atm, the ice begins to melt. As it melts, it absorbs some thermal energy from the system, causing the temperature to drop. The ice continues to melt until the temperature is low enough to stop the melting. Then a new equilibrium point is reached. Compare the "before" and "after" situation as shown in Table 17.1. This is an example of shifting the equilibrium position. The equilibrium position of a system can be changed by altering the *temperature* of the system, changing the *concentration* of one of the products or reactants, or sometimes by changing the *pressure* of the system. These factors affect the equilibrium point according to the following principle proposed by Le Chatlier in 1888. *When a stress is placed on a system in equilibrium, the equilibrium will shift so as to alleviate the stress.* The word "stress" can be interpreted to mean changes in temperature, pressure, or concentration. A catalyst has no effect on the equilibrium.

**Temperature**

If the temperature of a system in equilibrium is increased, the equilibrium will shift in such a way as to lower the temperature. Consider the hypothetical reaction:

$$A + B \rightleftharpoons C + \text{heat} \tag{17.12}$$

**TABLE 17.1**

*Le Chatelier's principle. When a stress is put on a system in equilibrium, the equilibrium will shift so as to alleviate the stress.*

| Factor | Original system | Final system |
|---|---|---|
| Temperature | 0.0 °C | −3 °C |
| Pressure | 1.00 atm | 400 atm |
| Mass of water | 100 g | 106 g |
| Mass of ice | 100 g | 94 g |

■ EXERCISE  *What effect would an increase of temperature have on the following equilibrium systems (shift to left, shift to right, or no change)?*
(a) $N_2O_4(g) + heat \rightleftharpoons 2NO_2(g)$
(b) $2SO_2(g) + O_2(g) \rightleftharpoons 2SO_3(g) + heat$

The reverse reaction is endothermic: it uses up thermal energy. If the temperature of this system is increased, the reverse reaction will occur to a greater extent and use up some of this increased thermal energy. More A and B and less C will exist than before the temperature was increased. ■

### Concentration

When the concentration of one reactant or product of a system in equilibrium is increased, the reaction occurs that uses up that particular substance. Consider the following reaction:

■ EXERCISE  *What effect would an increase in concentration have on the following equilibrium systems (shift to left, shift to right, or no change)?*
(a) $2NO_2(g) \rightleftharpoons 2NO(g) + O_2(g) +$ heat *(increase $O_2$)*
(b) $H_3O^+(aq) + OH^-(aq) \rightleftharpoons 2H_2O + heat$ *(increase $OH^-$)*

$$A(aq) + B(aq) \rightleftharpoons C(aq) \qquad (17.13)$$

If the concentration of B(aq) is increased by adding more of it to the solution, then the equilibrium will shift in such a way as to use up some of it. The forward reaction will occur to a greater extent, using up some A(aq) and B(aq) and forming some more C(aq). ■

### Pressure

Changes in pressure affect those systems in which reactants occupy a different volume than the products. Usually this involves gaseous equilibrium systems. Consider the following hypothetical system:

$$2A(g) + 3B(g) \rightleftharpoons 2C(g) \qquad (17.14)$$

In the gas phase the volume occupied by 2 molecules of A plus 3 molecules of B is greater than the volume occupied by 2 molecules of C. An increase in pressure will shift the equilibrium so as to lower the pressure. In this case A(g) and B(g) will react to form C(g). Each time the reaction occurs, 5 molecules become 2 molecules.

What effect will pressure increase have on the following equilibrium system?

$$A(g) + B(g) \rightleftharpoons 2D(g) \qquad (17.15)$$

*Hint:* In this reaction 2 molecules (A + B) react to form 2 molecules (2D).

Usually a pressure change does not have a large effect on equilibrium systems in the liquid or solid state. There may be some effect, however. Consider $H_2O(s)$ and $H_2O(l)$ in equilibrium again. $H_2O(s)$ has a larger volume than an equal mass of $H_2O(l)$. A pressure increase causes some $H_2O(s)$ to change to liquid:

■ EXERCISE  *What effect would an increase in pressure have on the following equilibrium systems (shift to left, shift to right, or no change)?*
*(a)* $N_2O_4(g) + heat \rightleftharpoons 2NO_2(g)$
*(b)* $2SO_2(g) + O_2(g) \rightleftharpoons 2SO_3(g) + heat$

$$\text{heat} + H_2O(s) \rightleftharpoons H_2O(l) \tag{17.16}$$

Pressure increase: shifts equilibrium to right
Pressure decrease: shifts equilibrium to left

In Section 17.5 we discuss some common equilibrium situations. ■

## 17.5  Common equilibrium situations

**Vapor pressure**

A common equilibrium situation is the vapor pressure in a closed system.

$$\text{heat} + H_2O(l) \rightleftharpoons H_2O(v) \tag{17.17}$$

An increase in temperature causes the endothermic reaction to occur to a greater extent (Table 8.3).

**Slightly soluble substances**

*Concentration*  Solid silver(I) chloride is slightly soluble in water. When excess AgCl is added to water, an equilibrium is established.

$$AgCl(s) \rightleftharpoons Ag^+(aq) + Cl^-(aq) \tag{17.18}$$

According to Le Chatelier's principle, the equilibrium can be shifted if the concentration of one of the species can be increased or decreased. If 0.100 mole of NaCl is added to a 1.00 liter solution, more AgCl(s) forms (Table 17.2). An increase in $Cl^-$ from NaCl(s) causes the equilibrium to shift to the left. The amount of aqueous silver ion in solution decreases from $1 \times 10^{-5}$ mole to $1 \times 10^{-9}$ mole, a ten thousandfold decrease.

Can we shift the reaction in the other direction and cause more AgCl(s) to dissolve? The $Ag^+(aq)$ ion concentration can be lowered by adding some $NH_3(aq)$ to the solution. The ammonia molecules react with $Ag^+(aq)$ ions and form $Ag(NH_3)_2^+(aq)$ (see Chapter 15).

$$Ag^+(aq) + 2NH_3(aq) \rightleftharpoons Ag(NH_3)_2^+(aq) \tag{17.19}$$

Could we have increased the solubility of AgCl(s) by adding more solid AgCl? This increases the volume and mass of AgCl(s) but not its concentration. Consequently the addition of solid AgCl to the solid already present has no effect on the equilibrium.

We could have removed $Ag^+(aq)$ ions by adding NaI(aq). The less soluble AgI forms and the AgCl dissolves.

**TABLE 17.2**

Methods of shifting the equilibrium of a slightly soluble substance: $AgCl(s) \rightleftharpoons Ag^+(aq) + Cl^-(aq)$

| Method | Reagent | Result |
|---|---|---|
| Increase $Cl^-$ concentration | solid NaCl | forms more AgCl(s) |
| Increase $Ag^+$ concentration | solid $AgNO_3$ | forms more AgCl(s) |
| Decrease $Ag^+$ concentration via complex information | aqueous $NH_3$ | AgCl(s) dissolves; $Ag(NH_3)_2^+$ forms |
| Decrease $Ag^+$ concentration via precipitation | aqueous NaI | AgCl(s) dissolves; AgI(s) forms |
| Decrease $Cl^-$ concentration via precipitation | aqueous $Hg_2(NO_3)_2$ | some AgCl(s) dissolves; some $Hg_2Cl_2(s)$ forms |

*Temperature* When cold water is added to $PbCl_2(s)$ it remains on the bottom of the test tube; however, if the water is heated the solid quickly dissolves:

$$PbCl_2(s) + \text{heat} \xrightleftharpoons{H_2O} Pb^{2+}(aq) + 2Cl^-(aq) \qquad (17.20)$$

Cooling causes the solid $PbCl_2$ to re-form. This method of shifting the equilibrium is often used in general chemistry to separate solid $PbCl_2$ from solid AgCl and $Hg_2Cl_2$.

**Weak electrolytes**

Ammonia gas dissolves in and reacts with water. The following equilibrium is established.

$$NH_3(aq) + H_2O(l) \rightleftharpoons NH_4^+(aq) + OH^-(aq) \qquad (17.21)$$

The amount of $NH_3(aq)$ in solution can be controlled by the amount of $OH^-(aq)$ ions in solution. A high concentration of $OH^-$ ions will drive the reaction to the left and give a high concentration of $NH_3(aq)$. A low concentration of $OH^-(aq)$ will drive the reaction to the right and give a low concentration of $NH_3(aq)$.

The $OH^-(aq)$ ion concentration can be increased by the addition of NaOH and decreased by the addition of sufficient amounts of a strong acid.

$$NH_3(aq) + H_2O(l) \rightleftharpoons NH_4^+(aq) + OH^-(aq) \qquad (17.22)$$
$$2H_2O \rightleftharpoons H_3O^+(aq) + OH^-(aq)$$

A useful application of this equilibrium is in the control of the concentration of $Ag^+(aq)$ ions. We saw in Section 17.5 that AgCl(s) dissolves in aqueous $NH_3$ by formation of $Ag(NH_3)_2^+$. If

■ EXERCISE State how the indicated change will affect the equilibrium system shown (shift to left, shift to right, no change). (a) Add KCN(aq); (b) Add HCl(aq); (c) Add NaOH(aq); (d) Add $Fe^{3+}$(aq) (forms complex with $CN^-$).

$HCN(aq) + H_2O(l) \rightleftharpoons H_3O^+(aq) + CN^-(aq)$

the concentration of $NH_3$ is lowered, the $Ag(NH_3)_2^+$ decomposes back into $Ag^+$(aq) and $NH_3$(aq). This can be shown as follows:

1. $AgCl(s) \rightleftharpoons Ag^+(aq) + Cl^-(aq)$
The dissolution of AgCl depends upon the concentration of $Ag^+$(aq) and $Cl^-$(aq).

2. $Ag^+(aq) + 2NH_3(aq) \rightleftharpoons Ag(NH_3)_2^+$
The amount of $Ag^+$(aq) is controlled by $NH_3$(aq).

3. $NH_3(aq) + H_2O(l) \rightleftharpoons NH_4^+(aq) + OH^-(aq)$
The amount of $NH_3$(aq) is controlled by $OH^-$(aq).

4. $2H_2O \rightleftharpoons H_3O^+ + OH^-$
The amount of $OH^-$(aq) is controlled by $H_3O^+$(aq). ■

## 17.6 The equilibrium constant

**The concept**

In studying the equilibrium of $H_2O$ in Chapter 16 we saw that at 20°C:

$$[H_3O^+] \times [OH^-] = K_w = 1.0 \times 10^{-14} \qquad (17.24)$$

The concentrations of reactants and products in all chemical systems at equilibrium are related to one another quantitatively through the equilibrium constant, $K_{eq}$. What is the equilibrium constant? Consider the following reaction at 360°C.

$$H_2(g) + I_2(g) \rightleftharpoons 2HI(g) \qquad (17.25)$$

Various amounts of hydrogen and iodine were mixed at this temperature and the systems were allowed to reach equilibrium. The final concentrations at equilibrium are shown in Table 17.3. The equilibrium constant, $K_{eq}$, for this reaction is also

TABLE 17.3

Various concentrations of $H_2(g)$ and $I_2(g)$ were mixed at 360°C and allowed to reach equilibrium.

| Initial concentration (moles/liter) | | Equilibrium concentrations (moles/liter) | | | Equilibrium constant |
|---|---|---|---|---|---|
| $[H_2]$ | $[I_2]$ | $[H_2]$ | $[I_2]$ | $[HI]$ | $\dfrac{[HI]^2}{[H_2][I_2]}$ |
| 1.0 | 1.0 | 0.19 | 0.19 | 1.6 | 71 |
| 2.0 | 2.0 | 0.39 | 0.39 | 3.2 | 71 |
| 5.0 | 5.0 | 0.96 | 0.96 | 8.1 | 71 |

**TABLE 17.4**

Equilibrium expressions of selected reactions.

| Equation | Equilibrium expression | Equilibrium constant |
|---|---|---|
| $2H_2(g) + I_2(g) \rightleftharpoons 2HI(g)$ | $\dfrac{[HI]^2}{[H_2][I_2]}$ | 71.3 at 360°C |
| $N_2(g) + 3H_2(g) \rightleftharpoons 2NH_3(g)$ | $\dfrac{[NH_3]^2}{[N_2][H_2]^3}$ | $6.0 \times 10^{-2}$ at 500°C |
| $N_2O_4(g) \rightleftharpoons 2NO_2(g)$ | $\dfrac{[NO_2]^2}{[N_2O_4]}$ | $6.4 \times 10^{-3}$ at 75°C |

shown. This term is calculated from the following equilibrium expression:

$$K_{eq} = \frac{[HI]^2}{[H_2][I_2]} \qquad (17.26)$$

The *equilibrium expression* for a reaction is determined by placing the concentration of each *product in the numerator* and the concentration of each *reactant in the denominator*. Each term is raised to a power as determined by the coefficient in the balanced equation. Some examples are shown in Table 17.4.

Once the equilibrium constant has been determined for one sample, it can be used to calculate the amounts of products that will be produced at equilibrium in other samples. The equilibrium constant changes with temperature; however, it is unaffected by the presence of a catalyst, pressure, or concentration. Simple calculations using equilibrium constants are shown in the next three sections. ■

■ EXERCISE  Write the equilibrium expression for the following reactions.
(a) $H_2(g) + Cl_2(g) \rightleftharpoons 2HCl(g)$
(b) $4HCl(g) + O_2(g) \rightleftharpoons 2Cl_2(g) + 2H_2O(g)$

**Gaseous phase equilibrium**

EXAMPLE 1  Calculate the equilibrium constant, $K_{eq}$, at 55°C for the reaction:

$$N_2O_4(g) \rightleftharpoons 2NO_2(g)$$

At equilibrium $[N_2O_4]$ is 2.00 M; $[NO_2]$ = 1.32 M.

Given: $[N_2O_4]$ = 2.00 M
       $[NO_2]$ = 1.32 M

Find: $K_{eq}$

Relationship: $K_{eq} = \dfrac{[NO_2]^2}{[N_2O_4]}$

Solution: $K_{eq} = \dfrac{[NO_2]^2}{[N_2O_4]} = \dfrac{[1.32]^2}{[2.00]} = \dfrac{1.74}{2.00} = 0.87$

In calculating and using $K_{eq}$, we will ignore units.

EXAMPLE 2   The $K_{eq}$ for the reaction $H_2(g) + I_2(g) \rightleftharpoons 2HI(g)$ is 50 at 730°C. What will be the concentration of HI(g) if 5.0 moles of $H_2(g)$ are in equilibrium with 4.0 moles of $I_2(g)$ in a 1.00 liter vessel?

Given: $[H_2] = 5.0$ moles per liter at equilibrium
$[I_2] = 4.0$ moles per liter at equilibrium
$K_{eq} = 50$

Find: Concentration of HI(g)

Relationship: $K_{eq} = \dfrac{[HI]^2}{[H_2][I_2]} = 50$

Solution: $K_{eq} = \dfrac{[HI]^2}{[H_2][I_2]} = \dfrac{[HI]^2}{[5.0][4.0]} = 50$

$[HI]^2 = 50 \times 20 = 1000$
$[HI] = (1000)^{1/2} = 31.6$

The concentration of HI(g) is 32 moles per liter.

EXAMPLE 3   The $K_{eq}$ for the reaction $COCl_2(g) \rightleftharpoons CO(g) + Cl_2(g)$ is 0.25. What are the concentrations of CO(g) and $Cl_2(g)$ in equilibrium with 6.0 M $COCl_2(g)$? Initially only $COCl_2$ was placed in the vessel.

Given: $[COCl_2] = 6.0\ M$
$K_{eq} = 0.25$

Find: Concentration of CO(g) and $Cl_2(g)$

Relationship: $K_{eq} = \dfrac{[CO][Cl_2]}{[COCl_2]} = 0.25$

Solution: $K_{eq} = \dfrac{[CO][Cl_2]}{[COCl_2]} = \dfrac{[CO)(Cl_2)}{[6.0]} = 0.25$

$[CO][Cl_2] = 1.5$

According to the ratio in the balanced equation, $[CO] = [Cl_2]$. Therefore,

$[CO][CO] = [CO]^2 = 1.5$
$[CO] = (1.5)^{1/2} = 1.22$
$[CO] = [Cl_2] = 1.2\ M$

■ EXERCISE  *What is the value of $K_{eq}$ at 76°C for the reaction: $H_2(g) + I_2(g) \rightleftharpoons 2HI(g)$? The equilibrium concentrations of a sample were: $[HI] = 13.5\ M$, $[H_2] = 2.00\ M$, $I_2 = 2.00\ M$.*

■ EXERCISE  *The $K_{eq}$ equals 51 at 25°C for the following reaction:*

$N_2O_3(g) \rightleftharpoons NO(g) + NO_2(g)$

*If an equilibrium mixture contains 2.50 M $N_2O_3$ and 10.00 M NO, what is the concentration of $NO_2$?*

## Weak electrolytes

The ratio of reactants to products in an equilibrium system containing a weak electrolyte is also given by the equilibrium expression. For an aqueous solution of acetic acid we have:

$$CH_3CO_2H(aq) + H_2O(l) \rightleftharpoons CH_3CO_2^-(aq) + H_3O^+(aq) \quad (17.27)$$

$$K_{eq} = \dfrac{[CH_3CO_2]^-[H_3O]^+}{[CH_3CO_2H][H_2O]} \quad (17.28)$$

In dilute aqueous solutions the concentration of water is large compared to the other species present. In the course of a reac-

**TABLE 17.5**
*The ionization constants of some weak acids and bases.*

| Acid | Reaction | $K_i$ |
|---|---|---|
| Sulfurous acid | $H_2SO_3(aq) + H_2O(l) \rightleftharpoons H_3O^+(aq) + HSO_3^-(aq)$ | $1.7 \times 10^{-2}$ |
| Hydrogen sulfate ion | $HSO_4^-(aq) + H_2O(l) \rightleftharpoons H_3O^+(aq) + SO_4^{2-}(aq)$ | $1.3 \times 10^{-2}$ |
| Phosphoric acid | $H_3PO_4(aq) + H_2O(l) \rightleftharpoons H_3O^+(aq) + H_2PO_4^-(aq)$ | $7.1 \times 10^{-3}$ |
| Hydrofluoric acid | $HF(aq) + H_2O(l) \rightleftharpoons H_3O^+(aq) + F^-(aq)$ | $6.6 \times 10^{-4}$ |
| Nitrous acid | $HNO_2(aq) + H_2O(l) \rightleftharpoons H_3O^+(aq) + NO_2^-(aq)$ | $5.1 \times 10^{-4}$ |
| Acetic acid | $CH_3COOH(aq) + H_2O(l) \rightleftharpoons H_3O^+(aq) + CH_3COO^-(aq)$ | $1.8 \times 10^{-5}$ |
| Carbonic acid | $H_2CO_3(aq) + H_2O(l) \rightleftharpoons H_3O^+(aq) + HCO_3^-(aq)$ | $4.4 \times 10^{-7}$ |
| Hydrogen sulfide | $H_2S(aq) + H_2O(l) \rightleftharpoons H_3O^+(aq) + HS^-(aq)$ | $1.0 \times 10^{-7}$ |
| Hydrocyanic acid | $HCN(aq) + H_2O(l) \rightleftharpoons H_3O^+(aq) + CN^-(aq)$ | $4.0 \times 10^{-10}$ |
| Ammonium ion | $NH_4^+(aq) + H_2O(l) \rightleftharpoons NH_3(aq) + H_3O^+(aq)$ | $5.6 \times 10^{-10}$ |
| Hydrogen carbonate ion | $HCO_3^-(aq) + H_2O(l) \rightleftharpoons H_3O^+(aq) + CO_3^{2-}(aq)$ | $4.7 \times 10^{-11}$ |
| **Base** | | |
| Ammonia | $NH_3(aq) + H_2O(l) \rightleftharpoons NH_4^+(aq) + OH^-(aq)$ | $1.8 \times 10^{-5}$ |
| Methylamine | $CH_3NH_2(aq) + H_2O(l) \rightleftharpoons CH_3NH_3^+(aq) + OH^-(aq)$ | $5.0 \times 10^{-4}$ |

tion, the concentration of water remains fairly constant and its value is incorporated with the equilibrium constant. This gives a second equilibrium constant called the *ionization constant*, $K_i$

$$K_{eq} \times [H_2O] = K_i = \frac{[CH_3CO_2]^-[H_3O]^+}{[CH_3CO_2H]} \tag{17.29}$$

Ionization constants are used for weak acids and weak bases. The size of the ionization constant indicates how extensively a weak acid (or base) ionizes and therefore how strong an acid (or base) the substance is (Table 17.5).

**EXAMPLE 4** The $K_i$ for hydrofluoric acid is $6.6 \times 10^{-4}$. What is the $H_3O^+$ ion concentration in an aqueous solution of hydrofluoric acid if the concentration of HF(aq) is originally 0.20 M?

Given: $K_i = 6.6 \times 10^{-4}$
HF(aq) = 0.20 M

Find: $H_3O^+$ concentration

Relationship: $HF(aq) + H_2O(l) \rightleftharpoons H_2O^+(aq) + F^-(aq)$

$$K_i = \frac{[H_3O^+][F^-]}{[HF]} = 6.6 \times 10^{-4}$$

*Solution:* The small value of the $K_i$ indicates that only a small percentage of HF ionizes. We will assume that the concentration of HF after ionization is still approximately 0.20 M.

$$K_i = \frac{[H_3O^+][F^-]}{[HF]} = \frac{[H_3O^+][F^-]}{0.20} = 6.6 \times 10^{-4}$$

According to the balanced equation, the concentration of $[H_3O^+]$ equals $[F^-]$. By substituting $[H_3O^+]$ for $[F^-]$ we obtain:

$$K_i = \frac{[H_3O^+][H_3O^+]}{[0.20]} = 6.6 \times 10^{-4}$$

$$[H_3O^+]^2 = 0.20 \times 6.6 \times 10^{-4} = 1.3 \times 10^{-4} \qquad [H_3O^+] = 1.1 \times 10^{-2}$$

The concentration of $H_3O^+$ is $1.1 \times 10^{-2}$ M.

## Precipitation

When solid AgBr is added to water, a small amount dissolves according to the equation:

$$AgBr(s) \rightleftharpoons Ag^+(aq) + Br^-(aq) \tag{17.30}$$

The equilibrium expression is:

$$K_{eq} = \frac{[Ag^+][Br^-]}{[AgBr]}$$

The concentration of solid AgBr is fixed by its density. It is a constant. In equilibrium expressions for slightly soluble substances, this constant is incorporated with the $K_{eq}$. The new expression is called the *solubility product* constant, $K_{sp}$. For silver bromide the $K_{sp}$ is:

$$K_{sp} = [Ag^+][Br^-] = 5.0 \times 10^{-13}$$

The product of the silver ion concentration and bromide ion concentration cannot exceed $5.0 \times 10^{-13}$ in a system in equilibrium. If it momentarily exceeds this value by the mixing of two solutions, a precipitate of AgBr will form. The precipitate continues to form until the product equals the $K_{sp}$. If the product of $[Ag^+]$ and $[Br^-]$ does not equal $5.0 \times 10^{-13}$ in an aqueous system, then a precipitate will not form.

EXAMPLE 5  A few drops of concentrated hydrochloric acid solution are added to a solution of silver nitrate. A white precipitate of silver chloride forms. What is the maximum concentration of $Ag^+$ ions left in solution, if the $Cl^-$ ion concentration is 0.10 M? $K_{sp} = 1.7 \times 10^{-10}$

■ EXERCISE  The $K_{sp}$ of radium sulfate is $4.0 \times 10^{-11}$. What is the maximum concentration of $Ra^{2+}$ ions in a solution that contains $0.50 \text{ M SO}_4^{2-}$ ions?

$RaSO_4(s) \rightleftharpoons Ra^{2+}(aq) + SO_4^{2-}(aq)$

Given: $[Cl^-] = 0.10$ M, $K_{sp} = 1.7 \times 10^{-10}$

Find: $[Ag^+]$

Relationship: $AgCl(s) \rightleftharpoons Ag^+(aq) + Cl^-(aq)$
$K_{sp} = [Ag^+][Cl^-]$

Solution: $K_{sp} = [Ag^+][Cl^-] = [Ag^+][0.10] = 1.7 \times 10^{-10}$

$[Ag^+] = \dfrac{1.7 \times 10^{-10}}{0.10} = 1.7 \times 10^{-9}$

The concentration of silver ions is $1.7 \times 10^{-9}$ M. ■

## Summary

The study of the factors which influence the rates of chemical reactions is called chemical kinetics. Before two particles can react with one another, the particles must (1) collide, (2) have the correct orientation, and (3) possess sufficient energy to break bonds.

The rate of collisions can be speeded up by increasing the concentration of the reactants. For gases, increasing the pressure will increase the concentration. If a reactant is solid, the collision rate depends on the surface area. For reactions that occur rapidly, the rate of collision may depend upon the rate of mixing.

The orientation of colliding molecules may be affected by some catalysts. The activation energy is an energy barrier that prevents the reactants from becoming products. This barrier is due in part to the bonds in the reactant that must be broken. The barrier can be overcome in two ways: (1) increasing the average kinetic energy of the reactants and (2) employing a catalyst to lower the activation energy.

Radioactive isotopes can be used to trace the whereabouts of certain elements in a system in equilibrium. Equilibrium can be recognized by constant macroscopic properties. In a system at equilibrium the forward and reverse reactions are occurring at the same rate. This is the principle of microscopic reversibility.

When a stress is placed on a system in equilibrium, the system undergoes a change in such a way as to relieve the stress and establish a new equilibrium point. If the temperature is increased, the endothermic reaction occurs to a greater extent. If the concentration of one reactant is increased, that reactant is partially removed by reaction. If the external pressure is increased, the equilibrium will shift so as to lower the volume of the system if this is possible.

The equilibrium constant relates the concentration of the reactants to products. It is useful in determining how much of a certain reactant or product will be present at equilibrium. In aqueous solutions of weak electrolytes the equilibrium constant is simplified and called the $K_i$. For slightly soluble substances that form precipitates, the equilibrium constant is simplified and called the $K_{sp}$.

*Objectives*  By the end of this chapter, you should be able to do the following:

1. State the conditions necessary for a reaction to take place. (Question 2)

2. List the factors that affect the rate of a chemical reaction. (Question 3)

3. Draw a diagram showing the kinetic energy distribution of particles in a gas at two different temperatures. Identify the particles in each sample that have sufficient kinetic energy to overcome the activation energy barrier. (Question 5)

4. Draw an energy diagram for a hypothetical reaction. Identify the energy of the reactants and the products and the activation energy. Show the effect of a catalyst on the activation energy. (Question 6)

5. From a given list, identify systems in equilibrium. (Questions 4)

6. Use Le Chatelier's principle to predict the effect of certain stresses on equilibrium systems. (Questions 3, 7, 16, 17, 18)

7. Given a chemical equation, write the equilibrium expression as either $K_{eq}$, $K_i$, or $K_{sp}$. (Questions 8, 19)

8. Given equilibrium concentrations of all reactants and products, calculate the $K_{eq}$, $K_i$, or $K_{sp}$. (Questions 9, 10, 11)

9. Given the equilibrium constant and all equilibrium concentrations but one, calculate the unknown concentration. (Questions 12, 13, 14, 15)

*Questions*  1. Define the following terms. How much does your definition rely on concepts and terms learned in previous chapters? kinetics; activation energy; catalyst; radioactive tracer; phase equilibrium; macroscopic properties; microscopic reversibility; Le Chatelier's principle; equilibrium expression; $K_{eq}$; chemical equilibrium; dynamic equilibrium; $K_i$; $K_{sp}$; equilibrium constant.

2. Consider the following hypothetical reaction:

A—B(g) + C(g) $\rightleftharpoons$ A—C(g) + B(g)

List the conditions necessary for a reaction to occur.

3. Ammonia is produced by the reaction of hydrogen with nitrogen:

$N_2(g) + 3H_2(g) \rightleftharpoons 2NH_3(g) + 22$ kcal

The reaction is very slow at room temperature. (a) List three methods that could be used to increase the rate of this reaction. (b) What effect does an increase in pressure have on the equilibrium of this system? (c) What effect does an increase in temperature have on the equilibrium of this system?

4. Which of the following systems are in a stage of equilibrium? (a) An unopened bottle of beer in a refrigerator; (b) a hot beaker of water on a

laboratory bench; (c) solid zinc dissolving in aqueous hydrogen chloride; (d) a growing precipitate of FeCO₃ in a hot water pipe; (e) a rusting automobile tail pipe; (f) a closed bottle of water at constant temperature; (g) a capped half-empty can of gasoline.

5. Draw a diagram showing the kinetic energy distribution of particles in a gas at two different temperatures. Identify the particles that have sufficient kinetic energy to overcome an arbitrary activation energy barrier.

6. The activation energy for the reaction

$$H_2(g) + I_2(g) \rightleftharpoons 2HI(g) + 2.2 \text{ kcal is about 40 kcal}$$

Draw an energy diagram for this reaction. Show the activation energy barrier and the relative difference in energy between reactants and products. What effect does a catalyst have on the activation energy barrier?

7. An important industrial reaction is the "water gas" reaction:

$$C(s) + H_2O(g) + 30 \text{ kcal} \rightleftharpoons CO(g) + H_2(g)$$

At 1000°C this reaction occurs to an appreciable extent. What effect on the equilibrium would lowering the temperature have? What effect would lowering the pressure have on the equilibrium?

8. Write the equilibrium expression for the following reactions.
(a) $4NH_3(g) + 3O_2(g) \rightleftharpoons 2N_2(g) + 6H_2O(g)$
(b) $CO(g) + H_2O(g) \rightleftharpoons CO_2(g) + H_2(g)$
(c) $3O_2(g) \rightleftharpoons 2O_3(g)$
(d) $PCl_3(g) + Cl_2(g) \rightleftharpoons PCl_5(g)$
(e) $SO_2(g) + NO_2(g) \rightleftharpoons SO_3(g) + NO(g)$
(f) $HNO_2(aq) + H_2O(l) \rightleftharpoons H_3O^+(aq) + NO_2^-(aq)$
(g) $FeS(s) \rightleftharpoons Fe^{2+}(aq) + S^{2-}(aq)$

9. In the atmosphere sulfur dioxide reacts with oxygen to form sulfur trioxide, $2SO_2(g) + O_2(g) \rightleftharpoons 2SO_3(g)$. A sample at equililbrium contained 0.40 M SO₂, 0.20 M O₂, and 4.5 M SO₃. Calculate the $K_{eq}$.

10. The solubility of NiS(s) is $1.0 \times 10^{-11}$ mole per liter. What is the $K_{sp}$?

$$NiS(s) \rightleftharpoons Ni^{2+}(aq) + S^{2-}(aq)$$

11. A solution contains $1.0 \times 10^{-2}$ M HNO₂, $2.1 \times 10^{-3}$ M H₃O⁺, and $2.1 \times 10^{-3}$ M NO₂⁻. What is the $K_i$ for HNO₂?

$$HNO_2(aq) + H_2O(l) \rightleftharpoons H_3O^+(aq) + NO_2^-(aq)$$

12. The equilibrium constant for the reaction

$$4NH_3(g) + 5O_2(g) \rightleftharpoons 4NO(g) + 6H_2O(g)$$

at a certain temperature is 400. From the following equilibrium concentrations, calculate the concentration of NO(g). [H₂O] = 1.0 M, [O₂] = 0.10 M, [NH₃] = 2.0 M.

13. The $K_{eq}$ for the reaction $CO(g) + Cl_2(g) \rightleftharpoons COCl_2(g)$ is 13 at a certain temperature. If there is 0.50 mole of CO and 0.50 mole of Cl₂, how many moles of COCl₂ are there at equilibrium in a 1.00 liter vessel?

14. A common eyewash is a solution of boric acid, $H_3BO_3$. If the $K_i = 6 \times 10^{-10}$ and the concentration of $H_3BO_3(aq)$ is 0.10 M, what is the concentration of $H_3O^+(aq)$?

$$H_3BO_3(aq) + H_2O(l) \rightleftharpoons H_3O^+(aq) + H_2BO_3^-(aq)$$

15. A solution of sodium sulfide was added to a solution of silver nitrate. A black precipitate of $Ag_2S$ formed. The $K_{sp}$ of silver sulfide is $1 \times 10^{-50}$. What is the silver ion concentration if the sulfide ion concentration is $1 \times 10^{-15}$?

$$Ag_2S(s) \rightleftharpoons 2Ag^+(aq) + S^{2-}(aq)$$

16. The hydrogen carbonate ion, $HCO_3^-(aq)$, in the blood stream serves as a buffer. It neutralizes excess $H_3O^+$ ions or $OH^-$ ions.

(a) $H_3O^+(aq) + HCO_3^-(aq) \rightleftharpoons CO_2(aq) + 2H_2O$
(b) $OH^-(aq) + HCO_3^-(aq) \rightleftharpoons H_2O(l) + CO_3^{2-}(aq)$

How does an excess of acid affect the equilibrium shown in Eq. (a)?
How does an excess of base affect the equilibrium shown in Eq. (b)?

17. A closed container of aqueous sodium sulfite is in equilibrium with a small amount of $H_2SO_3$.

$$SO_3^{2-}(aq) + 2H^+(aq) \rightleftharpoons H_2SO_3(aq) \rightleftharpoons H_2O(l) + SO_2(g)$$

As shown by the equation, the $H_2SO_3(aq)$ decomposes into $H_2O$ and $SO_2$. What would be the effect of adding concentrated aqueous HCl to a solution of $Na_2SO_3$?

18. A closed container of aqueous sodium carbonate is in equilibrium with $H_2O$ and $CO_2$.

$$CO_3^{2-}(aq) + 2H^+(aq) \rightleftharpoons H_2O(l) + CO_2(aq)$$

What would be the effect of adding concentrated aqueous HCl to a solution of $Na_2CO_3$?

19. Write the equilibrium expressions and the $K_i$ or $K_{sp}$ for the following reactions:
(a) $AgIO_3(s) \rightleftharpoons Ag^+(aq) + IO_3^-(aq)$
(b) $Al(OH)_3(s) \rightleftharpoons Al^{3+}(aq) + 3OH^-(aq)$
(c) $HCOOH(aq) + H_2O(l) \rightleftharpoons HCOO^-(aq) + H_3O^+(aq)$
(d) $HClO(aq) + H_2O(l) \rightleftharpoons ClO^-(aq) + H_3O^+(aq)$

# Oxidation-reduction

## 18.1 Introduction

What is oxidation-reduction? We learned in Chapter 5 that when elements combine to form compounds, valence electrons may be lost, gained, or shared between atoms. In many reactions, there is a competition between atoms for these valence electrons. Oxidation-reduction reactions result from the competition of the atoms for electrons.

Oxidation numbers are useful not only for naming compounds but also for tracing the electronic change which an atom undergoes in a chemical reaction. Any chemical change which involves an *increase in the oxidation number* of an atom is referred to as *oxidation*. Oxidation represents a net loss or apparent loss in electrons. *A decrease in oxidation number* is called *reduction*.

## 18.2 A review of oxidation-reduction reactions

Oxidation-reduction reactions—called simply *redox* reactions—occur when atoms compete with each other for valence electrons. Before redox reactions are considered further, review the concept of oxidation numbers in Chapter 7.

EXAMPLE 1 A white substance slowly forms on aluminum metal which is left in contact with the oxygen in the atmosphere. The chemical analysis of this substance indicates that it is aluminum oxide, $Al_2O_3$. What is the oxidation number of aluminum in aluminum metal? In $Al_2O_3$? What is the oxidation number of oxygen in elemental gaseous oxygen? In $Al_2O_3$?

Does the oxidation number of Al increase or decrease when aluminum reacts with oxygen? The oxidation number of Al in the metal is 0, and in $Al_2O_3$ it is 3+. Aluminum is oxidized by the loss of three electrons.

Oxidation: $Al \rightleftharpoons Al^{3+} + 3e^-$ (18.1)

The oxidation number of Al changes from 0 to 3+. This is an increase in oxidation number and is therefore oxidation. Al is said to be oxidized.

The oxidation number of oxygen changes from 0 to 2−. This is a decrease in oxidation number, and is therefore reduction. Oxygen has been *reduced*. The reduction of oxygen results from the *gain* of two electrons.

Reduction: $O_2 + 4e^- \rightleftharpoons 2O^{2-}$ (18.2)

Gain of electrons by one atom cannot occur without the resultant loss of electrons by another atom. *Oxidation and reduction must occur simultaneously.*

Since aluminum lost electrons during the reaction, Al can be viewed as being responsible for the fact that oxygen gained electrons. Al is the substance responsible for the reduction; thus it is called the *reducing agent*. *The element which is oxidized acts as the reducing agent.* Oxygen, which gains electrons during the reaction, can be viewed as being responsible for the loss of electrons by Al. Oxygen is called the *oxidizing agent* in this case. *The element which is reduced acts as the oxidizing agent.*

Oxidation: $Al \longrightarrow Al^{3+} + 3e^-$ (from Eq. 18.1)
Reduction: $O_2 + 4e^- \longrightarrow 2O^{2-}$ (from Eq. 18.2)

Species reduced: $O_2$  Species oxidized: Al
Oxidizing agent: $O_2$  Reducing agent: Al

Consider the half-reactions:

Oxidation: $Al \rightleftharpoons Al^{3+} + 3e^-$
Reduction: $O_2 + 4e^- \rightleftharpoons 2O^{2-}$

If these two half-reactions are summed, the number of electrons lost will *not be equal* to the number of electrons gained. If the entire oxidation half-reaction is multiplied by 4 and the reduction half-reaction multiplied by 3, the electrons lost will be equal to the electrons gained.

Oxidation: $\quad\quad\quad\quad\quad 4Al \longrightarrow 4Al^{3+} + 12e^-$
Reduction: $\quad\quad\quad 3O_2 + 12e^- \longrightarrow 6O^{2-}$
$$4Al + 3O_2 + 12e^- \longrightarrow 4Al^{3+} + 6O^{2-} + 12e^- \quad (18.3)$$

The equation can also be written as

$$4Al(s) + 3O_2(g) \longrightarrow 2Al_2O_3(s)$$

The number of electrons lost in this reaction is equal to the number of electrons gained. Whenever you balance an equation for mass, check to see that the net charge is the same on both sides of the equation. ■

■ EXERCISE  Write half-reactions for each of the following reactions. Use the half-reactions to write balanced equations.

(a) $Iron(s) + chlorine(g) \rightarrow$
$\quad\quad iron(III)\ chloride(s) + energy$
(b) $vanadium(s) + oxygen(g) \rightarrow$
$\quad\quad vanadium(V)\ oxide(s) + energy$

## 18.3  Half-cells

It has been observed experimentally that a reddish-brown solid forms when metallic zinc is placed in contact with a solution of copper(II) sulfate. The color of the copper(II) sulfate solution, which is originally bright blue, decreases in intensity as the reddish-brown solid forms, until it is finally colorless. The piece of metallic zinc is observed to decrease in size as the reddish-brown solid forms. Chemical analysis indicates that the reddish-brown solid is metallic copper and that zinc(II) sulfate is present in the colorless solution.

The blue color of a copper(II) sulfate solution indicates the presence of copper(II) ions, $Cu^{2+}(aq)$. Since the intensity of the blue color decreases when the solution is placed in contact with metallic zinc, there must be a decrease in $Cu^{2+}(aq)$ ions in the solution. The decrease in $Cu^{2+}(aq)$ ions is evidently involved in the formation of metallic copper, Cu. Metallic copper can be formed from $Cu^{2+}(aq)$ by the acceptance of two electrons.

Reduction: $\quad Cu^{2+}(aq) + 2e^- \rightleftharpoons Cu(s) \quad\quad\quad (18.5)$

Since $Cu^{2+}(aq)$ is reduced by the acceptance of two electrons, some other substance in the solution must release electrons and be oxidized. Since zinc metal, Zn, disappears while the zinc(II) ion, $Zn^{2+}(aq)$, is formed, zinc must be the substance undergoing oxidation.

Oxidation: $\quad Zn(s) \rightleftharpoons Zn^{2+}(aq) + 2e^- \quad\quad\quad (18.6)$

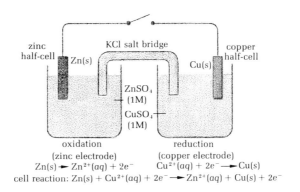

**FIGURE 18.1**
The Zn-Cu cell. The zinc half-cell is on the left; the copper half-cell on the right. The KCl salt bridge allows for flow of ions between the two solutions while the external circuit allows for flow of electrons.

The net equation can be obtained by adding the two half-reactions.

$$Cu^{2+}(aq) + Zn(s) \longrightarrow Zn^{2+}(aq) + Cu(s) \tag{18.7}$$

Note in Eq. 18.7 that the sulfate ion is not included since it has been neither oxidized nor reduced; in fact, it has undergone no change. The *net ionic equation* includes only those substances which have undergone a change. Note, however, that the net equation is still balanced for mass and charge. If it is desired, the $SO_4^{2-}$ ions can be included in the equation.

$$CuSO_4(aq) + Zn(s) \longrightarrow Cu(s) + ZnSO_4(aq) \tag{18.8}$$

An apparatus can be constructed which will allow each half-reaction of a redox reaction to be conducted in a separate container (Fig. 18.1). The apparatus consists of two separate containers called half-cells. The two half-cells are connected by a small piece of glass tubing as shown. The entire apparatus is called a cell. Each half-reaction can be conducted in a separate half-cell.

The previous reaction involving zinc and copper can be used to demonstrate the use of this apparatus. One of the containers is filled with a 1.0 M solution of $CuSO_4$ while the other is filled with a 1.0 M solution of $ZnSO_4$. The glass tubing is filled with a 1.0 M solution of KCl. This tubing is called a salt bridge. Its purpose is to prevent the mixing of the two solutions while allowing the migration of ions between them. A strip of copper metal is suspended into the $CuSO_4$ solution and a strip of zinc is suspended into the $ZnSO_4$ solution. A reddish-brown precipitate of metallic copper forms in the $CuSO_4$ chamber when the two metal strips are connected by a wire (Fig. 18.2). As the Cu metal appears, there is a corresponding decrease in the inten-

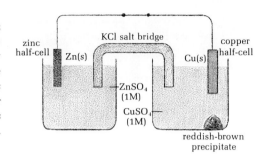

**FIGURE 18.2**
The zinc-copper cell. After the two half-cells have been connected by means of the external circuit for some time, a reddish-brown precipitate is noted in the bottom of the copper half-cell. The zinc electrode becomes smaller.

sity of the blue color indicating the disappearance of $Cu^{2+}(aq)$ from the solution. The piece of zinc in the $ZnSO_4$ chambers is observed to decrease in size. Chemical analysis of the solution in this chamber indicates an increase in the concentration of $ZnSO_4$.

What has happened in the cell to cause these changes? $Cu^{2+}(aq)$ is evidently being reduced to Cu by the acceptance of a pair of electrons.

Reduction: $Cu^{2+}(aq) + 2e^- \rightleftharpoons Cu$

What is the source of these electrons? The piece of zinc went into solution while $Zn^{2+}(aq)$ increased in concentration in the solution. Zn is evidently being oxidized to $Zn^{2+}(aq)$ by the loss of two electrons.

Oxidation: $Zn(s) \rightleftharpoons Zn^{2+}(aq) + 2e^-$

The observed changes do not occur unless the two metal strips are connected by the wire. We conclude that electron transfer must occur through the wire. Two electrons are given up by Zn. The two electrons pass through the wire where they are accepted by $Cu^{2+}(aq)$. For each pair of electrons transported through the wire, one sulfate ion, $SO_4^{2-}(aq)$, must pass through the salt bridge to prevent a buildup of charge Fig. 18.3.

The copper and zinc strips which are suspended in the two containers are called electrodes. *The electrode at which oxida-*

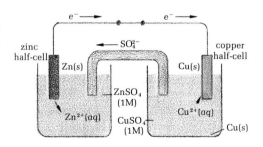

**FIGURE 18.3**
The zinc-copper cell. The zinc atoms from the strip of zinc metal release two electrons each. These electrons flow through the wire as indicated. The copper ions, $Cu^{2+}(aq)$, in contact with the copper electrode can accept two electrons each to form copper metal, $Cu(s)$.

tion occurs is called the anode. The zinc electrode is the anode in this case. Reduction occurs at the cathode; therefore the copper electrode is the cathode in this case.

Anode reaction: $\qquad Zn \rightleftharpoons Zn^{2+}(aq) + 2e^-$
Cathode reaction: $\underline{\qquad Cu^{2+}(aq) + 2e^- \rightleftharpoons Cu(s)\qquad}$
Cell reaction: $\qquad Cu^{2+}(aq) + Zn(s) \rightleftharpoons Zn^{2+}(aq) + Cu(s)$

(18.9)

## 18.4 The hydrogen electrode

The oxidizing tendency of a half-reaction can only be determined in relation to another half-reaction. For instance, it was determined that zinc had a greater oxidizing tendency than copper. It is convenient to chose one half-reaction as a reference point and then determine the oxidizing tendency of any half-reaction in relation to the reference point. The half-reaction which has been chosen as a reference point by chemists is the half-reaction for the oxidation of hydrogen from a solution containing 1.00 mole/liter of hydronium ion at a hydrogen pressure of 1.00 atm. Zinc has a greater tendency to be oxidized than does hydrogen because zinc will reduce hydrogen. Copper, however, has less tendency to be oxidized than does hydrogen. It will not reduce hydrogen. For these same reasons, zinc is placed above hydrogen in the activity series and copper is placed below hydrogen Table 18.1. The position of a metal in the activity series is a measure of its oxidizing tendency.

The cell in Fig. 18.4(a) contains a hydrogen half-cell and a zinc half-cell. As in the previous case, an electrode is dipped

FIGURE 18.4
(a) The zinc-hydrogen cell. (b) A voltmeter connected between the zinc half-cell and the hydrogen half-cell gives a reading of +0.76 volt.

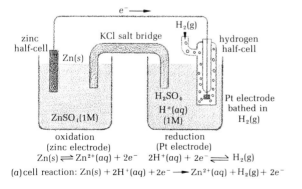
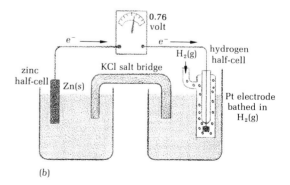

**TABLE 18.1**

Activity series. The more active metals, at top, are more easily oxidized than the less active metals, at bottom of table.

| Oxidizing tendency | Metal | Reaction |
|---|---|---|
| More active metals; best reducing agents | K<br>Ba<br>Sr<br>Ca<br>Na | react with cold water to give hydroxide + hydrogen |
| | Mg<br>Al<br>Mn<br>Zn<br>Cr<br>Fe | react with steam to give oxide + hydrogen |
| | Cd<br>Co<br>Ni<br>Sn<br>Pb | react with very hot steam to give oxide + hydrogen |
| Less active metals; best oxidizing agents | H<br>Cu<br>Hg<br>Ag<br>Pt<br>Au | won't react with steam |

↑ Increasing ease of oxidation ↑     ↓ Increasing ease of reduction ↓

into each of the half-cells. Since hydrogen is a gas, however, an inert metal strip such as platinum is suspended into the hydrogen half-cell and the hydrogen gas is bubbled over it. When the platinum electrode bathed in hydrogen gas and $H^+$(aq) is connected to the zinc electrode, there is a flow of electrons. The electron flow results from the difference in oxidizing tendency of the hydrogen half-cell and the zinc half-cell.

A voltmeter is used to measure the difference in oxidizing tendency between two half-cells [Fig. 18.4(b)]. When the temperature is 25 °C, the pressure is 1.0 atm, and the concentrations of $H^+$(aq) and $Zn^{2+}$(aq) are both 1.0 M, the needle of the voltmeter connected between the two electrodes moves to the right to give a reading of 0.76 volt [Fig. 18.4(b)]. This is called the *potential difference*, and is a measure of the tendency for the overall cell reaction to occur. The 0.76 volt is due to both half-cell reactions, that is, oxidation and reduction, but if the oxidizing tendency of the hydrogen half-cell is arbitrarily assigned a value of 0.00 volt, the oxidizing tendency of the zinc half-cell can be assigned a value of 0.76 volt relative to the hydrogen half-cell. The sign of the potential is positive, +0.76 volt, which indicates that zinc metal is more easily oxidized than hydrogen gas.

## 18.5 Oxidation potentials

The voltages which are assigned to half-cells in this manner are called *oxidation potentials*. When two half-cells are connected by an external circuit, electrons flow *from* the half-cell with the higher oxidation potential *to* the half-cell with the lower oxidation potential.

The oxidation potential of the copper electrode can be determined in the same way. When the copper electrode dips into a 1.0 M solution of $Cu^{2+}$(aq), and the inert platinum electrode dips into a 1.0 M solution of $H^+$(aq), the needle of the voltmeter connected between the two electrodes moves to the left giving a reading of $-0.34$ volt. When these two half-cells are connected by an external wire, electrons will flow through the wire from the hydrogen half-cell to the copper half-cell. A negative sign on an oxidation potential indicates that the half-cell undergoes oxidation *less* readily than hydrogen. A positive sign on an oxidation potential indicates that the half-cell undergoes oxidation *more* readily than hydrogen.

The oxidation potentials of a number of half-cells are included in Table 18.2. The double arrows used in this table indicate that any of these half-reactions can be made to go in either direction under appropriate conditions. This means that a half-reaction may be an oxidation in one case but may function as a reduction in another case. The sign of an oxidation potential is positive or negative depending on the direction of the reaction in the half-cell. Standard oxidation potentials apply when the half-reaction is written as an oxidation. The *standard oxidation potential*, $E^o$, of a half-cell is the potential difference in volts between the specific half-cell and the hydrogen half-cell when all concentrations are 1.0 M, the temperature is 25 °C, and the pressure is 1.0 atm.

What is the difference in potential between the copper half-cell and the zinc half-cell in the cell which was discussed previously? The standard oxidation potentials, $E^o$, for these two half-cells can be found in Table 18.2.

TABLE 18.2

The oxidation potentials and half-reactions for some selected half-cells.

| Half reaction | Oxidation potential (volt) |
|---|---|
| $Li(s) \rightleftharpoons Li^+(aq) + e^-$ | +3.05 |
| $Na(s) \rightleftharpoons Na^+(aq) + e^-$ | +2.71 |
| $Mg(s) \rightleftharpoons Mg^{2+}(aq) + 2e^-$ | +2.37 |
| $Al(s) \rightleftharpoons Al^{3+}(aq) + 3e^-$ | +1.66 |
| $Zn(s) \rightleftharpoons Zn^{2+}(aq) + 2e^-$ | +0.76 |
| $H_2(g) \rightleftharpoons 2H^+(aq) + 2e^-$ | 0.00 |
| $Cu(s) \rightleftharpoons Cu^{2+}(aq) + 2e^-$ | −0.34 |
| $2I^-(aq) \rightleftharpoons I_2(s) + 2e^-$ | −0.54 |
| $Ag(s) \rightleftharpoons Ag^+(aq) + e^-$ | −0.80 |
| $Hg(l) \rightleftharpoons Hg^{2+}(aq) + 2e^-$ | −0.85 |
| $2Cl^-(aq) \rightleftharpoons Cl_2(g) + 2e^-$ | −1.36 |
| $2F^-(aq) \rightleftharpoons F_2(g) + 2e^-$ | −2.87 |

| Reaction | Standard oxidation potential, $E^o$ (volt) |
|---|---|
| Zn(s) | $Zn^{2+}(aq) + 2e^- \rightleftharpoons$  +0.76 |
| Cu(s) | $Cu^{2+}(aq) + 2e^- \rightleftharpoons$  −0.34 |

We see that the zinc half-cell has the larger positive oxidation potential and we know already that zinc is more easily oxidized than copper. *For a pair of half-cells, the half-cell which has the largest positive oxidation potential will be the half-cell which undergoes oxidation.* Zinc will undergo oxidation in this case; thus copper must be reduced.

Anode reaction: $Zn(s) \rightleftharpoons Zn^{2+}(aq) + 2e^-$  $E^o = +0.76$ volt
Cathode reaction: $Cu^{2+}(aq) + 2e^- \rightleftharpoons Cu(s)$  $E^o = +0.34$ volt

Note that the sign of the $E^o$ for the copper half-cell has changed since this half-reaction is now written as a reduction. The potential difference between the two half-cells of a cell is called the *voltage* of the cell. The voltage of the cell is the sum of the standard potentials for each half-cell when the half-cell reactions are written in the direction in which they actually go. The voltage of the zinc-copper cell is calculated by summing the appropriate half-cell potentials.

Anode reaction: $Zn(s) \longrightarrow Zn^{2+}(aq) + 2e^-$  $E^o = +0.76$ volt
Cathode reaction: $Cu^{2+}(aq) + 2e^- \longrightarrow Cu(s)$  $E^o = +0.34$ volt
Cell reaction: $Zn(s) + Cu^{2+}(aq) \longrightarrow Zn^{2+}(aq) + Cu(s)$  $E^o = +1.10$ volt

The calculated voltage of the zinc-copper cell is +1.10 volts. This is in agreement with the experimental value as shown in Fig. 18.5. The positive value of the cell voltage indicates that the reaction is spontaneous as written. A *spontaneous reaction* proceeds by itself without any outside force. *A redox reaction which has a positive cell voltage tends to take place as it is written.* If a cell voltage is negative, the opposite reaction will spontaneously occur. The list of reactions and appropriate half-cell potentials listed in Table 18.2 is sometimes called the electromotive series or simply the EMF series.

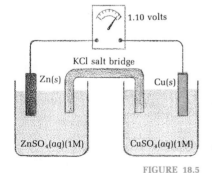

FIGURE 18.5
*The voltage measurement of the zinc-copper cell is +1.10 volts.*

EXAMPLE 1. If chlorine gas is bubbled into a solution of sodium bromide, will elemental bromine form? Write the balanced equation if the reaction occurs.

Given: (From Table 18.2)
$2Br^-(aq) \rightleftharpoons Br_2(g) + 2e^-$  $E^o = -1.09$ volts
$2Cl^-(aq) \rightleftharpoons Cl_2(g) + 2e^-$  $E^o = -1.36$ volts

Find: $Cl_2(g) + Na^+(aq) + Br^-(aq) \rightleftharpoons$ ?

*Solution:* Which is more easily oxidized, $Br^-$ or $Cl^-$? The $E^o$ value for the bromine half-cell, $-1.09$ volts, is a smaller negative value and thus a larger positive value than the $E^o$ value for the chlorine half-cell. Thus, the bromine half-cell will function as an oxidation in this case. The chlorine half-cell must undergo reduction.

| | | |
|---|---|---|
| Anode reaction: | $2Br^-(aq) \rightleftharpoons Br_2(g) + 2e^-$ | $E^o = -1.09$ volts |
| Cathode reaction: | $Cl_2(g) + 2e^- \rightleftharpoons 2Cl^-(aq)$ | $E^o = +1.36$ volts |
| Cell reaction: | $Cl_2(g) + 2Br^-(aq) \rightleftharpoons Br_2(g) + 2Cl^-(aq)$ | $E^o = +0.27$ volt |

The cell potential for this cell reaction is positive; thus the reaction proceeds spontaneously as written. Droplets of liquid bromine will form at the bottom of the container when chlorine gas is bubbled into a solution containing bromide ion. The complete equation for the reaction includes $2Na^+(aq)$ on both sides of the equation since $Na^+(aq)$ is a spectator ion which undergoes no change in this case.

■ EXERCISE *If chlorine gas is bubbled into a solution of sodium fluoride, does elemental fluorine form? Write the equation for any reaction that occurs. ($E^o$ for fluorine is $-2.87$ volts; for chlorine, $-1.36$ volts.)*

$F_2(g) + 2e^- \rightleftharpoons 2F^-(aq)$
$Cl_2(g) + 2e^- \rightleftharpoons 2Cl^-(aq)$
$Cl_2(g) + Na^+(aq) + F^-(aq) \rightleftharpoons ?$

$Cl_2(g) + 2Na^+(aq) + 2Br^-(aq) \rightleftharpoons$
$\qquad Br_2(g) + 2Cl^-(aq) + 2Na^+(aq)$ ■ (18.10)

EXAMPLE 3 If a piece of magnesium metal, Mg, is added to an aqueous solution of $AgNO_3$, will Ag metal be precipitated?

*Given:* $Ag(s) \rightleftharpoons Ag^+(aq) + e^-$ $\quad E^o = -0.80$
$Mg(s) \rightleftharpoons Mg^{2+}(aq) + 2e^-$ $\quad E^o = +2.37$

*Find:* Whether $Mg(s)$ is capable of reducing $Ag^+(aq)$ to $Ag(s)$.

*Solution:* Which has the higher positive oxidation potential? The oxidation potential for the magnesium half-cell, $+2.37$ volts, is a higher positive value than that of the silver half-cell, $-0.80$ volt; thus, the magnesium half-cell will undergo oxidation more readily than the silver half-cell.

| | | |
|---|---|---|
| Anode reaction: | $Mg(s) \rightleftharpoons Mg^{2+} + 2e^-$ | $E^o = +2.37$ volts |
| Cathode reaction: | $2Ag^+(aq) + 2e^- \rightleftharpoons 2Ag(s)$ | $E^o = +0.80$ volt |
| Cell reaction: | $2Ag^+(aq) + Mg(s) \rightleftharpoons 2Ag(s) + Mg^{2+}(aq)$ | $E^o = +3.17$ volts |

The positive value of the cell potential, $+3.17$ volts, indicates that the reaction is spontaneous as written. Note that the equation for the silver half-cell was multiplied by two in order to balance the electron change.

## 18.6 Balancing redox equations

**Redox equations**

A valid chemical equation satisfies three requirements. First, the equation must include all of the actual chemical species which have been experimentally shown to be involved in the chemical reaction. Second, the same mass must appear on both sides of the equation since mass is neither created nor destroyed in chemical reactions. Third, the electrical charge must be the same on both sides of the equation since electrons are neither created nor destroyed during a chemical reaction.

To be valid, a redox equation must satisfy the same three requirements:

1. Correct formulas must be written for all reactants and products.
2. The equation must be balanced for mass.
3. The equation must be balanced for charge.

Hydrogen gas is evolved when a piece of aluminum metal is placed in a 1 M solution of HCl. The aluminum metal is oxidized to $Al^{3+}(aq)$ while the $H^+(aq)$ is reduced to $H_2(g)$. The individual half-cell reactions for the reaction of hydrochloric acid with aluminum metal are:

Reduction: $2H^+(aq) + 2Cl^-(aq) + 2e^- \rightleftharpoons$
$$H_2(g) + 2Cl^-(aq) \quad (18.11)$$
Oxidation: $Al(s) \rightleftharpoons Al^{3+}(aq) + 3e^-$

Two electrons are gained in the reduction half-reaction but three electrons are lost in the oxidation half-reaction. The number of electrons gained in the reduction must equal the number of electrons lost in the oxidation since electrons are neither created nor destroyed in the process. An equation "balanced" for electron charge and for mass is obtained by multiplying all species in the reduction half-reaction by 3 and all of the species in the oxidation half-reaction by 2.

Reduction: $6H^+(aq) + 6Cl^-(aq) + 6e^- \rightleftharpoons 3H_2(g) + 6Cl^-(aq)$
Oxidation: $2Al(s) \rightleftharpoons 2Al^{3+}(aq) + 6e^-$
Cell reaction: $6H^+(aq) + 6Cl^-(aq) + 2Al(s) \rightleftharpoons$
$$3H_2(g) + 2Al^{3+}(aq) + 6Cl^-(aq) \quad (18.12)$$

The number of electrons gained, 6, is now equal to the number of electrons lost. Is the equation balanced for mass?

Dilute nitric acid reacts with metallic copper to form $Cu^{2+}(aq)$ and $NO(g)$. This reaction involves oxidation-reduction. For the reaction of dilute nitric acid with metallic copper, the skeletal equation is:

$$Cu + NO_3^-(aq) \rightleftharpoons Cu^{2+}(aq) + NO(g) \quad (18.13)$$

A skeletal equation indicates only the reacting species involved in a reaction. The individual half-cell reactions for the reaction of dilute nitric acid and of copper are:

Reduction: $NO_3^-(aq) + 4H^+(aq) + 3e^- \rightleftharpoons NO(g) + 2H_2O(l)$
Oxidation: $Cu(s) \rightleftharpoons Cu^{2+}(aq) + 2e^-$ (18.14)

The reactants and products of all species involved in the reaction are included in the two half-reactions. Two electrons are lost by each Cu but three electrons are gained by each $NO_3^-$. A balanced equation for the reaction is obtained by multiplying all species in the reduction half-reaction by 2, multiplying all the species in the oxidation half-reaction by 3, and then summing the two half-reactions.

Reduction: $2NO_3^-(aq) + 8H^+(aq) + 6e^- \rightleftharpoons 2NO(g) + 4H_2O(l)$
Oxidation: $3Cu(s) \rightleftharpoons 3Cu^{2+}(aq) + 6e^-$
Cell reaction: $2NO^-(aq) + 8H^+(aq) + 3Cu(s) \longrightarrow$
$2NO(g) + 3Cu^{2+}(aq) + 4H_2O(l)$ (18.15)

Note that the balanced equation is the equation for the net ionic reaction. The complete equation is obtained by adding enough $NO_3^-(aq)$ to each side to balance the $3Cu^{2+}(aq)$ formed. The $6NO_3^-(aq)$ are spectator ions.

Net equation: $2NO_3^-(aq) + 8H^+(aq) + 3Cu^0(s) \longrightarrow$
$2NO(g) + 3Cu^{+2}(aq) + 4H_2O(l)$
Complete equation: $8HNO_3(aq) + 3Cu^0(s) \longrightarrow$
$2NO(g) + 2Cu(NO_3)_2(aq) + 4H_2O(l)$ (18.16)

The equation for the redox reaction between copper and dilute nitric acid was balanced by multiplying each individual half-reaction by the appropriate coefficient necessary to balance the electrons and then adding the two half-reactions. A listing of half-reactions and their corresponding oxidation potentials can be found in reference texts. Many times, however, it is more convenient to write the necessary half-reactions yourself.

**Writing half-reactions**

The oxidation of molecular iodine, $I_2$, by potassium dichromate, $K_2Cr_2O_7$, in aqueous acidic solution to form iodate ion, $IO_3^-$ and chromium(III), $Cr^{3+}(aq)$, is a redox reaction which is of interest in analytical chemistry. The skeletal equation for this reaction is:

$$Cr_2O_7^{2-}(aq) + I_2(s) \rightleftharpoons Cr^{3+}(aq) + IO_3^-(aq) \quad (acidic) \quad (18.17)$$

Iodine is oxidized from 0 in $I_2$ to 5+ in $IO_3^-(aq)$. The oxidation of iodine from 0 to 5+ requires the loss of $5e^-$; thus oxidation of $I_2(s)$ to $2IO_3^-$ requires the loss of $10e^-$. A skeletal equation for the oxidation half-reaction can now be written:

$$\text{Oxidation:} \quad I_2(s) \rightleftharpoons 2IO_3^-(aq) + 10e^- \quad (acidic) \quad (18.18)$$

Note that there is a negative charge of $12e^-$ on the right side of the equation but zero charge on the left. *The charge must be balanced if the equation is valid.* What charged species is present in an acidic solution which could possibly be involved in the reaction? Hydrogen ion, which is present in acid solutions, is involved in the reaction; thus $H^+(aq)$ can be used to balance the charge.

$$\text{Oxidation:} \quad I_2(s) \rightleftharpoons 2IO_3^-(aq) + 12H^+(aq) + 10e^- \quad (acidic) \quad (18.19)$$

The equation is now balanced for charge but is it balanced for mass? There are 12 hydrogens on the right but none on the left. There are 6 oxygens on the right but none on the left. The solvent, water, is also involved in the reaction. Adding $6H_2O$ to the left side of the equation balances the equation for mass. Note that the equation can be balanced equally well using $H_3O^+$ instead of $H^+(aq)$.

$$\text{Oxidation:} \quad I_2(s) + 6H_2O(l) \rightleftharpoons 2IO_3^-(aq) + 12H^+(aq) + 10e^- \quad (acidic) \quad (18.20)$$

The method used for balancing the oxidation half-reaction can also be used for balancing the reduction half-reaction. What is the oxidation number of chromium in $Cr_2O_7^{2-}$? In $Cr^{3+}$? Chromium is reduced from an oxidation number of 6+ in $Cr_2O_7^{2-}$ to 3+ in $Cr^{3+}$. This reduction requires the gain of $3e^-$ by each chromium or $6e^-$ by the two chromiums in $Cr_2O_7^{2-}$.

$$\text{Reduction:} \quad Cr_2O_7^{2-}(aq) + 6e^- \rightleftharpoons 2Cr^{3+}(aq) \quad (acidic) \quad (18.21)$$

The charge on the left of the equation is $-8$ while that on the right is $+6$. Adding $14H^+(aq)$ to the left side balances the equation for charge.

$$\text{Reduction:} \quad Cr_2O_7^{2-}(aq) + 14H^+(aq) + 6e^- \rightleftharpoons 2Cr^{3+}(aq) \quad (acidic) \quad (18.22)$$

The equation is balanced for mass by placing $7H_2O$ on the right.

Reduction: $Cr_2O_7^{2-}(aq) + 14H^+(aq) + 6e^- \rightleftharpoons$
$$2Cr^{3+} + 7H_2O \quad \text{(acidic)} \quad (18.23)$$

The equations for both the oxidation and reduction half-reactions are "balanced." The two equations for the half-reactions can now be used to write the balanced equation for the reaction.

Oxidation: $I_2(s) + 6H_2O(l) \rightleftharpoons$
$$2IO_3^-(aq) + 10e^- + 12H^+(aq) \quad (18.24)$$
Reduction: $Cr_2O_7^{2-}(aq) + 14H^+(aq) + 6e^- \rightleftharpoons$
$$2Cr^{3+}(aq) + 7H_2O$$

Ten electrons are involved in the oxidation but only six in the reduction. Multiplying each of the species in the oxidation half-reaction by 3 and each of the species in the reduction half-reaction by 5 results in the same number of electrons, $30e^-$, involved in both half-reactions.

Oxidation: $3I_2(s) + 18H_2O(l) \rightleftharpoons$
$$6IO_3^-(aq) + 36H^+(aq) + 30e^-$$
Reduction: $5Cr_2O_7^{2-}(aq) + 70H^+(aq) + 30e^- \rightleftharpoons$
$$10Cr^{3+}(aq) + 35H_2O(l)$$

Cell reaction: $\overline{3I_2(s) + 5Cr_2O_7^{2-}(aq) + 18H_2O(l)}$
$+ 70H^+(aq) \rightleftharpoons 6IO_3^-(aq) + 10Cr^{3+}(aq)$
$$+ 35H_2O(l) + 36H^+(aq) \quad (18.25)$$

The equation can be simplified by subtracting $18H_2O$, $36H^+$, and $30e^-$ from both sides of the equation.

$$3I_2(s) + 5Cr_2O_7^{2-}(aq) + 34H^+(aq) \rightleftharpoons$$
$$6IO_3^-(aq) + 10Cr^{3+}(aq) + 17H_2O(l) \quad (18.26)$$

**Rules for balancing**

The procedure for balancing redox equations can be outlined as follows:

1. Use the skeletal equation for the redox reaction to write skeleton equations which represent the half-reactions.
2. Use the oxidation numbers of the elements in the various species to determine what is oxidized and what is reduced.
3. Add the appropriate number of electrons to the oxidation equation to bring about the oxidation of the oxidized species.
4. Add the appropriate number of electrons to the reduction equation to cause the reduction in oxidation number of the species reduced.

■ EXERCISE  Write balanced equations for the following redox reactions:
(a) $NO_2^-(aq) + Cr_2O_7^{2-}(aq) \rightleftharpoons NO_3^-(aq) + Cr^{3+}(aq)$ (acidic)
(b) $BrO_3^-(aq) + I^-(aq) \rightleftharpoons Br^-(aq) + I_2(s)$ (acidic)
(c) $Fe(s) + O_2(g) \rightleftharpoons Fe_2O_3(s)$

5. Balance each half-reaction separately by:
   (a) Balancing the charge. This is done by adding $H^+(aq)$ to the appropriate side of the equation for reactions occurring in acid solution. In basic solutions, the charge is balanced with $OH^-$.
   (b) Add $H_2O$ to the side necessary to balance H and O.
   (c) Check each half-reaction carefully to make sure that both charge and mass are balanced.
6. Multiply each half-reaction by the appropriate coefficient so that the number of electrons gained is equal to the number lost, then add the equations for the two half-reactions.
7. Duplications, if any, should be removed by adding or subtracting the appropriate number of the species concerned. ■

## 18.7 Stoichiometry of Redox reactions

**Equivalence**

How many moles of electrons are gained by one mole of $MnO_4^-$ during its reduction to $Mn^{2+}$? This question can be answered by considering the reduction half-reaction.

$$8H^+(aq) + MnO_4^-(aq) + 5e^- \rightleftharpoons Mn^{2+}(aq) + 4H_2O(l) \quad (18.27)$$

Mn(VII) is reduced to Mn(II) by the gain of $5e^-$. Thus, 1.00 mole of $MnO_4^-$ must gain 5.00 moles of electrons when 1.00 mole of $Mn^{2+}$ is produced.

It is convenient for the chemist to work with 1.00 equivalent of matter, that is, the quantity of matter which can accept or release 1.00 mole of electrons. The *equivalent weight* of a substance involved in a redox reaction is the weight of the substance which accepts or releases 1.00 mole of electrons in the reaction. What weight of permanganate ion, $MnO_4^-$, will accept 1.00 mole of $e^-$? The entire equation (18.28) is divided by 5.00 to obtain the weight of $MnO_4^-$ which will accept 1.00 mole of $e^-$.

$$\frac{1}{5.00} \text{ mole } MnO_4^- + 1.00 \text{ mole } e^- + \frac{8}{5.00} \text{ mole } H^+(aq) \longrightarrow \frac{1}{5.00} \text{ mole } Mn^{2+}(aq) + \frac{4}{5.00} \text{ mole } H_2O$$

The equivalent weight of $MnO_4^-$ for this reaction is the weight of 1.00 mole of $MnO_4^-$ divided by 5.00.

$$\frac{119 \text{ g MnO}_4^-}{1.00 \text{ mole MnO}_4^-} \times \frac{1.00 \text{ mole MnO}_4^-}{5.00 \text{ mole electrons}}$$
$$= \frac{23.8 \text{ g MnO}_4^-}{1.00 \text{ mole electrons}} = \frac{23.8 \text{ g MnO}_4^-}{\text{equivalent}}$$

The equivalent weight of a substance depends on the particular reaction in which the substance is involved. Three different equivalent weights for $MnO_4^-$ are listed below with their appropriate half-reactions.

$MnO_4^-(aq) + e^- \longrightarrow MnO_4^{2-}(aq)$  (18.28)
equivalent weight $MnO_4^- = 119 \text{ g}/1.00 = 119 \text{ g}$
$MnO_4^-(aq) + 3e^- + 2H_2O(l) \longrightarrow MnO_2(s) + 4OH^-(aq)$  (18.29)
equivalent weight $MnO_4^- = 119 \text{ g}/3.00 = 39.6 \text{ g}$
$MnO_4^-(aq) + 5e^- + 8H^+(aq) \longrightarrow Mn^{2+}(aq) + 4H_2O(l)$  (18.30)
equivalent weight $MnO_4^- = 119 \text{ g}/5.00 = 23.8 \text{ g}$

*Normality* is the unit of concentration used in stoichiometric calculations involving redox reactions. The amount of material equal to 1.00 equivalent weight is called an equivalent. The normality, N, of a solution is the number of equivalents which is dissolved in 1.00 liter of the solution. What would be the normality of a solution of $KMnO_4$ which contains 158.0 grams of $KMnO_4$ in 1.00 liter of solution?

There will be three possible normalities, depending on which half-reaction is considered. The weight of a mole of $KMnO_4$ is 158.0 grams. The equivalent weight of $KMnO_4$ in each of the three reactions considered previously can now be calculated.

$MnO_4^-(aq) + e^- \longrightarrow MnO_4^{2-}(aq)$
$$\text{eq. wt. } KMnO_4 = \frac{158 \text{ g KMnO}_4}{1.00 \text{ mole KMnO}_4} \times \frac{1.00 \text{ mole KMnO}_4}{1.00 \text{ mole electrons}}$$
$$= \frac{158 \text{ g KMnO}_4}{\text{equivalent}}$$

$MnO_4^-(aq) + 3e^- + 2H_2O(l) \longrightarrow MnO_2(s) + 4OH^-(aq)$
$$\text{eq. wt. } KMnO_4 = \frac{158 \text{ g KMnO}_4}{1.00 \text{ mole KMnO}_4} \times \frac{1.00 \text{ mole KMnO}_4}{3.00 \text{ mole electrons}}$$
$$= \frac{52.7 \text{ g KMnO}_4}{\text{equivalent}}$$

$MnO_4^-(aq) + 5e^- + 8H^+(aq) \longrightarrow Mn^{2+}(aq) + 4H_2O(l)$
$$\text{eq. wt. } KMnO_4 = \frac{158 \text{ g KMnO}_4}{1.00 \text{ mole KMnO}_4} \times \frac{1.00 \text{ mole KMnO}_4}{5.00 \text{ mole electrons}}$$
$$= \frac{31.6 \text{ g KMnO}_4}{\text{equivalent}}$$

■ EXERCISE  Calculate the equivalent weight of the indicated species in each of the following half-reactions.

(a) $\underline{ClO_3^-}$, $ClO_3^-(aq) + 6H^+(aq) + 6e^- \rightleftarrows Cl^- + 3H_2O$
(b) $\underline{Cu^{2+}}$, $Cu^{2+}(aq) + e^- \rightleftarrows Cu^+(aq)$
(c) $\underline{Cr_2O_7^{2-}}$, $Cr_2O_7^{2-}(aq) + 14H^+(aq) + 6e^- \rightleftarrows 2Cr^{3+}(aq) + 7H_2O$

**Redox titrations**

Redox titrations are widely used for quantitative analysis. Potassium permanganate, a powerful oxidizing agent, is often used in redox titrations. The intense purple color of $KMnO_4$ makes the end point visible since a slight excess of $KMnO_4$ gives a faint purple color which slowly fades away.

$Fe^{2+}$ is oxidized to $Fe^{3+}$ by $MnO_4^-$ in acidic solution. A solution of $KMnO_4$ can be used to titrate a sample of $Fe^{2+}$ which initially contains no $Fe^{3+}$.

Oxidation: $5Fe^{2+}(aq) \rightleftharpoons 5Fe^{3+}(aq) + 5e^-$ (acidic)

Reduction: $MnO_4^-(aq) + 8H^+(aq) + 5e^- \rightleftharpoons Mn^{2+}(aq) + 4H_2O(l)$ (acidic)

Cell reaction: $MnO_4^-(aq) + 5Fe^{2+}(aq) + 8H^+(aq) \rightleftharpoons 5Fe^{3+}(aq) + Mn^{2+}(aq) + 4H_2O(l)$ (18.32)

A 0.0258 gram sample of $KMnO_4$ was placed in a 100 ml volumetric flask and brought up to a volume of 100 ml with distilled water. What is the normality of this solution if it is used to oxidize $Fe^{2+}$ to $Fe^{3+}$ as indicated in the previous reaction?

$$\text{eq. wt. } KMnO_4 = \frac{158 \text{ g } KMnO_4}{1.00 \text{ mole } KMnO_4} \times \frac{1.00 \text{ mole } KMnO_4}{5.00 \text{ mole electrons}}$$

$$= \frac{31.6 \text{ g } KMnO_4}{1.00 \text{ mole electron}} = \frac{31.6 \text{ g } KMnO_4}{\text{equivalent}}$$

How many equivalents of $KMnO_4$ are there in 0.0258 gram of $KMnO_4$?

$$\text{number of eqs} = 0.0258 \text{ g} \times \frac{1.00 \text{ eq } KMnO_4}{31.6 \text{ g } KMnO_4}$$

$$= 0.00802 \text{ eq of } KMnO_4$$

What is the normality of a $KMnO_4$ solution which contains 0.00802 eq per 100 ml?

$$N_{KMnO_4} = \frac{0.0082 \text{ eq}}{0.100 \text{ liter}} = 0.0802 \text{ eq/liter}$$

What would be the number of equivalents of iron $Fe^{2+}$ in a 1.23 gram sample, which contains $Fe^{2+}$, plus inert substances, if 18.3 ml of 0.0802 N $KMnO_4$ is used to titrate the iron sample to a faint purple end point? The number of equivalents of $Fe^{2+}$ which have been oxidized at the end point is equal to the number of equivalents of $KMnO_4$ or $MnO_4^-$ which have been added.

eq $MnO_4^-$ added = (0.0802 eq/liter)(0.0183 liter) = 0.00147 eq

Since equivalents $KMnO_4$ = equivalents $Fe^{2+}$, the sample contained 0.00147 eq of $Fe^{2+}$.

What is the weight of iron in the 1.23 gram sample of iron and inert substances?

g Fe = (0.00147 eq) (55.8 g/eq) = $8.20 \times 10^{-2}$ g Fe

% Fe = $\dfrac{\text{g Fe}}{\text{g sample}} \times 100 = \dfrac{8.20 \times 10^{-2} \text{ g Fe}}{1.23 \text{ g}} \times 100 = 6.67\%$ Fe ∎

■ EXERCISE  Calculate the number of equivalents in 1.83 grams of $KMnO_4$ for each of the following half-reactions.

(a) $MnO_4^-(aq) + e^- \rightleftharpoons MnO_4^{2-}(aq)$
(b) $MnO_4^-(aq) + 3e^- + 2H_2O(l) \rightleftharpoons MnO_2(s) + 4OH^-(aq)$   (basic)

## 18.8  Redox reactions as a source of chemical energy

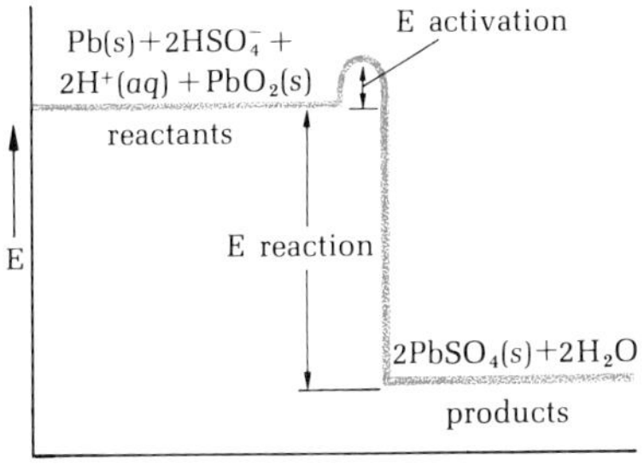

Redox reactions which occur spontaneously can be used as sources of chemical energy. A redox reaction in the lead storage battery, which is a chemical cell, furnishes the energy for starting your automobile. Flashlight batteries are also chemical cells. The fuel cells used on spacecraft are electrochemical cells. In electrochemical cells, chemical energy is converted to electrical energy. In other chemical reactions, chemical energy is converted primarily to heat energy.

**FIGURE 18.6**
*An energy diagram for the lead storage battery. Note the small energy of activation. An energy corresponding to the energy of reaction is given off when the reaction occurs. At least this much energy plus the energy of activation must be supplied to the battery in order to charge it.*

### The lead storage battery

Because of its convenience and dependability, the lead storage battery has become the most widely used chemical cell (Fig. 18.6). This battery consists of a spongy lead electrode, Pb, and a lead(IV) oxide electrode, $PbO_2$. The two electrodes dip into a solution of sulfuric acid. The half-reactions for these cells when the cell is producing electrical energy are:

Oxidation:  $Pb(s) + SO_4^{2-}(aq) \rightleftharpoons PbSO_4(s) + 2e^-$
Reduction:  $PbO_2(s) + SO_4^{2-}(aq) + 4H^+(aq) + 2e^- \rightleftharpoons PbSO_4(s) + 2H_2O(l)$
Cell reaction:  $\overline{Pb(s) + PbO_2(s) + 4H^+(aq) + 2SO_4^{2-}(aq) \rightleftharpoons 2PbSO_4(s) + 2H_2O(l)}$   (18.32)

What processes occur in the lead storage battery when the cell is producing an electrical current? The oxidation reaction indicates that Pb(s) is oxidized to $Pb^{2+}$ at the anode. The $Pb^{2+}$ formed at the anode then reacts with $SO_4^{2-}$ ion from the sulfuric acid to form solid $PbSO_4$. $Pb^{2+}$ is also formed by the reduction of $PbO_2$. This process occurs at the cathode. Because of these reactions, very slightly soluble $PbSO_4$ forms on both electrodes during the discharging or current-producing process. Note from the cell reaction that the concentration of

$H^+$(aq) and $SO_4^{2-}$(aq) must decrease during the discharge process. In a fully charged cell, the density of the sulfuric acid is usually about 1.25 grams per ml. The density of the sulfuric acid electrolyte in a completely discharged cell is about 1.10 grams per ml. Whether a battery needs to be charged can be determined by the density of sulfuric acid solution.

The utility of the lead storage battery is due to the fact that it can be recharged. Lead is formed at the anode when the cell reaction is reversed, while $PbO_2$ is formed at the cathode. The cell reaction for the charging process is:

$$2PbSO_4(s) + 2H_2O(l) \rightleftharpoons Pb(s) + PbO_2(s) + 4H^+(aq) + 2SO_4^{2-}(aq) \quad (18.33)$$

The cell reaction for the charging process occurs only when an external source of electrical energy is available to the cell. Three individual cells of 2.0 volts are connected in series in a 6-volt battery and six cells are connected in series in a 12-volt battery.


### Electrochemical cells and their uses

Rockets are propelled by the energy supplied from redox reactions. In this case chemical energy is transformed to heat energy. A rocket must have a fuel or reducing agent as well as an oxidizing agent. Many different fuels and many different oxidizing agents have been used in rockets. Liquid oxygen, a very strong oxidizing agent, is used to oxidize fuels such as hydrogen in many large rockets.

Chemical energy can be directly and efficiently converted to electrical energy with fuel cells. The fuel cell which is used to produce electrical energy on spacecraft is based on the combustion of hydrogen to form water. The half-reactions for the process are:

$$\begin{aligned}
\text{Oxidation:} \quad & 2H_2(g) + 4OH^-(aq) \rightleftharpoons 4H_2O(l) + 4e^- \\
\text{Reduction:} \quad & O_2(g) + 2H_2O(l) + 4e^- \rightleftharpoons 4OH^-(aq) \quad (18.34) \\
\text{Cell reaction:} \quad & 2H_2(g) + O_2(g) + 2H_2O(l) + 4OH^-(aq) \rightleftharpoons \\
& 4H_2O(l) + 4OH^-(aq) + \text{energy}
\end{aligned}$$

Removing duplications, the cell reaction becomes:

$$2H_2(g) + O_2(g) \rightleftharpoons 2H_2O(l) + \text{energy} \quad (18.35)$$

Astronauts use this electrochemical cell to produce the electrical energy which they need in outer space. They can also drink the $H_2O$ produced in the reaction.

*Summary*  Oxidation is any chemical change in which the oxidation number of an element is increased. This increase in oxidation number results from a loss or apparent loss of electrons. Reduction is any chemical change in which the oxidation number of an element is decreased or reduced. Reduction occurs with the gain or apparent gain of electrons.

Oxidation and reduction must occur simultaneously since electron gain by a substance cannot result without a corresponding loss of electrons by another substance. The number of electrons lost by one element must equal the number of electrons gained by another element; that is, charge is conserved in oxidation-reduction or redox reactions. The reducing agent is the substance that is oxidized, and the oxidizing agent is the substance which is reduced.

Each redox reaction may be divided into two separate half-reactions. One of the half-reactions represents the oxidation process; the other represents the process of reduction. Reaction vessels can be set up such that the individual half-reactions take place in a separate vessel or cell. The two half-reactions in the two separate cells must be connected by means of a salt bridge to allow a flow of ions. Electron flow results when the electrodes of the two cells are connected by a wire.

The driving force or potential of a half-reaction cannot be measured directly. Two half-cells are necessary for a cell reaction to occur; consequently any measurement of potential must involve two half-reactions. If a value is assumed for the potential of one half-cell, then the potentials of all other half-cells can be determined relative to the assumed potential. A potential of 0.00 volt was assumed for the hydrogen half-cell under a specified set of conditions. The potential for a half-reaction written as an oxidation is called an oxidation potential.

Whenever two half-reactions are combined, the cell reaction will be spontaneous as written only if the cell potential, $E^o$, for the over all reaction is positive. No appreciable reaction will occur if the cell potential for the over all reaction is negative.

The number of electrons given off in the oxidation half-reaction must equal the number of electrons gained in the reduction half-reaction before the two half-reactions are summed. Consequently, each half-reaction is multiplied by the proper coefficient such that the number of electrons gained will equal the number of electrons lost.

Redox reactions serve as useful sources of both electrical and heat energy. Electrical energy is supplied by electrochemical cells.

*Objectives*  By the end of this chapter, you should be able to do the following:

1. Assign oxidation numbers to any element in a compound. (Questions 2, 3)

2. Identify oxidation half-reactions, reduction half-reactions, oxidizing agents, and reducing agents. (Question 3)

3. Divide redox reactions into two separate half-reactions. (Question 4)

4. Write balanced equations for each half-reaction of a cell reaction from the skeletal equations. (Questions 3, 4, 7)

5. Use the balanced equations of the oxidation and reduction half-reactions to write the balanced equations for the cell reaction. (Questions 4, 7)

6. Determine whether a reaction is spontaneous given the cell reaction and the cell potential. (Questions 6, 7)

7. Calculate the cell potential for a given cell when the half-reactions and the half-cell potentials for each half-cell are given. (Questions 5, 7)

8. Use the half-cell potentials listed in the tables to predict whether cell reactions will be spontaneous. (Questions 5, 6, 7)

## Questions

1. Define and explain each of the following terms: *oxidation; reduction; oxidizing agent; reducing agent; half-cell reaction; cell reaction; redox reaction; lead storage battery; cell potential; half-cell potential*.

2. Assign oxidation numbers to each of the elements in the following: (a) $KMnO_4$; (b) $K_2Cr_2O_7$; (c) $H_3SbO_4$; (d) $MnO_4^{2-}$; (e) $MnO_2$; (f) $UO_2^{2+}$; (g) $Fe_2O_3$; (h) $ZnO_2^{2-}$; (i) $HCO_3^-$.

3. Write balanced equations for the half-reactions in the following skeletal equations. Which involve oxidation? Reduction? Identify the oxidizing or reducing agent.
 (a) $Cr_2O_7^{2-} \rightarrow Cr^{3+}$ (acidic)
 (b) $IO_3^- \rightarrow I_2$ (acidic)
 (c) $MnO_4^{2-} \rightarrow MnO_4^-$ (basic)
 (d) $Fe^{3+} \rightarrow Fe^{2+}$
 (e) $NO_3^- \rightarrow HNO_2$ (acidic)
 (f) $Mn(OH)_2 \rightarrow MnO_2$ (basic)
 (g) $V^{3+} \rightarrow VO^{2+}$ (acidic)
 (h) $MnO_2 \rightarrow MnO_4^-$ (acidic)
 (j) $H_2O_2 \rightarrow O_2$ (acidic)

4. Write balanced equations for the following cell reactions:
 (a) $ClO_3^-(aq) + H_2SO_3(aq) \rightarrow Cl^-(aq) + SO_4^{2-}(aq)$ (acidic)
 (b) $NO_3^-(aq) + H_2S(g) \rightarrow NO(g) + S(s)$ (acidic)
 (c) $AsO_3^{3-}(aq) + MnO_4^-(aq) \rightarrow AsO_4^{3-}(aq) + MnO_2(s)$ (basic)
 (d) $CrO_4^{2-}(aq) + S^{2-}(aq) \rightarrow S(s) + Cr(OH)_4^-(aq)$ (basic)
 (e) $Fe^{2+}(aq) + MnO_4^-(aq) \rightarrow Fe^{3+}(aq) + Mn^{2+}(aq)$ (acidic)
 (f) $I^-(aq) + NO_3^-(aq) \rightarrow I_2 + NO(g)$ (acidic)
 (g) $MnO_4^-(aq) + Sn(OH)_4(aq) \rightarrow MnO_2(s) + Sn(OH)^{2-}(aq)$ (basic)
 (h) $MnO_2(s) + Cl^-(aq) \rightarrow Mn^{2+}(aq) + Cl_2(g)$ (acidic)
 (i) $Al^0(s) + Cu^{2+}(aq) \rightarrow Al^{3+}(aq) + Cu^0(s)$
 (j) $Zn(s) + SO_4^{2-}(aq) \rightarrow Zn^{2+}(aq) + SO_2(g)$ (acidic)
 (k) $Sb^0(s) + NO_3^-(aq) \rightarrow Sb_2O_5(s) + NO(g)$ (acidic)

5. Calculate the standard potential, $E^o$, of the lead storage battery using the following half-cell potentials.

$$\text{PbSO}_4(s) + 2\text{H}_2\text{O}(l) \longrightarrow$$
$$\text{PbO}_2(s) + 4\text{H}^+(aq) + \text{SO}_4^{2-}(aq) + 2e^- \qquad E^o = -1.68 \text{ volts}$$
$$\text{Pb}^o(s) + \text{SO}_4^{2-}(aq) \longrightarrow \text{PbSO}_4(s) + 2e^- \qquad E^o = +0.36 \text{ volt}$$

6. Which of the following reactions will spontaneously occur? Rewrite all the nonspontaneous reactions in the direction in which they occur spontaneously.

(a) $\text{Zn}^o(s) + 2\text{Ag}^+(aq) \rightarrow \text{Zn}^{2+}(aq) + 2\text{Ag}^o(s)$ $\qquad E^o = -1.56$ volts
(b) $3\text{Fe}^{2+}(aq) + \text{Al}^{3+}(aq) \rightarrow 3\text{Fe}^{3+}(aq) + \text{Al}^o(s)$ $\qquad E^o = -2.43$ volts
(c) $\text{Cu}^o(s) + \text{Fe}^{2+}(aq) \rightarrow \text{Cu}^{2+}(aq) + \text{Fe}^o(s)$ $\qquad E^o = -0.78$ volt
(d) $\text{Sn}^o(s) + 2\text{H}^+(aq) \rightarrow \text{Sn}^{2+}(aq) + \text{H}_2(g)$ $\qquad E^o = -0.27$ volt
(e) $\text{Mg}^o(s) + \text{Ni}^{2+}(aq) \rightarrow \text{Mg}^{2+}(aq) + \text{Ni}^o(s)$ $\qquad E^o = +2.15$ volts

7. Use the half-cell potentials given in Table 18.2 to determine which of the following metals will replace hydrogen from a solution of HCl. Calculate the cell potential for the cell reaction. Write a balanced equation which occurs spontaneously. (a) Cu, (b) Ag, (c) Na, (d) Zn, (e) Mg, (f) Li, (g) Hg, (h) Al.

# 19

# Chemistry of periodic groups

## 19.1 Introduction

The models and concepts that have been developed in chemistry rely on observations and data obtained in the laboratory. In this chapter we shall summarize observations made on many elements in both the free and combined form. These elements will be studied by periodic groups.

The elements fall into two broad categories: metals and nonmetals. In general, the metals appear on the left side and near the bottom of the periodic chart. As we shall see in this chapter, the elements in any one group become more like metals as we progress down the group. The general properties of metals are contrasted with properties of nonmetals in Table 19.1.

In explaining the properties of different elements and compounds we must employ many of the concepts developed earlier: electronic structure, ionization energy, electronegativity, bond type, acid-base properties, solubility, and oxidation potential. For each group we discuss, we shall examine the general group properties as well as the individual behavior of the elements.

This chapter includes *inorganic chemistry*—the study of the chemistry of the elements and their compounds. The chemistry of most carbon compounds falls into a separate area called *organic chemistry*; this topic will be discussed in Chapter 20.

**TABLE 19.1**

*General properties of metals and nonmetals.*

| Property | Metals | Nonmetals |
|---|---|---|
| Charge on monoatomic ion | positive | negative |
| Ionization energy and electronegativity | low | high |
| Redox characteristics | reducing agents | oxidizing agents |
| Electrical conductivity | high | low |
| Luster, ductility, and malleability | high | low |
| Aqueous solutions of oxides | usually basic | usually acidic |

## 19.2 Group VIIIA—the noble gases

**Group characteristics**

The group VIIIA elements or *noble gases* are helium, neon, argon, krypton, xenon, and radon. Prior to 1960, no compound containing a noble gas had been identified, and therefore this group is sometimes referred to as the inert gases. They are also called the rare gases, but they are not rare. Argon alone makes up almost 1 percent by mass of the earth's atmosphere.

When Mendeleev proposed his periodic table, the noble gases had not yet been discovered. In 1894 Lord Rayleigh discovered argon and by 1900 the other members of the group had been discovered. Actually, the spectrum of helium had been detected in the sun in 1868, but it was assumed to be a metal. The discovery of this group did not upset Mendeleev's model. A new group was simply added to the right-hand end of the periodic table.

The valence electrons in the noble gases are not shielded much from the nucleus as compared with the elements with one higher atomic number. (Compare neon and sodium.) Consequently these elements have the highest ionization energies of each period and they do not lose electrons in chemical reactions. Because the noble gases have filled s and p orbitals (see Table 19.2), they have no room to gain electrons as the halogens do. This accounts for their general inertness.

As shown in Table 19.2 the size of the nobel gases increases down the group and the ionization energy decreases down the group. The only forces between atoms in a sample of a noble gas are weak van der Waals forces. This accounts for their low melting and boiling points.

| TABLE 19.2 | Property | Helium | Neon | Argon | Krypton | Xenon | Radon |
|---|---|---|---|---|---|---|---|
| Some properties of the noble gases. | Ionization energy (kcal/mole) | 566 | 495 | 364 | 323 | 280 | |
| | Valence electrons | $1s^2$ | $2s^2 2p^6$ | $3s^2 3p^6$ | $4s^2 4p^6$ | $5s^2 5p^6$ | $6s^2 6p^6$ |
| | Atomic radius (Å) | 0.93 | 1.12 | 1.54 | 1.69 | 1.90 | 2.2 |
| | Melting point (°C) | −270 | −249 | −189 | −157 | −112 | −71 |
| | Boiling point (°C) | −269 | −246 | −186 | −153 | −108 | −62 |

## Chemistry of the noble gases

Because of their general inertness, the noble gases are always found in the elemental form as monatomic gases. Helium is found in a number of oil fields. Neon, argon, krypton, and xenon are obtained from the atmosphere. Radon is radioactive and is formed as a product of radium disintegration.

In 1933 Linus Pauling proposed that xenon, because of its relatively low ionization energy, might easily form compounds with fluorine. The first attempts to do so were not successful, but by 1962 crystals of xenon tetrafluoride, $XeF_4$, were synthesized by mixing xenon and fluorine together at 400°C.

$$Xe(g) + 2F_2(g) \rightleftharpoons XeF_4(s) + 51.5 \text{ kcal} \tag{19.1}$$

Since then, over twenty other compounds of noble gases have been prepared.

FIGURE 19.1
The three fluorides of xenon.

The three known fluorides of xenon are stable compounds that are formed in exothermic reactions (Fig. 19.1). Several compounds of krypton have been formed, but they are unstable. Radon is difficult to work with because of its high radioactivity.

## Uses of noble gases

Both helium and hydrogen can be used to fill balloons, but helium is much safer because it does not burn. Chemical and mechanical operations such as the welding of metals and the heating of a light bulb filament are often carried out in an inert atmosphere of helium or some other noble gas. The noble gas does not react with the hot metal as oxygen or nitrogen might.

Temperatures close to 0°K can be reached using liquid

■ EXERCISE  Write a balanced equation for the reaction between xenon and fluorine to produce solid xenon hexafluoride.

helium. Near this temperature, many substances sho properties such as almost no resistance to electrica tivity.

Neon, argon, krypton, and xenon can be used in ..on" lights, fluorescent bulbs, and electronic flash bulbs. ■

## 19.3 Group VIIA— the halogens

**Oxyions and oxyacids**

The halogens have already been discussed in Chapter 11. Compounds whose negative ions contain a halogen and oxygen are called *halogen oxyions*. Since chlorine is the most common halogen, we shall examine its oxyions first.

When added to a cold solution of potassium hydroxide, chlorine gas reacts and forms hypochlorite ions and chloride ions.

$$2KOH(aq) + Cl_2(g) \rightleftharpoons KClO(aq) + KCl(aq) + H_2O(l) \quad (19.2)$$

If the same reactants are mixed at 70°C, the chlorine disproportionates into chlorate ions and chloride ions.

$$6KOH(aq) + 3Cl_2(g) \rightleftharpoons KClO_3(aq) + 5KCl(aq) + 3H_2O(l) \quad (19.3)$$

When potassium chlorate is heated carefully in the absence of a catalyst, the chlorine again disproportionates and forms chloride ions and perchlorate ions.

$$4KClO_3(s) \rightleftharpoons 3KClO_4(s) + KCl(s) \quad (19.4)$$

The stability of the chlorine oxyions increases as the number of oxygen atoms increases (Fig. 19.2). This is due in part to the dispersion of a negative charge on highly electronegative oxygen atoms and on increasingly larger ions.

Each of the chlorine oxyions can gain a proton and form an acid. The acids are not as stable as the salts, a property that makes them better oxidizing agents. The strength of the acid increases with the number of oxygens in the anion, perchloric acid being the strongest acid.

Oxyions of fluorine are unknown, perhaps because the electronegativity of fluorine is greater than oxygen. Bromine forms two oxyions: the hypobromite ion, $BrO^-$ and the bromate ion, $BrO_3^-$. Iodine forms three: $IO^-$, $IO_3^-$, and $IO_4^-$.

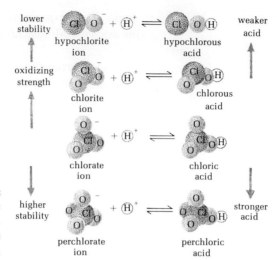

**FIGURE 19.2**
The stability of the chlorine oxyions increases with the number of oxygens. The strength of the acids increases from HClO to HClO$_4$.

### Interhalogens

Just as one chlorine atom can combine with a second chlorine atom to form a diatomic molecule of Cl$_2$, one chlorine atom can also combine with a fluorine, bromine, or iodine atom.

$$:\ddot{\underset{..}{Cl}}\cdot + \cdot\ddot{\underset{..}{Cl}}: \rightleftharpoons :\ddot{\underset{..}{Cl}}:\ddot{\underset{..}{Cl}}: \qquad (19.5)$$

$$:\ddot{\underset{..}{Cl}}\cdot + \cdot\ddot{\underset{..}{F}}: \rightleftharpoons :\ddot{\underset{..}{Cl}}:\ddot{\underset{..}{F}}: \qquad (19.6)$$

Compounds formed from two different halogen elements are called *interhalogens*, some samples are IF$_7$, BrF$_5$, and ClF$_3$. The smaller halogens are grouped about the larger one. Interhalogens are reactive compounds with many of the same chemical properties of pure halogens.

### Uses of the halogens

Fluorine compounds such as NaF, Na$_2$SiF$_6$, and SnF$_2$ are used in drinking water and in toothpastes to help prevent tooth decay. Small amounts are absorbed into the teeth and strengthen the structure of growing teeth. Compounds of carbon and fluorine, called *fluorocarbons*, form a wide variety of useful materials. The strong carbon–fluorine bonds make many fluorocarbons relatively inert. Freon, CCl$_2$F$_2$, is used for cooling in air conditioners and refrigerators. Teflon, C$_2$F$_4$, is a nonflammable plastic used in coating cooking utensils and in plastic tubing.

Chlorine is present in the body as chloride ions. Solutions of sodium hypochlorite, NaClO, are sold as household bleaching agents. Elemental chlorine is used to kill microorganisms in water treatment plants. Many insecticides, herbicides, and poison gases contain chlorine. Unfortunately these compounds often persist after they have done their job and may poison other life forms than those they were intended for.

Lead compounds are used in many gasoline fuels to prevent "knocking." To prevent an accumulation of solid lead compounds on the inside of the engine, bromine compounds are added to the fuel. Lead and bromine react to form volatile $PbBr_2$ which is vented into the atmosphere. Unfortunately, such lead compounds are highly toxic. Silver bromide, AgBr, reacts when exposed to light energy and consequently it is used in photography.

Iodine compounds, which are necessary for the proper functioning of the thyroid gland, may be taken into the body through seafood, drinking water in certain areas, or iodized salt. Iodized salt contains a small amount of KI mixed with NaCl. A solution of iodine, in the mixed solvent system of alcohol and water, is the common antiseptic, tincture of iodine. ■

■ EXERCISE  *Write equations for the preparation of a solution containing some chloric acid. Use $Cl_2$, $Ba(OH)_2$(aq), and $H_2SO_4$ as starting materials. Note that $BaSO_4$ is only slightly soluble.*

## 19.4 Group VIA— the oxygen family

**Group characteristics**

The group VIA elements are oxygen, sulfur, selenium, tellurium, and polonium (Table 19.3). Oxygen makes up about 49 percent of the earth's surface by weight. Sulfur is 0.05 percent selenium and tellurium about 0.0000001 or $10^{-7}$ percent. Sulfur is found both in compounds (usually metal sulfides) and in the elemental form. Trace amounts of selenium and tellurium which are chemically similar to sulfur are often found with sulfide ores. Polonium is a product of radioactive decay; because it is rare and highly radioactive, little is known about its chemistry.

Elemental oxygen occurs chiefly as a diatomic molecule, $O_2$. Sulfur occurs as a molecule of eight atoms in a ring, $S_8$. For the sake of simplicity, sulfur is written as a monoatomic atom in chemical equations. Selenium and tellurium and their compounds are poisonous and usually foul smelling.

In going down group VIA the electronegativities of the elements range from 3.5 for oxygen to about 2.0 for polonium. The

**TABLE 19.3** *Properties of the group VIA elements.*

| Property | Oxygen | Sulfur | Selenium | Tellurium | Polonium |
|---|---|---|---|---|---|
| Physical state at room temperature | gas | solid | solid | solid | solid |
| Ionization energy (kcal/mole) | 314 | 239 | 225 | 210 | 194 |
| Electronegativity | 3.5 | 2.5 | 2.4 | 2.1 | 2.0 |
| Valence electrons | $2s^2 2p^4$ | $3s^2 3p^4$ | $4s^2 4p^4$ | $5s^2 5p^4$ | $6s^2 6p^4$ |
| Atomic radius (Å) | 0.66 | 1.04 | 1.17 | 1.37 | |
| Melting point (°C) | −218 | 119 | 217 | 450 | 254 |
| Boiling point (°C) | −183 | 445 | 688 | 1390 | 965 |
| Oxidation state | 2− | 2−, 0 4+, 6+ | 2−, 0 4+, 6+ | 2−, 0 4+, 6+ | |
| Hydrogen compound | $H_2O$ | $H_2S$ | $H_2Se$ | $H_2Te$ | |
| Oxygen compounds | | $SO_2$ $SO_3$ | $SeO_2$ $SeO_3$ | $TeO_2$ $TeO_3$ | |

nonmetal properties of the elements decrease down the group. In fact polonium exhibits a luster, conducts electricity, and has a common oxidation state of 2+. Although tellurium is a poor conductor of electricity, it is $10^6$ times better than selenium.

How are the oxidation numbers of the group VIA elements related to their electronic configurations? Examine the configuration of sulfur:

$$_{16}S \quad [Ne] \quad \underset{3s^2}{(\uparrow\downarrow)} \quad \underset{3p^4}{(\uparrow\downarrow)(\uparrow)(\uparrow)}$$

By filling its two half-filled orbitals with electrons from metals or hydrogen, sulfur takes on a 2− oxidation number; examples are BaS or $H_2S$. By sharing the four valence electrons in the 3p orbitals with a nonmetal such as oxygen, sulfur takes on a 4+ oxidation number; examples are $SO_2$ or $H_2SO_3$. By sharing all six of its valence electrons (the 3s and 3p orbitals), sulfur takes on a 6+ oxidation number; examples are $SO_3$ or $H_2SO_4$.

**Reactions**

The first four elements in group VIA react with most metals to form the 2− anion.

$$2Ca(s) + O_2(g) \rightleftharpoons 2CaO(s) + \text{energy} \quad (19.7)$$
$$Zn(s) + S(s) \rightleftharpoons ZnS(s) + \text{energy} \quad (19.8)$$
$$Fe(s) + Se(s) \rightleftharpoons FeSe(s) + \text{energy} \quad (19.9)$$

The group VIA elements also react with nonmetals. The common oxidation states are 4+ and 6+ for sulfur, selenium, and tellurium.

$$S(s) + O_2(g) \rightleftharpoons SO_2(g) \quad (19.10)$$
$$S(s) + 3F_2(s) \rightleftharpoons SF_6(g) \quad (19.11)$$

**Oxycompounds**

The most commonly used compounds of sulfur, selenium, and tellurium are oxycompounds. Since sulfur can have oxidation numbers of 2−, 0, 4+, and 6+, much of the chemistry of sulfur involves oxidation-reduction reactions. If sulfur exists in the 6+ oxidation state it can be reduced to 4+, 0, or 2−. This is why $H_2SO_4$ and ionic sulfates are potential oxidizing agents.

*Sulfur 4+* Sulfur has a 4+ oxidation number in $SO_2$, $H_2SO_3$, and in sulfites. Sulfur dioxide is produced when sulfur or sulfur-containing compounds are burned. Some $SO_2$ is produced in the refining of metal sulfide ores.

$$2MnS(s) + 3O_2(g) \rightleftharpoons 2MnO(s) + 2SO_2(g) \quad (19.12)$$

When added to water, $SO_2(g)$ produces a weakly acidic solution.

$$H_2O(l) + SO_2(g) \rightleftharpoons H_2SO_3(aq) \quad (19.13)$$

If the $SO_2(g)$ is added to a basic solution, a sulfite salt forms.

$$2LiOH(aq) + SO_2(g) \rightleftharpoons Li_2SO_3(aq) + H_2O(l) \quad (19.14)$$

Sulfites, sulfur dioxide, and sulfurous acid are excellent reducing agents. For example, if any of these compounds is added to an acidic solutions of potassium permanganate, the purple color of the solution disappears. This is due to the reduction of the purple $MnO_4^-(aq)$ ion to the faint pink, almost colorless, $Mn^{2+}(aq)$ ion.

$$5H_2SO_3(aq) + KMnO_4(aq) \rightleftharpoons 2MnSO_4(aq) + K_2SO_4(aq)$$
$$+ 2SO_4^{2-}(aq) + 3H_2O(l) + 4H^+(aq) \quad (19.15)$$

Sulfurous acid is a weak acid, $K_i = 1.2 \times 10^{-2}$. It is a stronger acid than acetic acid, $K_i = 1.8 \times 10^{-5}$, or hydrofluoric acid, $K_i = 6.7 \times 10^{-4}$. Sulfurous acid solutions turn blue litmus to red, neutralize bases, and dissolve active metals. Many industrial areas have sulfurous acid present in the atmosphere from $SO_2$ gas reacting with water vapor in the air. This acid is a danger to living organisms because it is a strong reducing agent.

Selenium dioxide and tellurium dioxide are similar in behavior to sulfur dioxide. ■

■ EXERCISE  Use a table of oxidation potentials to look up two substances that an acidic aqueous solution of $SO_2$ will reduce. Write equations for these redox reactions.

*Sulfur 6+*  Sulfur dioxide gas can be oxidized to sulfur trioxide, a step in making sulfuric acid. The reaction occurs in the atmosphere in the presence of sunlight. In industry the oxidation is carried out by two different methods. In one case a gaseous mixture of $SO_2$ and $O_2$ is passed over a catalyst of $V_2O_5$. This is known as the *contact process*.

$$2SO_2(g) + O_2(g) \xrightleftharpoons{V_2O_5} 2SO_3(g) + 47.0 \text{ kcal} \tag{19.16}$$

When added to water, $SO_3$ forms sulfuric acid:

$$H_2O(l) + SO_3(g) \rightleftharpoons H_2SO_4(aq) \tag{19.17}$$

Over 58 billion pounds of sulfuric acid are produced annually in the United States. This compound is widely used to manufacture fertilizers, to clean metals, and to produce many other chemicals. Pure sulfuric acid is an oily liquid with a strong affinity for water and may be used as a drying agent.

The chemistry of sulfuric acid is based on two of its characteristics. First, it is a strong acid; second, it is a strong oxidizing agent. As an acid it can be used for neutralizing bases:

$$H_2SO_4(aq) + 2NaOH(aq) \rightleftharpoons Na_2SO_4(aq) + 2H_2O(aq) \tag{19.18}$$

Sulfuric acid acts best as an oxidizing agent when it is hot and concentrated. Even though copper is below hydrogen in the activity series, copper dissolves in hot $H_2SO_4$.

$$Cu(s) + 2H_2SO_4(s) \rightleftharpoons CuSO_4(aq) + SO_2(g) + 2H_2O(l) \tag{19.19}$$

The sulfur oxidizes the copper from 0 to 2+ and is itself reduced from 6+ to 4+.

Sulfates are salts that contain sulfur in the 6+ oxidation state. $CuSO_4$ is used to prevent or kill fungus; $MgSO_4 \cdot 7H_2O$ is sold as Epson salt; $CaSO_4 \cdot \frac{1}{2}H_2O$ is plaster of Paris. ■

■ EXERCISE  Write equations for the reaction of sulfuric acid with (a) $Mg(OH)_2$ and (b) MgO.

## 19.5 Group VA—the nitrogen family

**Group characteristics**

The group VA elements are nitrogen, phosphorus, arsenic, antimony, and bismuth. As in other groups, there is a trend from nonmetallic behavior to metallic behavior going down the group. Nitrogen, phosphorus, and arsenic are poor conductors of electricity and do not exhibit a metallic luster. Antimony and bismuth can be polished to a metallic luster and they conduct electricity (Table 19.4).

The ionization energy and electronegativity of the elements decreases down the group. The small nitrogen atom holds its electrons tightly; the larger bismuth atom has many shielding electrons and holds its electrons less tightly. Although nitrogen might be expected to be very reactive because of its small size and high electronegativity, it is almost inert at room temperature. This is because nitrogen occurs in the free state as a diatomic molecule. A strong triple bond holds the two atoms in the molecule together. The bond strength is 225 kcal/mole.

The general electronic configuration of the valence electrons

TABLE 19.4 Properties of the group VA elements.

| Property | Nitrogen | Phosphorus | Arsenic | Antimony | Bismuth |
|---|---|---|---|---|---|
| Physical state at room temperature | colorless gas | white to black solid | gray solid | solid metallic luster | solid, metallic luster |
| Ionization energy (kcal/mole) | 336 | 254 | 230 | 199 | 180 |
| Electronegativity | 3.0 | 2.1 | 2.0 | 1.9 | 1.9 |
| Valence electrons | $2s^2 2p^3$ | $3s^2 3p^3$ | $4s^2 4p^3$ | $5s^2 5p^3$ | $6s^2 6p^3$ |
| Atomic radius (Å) | 0.74 | 1.10 | 1.21 | 1.41 | 1.52 |
| Melting point (°C) | −210 | 44 | 814 | 630 | 271 |
| Boiling point (°C) | −196 | 280 | 610 | 1380 | 1470 |
| Oxidation state | 3− to 5+ | 3−, 3+, 5+ | 3+, 5+ | 3+, 5+ | 3+, 5+ |
| Hydrogen compound | $NH_3$ | $PH_3$ | $AsH_3$ | $SbH_3$ | $BiH_3$ |
| Chloride compound | $NCl_3$ | $PCl_3$ | $AsCl_3$ | $SbCl_3$ | $BiCl_3$ |

TABLE 19.5 Common oxidation states of nitrogen.

| Oxidation state | Formula | Name |
|---|---|---|
| 3− | $NH_3$ | Ammonia |
|    | $Li_3N$ | Lithium nitride |
| 2− | $N_2H_4$ | Hydrazine |
| 1− | $NH_2OH$ | Hydroxylamine |
| 0  | $N_2$ | Nitrogen |
| 1+ | $N_2O$ | Dinitrogen monoxide |
| 2+ | $NO$ | Nitrogen monoxide |
| 3+ | $N_2O_3$ | Dinitrogen trioxide |
| 4+ | $NO_2$ | Nitrogen dioxide |
| 5+ | $HNO_3$ | Nitric acid |

of the group VA elements is $ns^2np^3$. This configuration explains the common oxidation states of 3+ (the $np^3$ electrons are being used for bonding) and 5+ (the $ns^2np^3$ electrons are being used for bonding). Nitrogen has a high enough electronegativity to gain three electrons and take on a 3− oxidation state (Table 19.5).

**Hydrogen compounds**

Nitrogen reacts with hydrogen at elevated temperatures to form ammonia.

$$N_2(g) + 3H_2(g) \xrightleftharpoons{\text{catalyst}} 2NH_3(g) + 22 \text{ kcal} \quad (19.20)$$

Most nitrogen compounds are derived from ammonia made in this manner. The other group VA elements also form hydrogen compounds, but the compounds are progressively less stable down the group. The high electronegativity of nitrogen not only makes $NH_3$ stable but causes the molecule to be highly polar. Due to its polarity, ammonia has an abnormally high boiling point and great solubility in water.

**Oxides of group VA**

All the elements of group VA form oxides with the general formulas $X_2O_3$ and $X_2O_5$. Nitrogen also forms many other oxides.

The molecular formula of phosphorus(III) oxide is $P_4O_6$ and not $P_2O_3$ (Fig. 19.3). Phosphorus(V) oxide has the formula $P_4O_{10}$. This important chemical is made from naturally occurring phosphate rock.

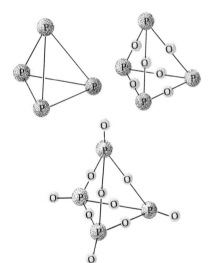

FIGURE 19.3 (a) Elemental phosphorus, $P_4$; (b) phosphorus(III) oxide, $P_4O_6$; (c) phosphorus(V) oxide, $P_4O_{10}$.

$$2Ca_3(PO_4)_2(s) + SiO_2(s) + 10C(s) \rightleftharpoons$$
$$6CaSiO_3(s) + 10CO(g) + P_4(g) \quad (19.21)$$
$$P_4(g) + 5O_2(g) \rightleftharpoons P_4O_{10}(s) \quad (19.22)$$

The oxide is shipped to chemical areas where phosphoric acid or phosphates are needed. Phosphoric acid is produced when water is added to $P_4O_{10}$.

$$P_4O_{10}(s) + 6H_2O(l) \rightleftharpoons 4H_3PO_4(aq) \quad (19.23)$$

Phosphates can be produced by adding a base to $P_4O_{10}$ or to $H_3PO_4$.

$$H_3PO_4(aq) + 3NH_3(aq) \rightleftharpoons (NH_4)_3PO_4(aq) \quad (19.24)$$

Large amounts of ammonium phosphate are used in fertilizers. This one compound supplies both nitrogen and phosphorus to plants.

### Nitric acid

Concentrated nitric acid is both a strong acid and a strong oxidizing agent. In the presence of $H_3O^+$ ions, $NO_3^-$ ions can oxidize most metals, nonmetals, and many compounds. The nitrogen can be reduced from 5+ to any lower oxidation state. The final products depend upon the reactants, their concentration, and the temperature.

$$3Cu(s) + 8HNO_3(aq) \rightleftharpoons$$
$$3Cu(NO_3)_2(aq) + 2NO(g) + 4H_2O(g) \quad (19.25)$$
$$4Zn(g) + 10HNO_3(aq) \rightleftharpoons$$
$$4Zn(NO_3)_2(aq) + NH_4NO_3(aq) + 3H_2O(l) \quad (19.26)$$
$$S(s) + 6HNO_3(aq) \rightleftharpoons$$
$$H_2SO_4(aq) + 6NO_2(g) + 2H_2O(l) \quad (19.27)$$

Mixed with hydrochloric acid, nitric acid forms a solution known as *aqua regia* or royal water. This solution is capable of dissolving gold. The nitric acid oxidizes the gold from 0 to 3+ and the chloride ion helps keep the gold(III) ion in solution by forming a complex ion.

$$Au(s) + 4HCl(aq) + 3HNO_3(aq) \rightleftharpoons$$
$$AuCl_4^-(aq) + 3NO_2(g) + H_3O^+(aq) + 2H_2O(l) \quad (19.28)$$

■ EXERCISE *Write the equation for the reaction between aqueous solutions of ammonia and nitric acid.*

Besides forming fertilizers, nitrates are widely used in making organic compounds and explosives. ■

## 19.6 Group IVA— the carbon family

**Group characteristics**

The elements in group IVA are carbon, silicon, germanium, tin, and lead. As in other groups, there is a gradual change from nonmetallic to metallic properties going down the group. Carbon and silicon are typical nonmetals. Carbon in the form of diamond does not conduct electricity; in the graphite form, however, it does. Silicon is a nonconductor and germanium is a semiconductor of electricity. Both tin and lead are good conductors (Table 19.6).

The high melting points of carbon and silicon are due to network bonding in their crystals. Elements in the group usually form covalent bonds with other elements by sharing two or four electrons. Common *oxidation states* are 2+ and 4+. Ions with a 4+ or 4− charge are rarely formed.

The sizes of the atoms of each element increase down the group. This is due to increasing valence-shell energy levels. The ionization energies and electronegativities decrease down the group. In tin and lead the valence electrons can easily move about in the solid, enabling them to conduct electricity easily.

**Hydrogen compounds**

Carbon forms many thousands of compounds with hydrogen. The simplest of these is methane, $CH_4$.

Carbon-hydrogen compounds are numerous. One carbon atom can form a strong bond with a second carbon atom and the number of carbon atoms that can bond together appears to be unlimited. Four simple examples are shown in Fig. 19.4. Carbon compounds are discussed in Chapter 20.

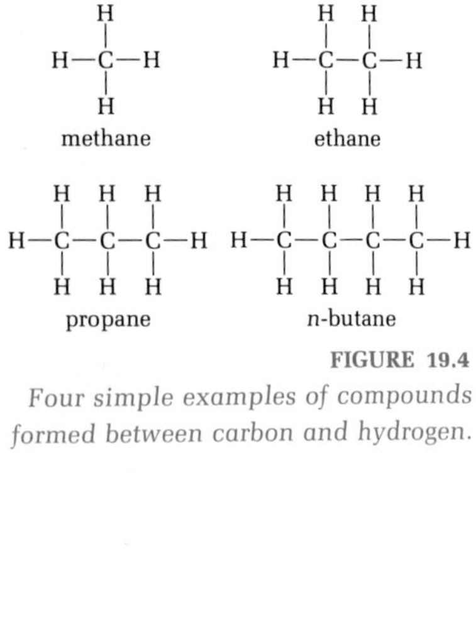

The hydrogen compounds of the other elements in group IV are progressively less stable going down the group. Silicon forms silane, $SiH_4$, with hydrogen. Germanium forms germane, $GeH_4$.

FIGURE 19.4
Four simple examples of compounds formed between carbon and hydrogen.

**Oxides**

All the elements in group IVA react with oxygen at elevated temperatures. Carbon, both as graphite and diamond, forms $CO_2(g)$ when there is an ample supply of oxygen:

$$C(s) + O_2(g) \rightleftharpoons CO_2(g) + \text{energy} \quad (19.29)$$

**TABLE 19.6**

*Properties of the group IV elements.*

| Property | Carbon | Silicon | Germanium | Tin | Lead |
|---|---|---|---|---|---|
| Physical state at room temperature | solid | solid | solid | solid | solid |
| Ionization energy (kcal/mole) | 260 | 188 | 182 | 169 | 171 |
| Electronegativity | 2.5 | 1.8 | 1.8 | 1.8 | 1.8 |
| Valence electrons | $2s^2 2p^2$ | $3s^2 3p^2$ | $4s^2 4p^2$ | $5s^2 5p^2$ | $6s^2 6p^2$ |
| Atomic radius (Å) | 0.77 | 1.17 | 1.22 | 1.41 | 1.54 |
| Melting point (°C) | 3570 | 1420 | 940 | 232 | 327 |
| Boiling point (°C) | 4200 | 2680 | 2830 | 2687 | 1750 |
| Oxidation state | 2+, 4+, 4− | 4+ | 2+, 4+ | 2+, 4+ | 2+, 4+ |
| Oxygen compound | CO, $CO_2$ | $SiO_2$ | GeO, $GeO_2$ | SnO, $SnO_2$ | PbO, $PbO_2$ |
| Chlorine compound | $CCl_4$ | $SiCl_4$ | $GeCl_4$ | $SnCl_2$, $SnCl_4$ | $PbCl_2$ |

Carbon dioxide is a colorless, odorless gas. Since it does not burn or support combusion, it is frequently used to extinguish fires. Being denser than air, the $CO_2$ blankets the fire area, keeping oxygen out.

$CO_2(g)$ is constantly vented into the air by animals. It is removed from the air by plants and used by them in photosynthesis. Some industrial areas are threatened by too much $CO_2$ produced by burning fuels. Too much or too little $CO_2$ in the air changes the pH of the blood and this may lead to unconsciousness or death.

$$CO_2(g) + 2H_2O(l) \rightleftharpoons HCO_3^-(aq) + H_3O^+(aq) \quad (19.30)$$

When the supply of oxygen is limited, carbon burns to form carbon monoxide:

$$2C(s) + O_2(g) \rightleftharpoons 2CO(g) + \text{energy} \quad (19.31)$$

This product will burn further in oxygen:

$$2CO(g) + O_2(g) \rightleftharpoons 2CO_2(g) + \text{energy} \quad (19.32)$$

Carbon monoxide gas is colorless, odorless, and poisonous. It combines with hemoglobin in the blood and prevents the hemoglobin from carrying oxygen to cells. As a result, the cells and the organism die. Carbon monoxide is produced in small amounts by most fires and in automobile engines.

Silicon dioxide, $SiO_2$, is the chief component of sand, quartz, flint, and most rocks. The silicon and oxygen atoms are bound together in a network. As a result, most silicon-oxygen compounds have high melting points of approximately 1700°C. Many rocks and clays contain silicon in complex formulas, such as $Al_2Si_2O_5(OH)_4$.

Lead and tin each form two oxides: $PbO$, $PbO_2$, $SnO$, and $SnO_2$. Lead(IV) oxide is used in automobile batteries. Despite its poisonous nature, lead is used in gasolines and plumbing. Lead was once used in paint, but laws have limited this practice.

## Carbonates

Large amounts of carbonate rocks are found in nature. Limestone and marble are calcium carbonate, $CaCO_3$. This naturally occurring chemical decomposes upon heating and produces calcium oxide and carbon dioxide.

$$CaCO_3(s) \rightleftharpoons CaO(s) + CO_2(g) \tag{19.33}$$

CaO produced in this manner is inexpensive and is a widely used base known as lime.

Carbonate ions and hydrogen carbonate ions are present in all naturally occurring waters and in body fluids. Carbon dioxide in the air reacts with water to form a weakly acidic solution:

$$CO_2(g) + 2H_2O(l) \rightleftharpoons H_3O^+(aq) + HCO_3^-(aq) \tag{19.34}$$

Rainfall is slightly acid due to the presence of $CO_2$; therefore, rain water dissolves limestone, $CaCO_3$:

$$CaCO_3(s) + H_3O^+(aq) \rightleftharpoons Ca^{2+}(aq) + HCO_3^-(aq) + H_2O(l) \tag{19.35}$$

Many areas of the United States contain caves, sink holes, and lakes formed by the dissolution of limestone.

## Radioactive carbon dating

Small amounts of radioactive carbon-14 are constantly being formed in the upper part of the earth's atmosphere by cosmic rays from space. The carbon-14 is formed by a nuclear reaction when a neutron collides with a nitrogen-14 nucleus:

$$^{14}_{7}N + ^{1}_{0}n \longrightarrow ^{14}_{6}C + ^{1}_{1}H \tag{19.36}$$

The carbon-14 thus formed mixes with the rest of the atmosphere, where small amounts of it are absorbed by all living species. When a living organism dies, no further carbon-14 is absorbed. The radioactive carbon present in the organism at death begins to decrease via the following reaction.

$$^{14}_{6}C \longrightarrow {}^{14}_{7}N + e^- \tag{19.37}$$

It takes 5580 years for half of the carbon-14 to disappear. Using this knowledge, scientists such as archeologists are able to determine approximately how long ago a living organism died.

## 19.7 Group IIIA— the boron family

**Group characteristics**

The group IIIA elements are boron, aluminum, gallium, indium, and thallium. Boron has nonmetallic properties and the other elements exhibit metallic properties. The general configuration for the elements is $ns^2np^1$. The three valence electrons are usually shared or lost, giving the elements a common oxidation state of 3+. The 1+ oxidation state is also common for thallium.

Pure crystalline boron is almost as hard as diamond. This is due to the network bonds in the solid. Boron is a poor conductor of electricity at room temperature. At higher temperatures its conductivity increases. This is a behavior typical of semiconductors.

Aluminum, gallium, indium, and thallium are typical metals. They all exhibit a metallic luster and conduct electricity. In fact, because of its light weight, aluminum has replaced copper in high voltage transmission lines. Gallium is solid at room temperature (23°C) but when held in the hand it melts into a silvery liquid resembling mercury.

The trends in ionization energy and electronegativity are irregular going down the group (Table 19.7). This is due in part to the fact that gallium, indium, and thallium are preceded by B group elements. Electrons are filling the $d$ orbitals of the B groups and these electrons do not shield the valence shell very well from the nucleus.

**Reactions with acids**

The four metals of group IIIA react with nonoxidizing acids such as HCl. They form aqueous salts and hydrogen gas.

TABLE 19.7

Properties of the group IIIA elements.

| Property | Boron | Aluminum | Gallium | Indium | Thallium |
|---|---|---|---|---|---|
| Physical state at room temperature | solid | solid | solid | solid | solid |
| Ionization energy (kcal/mole) | 191 | 138 | 138 | 133 | 141 |
| Electronegativity | 2.0 | 1.5 | 1.6 | 1.7 | 1.9 |
| Valence electrons | $2s^2 2p^1$ | $3s^2 3p^1$ | $4s^2 4p^1$ | $5s^2 5p^1$ | $6s^2 6p^1$ |
| Atomic radius (Å) | 0.80 | 1.25 | 1.25 | 1.50 | 1.55 |
| Melting point (°C) | 2300 | 660 | 29.8 | 156 | 449 |
| Boiling point (°C) | 2500 | 2500 | 2070 | 2100 | 1390 |
| Oxidation state | 3+ | 3+ | 3+ | 3+ | 3+, 1+ |
| Oxygen compounds | $B_2O_3$ | $Al_2O_3$ | $Ga_2O_3$ | $In_2O_3$ | $Tl_2O_3$, $Tl_2O$ |
| Chlorine compounds | $BCl_3$ | $Al_2Cl_6$ | $GaCl_3$ | $InCl_3$ | $TlCl_3$, $TlCl$ |

$$2Al(s) + 6HCl(aq) \rightleftharpoons 2AlCl_3(aq) + 3H_2(g) \qquad (19.38)$$
$$2Ga(s) + 6HCl(aq) \rightleftharpoons 2GaCl_3(aq) + 3H_2(g) \qquad (19.39)$$

Boron does not dissolve in HCl but it does dissolve in an oxidizing acid such as $HNO_3$ to produce boric acid.

$$B(s) + HNO_3(aq) + H_2O(l) \rightleftharpoons H_3BO_3(aq) + NO(g) \qquad (19.40)$$

Aluminum does not react with nitric acid. This is because, even in air, aluminum forms a tough protective coating of $Al_2O_3$. This oxide coating is thought to be formed in nitric acid also thus protecting the metal from further oxidation. Many other normally active metals form a protective oxide coating in nitric acid.

## Oxides

All elements in group IIIA react with oxygen at sufficiently high temperatures.

$$4B(s) + 3O_2(g) \rightleftharpoons 2B_2O_3(s) \qquad (19.41)$$
$$4Al(s) + 3O_2(g) \rightleftharpoons 2Al_2O_3(s) \qquad (19.42)$$

Gallium, indium, and thallium form $Ga_2O_3$, $In_2O_3$, and $Tl_2O_3$; thallium also forms $Tl_2O$. The oxides of gallium, indium, and thallium are, like other metallic oxides, basic. They can neutralize strong acids:

$$Ga_2O_3(s) + 6HCl(aq) \rightleftharpoons 2GaCl_3(aq) + 3H_2O(l) \quad (19.43)$$

Aluminum oxide can neutralize acids or bases. A substance that can neutralize both acids and bases is *amphoteric*.

$$Al_2O_3(s) + 6HCl(aq) \rightleftharpoons 2AlCl_3(aq) + 3H_2O(l) \quad (19.44)$$
$$Al_2O_3(s) + 2NaOH(aq) + 3H_2O(l) \rightleftharpoons$$
$$2Al(OH)_4^-(aq) + 2Na^+(aq) \quad (19.45)$$

Boron oxide is acidic. It reacts with water to form boric acid:

$$B_2O_3(s) + 3H_2O(l) \rightleftharpoons 2H_3BO_3(aq) \quad (19.46)$$

Boric acid is a weak acid; $K_i = 6 \times 10^{-10}$. It is commonly used as an eye wash. When added to a strong base, $H_3BO_3$ forms sodium tetraborate:

$$H_3BO_3(aq) + 2NaOH(aq) \rightleftharpoons Na_2B_4O_7(aq) + H_2O(l) \quad (19.47)$$

The hydrated form of $Na_2B_4O_7$ is known as borax, a useful cleansing agent. The tetraborate reacts with water (the reverse of Eq. 19.47) to form a low concentration of $OH^-$ ions in solution. Hydroxide ions help disperse dirt.

**Aluminum solutions**

When added to water, aluminum salts produce rather acidic solutions. Depending upon the concentration, the pH may be as low as 2. The hydronium ions are produced in a hydrolysis reaction between the hydrated aluminum ion $Al(H_2O)_6^{3+}$ and water molecules (Fig. 19.5).

$$Al(H_2O)^{3+}(aq) + H_2O(l) \rightleftharpoons$$
$$Al(H_2O)_5(OH)^{2+}(aq) + H_3O^+(aq) \quad (19.48)$$

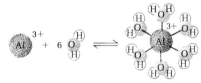

$$Al^{3+} + 6H_2O \rightleftharpoons Al(H_2O)_6^{3+}$$

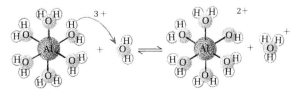

$$Al(H_2O)_6^{3+} + H_2O \rightleftharpoons Al(H_2O)_5(OH)^{2+} + H_3O^+$$

**FIGURE 19.5**
*Aqueous aluminum salts are acidic solutions. The hydrated aluminum ion reacts with water to produce $H_3O^+$ ions.*

When three moles of hydroxide are added per mole of aluminum ion in solution, a clear, gelatinous precipitate of $Al(OH)_3$ forms. Aluminum hydroxide is amphoteric; it reacts with and dissolves in either strong acids or strong bases.

$$Al(OH)_3(s) + 3HCl(aq) \rightleftharpoons AlCl_3(aq) + 3H_2O(l) \qquad (19.49)$$
$$Al(OH)_3(s) + NaOH(aq) \rightleftharpoons Al(OH)_4^-(aq) + Na^+(aq) \qquad (19.50)$$

### Boranes

Boron forms with hydrogen a number of highly combustible compounds called boranes. Although we might expect the simplest borane compound to have the formula $BH_3$, the observed molecule has the formula $B_2H_6$ (Fig. 19.6). Two of the hydrogen atoms appear to have two bonds. Other boranes are $B_5H_9$ and $B_6H_{10}$. These compounds are unstable and are potential high-energy rocket fuels. ■

■ EXERCISE  Aluminum and chlorine react to form a compound with the formula $Al_2Cl_6$. Write the equation for the reaction.

■ EXERCISE  Aluminum is a highly reactive metal. Explain why railroad tank cars made of aluminum can be used to ship nitric acid.

## 19.8 Group IIA—the alkaline earth family

### Group characteristics

The group IIA elements are beryllium, magnesium, calcium, strontium, barium, and radium. They are reactive metals and are always found in a combined state in nature.

The general electronic configuration for group IIA is $ns^2$. The elements readily lose their two valence electrons and form ions with a $2+$ charge. Because of its smaller size, beryllium tends to form covalent bonds. The chemical behavior of beryllium more closely resembles aluminum than the other group IIA elements.

Because the valence energy level increases down the group, the valence electrons are more easily removed as one progresses down the group. Both ionization energies and electronegativities decrease down the group (Table 19.8).

Radium is a product of radioactive disintegration and is itself radioactive. The radiation emitted by this element is used to help destroy cancerous tissues. Radium is found in small amounts in uranium ores.

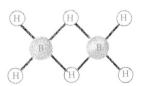

FIGURE 19.6  Diborane, $B_2H_6$. Two of the six hydrogens in the molecule appear to be bonded to two boron atoms. The electrons in these bonds are believed to travel about three different nuclei—the hydrogen nucleus and the nuclei of the two boron atoms.

### Activity

Except for beryllium, the group IIA elements react with water, releasing hydrogen. Calcium, strontium, barium, and radium

**TABLE 19.8** Properties of the group IIA elements.

| Property | Beryllium | Magnesium | Calcium | Strontium | Barium | Radium |
|---|---|---|---|---|---|---|
| Physical state at room temperature | solid | solid | solid | solid | solid | solid |
| Ionization energy (kcal/mole) | 215 | 176 | 141 | 131 | 120 | |
| Electronegativity | 1.5 | 1.2 | 1.0 | 1.0 | 0.9 | 0.9 |
| Valence electrons | $2s^2$ | $3s^2$ | $4s^2$ | $5s^2$ | $6s^2$ | $7s^2$ |
| Atomic radius (Å) | 0.89 | 1.36 | 1.74 | 1.91 | 1.98 | |
| Melting point (°C) | 1280 | 650 | 850 | 770 | 700 | 700 |
| Boiling point (°C) | 1500 | 1120 | 1490 | 1380 | 1640 | 1500 |
| Oxidation state | 2+ | 2+ | 2+ | 2+ | 2+ | 2+ |
| Oxygen compound | BeO | MgO | CaO | SrO | BaO | RaO |
| Chlorine compound | $BeCl_2$ | $MgCl_2$ | $CaCl_2$ | $SrCl_2$ | $BaCl_2$ | $RaCl_2$ |

react with cold water to produce basic solutions. Magnesium reacts only with hot water or steam:

$$Mg(s) + H_2O(g) \rightleftharpoons MgO(s) + H_2(g) \tag{19.51}$$
$$Ca(s) + 2H_2O(l) \rightleftharpoons Ca(OH)_2(aq) + H_2(g) \tag{19.52}$$

The elements in group IIA react with most nonmetals to produce ionic salts:

$$Ca(s) + S(s) \rightleftharpoons CaS(s) \tag{19.53}$$
$$Mg(s) + Cl_2(g) \rightleftharpoons MgCl_2(s) \tag{19.54}$$

The metals are strong reducing agents and replace less active metals from their compounds. The reaction involves electron transfer:

$$Mg(s) + Cu(NO_3)_2(aq) \rightleftharpoons Cu(s) + Mg(NO_3)_2(aq) \tag{19.55}$$

### Oxides

Magnesium oxide reacts to a slight degree with water to produce slightly soluble magnesium hydroxide. A suspension of magnesium hydroxide, known as milk of magnesia, is used as a laxative and for treating excess stomach acid.

$$MgO(s) + H_2O(l) \rightleftharpoons Mg(OH)_2(s) \quad (19.56)$$
$$Mg(OH)_2(s) + 2HCl(aq) \rightleftharpoons MgCl_2(aq) + 2H_2O(l) \quad (19.57)$$

Calcium hydroxide is soluble enough in water to make a weakly basic solution. If the hydroxide ions in equilibrium with solid calcium hydroxide are neutralized by an acid, the equilibrium shifts and more hydroxide ions form.

$$Ca(OH)_2(s) \rightleftharpoons Ca^{2+}(aq) + 2OH^-(aq) \quad (19.58)$$
$$+$$
$$H_3O^+(aq) \longrightarrow 2H_2O(l) \quad (19.59)$$

Solutions of calcium hydroxide, known as lime water, can be used for the detection of carbon dioxide gas. Consider the apparatus shown in Fig. 19.7. The glowing light bulb indicates that the dilute solution of calcium hydroxide contains ions. When carbon dioxide, from a person's lungs, is passed through the solution for a few minutes, the light goes out and a white precipitate forms. As more carbon dioxide is passed through the solution, the light goes on again and the white precipitate dissolves. What reactions are occurring?

The carbon dioxide gas neutralizes the calcium hydroxide and forms solid calcium carbonate and water:

$$Ca^{2+}(aq) + 2OH^-(aq) + CO_2(g) \rightleftharpoons CaCO_3(s) + H_2O(l) \quad (19.60)$$

The removal of ions from the solution causes the light to go out. The addition of more $CO_2$ causes the formation of soluble calcium hydrogen carbonate.

$$CaCO_3(s) + CO_2(g) + H_2O(l) \rightleftharpoons Ca(HCO_3)_2(aq) \quad (19.61)$$

The presence of calcium ions and hydrogen carbonate ions in the solution causes the light bulb to glow again.

The reaction between calcium hydroxide and carbon dioxide occurs in the formation of mortar in building construction. A watery mixture of calcium hydroxide is mixed with sand and other materials and left to set. Carbon dioxide from the air reacts with the hydroxide to form the solid carbonate.

$$Ca(OH)_2(s) + CO_2(g) \rightleftharpoons CaCO_3(s) + H_2O(l) \quad (19.62)$$

The oxides of strontium and barium form solutions of strong bases when added to water. The hydroxides of these elements react with $CO_2$ in a manner similar to $Ca(OH)_2$.

$$BaO(s) + H_2O(l) \rightleftharpoons Ba(OH)_2(aq) \quad (19.63)$$
$$Ba(OH)_2(aq) + CO_2(aq) \rightleftharpoons BaCO_3(s) + H_2O(l) \quad (19.64)$$

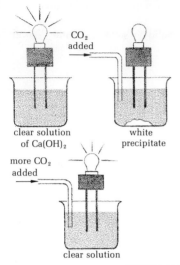

FIGURE 19.7

When $CO_2(g)$ is passed into $Ca(OH)_2$ solution, a precipitate forms and the light goes out. As more $CO_2$ is passed through the solution, the precipitate redissolves and the light goes on again.

■ EXERCISE  *Calcium oxide removes $CO_2$ from the air and forms $CaCO_3$. Calcium oxide can also be used to remove $SO_2$ and $SO_3$ from the air, the products being calcium sulfite and calcium sulfate. Write balanced equations for the reaction of CaO with $SO_2$ and $SO_3$.*

**Hard water**

Deposits of calcium carbonate, calcium sulfate, and calcium phosphate are widely distributed on the earth's surface. Although these compounds are only slightly soluble, enough calcium ions do dissolve in natural waters to produce what is known as hard water. The calcium ions in hard water react with negative ions in soaps and form precipitates or curds:

$$Ca^{2+}(aq) + 2C_{17}H_{35}CO_2^-(aq) \rightleftharpoons Ca[C_{17}H_{35}O]_2(s) \qquad (19.65)$$

The formation of these precipitates uses up some of the soap; thus more soap is required in hard water. These precipitates cling to surfaces and cause what is known as "soap scum." ■

## 19.9 Group B elements—the transition elements

**General characteristics**

The B group elements are known as *transition elements*. We will include the copper and zinc families in this group (Table 19.9). Elements 21 through 30 belong to the *first transition series*.

In this series electrons are filling the 3d orbitals. Elements 39 through 48 belong to the *second transition series*; here electrons are filling the 4d orbitals. Elements 72 through 80 belong to the *third transition series*; here electrons are filling the 5d orbitals. Elements 105 and higher are presently being formed in nuclear laboratories. Element 104 is expected to start a new transition series beginning below hafnium on the periodic table. What orbitals are probably being filled by this series?

All the transition elements are metals. They are good conductors of electricity and have a metallic luster. Copper and gold have their own characteristic color.

In going from scandium through zinc in the first transition series, electrons enter the inner 3d orbitals. These electrons do

| TABLE 19.9 | | IIIB | IVB | VB | VIB | VIIB | VIIIB | | | IB | IIB |
|---|---|---|---|---|---|---|---|---|---|---|---|
| The B group elements. | First transition series | $_{21}$Sc | $_{22}$Ti | $_{23}$V | $_{24}$Cr | $_{25}$Mn | $_{26}$Fe | $_{27}$Co | $_{28}$Ni | $_{29}$Cu | $_{30}$Zn |
| | Second transition series | $_{39}$Y | $_{40}$Zr | $_{41}$Nb | $_{42}$Mo | $_{43}$Tc | $_{44}$Ru | $_{45}$Rh | $_{46}$Pd | $_{47}$Ag | $_{48}$Cd |
| | Third transition series | $_{57}$La | $_{72}$Hf | $_{73}$Ta | $_{74}$W | $_{75}$Re | $_{76}$Os | $_{77}$Ir | $_{78}$Pt | $_{79}$Au | $_{80}$Hg |

**TABLE 19.10**

Properties of the elements of the first transition series.

| Element | Electronic configuration | Common oxidation state | Ionization energy (kcal/mole) | Electro-negativity | Density (g/ml) |
|---|---|---|---|---|---|
| Scandium | $3d^1 4s^2$ | 3+ | 150 | 1.3 | 2.5 |
| Titanium | $3d^2 4s^2$ | 4+ | 158 | 1.5 | 4.5 |
| Vanadium | $3d^3 4s^2$ | 2+, 3+, 4+, 5+ | 156 | 1.6 | 6.0 |
| Chromium | $3d^5 4s^1$ | 3+, 6+ | 156 | 1.6 | 7.1 |
| Manganese | $3d^5 4s^2$ | 2+, 3+, 4+, 7+ | 171 | 1.5 | 7.2 |
| Iron | $3d^6 4s^2$ | 2+, 3+ | 182 | 1.8 | 7.9 |
| Cobalt | $3d^7 4s^2$ | 2+, 3+ | 180 | 1.8 | 8.9 |
| Nickel | $3d^8 4s^2$ | 2+ | 176 | 1.8 | 8.9 |
| Copper | $3d^{10} 4s^1$ | 1+, 2+ | 178 | 1.9 | 8.9 |
| Zinc | $3d^{10} 4s^2$ | 2+ | 216 | 1.6 | 7.1 |

not shield the 4s electrons completely from the nucleus. Consequently in moving across this series the increasing nuclear change holds the valence electrons more tightly. See Table 19.10 for ionization energies and electronegativities of this series.

The melting points, boiling points, and densities of the transition elements vary greatly. Mercury melts at $-38.9°C$ but tungsten does not melt until a temperature of $3380°C$ is reached. Titanium, a light weight structural metal, has a density of 4.5 grams per ml. Scandium is lighter than aluminum; its density is 2.5 grams per ml. Osmium, on the other hand, has a density of 22.5 grams per ml, the highest of any known material.

**Oxidation states**

Many of the transition metals have several stable oxidation states. For instance, vanadium has oxidation states of 2+, 3+, 4+, and 5+. What are the oxidation states of manganese in the following compounds: $MnCl_2$, $Mn_2O_3$, $MnO_2$, $K_2MnO_4$, $KMnO_4$?

In aqueous solution or as solid hydrated salt the particular oxidation state of each element often has a characteristic color. A solution containing $Mn^{2+}(aq)$ has a faint pink color; $MnO_4^-(aq)$ ions have a deep purple color; $Fe^{2+}(aq)$ is pale green, and $Fe^{3+}(aq)$ is pale yellow. These characteristic colors of transition metals ions are due to electron transitions in the $d$ orbitals. Energy, in the form of visible light, is given off when electrons move from one $d$ orbital to a lower energy $d$ orbital.

When iron is acted upon by $HCl(aq)$ it dissolves and forms the $Fe^{2+}(aq)$ ion:

$$Fe(s) + 2HCl(aq) \rightleftharpoons FeCl_2(aq) + H_2(g) \qquad (19.66)$$

When iron reacts with a stronger oxidizing agent than $H_3O^+(aq)$ the 3+ may be formed. For instance, the action of chlorine on iron produces iron(III) chloride.

$$2Fe(s) + 3Cl_2(g) \rightleftharpoons 2FeCl_3(s) \tag{19.67}$$

**Oxides**

Nonmetal oxides are usually acidic and metal oxides are usually basic. When a metal has a high oxidation state, it has a greater attraction for electrons than might be expected. Such is the case of chromium in $CrO_3$. Instead of being basic, chromium (VI) oxide is acidic. Chromium acts as if it were a nonmetal. When placed in water, $CrO_3$ reacts to form chromic acid:

$$CrO_3(s) + H_2O(l) \rightleftharpoons H_2CrO_4(aq) \tag{19.68}$$

The change in acid-base character with oxidation number is well illustrated by the vanadium oxides:

| VO | basic | $VO_2$ | amphoteric |
|---|---|---|---|
| $V_2O_3$ | basic | $V_2O_5$ | acidic |

**Inner transition elements**

Elements that have electrons filling their $4f$ or $5f$ orbitals are known as *inner transition elements*. Elements 58 through 71 belong to the lanthanide series and elements 90 through 103 belong to the actinide series. The lanthanide elements are also called the *rare earths* although many of them are not rare. Many of the actinide elements are man-made. Because of the similarity of their valence shells (see Table 19.11) the elements in the lanthanide series have very similar chemical properties. The electronic configurations of these elements are difficult to determine from data and are not definitely known.

The nuclei of many of the actinide elements are radioactive. For example, $^{238}_{92}U$ decays into $^{234}_{90}Th$ by emitting an alpha particle (helium nucleus):

■ EXERCISE  Tungsten has a high melting point. What is the pratical application of this property?

$$^{238}_{92}U \longrightarrow {}^{234}_{90}Th + {}^{4}_{2}He \tag{19.69}$$

The thorium produced is radioactive and it decays through a series of nuclear reactions to $^{206}_{82}Pb$. The time required for half of a given amount of a radioactive isotope to decompose is called its *half-life*. The half-life of $^{238}_{92}U$ is 4.5 billion years. The half-life of one isotope of element 105 is about 1.6 seconds. ■

■ EXERCISE  Assume that copper and gold had the same color and were equally scarce. What chemical property would make gold more valuable?

Some nuclei split into two smaller nuclei when struck by other particles. This process is called *fission*. For example, $^{235}_{92}U$ can react as follows with a neutron:

TABLE 19.11 Electronic configuration of the lanthanide series.

| Element | Atomic number | Electronic configuration | Electronegativity | Common oxidation state |
|---|---|---|---|---|
| Lanthanium | 57 | [Xe] $5d^1 6s^2$ | 1.1 | 3+ |
| Cerium | 58 | [Xe] $4f^2 6s^2$ | 1.1 | 3+, 4+ |
| Praseodymium | 59 | [Xe] $4f^3 6s^2$ | 1.1 | 3+, 4+ |
| Neodymium | 60 | [Xe] $4f^4 6s^2$ | 1.2 | 3+ |
| Promethium | 61 | [Xe] $4f^5 6s^2$ |  | 3+ |
| Samarium | 62 | [Xe] $4f^6 6s^2$ | 1.2 | 2+, 3+ |
| Europium | 63 | [Xe] $4f^7 6s^2$ |  | 2+, 3+ |
| Gadolinium | 64 | [Xe] $4f^7 5d^1 6s^2$ | 1.1 | 3+ |
| Terbium | 65 | [Xe] $4f^9 6s^2$ | 1.2 | 3+, 4+ |
| Dysprosium | 66 | [Xe] $4f^{10} 6s^2$ |  | 3+ |
| Holmium | 67 | [Xe] $4f^{11} 6s^2$ | 1.2 | 3+ |
| Erbium | 68 | [Xe] $4f^{12} 6s^2$ | 1.2 | 3+ |
| Thubium | 69 | [Xe] $4f^{13} 6s^2$ | 1.2 | 2+, 3+ |
| Ytterbium | 70 | [Xe] $4f^{14} 6s^2$ | 1.1 | 2+, 3+ |
| Lutetium | 71 | [Xe] $4f^{14} 5d^1 6s^2$ | 1.2 | 3+ |

$$^{235}_{92}\text{U} + ^{1}_{0}\text{n} \longrightarrow ^{93}_{36}\text{Kr} + ^{140}_{56}\text{Ba} + 3\,^{1}_{0}\text{n} \tag{19.70}$$

The large amounts of energy released by such reactions can be used for the advancement or destruction of mankind.

## Summary

Elements in any one group of the periodic table have many similar properties. Starting at the top of a group and going down, the difference in physical and chemical behavior is usually gradual. In going down a group the valence shells are usually farther from the nucleus. This gives elements at the bottom of groups lower ionization energies and lower electronegativities. The most metallic element in each group is found at the bottom; the most nonmetallic element in each group is found at the top.

The group VIIIA elements are known as the noble gases. These elements have high ionization energies and no half-empty orbitals in their valence energy levels. Consequently, these elements do not readily form compounds with many elements. Xenon does, however, react readily with fluorine to give either $XeF_2$, $XeF_4$, or $XeF_6$.

The group VIIA elements or halogens are reactive nonmetallic elements. The halogens react readily with most metals, forming salts. Halogens form a number of ternary compounds containing (1) a halogen, (2) oxygen, and (3) a metal or hydrogen. The hydrogen-containing compounds are acids. The more oxygen these acids contain the stronger they are.

The group VIA elements, except for polonium, are nonmetals. They react with metals to produce binary salts. Sulfur reacts with oxygen to form two different compounds, $SO_2$ and $SO_3$. Both are common causes of air pollution.

Nitrogen and phosphorus from group VA are two elements that make up a large part of the soil nutrients essential for life. Nitrogen alone is unreactive and must be added to the soil in the form of ammonia or ammonium compounds, or as nitrate. Phosphorus is a reactive nonmetal. Oxides of both nitrogen and phosphorus are acidic. Nitric acid is a strong oxidizing agent.

The group IVA elements illustrate well the change from nonmetallic to metallic in going down a group. Carbon and silicon are nonmetals. Germanium is a metalloid. Tin and lead are primarily metallic. All known living tissue consists primarily of carbon compounds. The 1 percent of $CO_2$ present in the atmosphere is essential for plant life.

The group IIIA elements are mostly metallic. Boron and aluminum are commonly used elements in group IIIA. Aluminum metal is a light weight but strong material. Although it is a reactive element, its strong oxide coating protects it from corrosion.

The group IIA elements are reactive metals. They react with most nonmetals, producing ionic compounds. Calcium, strontium, barium, and radium behave much like the group IA elements: they react in cold water, producing basic solutions and liberating hydrogen. The group IIA hydroxides remove $CO_2$ from the air by forming carbonates.

The group B elements are all metals with widely different properties. Most transition elements have many possible oxidation states. The acid-base character of the oxides of transition elements depends in part on the oxidation state of the metal. The oxides are basic when the oxidation state is low; the oxides are acidic when the oxidation state is high.

## Objectives

By the end of this chapter, you should be able to do the following:

1. List the general electronic configuration of the valence level for groups IA through VIIIA. (Question 2)

2. Use equations to illustrate the reaction of water with an oxide of a metal from groups IA and IIA. (Question 3)

3. Use equations to illustrate the reaction of water with an oxide of a nonmetal from groups IVA, VA, and VIA. (Question 4)

4. List the most metallic and the most nonmetallic element in each A group. (Questions 5, 6)

5. Use equations to illustrate the behavior of an amphoteric hydroxide with an acid and with a base. (Question 7)

6. Use equations to illustrate the behavior of one element from each A group with oxygen and fluorine. (Questions 8, 9)

7. List the general trends in ionization energy, atomic size, and the electronegativity for each A group of elements. Which group or groups do not show any trends? (Question 10)

## Questions

1. Define the meaning of each of the following terms. How does your definition depend upon concepts learned earlier? *inorganic chemistry; organic chemistry; noble gases; halogens; halogen oxyions; interhalogens; contact process; amphoteric; half-life; fission; transition elements; lanthanide series; actinide series.*

2. List the general electronic configuration of the valence level for groups IA through VIIIA.

3. Complete the following equations:
   (a) $Li_2O(s) + H_2O(l) \rightleftharpoons$
   (b) $SrO(s) + H_2O(l) \rightleftharpoons$
   (c) $BaO(s) + H_2O(l) \rightleftharpoons$
   (d) $Na_2O(s) + H_2O(l) \rightleftharpoons$

4. Complete the following equations:
   (a) $CO_2(g) + H_2O(l) \rightleftharpoons$
   (b) $SO_2(g) + H_2O(l) \rightleftharpoons$
   (c) $P_4O_6(s) + H_2O(l) \rightleftharpoons$
   (d) $SeO_3(g) + H_2O(l) \rightleftharpoons$

5. If a metal is an element that (a) tends to lose electrons to a nonmetal and (b) forms a basic oxide, what is the most metallic element in each A group?

6. If a nonmetal is an element that (a) tends to gain electrons from a metal and (b) forms an acidic oxide, what is the most nonmetallic element in each A Group?

7. Zinc hydroxide is amphoteric. It reacts with strong bases to form soluble $Zn(OH)_4^{2-}$ ions. It reacts with acids and neutralizes them. Complete the following equations:
   (a) $Zn(OH)_2(s) + KOH(aq) \rightleftharpoons$
   (b) $Zn(OH)_2(s) + RbOH(aq) \rightleftharpoons$
   (c) $Zn(OH)_2(s) + HCl(aq) \rightleftharpoons$
   (d) $Zn(OH)_2(s) + HNO_3(aq) \rightleftharpoons$

8. Complete the following equations:
   (a) $Li(s) + O_2(g) \rightleftharpoons$
   (b) $Be(s) + O_2(g) \rightleftharpoons$
   (c) $Sn(s) + O_2(g) \rightleftharpoons$
   (d) $Ga(s) + O_2(g) \rightleftharpoons$
   (e) $As(s) + O_2(g) \rightleftharpoons$

9. Complete the following equations:
   (a) $K(s) + F_2(g) \rightleftharpoons$
   (b) $Ca(s) + F_2(g) \rightleftharpoons$
   (c) $In(s) + F_2(g) \rightleftharpoons$
   (d) $Si(s) + F_2(g) \rightleftharpoons$
   (e) $Sb(s) + F_2(g) \rightleftharpoons$
   (f) $O_2(g) + F_2(g) \rightleftharpoons$
   (g) $Cl_2(g) + F_2(g) \rightleftharpoons$
   (h) $Xe(g) + F_2(g) \rightleftharpoons$

10. List the general trends in ionization energy, atomic size, and electronegativity for each A group of elements. Which group or groups do not follow these trends?

11. Consider the areas of biology, medicine, engineering, agriculture, home economics, oceanography, astronomy, and space exploration. Pick one of these areas and, using a reference, determine what role the chemistry of elements from any three groups has in this area.

12. Ask an official of your nearest pollution control board what chemicals may be present in the air and water of your locality. What reactions put these chemicals there? What reactions would help remove them?

# Organic chemistry

## 20.1 Introduction

Organic chemistry is the chemistry of compounds which contain carbon. Most compounds which contain carbon are called *organic* compounds and those which do not contain carbon are called *inorganic*. Common organic chemicals include gasoline, plastics, vitamins, and aspirin.

The term "organic" is somewhat misleading since it implies a direct relationship between organic chemistry and living organisms. Until 1828, it was thought that all organic compounds did indeed originate from living organisms. In 1828, however, Friedrich Wohler synthesized an organic compound, urea, from an inorganic compound, ammonium isocyanate. With the synthesis of an organic compound from an inorganic source, the division between organic and inorganic chemistry became less clear. Today, the division is purely arbitrary and has been maintained primarily because of convenience. The number of organic compounds is extremely large; since they have common properties, they are best studied as a group.

Thousands of organic compounds, including aspirin, nylon, and proteins, have been synthesized in organic chemistry laboratories. New organic compounds are synthesized almost every day. Simple organic compounds serve as building blocks for the synthesis of larger and more complex organic compounds such as enzymes and genes. The manufacture of drugs, dyes, inks, paint, plastics, and rubber involves organic chemistry. The food we eat and the clothes we wear are organic compounds. Because the chemistry of all living organisms includes much organic chemistry, this field is of special significance to the medical professions.

## 20.2 Chemical bonds in organic chemistry

### Ionic bonds

Chemical bonds result from the competition between atoms for electrons. An *ionic bond* results from a *transfer of electrons*. Lithium fluoride is an ionic compound which is formed by the transfer of an electron from a lithium atom to a fluorine atom:

$$\text{LiF} \longrightarrow \text{Li}^+ + \text{F}^- + \text{energy}$$

### Covalent bonds

The bond of primary importance in organic chemistry is the covalent bond since carbon typically forms these bonds. Covalent bonds result from the sharing of electrons. Covalent bonds are responsible for binding the atoms together to form molecules in the following:

$$H\cdot + \cdot H \longrightarrow H\!:\!H$$

$$:\!\ddot{\underset{..}{Cl}}\!\cdot + \cdot\ddot{\underset{..}{Cl}}\!: \longrightarrow \;:\!\ddot{\underset{..}{Cl}}\!:\!\ddot{\underset{..}{Cl}}\!:$$

$$2H\cdot + \cdot\ddot{\underset{\cdot}{O}}\!: \longrightarrow H\!:\!\ddot{\underset{..}{O}}\!: \atop H$$

$$4H\cdot + \cdot\dot{\underset{\cdot}{C}}\cdot \longrightarrow {H \atop H\!:\!\ddot{C}\!:\!H \atop H}$$

### Bond lengths and bond angles

The most stable distance between the two nuclei of a covalent bond is known as the *bond length*. The bond length is the distance at which the stabilizing effect of overlap of atomic orbitals is exactly balanced by the repulsion between the two nuclei.

The angle between any two bonds in a molecule is known as

FIGURE 20.1

Bond angles for the three H—N—H bonds in ammonia, NH$_3$. Note that all three bonds have the same angle, 107°.

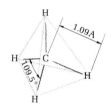

FIGURE 20.2

The four C—H bonds of methane. All four C—H bonds have the same bond length, 1.09 A. The angle between the bonds, the H—C—H bond angle, is 109.5°.

the *bond angle*. The bond angles for the three bonds in the ammonia molecule are shown in Fig. 20.1.

*Structure of methane* Methane, CH$_4$, which is the primary component of natural gas, is one of the simplest organic molecules. The bond angle between any two carbon-hydrogen bonds has been experimentally determined to be 109.5°. Further, experiments indicate that the methane molecule has the tetrahedral structure shown in Fig. 20.2. All four H—C—H bond angles are the same, 109.5°, and all four C—H bond lengths are the same 1.09 Å, From this, we conclude that all four C—H bonds of the methane molecule are equivalent.

The electronic configuration of carbon is:

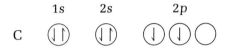

On the basis of electronic structure, carbon might be expected to combine with two hydrogen atoms to form CH$_2$ by the overlap of the available *p* orbitals with the *s* orbitals of two hydrogen atoms. Instead, a carbon atom combines with four hydrogen atoms to form methane, CH$_4$, presumably with the use of four bonding orbitals.

## Polarity of bonds

Two nuclei may share the electrons equally but in most cases the sharing is unequal. A covalent bond will be *polar* if the bond is between atoms which differ in electronegativity, that is, in their ability to attract electrons. The electron cloud which corresponds to a molecular orbital will be denser about the more electronegative atoms. The end of the bond adjacent to the more electronegative atom is relatively negatively charged and the other end of the bond is relatively positively charged. The bond has a negative pole and a positive pole. This separation of charge is indicated by δ+, or partial +, and δ−, or partial −. For example:

$$\overset{\delta+}{H}-\overset{\delta-}{Cl} \qquad \overset{\delta-}{\underset{\underset{\underset{H}{\delta+}}{H}}{\overset{\nearrow N \nwarrow}{H\ \ \ \ H}}}_{\delta+ \ \ \ \delta+} \qquad \overset{\delta-}{\underset{\overset{\delta+}{H}\ \ \ \ \overset{\delta+}{H}}{\overset{\nearrow O \nwarrow}{}}}$$

The greater the difference in electronegativity between the bonded atoms, the greater will be the polarity of the bond. Physical and chemical properties are determined by the polarities of their bonds. Whether a reaction will take place at a bond is influenced by its polarity.

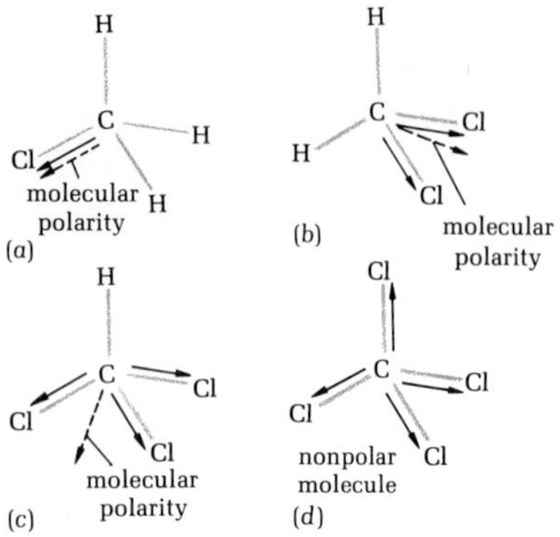

## Polarity of molecules

A molecule is polar if its centers of positive and negative charge do not exactly coincide. Since the H—Cl bond is polar, hydrogen chloride is a polar molecule. The HCl molecule constitutes a *dipole* which consists of two equal but opposite charges separated in space. The symbol ↔ is used to designate the dipole where the arrow points from positive to negative. The *dipole moment* of a molecule is an experimental measure of the polarity of the molecule. Methane has no dipole moment, which indicates that $CH_4$ is a nonpolar molecule. Why?

Chlorine is considerably more electronegative than carbon, and thus the C—Cl bond of $CH_3Cl$, chloromethane, is a polar bond. Carbon tetrachloride, $CCl_4$, contains four polar C—Cl bonds but the $CCl_4$ molecule is nonpolar as evidenced by no dipole moment. Why? Because of the symmetrical tetrahedral arrangement of the four C—Cl bonds in $CCl_4$, each of the C—Cl bonds opposes the other three bonds and they cancel each other; thus, $CCl_4$ is nonpolar. In $CH_3Cl$, however, the polar C—Cl bond is symmetrically opposed by the three C—H bonds, which are much less polar. The polarity of the C—Cl bond in $CH_3Cl$ is *not* exactly canceled by the three C—H bonds as evidenced by the fact that $CH_3Cl$ is experimentally found to be a polar molecule (Fig. 20.3).

*Similarities in properties* The polarity of molecules has a pronounced effect on such properties as melting point, boiling point, and solubility. The relatively negative end of one H—Cl molecule is attracted to the relatively positive end of another H—Cl molecule. This attraction is called *dipole-dipole interaction*. Polar molecules are generally quite strongly attracted to each other because of dipole-dipole interaction. As a consequence, the melting and boiling points of polar compounds are generally higher than those of nonpolar compounds of comparable molecular weight.

In Chapter 14, we made the generalization that "like dissolves like." Nonpolar methane is only very slightly soluble in the polar liquid, water. Methanol, $CH_3OH$, a highly polar organic compound, is quite soluble in polar water. From this observation we derive the generalization that polar dissolves polar or like dissolves like. Methanol dissolves in water because the positive end of a methanol molecule is attracted to the negative ends of water molecules (Fig. 20.4). The negative

FIGURE 20.3
The relationship between bond polarities and polarity of molecules for $CH_3Cl$, $CH_2Cl_2$, $CHCl_3$, and $CCl_4$. (a) The $CH_3Cl$ molecule is polar because of the polar C—Cl bond. The direction of the molecular polarity is the same as the direction of the bond polarity. (b) and (c) The direction of the molecular polarity is as shown. (d) The $CCl_4$ molecule is nonpolar because the four polar C—Cl bonds oppose and cancel each other.

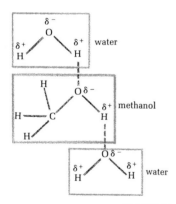

FIGURE 20.4
The attraction of polar $CH_3OH$, methanol molecules, for polar $H_2O$ molecules.

■ EXERCISE  Consider the molecule of sodium acetate. Which bonds are nonpolar? Polar? Ionic?

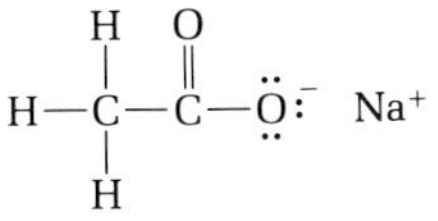

ends of the methanol molecules are likewise attracted to the positive ends of water molecules. ■

## 20.3 Saturated and unsaturated hydrocarbons

### The alkanes

Compounds which contain *only* carbon and hydrogen are called hydrocarbons. The simplest known hydrocarbon is methane, which is the first member of the homologous series called *alkanes*. A *homologous series* is a series of compounds in which each member of the series differs from the next member by a constant amount.

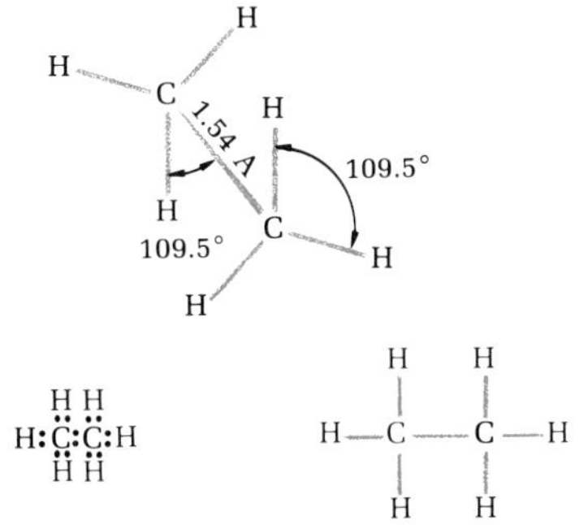

Methane, $CH_4$, ethane, $C_2H_6$, propane, $C_3H_8$, and butane, $C_4H_{10}$, all belong to the alkane family of compounds. This family forms a homologous series in which the constant difference between successive members of the series is $CH_2$; the general formula is $C_nH_{2n+2}$. For example, ethane, $C_2H_6$, is methane, $CH_4$, plus $CH_2$. There are single bonds between each carbon atom (Fig. 20.5). The first ten members of this homologous series are listed in Table 20.1. Alkanes are often used as fuels; they are relatively inert in the presence of acids, bases, and oxidizing agents. Structure, properties, and reactions of the alkanes will be discussed in Section 20.4.

FIGURE 20.5
The structure of ethane.

### The alkenes

Another homologous series is the family of hydrocarbons known as the *alkenes*, whose general formula is $C_nH_{2n}$. The

TABLE 20.1

The alkanes make up the homologous series of compounds whose formulas differ from each other by a multiple of $CH_2$. The general formula for the alkanes is $C_nH_{2n+2}$.

| Alkane ($C_nH_{2n+2}$) | n | Formula |
|---|---|---|
| Methane | 1 | $CH_4$ |
| Ethane | 2 | $C_2H_6$ |
| Propane | 3 | $C_3H_8$ |
| Butane | 4 | $C_4H_{10}$ |
| Pentane | 5 | $C_5H_{12}$ |
| Hexane | 6 | $C_6H_{14}$ |
| Heptane | 7 | $C_7H_{16}$ |
| Octane | 8 | $C_8H_{18}$ |
| Nonane | 9 | $C_9H_{20}$ |
| Decane | 10 | $C_{10}H_{22}$ |

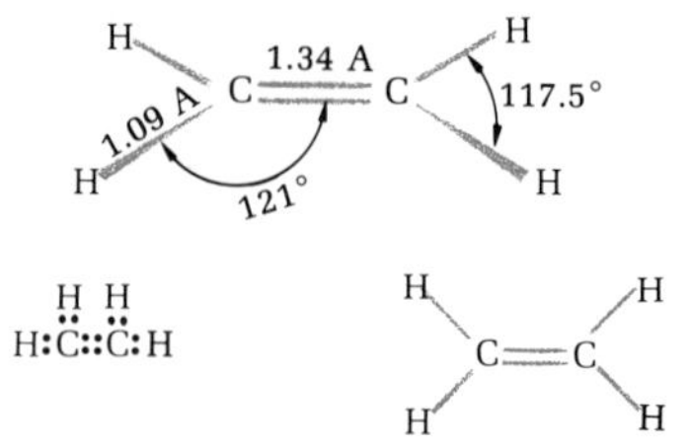

**FIGURE 20.6**
The structure of ethylene.

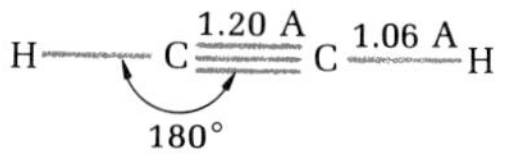

**FIGURE 20.7**
The structure of acetylene.

■ **EXERCISE** Classify the following as alkanes, alkenes, or alkynes.

(a) H—C(H)(H)—C(H)=C(H)—H

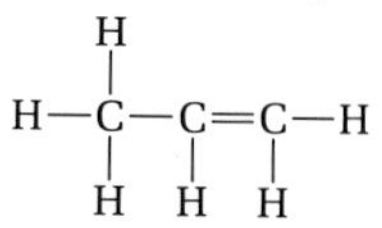

(b) H—C(H)(—C(H)—H)(—C(H)—H)—H

(c) H—C(H)(H)—C≡C—H

simplest member of the alkene family is ethylene, $C_2H_4$. Ethylene contains the same number of carbon atoms as ethane but fewer hydrogen atoms. The alkenes are called *unsaturated* hydrocarbons because they contain less than the maximum quantity of hydrogen. Alkanes are referred to as *saturated* hydrocarbons since they contain the maximum quantity of hydrogen. The distinguishing feature of the alkene structure is the presence of a carbon-carbon double bond. The structural formula is shown in Fig. 20.6. Carbon-carbon double bonds result from the sharing of *two pairs* of electrons. The double bond in an alkene is a potential source of an electron pair. This makes alkenes more reactive than alkanes.

Experimental evidence indicates that the four C—H bonds of ethylene are equivalent and that the double bond consists of two types of bonds. The H—C—H bond angle in ethylene is 120° (Fig. 20.6). Structure, properties, and reactions of the alkenes will be discussed in Section 20.5.

### The alkynes

Another family of unsaturated hydrocarbons are the *alkynes*, whose general formula is $C_nH_{2n-2}$. The simplest member of this family is acetylene, $C_2H_2$ or H—C≡C—H. Perhaps you are familiar with the use of acetylene in cutting torches and for welding. The most distinguishing feature of the alkyne structure is the carbon-carbon triple bond, —C≡C—. Like alkenes, alkynes are relatively reactive. The —C≡C— bond angle is experimentally found to be 180°; that is, H—C≡C—H is a linear molecule (Fig. 20.7).

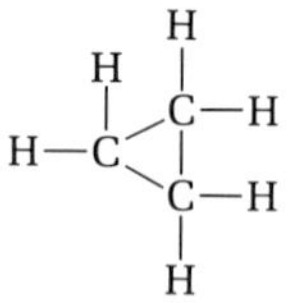

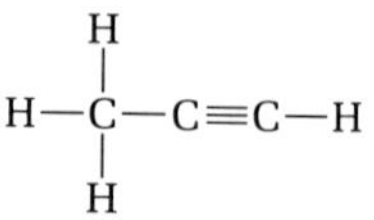

### Benzene

Benzene, $C_6H_6$, and compounds derived from it are called *aromatic compounds*. Although benzene is unsaturated, its behavior is different from that of the alkenes. Experimental evidence indicates that benzene consists of a closed ring structure with electron density above and below the plane of the ring. (Fig. 20.8).

Benzene is more stable than other compounds with double bonds such as alkenes. This is a consequence of its particular structure. A detailed study of benzene and benzene-like compounds is made in organic chemistry courses. ■

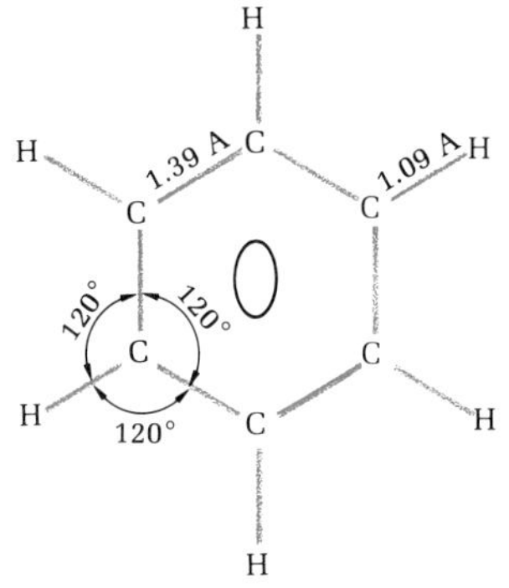

## 20.4 Alkanes

**Structure and properties**

There are two kinds of covalent bonds in alkane molecules, C—C and C—H. The C—C bonds are nonpolar since the covalent bond is between two atoms of the same element. The C—H bond is only slightly polar but even this small polarity is cancelled because of the tetrahedral symmetry of the carbon atoms; thus, the alkane molecules are nonpolar. For this reason natural gases such as methane, ethane, propane, and butane are only slightly soluble in water.

The alkanes have low melting and boiling points compared to polar molecules of equivalent molecular weight. Methane, ethane, propane, and butane are all gases at room temperature and 1.00 atm pressure. The next thirteen alkanes, $C_5H_{12}$ to $C_{17}H_{36}$, are liquids at room temperature. Only those alkanes which contain 18 or more carbons are solids at room temperature.

The relatively low boiling and melting points of the alkanes indicate that there is very little attraction of one alkane molecule for another as we might expect. Weak intermolecular forces are related to low polarities.

The primary source of alkanes is petroleum and the accompanying natural gas. Petroleum is the result of the decay of complex molecules from living animals under the stress of geologic changes.

FIGURE 20.8
*The structure of benzene. The benzene molecule is flat with all bond angles 120°.*

**Reactions**

Alkanes are saturated hydrocarbons; that is, they contain the maximum possible number of hydrogen atoms. Alkanes may undergo reactions such as the following:

1. Some other group may be *substituted* for one or more of the hydrogens of an alkane.
2. The alkanes can be burned in the presence of oxygen (*combustion*).
3. A molecule of an alkane may be broken apart by a process called *cracking*.

*Substitution* The chlorination of methane is typical of the substitution reactions of alkanes.

$$CH_4(g) + Cl_2(g) \xrightarrow{\text{heat or light}} CH_3Cl(g) + HCl(g) \qquad (20.1)$$

The first product of the reaction of chlorine with methane is monochloromethane, $CH_3Cl$. Note that the name was formed by indicating the number and type of atom which has been substituted for hydrogen. Further chlorination yields, in addition to $CH_3Cl$, the following:

$$\begin{array}{c} H \\ | \\ H-C-Cl \\ | \\ Cl \end{array} \quad CH_2Cl \quad \text{dichloromethane}$$

$$\begin{array}{c} Cl \\ | \\ H-C-Cl \\ | \\ Cl \end{array} \quad CH_2Cl_3 \quad \begin{array}{l}\text{trichloromethane}\\ \text{(chloroform)}\end{array}$$

$$\begin{array}{c} Cl \\ | \\ Cl-C-Cl \\ | \\ Cl \end{array} \quad CCl_4 \quad \begin{array}{l}\text{tetrachloromethane}\\ \text{(carbon tetrachloride)}\end{array}$$

The reaction occurs slowly, if at all, in the dark. It readily takes place either in the presence of light or with heat at 250–400°C.

*Combustion* Many alkanes burn rather easily. Gasoline, which is largely a mixture of alkanes, undergoes combustion readily in automobile engines. The burning of propane serves as an example:

$$C_3H_8(g) + 5O_2(g) \longrightarrow 3CO_2(g) + 4H_2O(g) + \text{energy} \qquad (20.2)$$

If conditions for burning are not ideal, other products may result. For instance, if the oxygen supply is limited or if proper mixing does not occur, combustion may be incomplete and CO or carbon may be produced.

*Cracking* A reaction of importance to the petroleum industry is the cracking of high molecular weight alkanes. The reaction produces hydrogen, alkenes, and smaller alkanes.

■ EXERCISE  *Octane, $C_8H_{18}$ burns in the presence of oxygen. Write a balanced equation for this reaction.*

$$\text{alkane} \xrightarrow{600°C} H_2 + \text{alkenes} + \text{smaller alkanes} \qquad ■ \quad (20.3)$$

## 20.5 Alkenes

**Structure and properties**

The distinguishing feature of the alkenes is the carbon-carbon double bond. The melting and boiling points for the alkenes are relatively low, as is the case for the alkanes. The alkenes are soluble in nonpolar solvents. Because of their very low polarities, or nonpolarity, the alkenes are only slightly soluble in polar solvents like water.

Alkenes can undergo substitution reactions just as the alkanes do, but the reaction which characterizes the alkenes is *addition* of various reagents to the carbon-carbon double bond. In fact, the chemistry of the alkenes is the chemistry of the double bond.

**Reactions**

*Addition* The combining of two molecules to form a single molecule of product is called an *addition reaction*. This can be contrasted to the substitution reactions in which one atom or group is substituted for another atom on a molecule. The addition of various reagents to the carbon-carbon double bond is demonstrated by the following:

Addition of hydrogen:

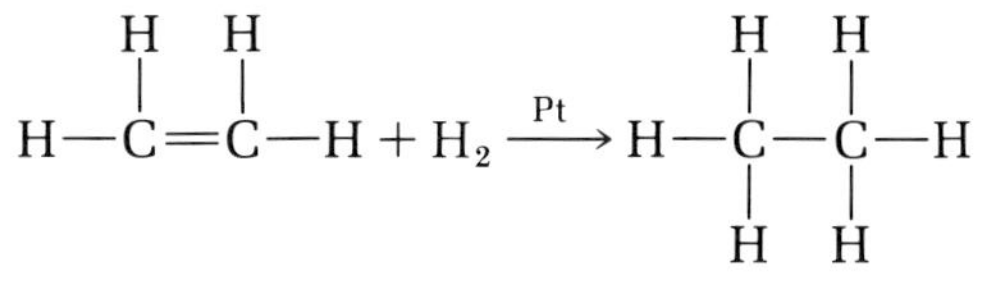

$$H-\underset{H}{\overset{H}{C}}=\underset{H}{\overset{H}{C}}-H + H_2 \xrightarrow{Pt} H-\underset{H}{\overset{H}{C}}-\underset{H}{\overset{H}{C}}-H \quad (20.4)$$

ethene → ethane

Addition of hydrogen halides:

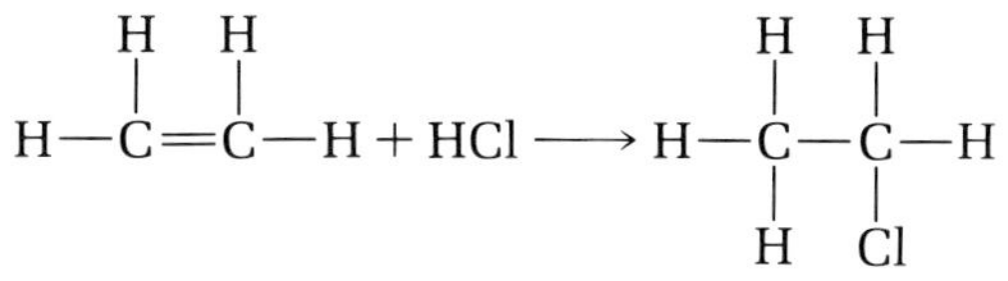

$$H-\underset{H}{\overset{H}{C}}=\underset{H}{\overset{H}{C}}-H + HCl \longrightarrow H-\underset{H}{\overset{H}{C}}-\underset{Cl}{\overset{H}{C}}-H \quad (20.5)$$

ethene → monochloroethane

Addition of water:

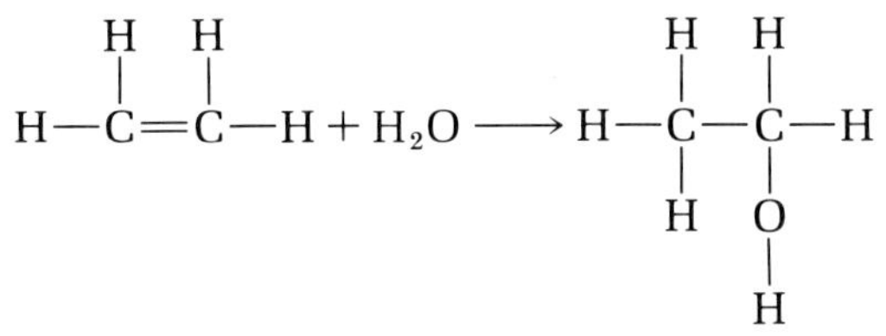

$$H-\underset{H}{\overset{H}{C}}=\underset{H}{\overset{H}{C}}-H + H_2O \longrightarrow H-\underset{H}{\overset{H}{C}}-\underset{\underset{H}{\overset{|}{O}}}{\overset{H}{C}}-H \quad (20.6)$$

ethene → ethyl alcohol

Alkenes may react with one another to form long chains or polymers (Section 20.7). Polyethylene contains thousands of ethylene molecules linked together.

$$CH_2{=}CH_2 + CH_2{=}CH_2 \xrightarrow{(cat)} -\underset{\underset{H}{|}}{\overset{\overset{H}{|}}{C}}-\underset{\underset{H}{|}}{\overset{\overset{H}{|}}{C}}-\underset{\underset{H}{|}}{\overset{\overset{H}{|}}{C}}-\underset{\underset{H}{|}}{\overset{\overset{H}{|}}{C}}- \quad (20.7)$$

Reactions 20.4–20.7 are general for the alkenes. There are many more addition reactions of alkenes which are considered in courses in organic chemistry. Note that the -ane ending of the alkane was dropped and -ene added to obtain the name ethene for $H_2C{=}CH_2$. The names of the other alkenes are formed in the same way.

**Substitution** One or more of the hydrogens on an alkene can be replaced by another group such as —Cl, although substitution reactions are much less important than addition to the double bond.

$$H-\underset{\underset{H}{|}}{\overset{\overset{H}{|}}{C}}-\overset{\overset{H}{|}}{C}{=}\overset{\overset{H}{|}}{C}-H + Cl_2 \xrightarrow[\substack{\text{low} \\ \text{concentration} \\ \text{of } Cl_2}]{\text{heat}}$$

$$Cl-\underset{\underset{H}{|}}{\overset{\overset{H}{|}}{C}}-\overset{\overset{H}{|}}{C}{=}\overset{\overset{H}{|}}{C}-H + HCl \quad \blacksquare \quad (20.8)$$

■ EXERCISE  *Draw the structures of the products in the following addition reactions.*

(a) $H-\underset{\underset{H}{|}}{\overset{\overset{H}{|}}{C}}-\overset{\overset{H}{|}}{C}{=}\overset{\overset{H}{|}}{C}-H + Br_2 \rightleftharpoons$

(b) $H-\underset{\underset{H}{|}}{\overset{\overset{H}{|}}{C}}-\overset{\overset{H}{|}}{C}{=}\overset{\overset{H}{|}}{C}-H + H_2 \rightleftharpoons$

## 20.6 The effect of structure on properties— alcohols and acids

**Alcohols**

The atom or group of atoms which determines the structure and properties of a particular family of organic compounds is called the *functional group*. The functional group for the alkenes is the characteristic carbon-carbon double bond. For the alkynes, the functional group is the carbon-carbon triple bond.

Alcohols are the family of compounds whose functional group is —O—H, the hydroxyl group. The general formula for alcohols is ROH where R is any group which has the general structure of either an alkane, an alkene, or an alkyne. The alcohol structure may be viewed as being the result of substitution of a hydrocarbon group for one of the hydrogens of water. Some examples of alcohols are:

$$\begin{array}{c} H \\ | \\ H-C-O-H \\ | \\ H \end{array} \qquad \begin{array}{cc} H & H \\ | & | \\ H-C-C-O-H \\ | & | \\ H & H \end{array} \qquad \begin{array}{ccc} H & H & H \\ | & | & | \\ H-C-C-C-H \\ | & | & | \\ H & O & H \\ & | \\ & H \end{array}$$

methyl alcohol (wood alcohol)     ethyl alcohol (grain alcohol)     isopropyl alcohol (rubbing alcohol)

**Properties of alcohols**

Because of the polar —O—H group, methyl alcohol, $CH_3$—O—H, is a polar compound. Its polarity makes methyl alcohol soluble in all proportions in water, a polar compound. Like methyl alcohol, all of the alcohols which contain less than ten carbon atoms are soluble in water at least to some extent.

The polarities of the low molecular weight alcohols result in strong intermolecular attractions between the molecules. Strong intermolecular attractions give rise to high boiling points. The molecular weights of ethyl alcohol, $C_2H_5OH$, and propane, $C_3H_8$, are similar, 46 for ethyl alcohol, and 44 for propane. Their boiling points, however, are quite different. The boiling point for propane is $-42°C$, the boiling point for ethyl alcohol is $+78°C$.

Because of their high polarities, the lower molecular weight alcohols make excellent solvents. All alcohols are toxic but they vary greatly in toxicity from methyl alcohol, which can cause blindness and death, to the intoxicating effect of ethyl alcohol. The low toxicity of ethyl alcohol has permitted its use as a beverage in some cases.

**Reactions of alcohols**

The chemical reactions which alcohols can undergo are determined by the characteristic functional group, —O—H, the hydroxyl group. The reactions of alcohols can involve the breaking of the carbon-oxygen bond or the oxygen-hydrogen bond. Reactions of this type are substitutions in which some group is substituted either for hydrogen or for the hydroxyl group. Alcohols can also undergo elimination reactions in which a molecule of water is lost. Examples of these reactions are listed below:

*Carbon-oxygen bond cleavage and replacement by halide*

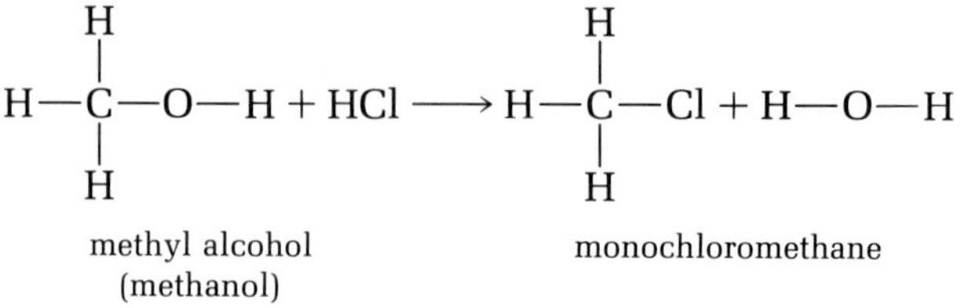

$$H-\underset{\underset{H}{|}}{\overset{\overset{H}{|}}{C}}-O-H + HCl \longrightarrow H-\underset{\underset{H}{|}}{\overset{\overset{H}{|}}{C}}-Cl + H-O-H \quad (20.9)$$

methyl alcohol (methanol)      monochloromethane

*Oxygen-hydrogen bond cleavage, and replacement by an active metal*

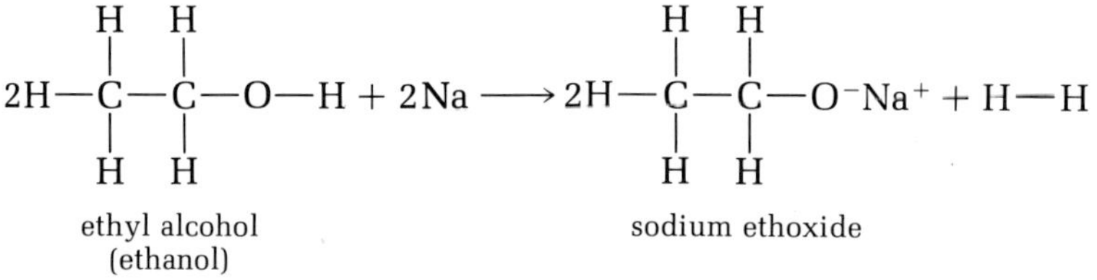

$$2H-\underset{\underset{H}{|}}{\overset{\overset{H}{|}}{C}}-\underset{\underset{H}{|}}{\overset{\overset{H}{|}}{C}}-O-H + 2Na \longrightarrow 2H-\underset{\underset{H}{|}}{\overset{\overset{H}{|}}{C}}-\underset{\underset{H}{|}}{\overset{\overset{H}{|}}{C}}-O^-Na^+ + H-H \quad (20.10)$$

ethyl alcohol (ethanol)      sodium ethoxide

In this reaction, the alcohol plays the role of an acid. Recall that when an active metal is added to an acid, the metal replaces the hydrogen.

*Elimination of $H_2O$: dehydration*

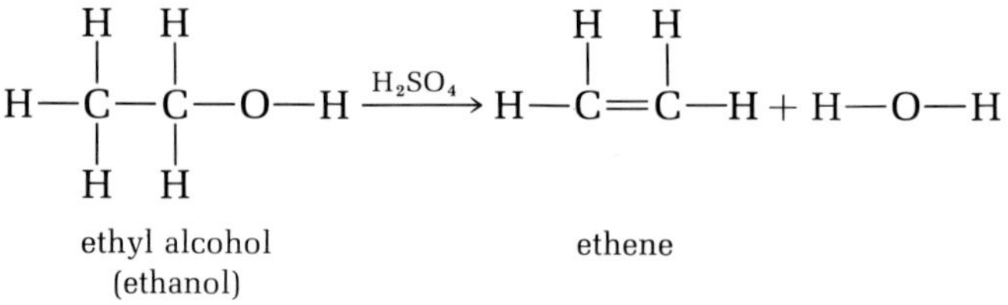

$$H-\underset{\underset{H}{|}}{\overset{\overset{H}{|}}{C}}-\underset{\underset{H}{|}}{\overset{\overset{H}{|}}{C}}-O-H \xrightarrow{H_2SO_4} H-\overset{\overset{H}{|}}{C}=\overset{\overset{H}{|}}{C}-H + H-O-H \quad (20.11)$$

ethyl alcohol (ethanol)      ethene

Note that the dehydration of alcohols can be used to prepare alkenes.

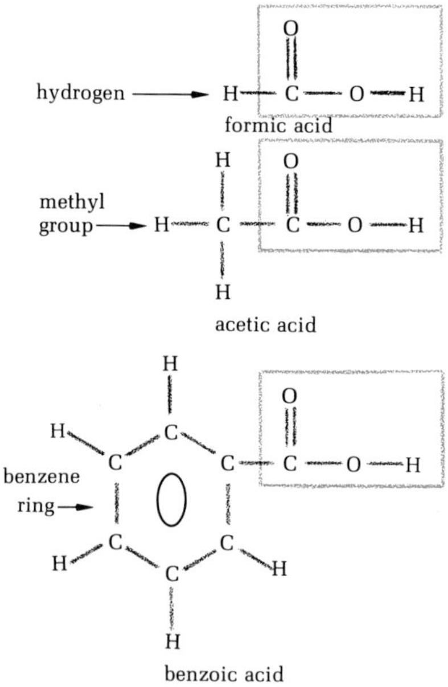

**Organic acids**

Organic acids are the family of compounds which contain the carboxyl group, $O{=}\underset{|}{C}{-}O{-}H$ as their functional group. The carboxyl group may be attached to hydrogen, a straight chain, or a benzene ring (Fig. 20.9). Formic acid, shown in the figure, is responsible for the sting of ants; acetic acid is present in vinegar.

FIGURE 20.9
*Organic acids differ in the group which is attached to the functional carboxyl group $-\overset{\overset{O}{\|}}{C}-O-H$.*

formic acid: hydrogen — H—C(=O)—O—H
acetic acid: methyl group — H—C(H)(H)—C(=O)—O—H
benzoic acid: benzene ring — C₆H₅—C(=O)—O—H

**Properties of organic acids**

The properties of organic acids are determined by their characteristic functional group, $O{=}\underset{|}{C}{-}O{-}H$, the carboxyl group.

Because they contain the carboxyl group, the organic acids are called *carboxylic acids*.

Carboxylic acid molecules are very polar, even more polar than the alcohols. Therefore, the lower molecular weight acids are soluble in water. The boiling points of the carboxylic acids as a class are even higher than the boiling points of the alcohols. The lower molecular weight acids have sharp, penetrating unpleasant odors. Perhaps you are familiar with the sharp odor of acetic acid from its presence in vinegar. Butyric acid, one of the higher molecular weight acids, gives butter its characteristic odor and is used as a seasoning to give a butter flavor.

**Reactions of organic acids**

The carboxyl group consists of a hydroxyl group, —O—H, attached to a carbonyl group, $O{=}\underset{|}{C}{-}$:

$$\underset{\text{hydroxyl group}}{-O-H} + \underset{\text{carbonyl group}}{\overset{\overset{O}{\|}}{-C-}} \Big\} \underset{\text{carboxyl group}}{\overset{\overset{O}{\|}}{-C-O-H}}$$

The —O—H group of carboxylic acids undergoes reactions which are possible only because it is attached to the carbonyl group. Some examples of the reactions of carboxylic acids are:

Replacement of the H of $O{=}\underset{|}{C}{-}O{-}H$

$$\underset{\text{acetic acid}}{CH_3-\overset{\overset{O}{\|}}{C}-O-H} + NaOH \longrightarrow \underset{\text{sodium acetate}}{CH_3-\overset{\overset{O}{\|}}{C}-O^-Na^+} + H_2O \quad (20.12)$$

$$\text{acid} \quad + \quad \text{base} \quad \longrightarrow \quad \text{salt}$$

Reaction with alcohols; formation of esters:

$$\underset{\text{acetic acid}}{CH_3-\overset{\overset{O}{\|}}{C}-O-H} + \underset{\text{ethyl alcohol}}{CH_3-CH_2-O-H} \xrightarrow{H^+} \underset{\text{ethyl acetate (an ester)}}{CH_3-\overset{\overset{O}{\|}}{C}-O-CH_2-CH_3} + H_2O \quad (20.13)$$

*Reduction to alcohols:*

$$CH_3-\overset{\overset{O}{\|}}{C}-O-H \xrightarrow{\text{LiAlH}_4} CH_3-CH_2-OH \qquad (20.14)$$

acetic acid ethyl alcohol

(This equation has *not* been balanced.)

*Reaction with $NH_3$ to form amides:*

$$CH_3-\overset{\overset{O}{\|}}{C}-O-H + NH_3 \xrightarrow{\text{heat}} CH_3-\overset{\overset{O}{\|}}{C}-N\overset{H}{\underset{H}{\diagdown}} + H_2O \qquad (20.15)$$

acetic acid acetamide (an amide)

### The relation of structure to properties

Carboxylic acids are usually much weaker than the strong mineral acids ($HCl$, $H_2SO_4$, and $HNO_3$) with which you are already familiar. The carboxylic acids are, however, much more acidic than the very weakly acidic alcohols. Both carboxylic acids and alcohols are sufficiently acidic to react with active metals. The metal can replace the active hydrogen of the alcohol or acid.

$$2CH_3-\overset{\overset{O}{\|}}{C}-O-H + 2Na(\text{metal}) + 2CH_3-\overset{\overset{O}{\|}}{C}-O^-Na^+ + H_2 \qquad (20.16)$$

acetic acid sodium acetate

$$2CH_3-CH_2-O-H + 2Na \longrightarrow 2CH_3-CH_2-O^-Na^+ + H_2 \qquad (20.17)$$

ethyl alcohol sodium ethoxide

The weakly acidic hydrogen of an alcohol is insufficiently acidic to react with sodium hydroxide. Carboxylic acids, on the other hand, react with sodium hydroxide to form the corresponding salt.

$$CH_3-CH_2-O-H + NaOH \longrightarrow \text{No reaction} \qquad (20.18)$$

ethyl alcohol

$$CH_3-\overset{\overset{O}{\|}}{C}-O-H + NaOH \longrightarrow CH_3-\overset{\overset{O}{\|}}{C}-O^-Na^+ + H_2O \qquad (20.19)$$

acetic acid sodium acetate

Why isn't the hydrogen of the alcohol donated to the hydroxide ion? The answer involves the concept of electronegativity. In an alcohol, the OH group is attached to carbon, an element of intermediate electronegativity. In an acid, the OH

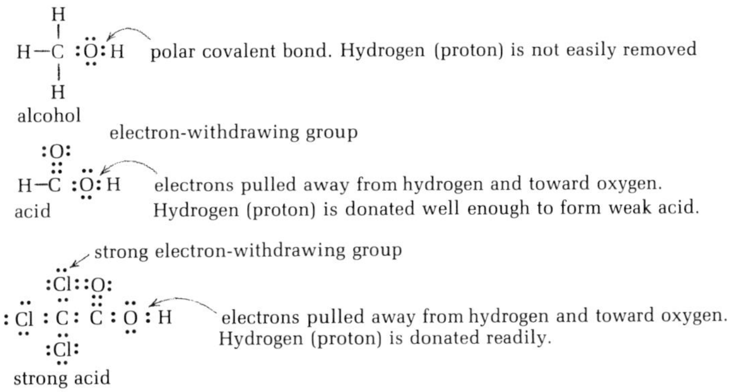

**FIGURE 20.10**
*Electron-withdrawing groups enhance the ease with which the proton is donated in the* $-\overset{|}{\underset{|}{C}}-O-H$ *group.*

$$H-\overset{\overset{H}{|}}{\underset{\underset{H}{|}}{C}}-\overset{..}{\underset{..}{O}}:H \quad \text{polar covalent bond. Hydrogen (proton) is not easily removed}$$

alcohol

electron-withdrawing group

$$H-\overset{:O:}{\overset{||}{C}}:\overset{..}{\underset{..}{O}}:H \quad \text{electrons pulled away from hydrogen and toward oxygen.}$$
$$\text{Hydrogen (proton) is donated well enough to form weak acid.}$$

acid

strong electron-withdrawing group

$$:\overset{..}{\underset{..}{Cl}}:\overset{\overset{:\overset{..}{Cl}:}{|}}{\underset{\underset{:\overset{..}{Cl}:}{|}}{C}}:\overset{:O:}{\overset{||}{C}}:\overset{..}{\underset{..}{O}}:H \quad \text{electrons pulled away from hydrogen and toward oxygen.}$$
$$\text{Hydrogen (proton) is donated readily.}$$

strong acid

group is attached to a carbonyl group, $-\overset{|}{C}=O$. This group has a high electronegativity and withdraws electrons from the O—H bond (Fig. 20.10). This weakens the O—H bond and the hydrogen comes off with greater ease. Therefore the $O=\overset{|}{\underset{|}{C}}-O-H$ group is more acidic than the OH group. Adding other electron-withdrawing groups to a molecule may make the $O=\overset{|}{\underset{|}{C}}-O-H$ even more acidic. For instance, the replacement of the hydrogens in acetic acid with highly electronegative chlorines makes trichloracetic acid a strong acid.

Carboxylic acids, although usually weaker than mineral acids such as HCl, function in much the same manner as mineral acids. The carboxylic acids can donate a proton to strong bases like NaOH to form salts. Salts of carboxylic acids are nonvolatile crystalline solids made up of positive and negative ions. The physical properties of these salts are similar to those of the inorganic salts. The alcohols have such slight acidities that they do not react with strong bases like NaOH. A seemingly subtle difference in structure, $-\overset{|}{\underset{|}{C}}-O-H$ versus $O=\overset{|}{\underset{|}{C}}-O-H$, causes a large difference in properties.

## 20.7 Polymers

Polymers are very large molecules which have been formed by linking together many small molecules. The polymerization of ethylene noted earlier involves the linking together of a large number of ethylene molecules to form the polymer, polyethylene.

$$-\underset{\underset{H}{|}}{\overset{\overset{H}{|}}{C}}=\underset{\underset{H}{|}}{\overset{\overset{H}{|}}{C}}- \rightleftharpoons -\underset{\underset{H}{|}}{\overset{\overset{H}{|}}{C}}-\underset{\underset{H}{|}}{\overset{\overset{H}{|}}{C}}-\underset{\underset{H}{|}}{\overset{\overset{H}{|}}{C}}-\underset{\underset{H}{|}}{\overset{\overset{H}{|}}{C}}-\underset{\underset{H}{|}}{\overset{\overset{H}{|}}{C}}-\underset{\underset{H}{|}}{\overset{\overset{H}{|}}{C}}-\underset{\underset{H}{|}}{\overset{\overset{H}{|}}{C}}-\underset{\underset{H}{|}}{\overset{\overset{H}{|}}{C}}- \quad (20.20)$$

ethylene unit     polyethylene

Polyethylene is widely used for packaging films and for household food wraps.

A molecule of polymer is very large; that is, it has a long chain of molecules linked together by chemical bonds. These long chains give strength to the polymer just as small strands give strength to a cable or rope. Chemists have learned to modify the properties of polymers so that strong and durable yet flexible polymers can be made.

Polyvinyl chloride, PVC, is a polymer which is used to make flexible plastic pipe as well as hard phonograph records. When a naturally occurring polymer, "natural rubber," became very scarce during World War II, chemists developed several rubber substitutes. These included isoprene and polychloroprene (Neoprene). Today automobile tires and most other rubber articles are made of synthetic polymers.

Silk is a naturally occurring polymer which was once needed in the manufacture of parachutes. Today a man-made polymer, nylon, is used for this and many other purposes. Nylon is strong enough to be used in making tire cord and cables yet soft and light enough to make parachutes and lingerie. Polyesters, another polymer, are now used for making tire cord.

**TABLE 20.2**

*The functional groups of several families of organic compounds.*

| Family | Functional group |
|---|---|
| Alkanes | —H |
| Alkenes | —C=C— |
| Alkynes | —C≡C— |
| Alcohols | —O—H |
| Organic acids | —C(=O)—O—H |
| Aldehydes | —C(=O)—H |
| Ketones | —C—C(=O)—C— |
| Amines | —NH₂ |
| Amides | —C(=O)—NH₂ |
| Amino acids | —C(NH₂)(H)—C(=O)—O—H |

## 20.8 Organic compounds and their functional groups

**Aldehydes**

As we saw in Section 20.7, the structure of an organic compound has a profound influence on its properties. The properties of a family of organic compounds are determined by their functional group(s). Table 20.2 contains a listing of the families of organic compounds discussed in this chapter, with their functional groups.

The aldehydes are the family of compounds whose functional group is the carbonyl group attached to hydrogen,

$-\overset{\overset{\displaystyle O}{\|}}{C}-H$. A common aldehyde is formaldehyde, $H-\overset{\overset{\displaystyle O}{\|}}{C}-H$, a primary constituent of embalming fluid. Aldehydes are polar compounds which have high boiling points; the temperatures are lower, however, than the boiling points of the alcohols of similar molecular weight.

Oxidation of organic compounds results in the removal of hydrogen atoms and/or the addition of oxygen atoms. An aldehyde may be formed by the oxidation of an alcohol, provided the —O—H group of the alcohol in question is attached to a carbon atom which bears two hydrogen atoms.

$$H-\underset{\underset{H}{|}}{\overset{\overset{H}{|}}{C}}-\underset{\underset{H}{|}}{\overset{\overset{H}{|}}{C}}-O-H \xrightarrow{\text{hot Cu}} H-\underset{\underset{H}{|}}{\overset{\overset{H}{|}}{C}}-\overset{\overset{O}{\|}}{C}-H + H_2 \qquad \text{oxidation: removal of two hydrogens} \qquad (20.21)$$

ethyl alcohol → acetaldehyde

Oxidation of an aldehyde produces the corresponding carboxylic acid.

$$5H-\underset{\underset{H}{|}}{\overset{\overset{H}{|}}{C}}-\overset{\overset{O}{\|}}{C}-H + 2MnO_4^- + 6H^+(aq) \longrightarrow 5H-\underset{\underset{H}{|}}{\overset{\overset{H}{|}}{C}}-\overset{\overset{O}{\|}}{C}-OH + 2Mn^{2+} + 3H_2O \quad (20.22)$$

oxidation: addition of one oxygen

### Ketones

Oxidation of alcohol in which the —O—H group is attached to a carbon bearing less than two hydrogen atoms produces a ketone. The family of compounds containing the carbonyl group attached to two hydrocarbon groups is called the ketones. Dimethyl ketone, $CH_3-\overset{\overset{\displaystyle O}{\|}}{C}-CH_3$, is an example. This ketone, also called acetone, is a widely used solvent. It is produced by the oxidation of isopropyl alcohol.

$$H-\underset{\underset{H}{|}}{\overset{\overset{H}{|}}{C}}-\underset{\underset{O}{|}\phantom{|}\underset{H}{|}}{\overset{\overset{H}{|}}{C}}-\underset{\underset{H}{|}}{\overset{\overset{H}{|}}{C}}-H \xrightarrow{\text{hot Cu}} H-\underset{\underset{H}{|}}{\overset{\overset{H}{|}}{C}}-\underset{\underset{O}{\|}}{C}-\underset{\underset{H}{|}}{\overset{\overset{H}{|}}{C}}-H \qquad \text{oxidation: removal of hydrogen} \qquad (20.23)$$

isopropyl alcohol → acetone (dimethyl ketone)

### Reactivities of aldehydes and ketones

Since both aldehydes and ketones contain the carbonyl group, $O=C-$, they are collectively called *carbonyl* compounds. The chemistry of aldehydes and ketones is largely the chemistry of the carbonyl group.

The properties of aldehydes and ketones are quite similar because of their similarity in structure. There is, however, a slight difference in structure between the two families. There must be at least one hydrogen attached to the carbonyl group in aldehydes. Aldehydes form at the *end* of a carbon chain. In ketones, there are two hydrocarbon groups and no hydrogen attached to the carbonyl group. Ketones form in the *middle* of carbon chains. This difference in structure affects the reactivities of these two compounds. Aldehydes are quite easily oxidized to their corresponding acids because the acid group must form at the end of the molecule. Ketones are very difficult to oxidize. In general, aldehydes are more reactive than ketones.

### Amines

Organic compounds whose functional group is $-NH_2$, the amino group, are called amines. The general formula for amines is $RNH_2$, $R_2NH$, or $R_3N$, where R stands for a hydrocarbon group. Amines may be viewed as being formed by the replacement of one or more of the hydrogen atoms of ammonia with one or more hydrocarbon groups. Like ammonia, the amines are basic compounds. They are proton acceptors and their solution will turn litmus blue. Volatile amines have a fishlike odor. Many amines are widely used as medicines.

Amines react with $H^+(aq)$ to form salts in the same manner as ammonia:

$$NH_3(g) + H^+(aq) + Cl^-(aq) \longrightarrow NH_4^+(aq) + Cl^-(aq) \quad (20.24)$$
$$\text{ammonia} \hspace{4cm} \text{ammonium chloride}$$

$$\underset{\text{methyl amine}}{H_3C-NH_2} + H^+(aq) + Cl^-(aq) \longrightarrow \underset{\text{methyl ammonium chloride}}{H_3C-NH_2-H^+}(aq) + Cl^-(aq) \quad (20.25)$$

454   20: Organic chemistry

### Amino acids

All amino acids contain the two functional groups: the amino group, $-NH_2$, and the carboxyl group, $O=C-O-H$. The properties of the amino acids are therefore determined by two groups instead of one. An example of an amino acid is glycine, whose structural formula is:

$$H-\underset{\underset{\underset{H\;\;H}{\diagdown\!\diagup}}{N}}{\overset{\overset{H}{|}}{C}}-\overset{\overset{O}{\|}}{C}-O-H$$

glycine

The amino acids are nonvolatile crystalline solids which decompose at high temperatures. They are soluble in polar solvents such as water but only slightly soluble in nonpolar solvents.

Amino acids have the ability to bond together. Compounds which are formed by the bonding of two or more amino acids together are *peptides*.

$$\underset{\text{glycine}}{\overset{H}{\underset{H}{\diagdown}}N-\overset{\overset{H}{|}}{\underset{\underset{H}{|}}{C}}-\overset{\overset{O}{\|}}{C}-O-H} + \underset{\text{glycine}}{\overset{H}{\underset{H}{\diagdown}}N-\overset{\overset{H}{|}}{\underset{\underset{H}{|}}{C}}-\overset{\overset{O}{\|}}{C}-O-H} \rightleftharpoons \underset{\substack{\text{glycyl glycine,}\\ \text{a dipeptide}}}{\overset{H}{\underset{H}{\diagdown}}N-\overset{\overset{H}{|}}{\underset{\underset{H}{|}}{C}}-\overset{\overset{O}{\|}}{C}-\overset{}{\underset{\underset{H}{|}}{N}}-\overset{\overset{H}{|}}{\underset{\underset{H}{|}}{C}}-\overset{\overset{O}{\|}}{C}-OH} + HOH \quad (20.26)$$

The group $-\overset{\overset{O}{\|}}{C}-\underset{\underset{H}{|}}{N}-$ is called the amide linkage. Compounds which contain only this functional group are called *amides*.

### Proteins

Amino acids can bond together to form polymers of very high molecular weight. These polymers, called proteins, make up the muscles and enzymes in your body. There are twenty naturally occurring protein amino acids; that is, all protein is thought to be constructed of the same twenty amino acid molecules. Consider the general formula of amino acids to be:

$$R-\overset{\overset{\displaystyle H}{|}}{\underset{\underset{\displaystyle N}{|}}{C}}-\overset{\overset{\displaystyle O}{\|}}{C}-O-H$$
$$\phantom{R-C-C-}H\quad H$$

The differences between the twenty protein amino acids are due to the nature of the R groups.

At least ten of these naturally occurring amino acids are required in the diet of man; these are called the essential amino acids. The physiological development of the human body cannot take place without these amino acids. Meat is an excellent source of protein, and of course amino acids, hence the wide use of meat as a food. Much research has been conducted to find new ways of producing more protein to feed the people of tomorrow.

## Summary

Organic chemistry is the chemistry of compounds that contain carbon. The bond which is most important in organic chemistry is the covalent bond. Overlap of a half-filled atomic orbital of one atom with a half-filled atomic orbital of another atom results in a covalent bond. The most stable distance between these two nuclei is known as the bond length. The angle between any two bonds in a molecule is known as the bond angle.

A covalent bond formed between two atoms which differ in electronegativity will be polar. A molecule will be polar if its centers of positive and negative charge do not exactly coincide. The symmetry of a molecule may result in a cancellation of the polarities of bonds. For example, a molecule such as $CCl_4$, which contains four polar bonds, is a nonpolar molecule.

Methane, $CH_4$, is one of the simplest organic compounds. Those compounds, like methane, which contain only carbon and hydrogen are known as hydrocarbons. Methane is the first member of the homologous series called alkanes. Each member of a homologous series differs from each other member by a definite amount.

The general formula for the alkanes is $C_nH_{2n+2}$; for alkenes it is $C_nH_{2n}$; and for alkynes it is $C_nH_{2n-2}$.

A group which defines the structure and determines the properties of a family of organic compounds is called a functional group. The functional group for alkenes is the carbon-carbon double bond, that for the alkynes is the carbon-carbon triple bond.

Alkanes can undergo substitution reactions, in which some group is substituted for one or more of the hydrogen atoms. Alkenes can undergo addition reactions, in which a reagent adds to the carbon-carbon double bond causing it to break.

The hydrogen of a carboxyl group is a much more acidic hydrogen than the hydrogen of the alcohol groups.

The differences in properties between alcohols and carboxylic acids demonstrate again the effect of the structure of a molecule upon its properties.

## Objectives

By the end of this chapter, you should be able to do the following:

1. Write a structural formula for each of the following: an alkane, an alkene, an alkyne, an alcohol, a carboxylic acid, a ketone, and aldehyde, an amine, an amino acid, and benzene. (Questions 2, 3, 8)

2. Give examples of both addition and substitution reactions. (Questions 5, 6)

3. Write an equation which demonstrates the preparation of an aldehyde. (Questions 10, 11)

4. Write an equation for the preparation of an acid. (Questions 10, 11)

5. Draw a diagram of a molecule which contains polar bonds but is nonpolar. (Question 9)

6. Write balanced equations for the combustion of alkanes and alkenes. (Question 4)

## Questions

1. Define or explain the following terms: *alkane; alkene; alkyne; alcohol; carboxylic acid; functional group; hydrocarbon; saturated; unsaturated; homologous series; double bond; triple bond; aldehyde; ketone; amide linkage; amino acid; tetrahedral angle; bond angle; bond length; bond polarity; polar molecule.*

2. Write structural formulas for ethene, ethane, acetylene, and benzene.

3. Use structural formulas to demonstrate the functional group of a representative example of each of the following families of compounds: (a) alkanes; (b) alkenes; (c) alkynes; (d) alcohols; (e) organic acids; (f) aldehydes; (g) ketones; (h) amides; (i) amino acids; (j) amines.

4. Write balanced equations for the combustion of: (a) ethane; (b) propane; (c) ethene; (d) ethyne; (e) ethyl alcohol.

5. Write balanced equations for the following: (a) the reaction of chlorine with ethane and (b) the reaction of ethene with water in an acidic medium.

6. Which of the reactions in Question 5 is an addition? Which is a substitution? Contrast addition with substitution. Which classes of compounds would you expect to undergo substitution? Addition? How is structure related to properties in this case?

7. Which compound in each of the following pairs would you expect to have the *highest* boiling point?

(a) $CH_3-\overset{\overset{O}{\|}}{C}-H$ or $CH_3-CH_3$

(b) $CH_3-O-H$ or $CH_4$

(c) $CH_4$ or $H_2O$

(d) $H-\overset{\overset{O}{\|}}{C}-H$ or $CH_4$

8. Identify the family of compounds to which each of the following belongs:

(a) $CH_3-O-H$

(b) $CH_3-\overset{\overset{O}{\|}}{C}-O-H$

(c) $CH_3-\overset{\overset{O}{\|}}{C}-H$

(d) $CH_3-N\overset{H}{\underset{H}{\diagdown}}$

(e) $CH_3-\overset{\overset{O}{\|}}{C}-N\overset{H}{\underset{H}{\diagdown}}$

(f) $CH_3-C\equiv C-H$

9. The polar $-\overset{|}{\underset{|}{C}}-Cl$ bonds in $CHCl_3$, $CH_2Cl_2$, and $CH_3Cl$ make these compounds very polar. Why, then, are $CCl_4$ and $CH_4$ nonpolar compounds? Use the concepts of bond polarity and polarity of molecules in your discussion.

10. Write balanced equations for the following:

(a) formation of $CH_3-\overset{\overset{O}{\|}}{C}-$ from an alcohol

(b) formation of an alcohol from $H-\overset{\overset{H}{|}}{C}=\overset{\overset{H}{|}}{C}-H$

(c) formation of an aldehyde from $CH_3OH$
(d) reaction of ethane with chlorine in the presence of light
(e) reaction of sodium metal with acetic acid
(f) reaction of sodium metal with ethanol
(g) reaction of sodium hydroxide with ethanol
(h) reaction of sodium hydroxide with acetic acid

11. Which family of compounds is more readily oxidized, aldehydes or ketones? What compound would you select for the preparation of acetic acid?

12. Why are ketones oxidized only with great difficulty?

13. Look at the labels on your clothing and other labels around your house and make a list of the polymers you find.

14. There is some disagreement as to the concentration of amino acids in lunar soil but it appears that the total amino acid concentration is less than 100 nanograms (ng) of amino acids per gram of lunar soil (1 ng = 1 × $10^{-9}$ g). Assuming that this is the approximate amino acid

concentration of the lunar surface, calculate the approximate percentage of amino acids. What would the presence of amino acids in the lunar soil suggest?

15. LSD is a hallucinogenic drug which has drastic and damaging effects on the human body. LSD is prepared from lysergic acid, an organic acid whose structure is

Why do the properties of lysergic acid differ from those of acetic acid?

16. Develop a model to explain why carbon forms four bonds in $CH_4$ and no compounds with two bonds. Compare your model to the model in an organic chemistry text.

# Appendix I

# A review of mathematics

## 1 Significant figures

A significant figure is a number in a measurement which we believe to be correct. Whenever a measurement is reported, only one figure with any uncertainty may be written.

Using a balance which could be read to within 0.00001 gram, a chemist carefully measured and recorded 0.01438 gram as the mass of a sample of an amino acid. How many significant figures are there in his reported value, 0.01438 gram? The 8 contains some uncertainty but is still a significant figure. The other figures, 1, 4, and 3, contain no uncertainty. Thus, the measurement 0.01438 gram contains four significant figures. The two zeros have no measurement significance and are used merely to place the decimal. The measurement could have been reported as 14.38 milligrams, mg, without using zeros. Examples of the correct procedure for reporting significant figures are as follows:

| Measurement | Accuracy possible with instrument used in measurement | Number of significant figures |
|---|---|---|
| 20.7 ml | ±0.1 ml | 3 |
| 84.72 ml | ±0.01 ml | 4 |
| 1891.0 g | ±0.1 g | 5 |
| 158,000 gal | ±1,000 gal | 3 |
| 4800 mi | ±100 mi | 2 |
| 5100 mi | ±10 mi | 3 |
| 3700 g | ±1 g | 4 |

Zeros may not only be used to place the decimal point; they may be significant figures. Consider the mass measurement 5.0073 grams made on a balance which is accurate to ±0.0001 gram. In this case the two zeros in 5.0073 grams are significant; thus the measurement has five significant figures. The mass measurement 0.0053 gram made on the balance (±0.0001 gram) contains only two significant figures. The two zeros in front of the 5 and 3 are used only to place the decimal and are not significant.

We must know the accuracy with which a measurement was made in order to determine the number of significant figures in the measurement. Consider the mass measurement 8.7190 grams made on a balance which is accurate to ±0.0001 gram. Any error in a mass measurement made with this balance would be in the fourth figure to the right of the decimal point. Thus, the first figure with any uncertainty in 8.7190 grams is the zero; it is significant.

Whenever a calculation is made using measured values, the answer obtained can be no more significant than the least accurate value used in the calculation. For instance, consider that you measure your height and report the value as 63 inches but you want to know the value in centimeters.

$$63 \text{ in.} \times \frac{2.54 \text{ cm}}{1.00 \text{ in.}} = 160.02 \text{ cm} = 160 \text{ cm} = 1.6 \times 10^2 \text{ cm}$$

Note that the measurement 63 in. contains only two significant figures; thus the answer must be rounded to two significant figures or 160 cm. The zero places the decimal and is not significant.

## 2  Scientific notation

Frequently it is convenient to write ordinary numbers using scientific notation. This method is especially useful for writing very large or very small numbers. It is also handy when one is using a slide rule for calculations.

To rewrite a number in scientific notation, follow these rules:

1. Replace the decimal point to the right of the first significant figure.

2. Maintain the value of the original number by multiplying the new number by a power of ten. The correct power of ten is determined by the number of places the decimal was moved and the direction of the move.

a. If the decimal was moved from right to left, the power of ten is positive.
b. If the decimal was moved from left to right, the power of ten is negative.
c. If the decimal was moved two places, the power of ten is 2. If it was moved three places, the power of ten is 3, and so on.

EXAMPLE 1   Rewrite 6425 in scientific notation.
1. Place the decimal to the right of the first significant figure.

$6425 \longrightarrow 6.425$

2. Multiply the new number by the correct power of ten.

$6425 \longrightarrow 6.425 \times 10^3$

Note that the power of ten is a positive 3: the decimal was moved three places from *right to left*.

EXAMPLE 2   Rewrite 0.000261 in scientific notation.
1. Place the decimal to the right of the first significant figure.

$0.000261 \longrightarrow 2.61$

2. Multiply the new number by the correct power of ten.

$0.000261 \longrightarrow 2.61 \times 10^{-4}$

Note that the power of ten is a negative 4: the decimal was moved four places from *left to right*.

The following equivalencies demonstrate the use of scientific notation.

$$10 = 1 \times 10^1 \qquad\qquad 1 = 1 \times 10^0$$
$$100 = 1 \times 10^2 \qquad\qquad 0.1 = 1 \times 10^{-1}$$
$$1000 = 1 \times 10^3 \qquad\qquad 0.01 = 1 \times 10^{-2}$$
$$10{,}000 = 1 \times 10^4 \qquad\qquad 0.001 = 1 \times 10^{-3}$$
$$100{,}000 = 1 \times 10^5 \qquad\qquad 0.0001 = 1 \times 10^{-4}$$

**Multiplication**

To multiply terms written in scientific notation:

1. Multiply the digit terms.
2. Multiply the exponential terms by algebraically adding together the powers of ten.
3. Write the answer in scientific notation.

The following examples demonstrate the process.

EXAMPLE 3  Multiply $3.7 \times 10^{-9}$ by $4.8 \times 10^{7}$.
1. Multiply the digit terms:

$3.7 \times 4.8 = 17$

2. Add the powers of 10:

$10^{-9} \times 10^{+7} = 10^{(-9+7)} = 10^{-2}$

(Note that the powers of ten were added algebraically.)
3. Rewrite the terms:

$17 \times 10^{-2} = 1.7 \times 10^{-1}$

EXAMPLE 4  Multiply $5.70 \times 10^{3}$ by $4.00 \times 10^{5}$.
1. Multiply the digit terms:

$5.07 \times 4.00 = 22.8$

2. Add the powers of ten:

$10^{3} \times 10^{5} = 10^{8}$

3. Rewrite the terms:

$22.8 \times 10^{8} = 2.28 \times 10^{9}$

**Division**

To divide terms written in scientific notation:

    1. Divide the digit terms as usual.
    2. Algebraically subtract the power of ten in the denominator (on bottom) from the power of ten in the numerator (on top).
    3. Write the quotient (answer) in scientific notation.

The following examples demonstrate division using scientific notation.

EXAMPLE 5  Divide $6.02 \times 10^{23}$ by $3.0 \times 10^{6}$.
1. Divide the digit terms:

$$\frac{6.02}{3.0} = 2.0$$

$$\frac{6.02 \times 10^{23}}{3.0 \times 10^{6}} = 2.0 \times \frac{10^{23}}{10^{6}}$$

2. Algebraically subtract the power of ten on the bottom from the one on top:

$$\frac{6.02 \times 10^{23}}{3.0 \times 10^{6}} = 2.0 \times \frac{10^{23}}{10^{6}} = 2.0 \times 10^{(23-6)}$$

3. Write the answer in scientific notation:

$$\frac{6.02 \times 10^{23}}{3.0 \times 10^{6}} = 2.0 \times 10^{17}$$

**EXAMPLE 6** Determine the density of a gas if 22,400 ml of the gas has a mass of 44.01 grams.

Note: First, write the two numbers in scientific notation:

22,400 ml = $2.24 \times 10^4$ ml
44.01 g = $4.401 \times 10^1$ g

$$\text{density} = \frac{\text{mass}}{\text{volume}} = \frac{4.401 \times 10^1 \text{ g}}{2.24 \times 10^4 \text{ ml}}$$

$$\text{density} = \frac{4.401}{2.24} \times \frac{10^1}{10^4} = 1.96 \times \frac{10^1 \text{ g}}{10^4 \text{ ml}}$$

$$\text{density} = 1.96 \times 10^{(1-4)} = 1.96 \times 10^{-3} \text{ g/ml}$$

**Addition**

To add numbers in exponential form, first move the decimal point to the right or left in one of the numbers so as to adjust to the same power of ten as the other number. Next add as usual.

**EXAMPLE 7** $3.78 \times 10^{-8} + 4.01 \times 10^{-7}$

$4.01 \times 10^{-7} = 40.1 \times 10^{-8}$

$$\begin{array}{r} 3.78 \times 10^{-8} \\ +40.1\phantom{0} \times 10^{-8} \\ \hline 43.88 \times 10^{-8} \end{array}$$ or $4.39 \times 10^{-7}$ (three significant digits)

# Appendix II
## Answers to selected odd-numbered questions

## Chapter 2

3. 1.00 g/ml
5. 76 liters or 76,000 ml
7. 0.555 lb or 252 g
9. 2.60 g/ml
11. 0.625 g
13. The mass of 1.00 liter of Hg is 29.7 lb.
15. 1.00 oz = 28.4 g
17. Approximately 45 liters for each 100 lb of body weight

## Chapter 3

3. All of the elements listed occur in sea water.
5. (a) $FePO_4$
   (b) $PbCl_2$
   (c) $PbO$
   (d) $CaCl_2$
   (e) $CaO$
7. (a) $^{131}_{53}I$, 53 electrons
       53 protons
       78 neutrons
   (b) $^{24}_{11}Na$, 11 electrons
       11 protons
       13 neutrons
   (c) $^{60}_{27}Co$, 27 electrons
       27 protons
       33 neutrons
   (d) $^{99}_{43}Tc$, 43 electrons
       43 protons
       56 neutrons
   (e) $^{51}_{24}Cr$, 24 electrons
       24 protons
       27 neutrons
   (f) $^{130}_{54}Xe$, 54 electrons
       54 protons
       76 neutrons
   (g) $^{197}_{80}Hg$, 80 electrons
       80 protons
       117 neutrons
   (h) $^{198}_{79}Au$, 79 electrons
       79 protons
       119 neutrons
9. Dalton's model assumes that one atom of Fe combines with one atom of O to form FeO.
15. (a) 1.01 g
    (b) 2.02 g
    (c) 16.0 g
    (d) 32.0 g
    (e) 19.0 g
    (f) 38.0 g
    (g) 23.0 g
    (h) 39.1 g
17. (a) Diamond—100% carbon
    (b) Corundum—52.9% aluminum
                  47.1% oxygen
    (c) Fluorite—51.3% calcium
                 48.7% fluorine
    (d) Gypsum—29.4% calcium
               23.6% sulfur
               47.0% oxygen
    (e) Talc—19.2% magnesium
             29.6% silicon
             50.6% oxygen
             0.53% hydrogen
19. (a) 5.93% H         (b) 1.60% H
        94.07% S            22.2 % N
                            76.2 % O

(c) 4.45% H  (d) 50.1 % S
95.6 % C   49.9 % O
21. (a) 0.4625 mole B atoms
(b) 0.4163 mole C atoms
(c) 0.2175 mole Na atoms
(d) 0.1614 mole P atoms
(e) 0.08516 mole Ni atoms
23. (a) CH            (e) NaO
(b) $CH_2$         (f) $Na_2O$
(c) HO             (g) $Zn_2O_6$
(d) $H_2O$         (h) $CuSO_4$
33. LiOH removes more $CO_2$ per unit weight.
35. Some mass must be lost, that is, transformed to energy when the 6 protons and 6 neutrons are brought together into the carbon-12.

## Chapter 4

3. (a) $_4$Be—2 valence electrons
(b) $_{12}$Mg—2 valence electrons
(c) $_{20}$Ca—2 valence electrons
(d) $_8$O—6 valence electrons
(e) $_{16}$S—6 valence electrons
(f) $_9$F—7 valence electrons
(g) $_{17}$Cl—7 valence electrons
(h) $_{10}$Ne—8 valence electrons
(i) $_{18}$Ar—8 valence electrons
5. The elements with unpaired electrons are paramagnetic (that is, O, S, F, and Cl) while the elements with no unpaired electrons are diamagnetic.
7. 10 d electrons—maximum
14 f electrons—maximum
9. The electron configuration of $_{14}$Si,

| ⇅ | ⇅ | ⇅ ⇅ ⇅ | ⇅ | ↓ ↓ ○ |
|---|---|---|---|---|
| 1s | 2s | 2p | 3s | 3p |

indicates that $_{14}$Si has *two* unpaired electrons.
11. The smallest unit of energy is a quantum of energy. The smallest unit of people is one person. The smallest unit of electricity is the electron and the atom is the smallest unit of matter. A quantum of anything is the smallest possible unit of that thing.
13. Energy is given off; thus the electrons must move from higher energy levels to lower energy levels. Energy is neither created nor destroyed in the process.
15. Maximum number of $e^-$ for $n = 1$ is 2; filled at He.

17. $n = 3$, 18 electrons maximum
$n = 4$, 32 electrons maximum
$n = 5$, 50 electrons maximum
19. (a) $n = 1$, s only
(b) $n = 2$, s and p
(c) $n = 3$, s, p, and d
(d) $n = 4$, s, p, d, and f
21. $1s < 2s < 2p < 3s < 3p < 4s < 3d < 4p$
25. (a) Second shell, $n = 2$
(b) Fourth shell, $n = 4$
(c) Fifth shell, $n = 5$
(d) Fourth shell, $n = 4$

## Chapter 6

9. (a) 10.1 g         (d) 1.43 g
(b) 260 g          (e) 4.77 g
(c) 2060 g         (f) 2250 g
11. (a) $4.35 \times 10^{24}$ $XeF_4$ molecules
(b) $4.35 \times 10^{24}$ $NO_2$ molecules
(c) $1.205 \times 10^{26}$ $CH_4$ molecules
(d) $2.29 \times 10^{22}$ Xe atoms
(e) $2.10 \times 10^{22}$ $XeF_4$ molecules
(f) $9.46 \times 10^{22}$ $NO_2$ molecules
13. (a) 1.00 mole $Cl_2$
(b) $1.99 \times 10^{-2}$ mole HCl
(c) $1.51 \times 10^{-13}$ mole $SO_2$
15. The elements increase in nuclear charge, that is, an increased number of protons, which exerts a progressively stronger force of attraction on the valence electrons.
19. (a) Nonpolar      (c) Polar
(b) Nonpolar      (d) Polar

## Chapter 8

5. $7.23 \times 10^2$ mi/hr = 723 mi/hr
7. (a) 48.5 lb/in²
(b) 2510 mm Hg
(c) 3.30 atm
9. 1128 mm Hg
11. (a) 0°F = −17.8°C
(b) 0°F = 255°K
(c) Absolute zero = −459.69°F
(d) Mr. Farenheit had a fever on the day he calibrated his thermometer.
15. 4080 cal
17. (a) 463 cal
(b) 7120 cal = $7.12 \times 10^3$ cal = 7.12 kcal
(c) 21,000 cal = $2.10 \times 10^4$ cal = 21.0 kcal

## Chapter 9

9. (a) $2\,Li(s) + 2H_2O(l) \rightleftharpoons 2LiOH(aq) + H_2(g)$
   (b) $2Na(s) + 2H_2O(l) \rightleftharpoons 2NaOH(aq) + H_2(g)$
11. (a) $Na_2O(s) + H_2O(l) \rightleftharpoons 2NaOH(aq)$
    (b) $Li_2O(s) + H_2O(l) \rightleftharpoons 2LiOH(aq)$
    (c) $KH(s) + H_2O(l) \rightleftharpoons KOH(aq) + H_2(g)$
19. (a) 0.813 mole $SiO_2$
    (b) 0.813 mole $SiF_4$
    1.63 moles $H_2O$
21. (a) 28,000 moles S
    (b) $1.0 \times 10^7$ moles S per year
    (c) 7,000 moles $CH_4$, $1.6 \times 10^5$ liters $CH_4$ at STP
23. 62.9% $H_2O$

## Chapter 10

5. (a) 4 $BrF_3$ molecules
   (b) 12 $TiF_4$ molecules
   (c) 16 $Br_2$ molecules
   (d) 127 $O_2$ molecules
   (e) $1.5 \times 10^{23}$ $O_2$ molecules
7. (a) 1.00 mole $NaIO_3 = 198$ g
      1.00 mole $NaHSO_3 = 104$ g
      1.00 mole $I_2 = 254$ g
      1.00 mole $Na_2SO_4 = 142$ g
      1.00 mole $NaHSO_4 = 120$ g
      1.00 mole $H_2O = 18.0$ g
   (b) 254 g $I_2$
   (c) 520 g $NaHSO_3$
   (d) 781 g $Na_2SO_4$
   (e) 1980 g $NaIO_3$
   (f) 1430 g $NaHSO_3$
9. (a) 150 cm²     (b) 750 cm²
11. The concentration of ethanol, the intoxicating component of beer and whiskey, is greater in whiskey.
17. (a) Decomposition    (f) Replacement
    (b) Combination      (g) Decomposition
    (c) Replacement      (h) Combination
    (d) Displacement     (i) Displacement
    (e) Replacement      (j) Displacement

## Chapter 11

13. $2.0 \times 10^{-4}$ mole iodine atoms
15. (a) 3.22 g $Na^+$     (b) 28 $Na^+$ to 1 $K^+$
    0.2 g $K^+$
    3.65 g $Cl^-$
17. (a) 1.00 mole $KMnO_4$   (b) 2.50 moles $Cl_2$

## Chapter 12

7. 1.0 atm
9. 12.3 moles of gas, $7.39 \times 10^{24}$ molecules
11. The lighter $CH_4$ will diffuse about 1.7 times faster than $CO_2$.
13. 63.8 g/mole
15. 16.7 ml $N_2$ at 373°K and 1.00 atm
17. P total = 2.91 atm
    $P_{O_2} = 0.45$ atm
    $P_{N_2} = 0.22$ atm
    $P_{He} = 2.2$ atm
19. $P_{He} = 2.36$ atm
    $P_{Ar} = 0.94$ atm
21. The lighter neon atoms will diffuse about 3.15 times faster than the mercury atoms.

## Chapter 13

9. (a) 88.0 moles of ethane
   (b) 2.00 moles of ethane
11. (a) 8.0            (d) 20.0
    (b) 12             (e) 12
    (c) 2.00
13. 5.64 moles NaOH
15. 6.51 g Na
17. 179 liters $CO_2(g)$
    (a) 10.0 moles $H_2O(g)$   (d) 20 moles $H_2O(g)$
    (b) 20.0 moles $CO_2(g)$   (e) 80 moles $CO_2(g)$
    (c) 8.0 moles $CO_2(g)$
19. 2.75 g $N_2O(g)$
21. 3.00 liters $CO(g)$ will produce 3.00 liters $CO_2(g)$
23. 381 g $H_2SO_4$
25. 0.038 g baking soda ($NaHCO_3$)
27. (a) $C_3H_8(g)$ is in excess
    (b) 0.060 mole $CO_2(g)$
    (c) 1.4 g $H_2O(g)$

## Chapter 14

7. (a) 6.8 g $H_2O_2$ (b) 220 g $H_2O$ (c) 0.20 mole $H_2O_2$
9. (a) 0.992% sugar
      0.0293 m, 0.0292 M
   (b) 0.992% NaCl
      0.171 m, 0.171 M
   (c) 0.992% $CuSO_4 \cdot 5H_2O$
      0.0401 m, 0.0400 M
   (d) 0.992% $KAl(SO_4)_2 \cdot 12H_2O$
      0.0211 m, 0.0210 M
11. 140 g alcohol, 3.0 M

13. (a) 22.3% urea, 77.7% $H_2O$
    (b) 4.78 m
15. $5 \times 10^{-6}$ mole HCl
17. (a) 420 g KBr   (c) 4.2 g KBr
    (b) 42 g KBr    (d) 0.42 g KBr
19. 1.4 m
21. (a) 0.742 m     (c) 0.476 moles
    (b) 0.742 moles (d) 74.0 g/mole
23. 0.900 M
25. 0.490 M

## Chapter 15

7. (a) $3.7 \times 10^{-4}$°C   (c) $7.4 \times 10^{-2}$°C
   (b) $3.7 \times 10^{-3}$°C   (d) 0.11°C
9. The $1.00 \times 10^{-4}$ m sugar solution has the higher boiling point. (A $3.00 \times 10^{-4}$ m AgCl solution cannot be prepared due to the lack of solubility of AgCl.)
11. $Pb^{2+}$, $Cu^+$, $Hg_2^{2+}$, $Au^+$
13. $Ca^{2+}$, $Mg^{2+}$, $Fe^{2+}$, $Cu^{2+}$
15. (a) $KCl(s) \xrightarrow{H_2O} K^+(aq) + Cl^-(aq)$
    (b) $H_2S(g) + H_2O(l) \rightleftharpoons H_3O^+(aq) + HS^-(aq)$
        $HS^-(aq) + H_2O(l) \rightleftharpoons H_3O^+(aq) + S^{2-}(aq)$
    (c) $NH_3(g) + H_2O(l) \rightleftharpoons NH_4^+(aq) + OH^-(aq)$
    (d) $HI(g) + H_2O(l) \rightleftharpoons H_3O^+(aq) + I^-(aq)$
    (e) $H_2SO_4(l) + H_2O(l) \rightleftharpoons$
        $H_3O^+(aq) + HSO_4^-(aq)$
        $HSO_4^-(aq) + H_2O(l) \rightleftharpoons$
        $H_3O^+(aq) + SO_4^{2-}(aq)$
    (f) $CoCl_2(s) \xrightarrow{H_2O} Co^{2+}(aq) + 2Cl^-(aq)$
    (g) $Zn(NO_3)_2(s) \xrightarrow{H_2O} Zn^{2+}(aq) + 2NO_3^-(aq)$
    (h) $Na_3PO_4(s) \xrightarrow{H_2O} 3Na^+(aq) + PO_4^{3-}(aq)$
        $PO_4^{3-}(aq) + H_2O(l) \rightleftharpoons$
        $HPO_3^{2-}(aq) + OH^-(aq)$
        $HPO_3^{2-}(aq) + H_2O(l) \rightleftharpoons$
        $H_2PO_3^{2-}(aq) + OH^-(aq)$ (slight)
17. (a) $SnS(s) + S^{2-}(aq) \rightleftharpoons SnS_2^{2-}(aq)$
    (b) $Ti(OH)_2(s) + 2H_3O^+(aq) \rightleftharpoons$
        $Ti^{2+}(aq) + 4H_2O(l)$
    (c) $BaSO_3(s) + 2H_3O^+(aq) \rightleftharpoons$
        $Ba^{2+}(aq) + H_2SO_3(aq) + 2H_2O(l)$
    (d) $Cd(OH)_2(s) + 2H_3O^+(aq) \rightleftharpoons$
        $Cd^{2+}(aq) + 4H_2O$
    (e) $Zn(OH)_2(s) + 2OH^-(aq) \rightleftharpoons Zn(OH)_4^{2-}(aq)$
19. (a) 1340 g
    (b) There are about 1280 g of water in 1340 g urine.

(c) 0.20 m, 0.20 M
(d) 0.39 m, 0.38 M
21. (a) $AgNO_3(aq) + HCl(aq) \rightleftharpoons$
        $AgCl(s) + HNO_3(aq)$
        $Ag^+(aq) + Cl^-(aq) \rightleftharpoons AgCl(s)$
    (b) $AgNO_3(aq) + HBr(aq) \rightleftharpoons$
        $AgBr(s) + HNO_3(aq)$
        $Ag^+(aq) + Br^-(aq) \rightleftharpoons AgBr(s)$
    (c) $BaCl_2(aq) + H_2SO_4(aq) \rightleftharpoons$
        $BaSO_4(s) + 2HCl(aq)$
        $Ba^{2+}(aq) + SO_4^{2-}(aq) \rightleftharpoons BaSO_4(s)$
    (d) $Ba(ClO_3)_2(aq) + H_2SO_4(aq) \rightleftharpoons$
        $BaSO_4(s) + 2HClO_3(aq)$
        $Ba^{2+}(aq) + SO_4^{2-}(aq) \rightleftharpoons BaSO_4$
    (e) $Pb(NO_3)_2(aq) + 2HCl(aq) \rightleftharpoons$
        $PbCl_2(s) + 2HNO_3$
        $Pb^{2+}(aq) + 2Cl^- \rightleftharpoons PbCl_2(s)$

## Chapter 16

11. (a) 10.0 N    (c) 2.10 N
    (b) 4.0 N     (d) 0.0106 N
13. $4.0 \times 10^{-4}$ M in $H_3O^+$
15. (a) Weakly acidic medium
    (b) $1.0 \times 10^{-6}$ M in $H_3O^+$
17. 403 liters or $4.03 \times 10^5$ ml

## Chapter 17

9. $K_{eq} = 6.3 \times 10^2$
11. $K_i = 4.4 \times 10^{-4}$
13. 3.3 mole $CoCl_2$
15. $[Ag^+(aq)] = 3 \times 10^{-18}$ M

## Chapter 18

3. (a) $14H_3O^+ + Cr_2O_7^{2-} + 6e^- \rightleftharpoons 2Cr^{3+} + 21H_2O$
   (b) $12H_3O^+ + 2IO_3^- + 10e^- \rightleftharpoons I_2 + 18H_2O$
   (c) $MnO_4^{2-} \rightleftharpoons MnO_4^- + e^-$
   (d) $Fe^{3+} + e^- \rightleftharpoons Fe^{2+}$
   (e) $3H_3O^+ + NO_3^- + 2e^- \rightleftharpoons HNO_2 + 4H_2O$
   (f) $Mn(OH)_2 + 2OH^- \rightleftharpoons MnO_2 + 2H_2O + 2e^-$
5. 2.04 volts

## Chapter 19

5. The element at the bottom of each one of the A groups is the most metallic element in that group.

7. (a) $Zn(OH)_2(s) + 2KOH(aq) \rightleftharpoons$
    $Zn(OH)_4^{2-}(aq) + 2K^+(aq)$
   (b) $Zn(OH)_2(s) + 2RbOH(aq) \rightleftharpoons$
    $Zn(OH)_4^{2-}(aq) + 2Rb^+(aq)$
   (c) $Zn(OH)_2(s) + 2HCl(aq) \rightleftharpoons$
    $Zn^{2+}(aq) + 2Cl^-(aq) + 2H_2O(l)$
   (d) $Zn(OH)_2(s) + 2HNO_3(aq) \rightleftharpoons$
    $Zn^{2+}(aq) + 2NO_3^-(aq) + 2H_2O(l)$
9. (a) $2K(s) + F_2(g) \rightleftharpoons 2KF(s)$
   (b) $Ca(s) + F_2(g) \rightleftharpoons CaF_2(s)$
   (c) $2In(s) + 3F_2(g) \rightleftharpoons 2InF_3(s)$
   (d) $Si(s) + 2F_2(g) \rightleftharpoons SiF_4(g)$
   (e) $2Sb(s) + 3F_2(g) \rightleftharpoons 2SbF_3(s)$
   (f) $O_2(g) + 2F_2(g) \rightleftharpoons 2OF_2(g)$
   (g) $Cl_2(g) + F_2(g) \rightleftharpoons 2ClF(g)$
   (h) $Xe(g) + 2F_2(g) \rightleftharpoons XeF_4(s)$

# Chapter 20

5. (a) 

$$H-\underset{\underset{H}{|}}{\overset{\overset{H}{|}}{C}}-\underset{\underset{H}{|}}{\overset{\overset{H}{|}}{C}}-H(g) + Cl_2(g) \rightleftharpoons$$

$$H-\underset{\underset{H}{|}}{\overset{\overset{H}{|}}{C}}-\underset{\underset{H}{|}}{\overset{\overset{H}{|}}{C}}-Cl + HCl$$

ethyl chloride

(b) 

$$H-\overset{\overset{H}{|}}{C}=\overset{\overset{H}{|}}{C}-H(g) + H_2O(l) \xrightarrow{H_3O^+}$$

$$H-\underset{\underset{H}{|}}{\overset{\overset{H}{|}}{C}}-\underset{\underset{H}{|}}{\overset{\overset{H}{|}}{C}}-O-H$$

ethyl alcohol or ethanol

7. (a) $CH_3-\overset{\overset{O}{\|}}{C}-H$  (c) $H_2O$  (d) $H-\overset{\overset{O}{\|}}{C}-H$
   (b) $CH_3-O-H$

9. $CCl_4$ and $CH_4$ are both symmetric molecules. The polarity caused by one group is exactly negated by the forces of the three other identical groups.

11. Aldehydes are more easily oxidized than are ketones. Oxidation of ethyl alcohol, $CH_3CH_2OH$,

 yields $CH_3-\overset{\overset{O}{\|}}{C}-H$, acetaldehyde, while

 oxidation of $CH_3-\overset{\overset{O}{\|}}{C}-H$ yields acetic acid,

 $CH_3-\overset{\overset{O}{\|}}{C}-O-H$.

15. Lysergic acid differs considerably in the group that is attached to the acid group, $-\overset{\overset{O}{\|}}{C}-O-H$.

# Index

Absolute zero, 167
Acid-base reactions, 345–365
Acids, amino, 455
   aqueous solutions, 242–245, 346
   Arrhenius, preparation, 355–356
   Arrhenius model, 346–347
   Brønsted, 220, 363–365
   Brønsted-Lowry model, 349
   conjugate, 351–352
   electron donor reactions, 353–355
   ionization, 349–350
   Lewis model, 353–355
   neutralization, 216–217, 347–348
   Brønsted-Lowry model, 352
   stoichiometry, 361–362
   normality, 357
   organic, 448–451
   pH concept, 357–361
   pH scale, 360–361
   proton transfer reactions, 348–353
   reactions of boron family with, 425–426
   reactions of metals with, 242–245, 249
   weak, 245
Activation energy, 371
Activity series, 189, 243, 394
Addition, 464
Addition reactions, 445

Air, components, 15
Alcohols, 446–448
Aldehydes, 452–453, 454
Alkali metal halide solutions, 238–242
Alkali metals, 229–235
   chemical properties, 229–230
   electrolysis, 233–234
   physical properties, 229
   reaction with halogens, 230–233
   reaction with water, 230
   solubility, 233
Alkaline earth family of elements, 428–431
   activity, 428–429
   oxidation numbers, 143
   oxides, 429
   properties, 428, 429
Alkanes, 441, 443–444
Alkenes, 441–442, 444–446
Alkynes, 442
Aluminum solutions, 427–428
Amides, 455
Amines, 454
Amino acids, 455
Ammonia, 420
Amphiprotic substances, defined, 351
Anhydrous, defined, 194
Anode, defined, 234, 292–293
Aqueous, defined, 9
Argon, 411
Aromatic compounds, 442
Arrhenius, Svante August, 346
Arrhenius acids and bases, 355–356
Arrhenius model, 346–347
Atmosphere (air), composition, 15
Atmosphere (unit of pressure), 164
Atomic mass units, 54, 57
Atomic numbers, 55, 100–101
Atomic orbital diagrams, 87–89
Atomic weight, 57–58
Atoms, 50–52
   defined, 23
   electron-dot structures, 92–93, 102, 103
   electronic structure, 76–93
   model, 55, 79–80
   nucleus, 55
Auto-oxidation reduction, 237

Avogadro, Amedeo, 52, 58–59, 258
Avogadro's number, 58–59, 258

Balance (instrument), 35
Balanced equations,
   interpretation, 208–210
   redox reactions, 397–402
   writing, 62–64
   see also Equations
Barometer, 164
Bases, aqueous solutions, 346
   Arrhenius, preparation, 355–356
   Arrhenius model, 347
   Brønsted, 220, 363–356
   Brønsted-Lowry model, 349
   conjugate, 351–352
   electron-donor reactions, 353–355
   ionization, 350–351
   Lewis model, 353–355
   neutralization, 216–217, 347–348
      Brønsted-Lowry model, 352
      stoichiometry, 361–362
   normality, 357
   pH concept, 357–361
   pH scale, 360–361
   proton-transfer reactions, 348–353
   solutions, 346
Battery, lead storage, 405–406
Benzene, 442
Binary compounds, defined, 125
   ionic, 129–130
   naming, 125, 129–130, 149–153
Bohr, Niels Henrik David, 78–79
Boiling points, 176
   electrolytes, 328–329
   solutions, 307–308, 320–321
   trends in groups, 104
Bond angle, 438–439
Bond length, 115, 438
Bonds, chemical, 114–118
   classification, 116–117
   covalent, 120–124
      and electronegativity, 124
      organic chemistry, 438
   defined, 51, 114
   hydrogen, 135, 185–186

   ionic, 117–118, 125–129, 135
      organic chemistry, 438
   metallic, 135–136
   nonpolar covalent, 116
   in organic chemistry, 438–441
   polar covalent, 116–117, 130–133
   polarity of, 439–440
Borances, 428
Boron family of elements, 425–428
   oxides, 426–427
   properties, 425, 426
   reactions with acids, 425–426
Boyle, Robert, 261, 345
Boyle's law, 261, 264
Brønsted, J. N., 349
Brønsted acids and bases, 220
   salts, 363–365
Brønsted-Lowry model, 349, 352

Calorie, defined, 169
Carbon family of elements, 422–425
   hydrogen compounds, 422;
      see also Organic chemistry
   oxides, 422–424
   properties, 422, 423
Carbonates, 424
Carbon-14, 424–425
Catalysts, 212, 372
Cathode, defined, 234, 393
Celsius scale, 167
   conversion, 168–169
Centrifuge, 18
Changes, chemical, defined, 4–5
   and physical distinguished, 24
Charles, Jacques Alexandre César, 263
Charles' law, 263, 266, 267
Chemist, activities, 1–2
Chemistry
   defined, 1
   inorganic, defined, 410
   organic, 437–456
      chemical bonds, 438–441
      defined, 437
      functional groups, 446, 452
   relationship to world, 7–8
Chlorine, see Halogens
Closed system, defined, 7

Colligative properties, 308, 329
Collision theory, 210–214
Combination reactions, 214–215
Combined gas laws, 269–271
Combining ratios, 128–129, 130
Complex ion, 340
Composition, defined, 3
Compounds
  aromatic, 442
  binary
    defined, 125
    ionic, 129–130
    naming, 125, 129–130, 149–153
  covalent, 147
  defined, 14–15
  ionic, 125–130
    combining ratios, 128–129
    names and formulas, 129–130
  ternary, 147–148
    naming, 153–156
  transition, 92
Compressibility, 178
Concentration, defined, 213
  solutions, 313–317
Condensation, 178
Conductivity, electrical, 307, 326–328, 333–334
Configurations, electronic, 83–84
  and periodic chart, 102–103
Conjugate acid-base pairs, 351–352
Conservation of mass, law of, 62
Constant, equilibrium, 379–384
  ionization, 359, 382–383
  solubility product, 383–384
Covalent bonds, 120–124
  and electronegativity, 124
  nonpolar, 116
  organic chemistry, 438
  polar, 116–117, 130–133
Covalent compounds, 147
Cracking, 444
Crust, earth's, distribution of elements, 48
Crystals, 170–171
Cylinder, graduated, 33, 34

Dalton, John, 51–52, 62, 77, 214
Dalton's law of partial pressures, 275

Dating, radioactive carbon, 424–425
Davy, Sir Humphrey, 346
Decantation, defined, 18
Decomposition reactions, 215
Definitions, operational, defined, 6
  theoretical, defined, 7
Democritus, 50–51
Density, 40–42
  of gases, 256–258
Derived property, defined, 28
Diamagnetic, 86
Diatomic molecules, 53
Diffusion, of gases, 178
  law of, 279
  relative rates, 277–280
  of liquids, 174
Dipole moment, 440
Dipole-dipole interactions, 134, 440
Displacement reactions, 215–217
Disproportionation, 237
Division, 463–464
Döbereiner, Johann Wolfgang, 99
Dynamic equilibrium, 374–375
  defined, 172, 351
  liquid-gas, 176, 373–374
  solid-gas, 173
  solid-liquid, 172–173

Earth's crust, distribution of elements, 48
Effusion, law of, 279
Eight, rule of, 122
  combining ratios and, 130
Electrical conductivity, 307, 326–328, 333–334
Electrical energy, 23
Electrical nature of matter, 19–21
Electrochemical cells, 406
Electrolysis, 233–234
Electrolytes, 326–340
  boiling points, 328–329
  conductivity, 307, 326–328, 333–334
  defined, 307
  freezing points, 328–329
Electromotive series, 395, 396
Electron donor reactions, 221–222

Lewis model, 353–355
Electron transfer reactions, 218–220
Electron-dot structures, 92–93, 102, 103
Electronegativity, 118–120
  covalent bonds and, 124
Electronic configurations, 83–84
  and periodic chart, 102–103
Electronic structure, 76–93
Electrons, 53–54, 76–93
  defined, 20
  energy levels, 78–79
  energy sublevels, 81–84
  and magnetism, 86
  orbitals, 84–86
  valence, 90–92
Elements, alkali metals, 229–235
  chemical properties, 229–230
  electrolysis, 233–234
  physical properties, 229
  reaction with halogens, 230–233
  reaction with water, 230
  solubility, 233
  alkaline earth family, 428–431
    activity, 428–429
    oxidation numbers, 143
    oxides, 429
    properties, 428, 429
  atomic numbers, 55, 100–101
  atomic weights, 57–58
  boron family, 425–428
    oxides, 426–427
    properties, 425, 426
    reaction with acids, 425–426
  carbon family, 422–425
    hydrogen compounds, 422; see also Organic chemistry
    oxides, 422–424
    properties, 422, 423
  defined, 14–15
  distribution, 47–48
  group IA, see alkali metals
  group IIA, see alkaline earth family
  group IIIA, see boron family
  group IVA, see carbon family

Elements (Continued)
  group VA, see nitrogen family
  group VIA, see oxygen family
  group VIIA, see halogens
  group VIIIA, see noble gases
  group B, see transition
  halogens, 235–242, 413–415
    chemical properties, 236–238
    interhalogens, 414
    oxidation numbers, 144
    physical properties, 235–236
    reaction with alkali metals, 230–233
    reaction with halogens, 239–242
    reaction with hydrogen, 237
    reaction with silver nitrate solutions, 238–239
    reaction with water, 236–237
    solubility, 235, 236
    uses, 414–415
  inner transition, 103, 433–434
  names, 48
  nitrogen family, 419–422
    characteristics, 419–420
    hydrogen compounds, 420
    oxides, 420–421
  noble gases, 106–107, 411–413
    characteristics, 411–412
    chemistry, 412
    uses, 412–413
  oxygen family, 415–418
    characteristics, 415–416
    reactions, 416–417
  periodic chart, see Periodic chart
  representative, defined, 103
  symbols, 48–49
  transition, 431-434
    defined, 103
    inner, 103, 433–434
    oxides, 433
    properties, 431–432
Empirical formulas, 67–70
Endothermic reactions, 233
Energy, activation, 371
  electrical, 23
    redox reactions as source, 405–406
  forms, 21–23
  heat, 22, 167
  ionization
    defined, 106
    electronegativity and, 120
    metals, 107
    as periodic property, 105–106
  kinetic, 21–22, 161–162
    formula, 22
    types, 161
  light, 22–23, 77–78
  potential, 22
  thermal, 22, 167
    defined, 167
Energy levels, 78–84
Enthalpy of formation, 232–233
Enzymes, 371
Equations, chemical, balanced, interpretation, 208–210
  writing, 62–65, 397–402
  calculations, 294–300
  mole relationships, 209, 292–294
  redox, 397–402
  symbols in, 11, 49–50
  word, 9–10
Equilibrium, chemical, 374–375, 377–379
  defined, 369
  dynamic, defined, 172, 351
  liquid-gas, 176, 373–374
  phase, defined, 373
  shifting, 375–377
  solid-gas, 173
  solid-liquid, 172–173
Equilibrium constant, 379–384
Evaporation, 175
Excited state, defined, 81
Exothermic reactions, 233
Exponents, 461–464

Fahrenheit scale, 167
  conversion, 168–169
Ferromagnetic, 86
Figures, significant, 31, 42–43, 460–461
Fission, 433–434
Fluidity, 174
Fluorocarbons, 414
Formation, heat of, 232–233
Formula unit, defined, 61
Formula weight, defined, 61
Formulas, 52–53
  defined, 50
  empirical, 67–70
  percent weight from, 65–66
Freezing points, 175
  electrolytes, 328–329
  solutions, 307–308, 321–322
  see also Melting points
Fuel cells, 406
Functional groups, defined, 446
  table, 452
Fusion, heat of, 171–172

Gases, 255–281
  Boyle's law, 261, 264
  calculations, 263–269
  Charles' law, 263, 266, 267
  combined gas laws, 269–271
  compressibility, 178
  defined, 160
  densities, 256–258
  diffusion, 178
    law of, 279
    relative rates of, 277–280
  effusion, law of, 279
  ideal, defined, 257
  ideal gas law, 271–273
  ideal and real, 280–281
  kinetic molecular model, 162–163, 255–256
  noble, 106–107, 411–413
    characteristics, 411–412
    chemistry, 412
    uses, 412–413
  partial pressures, 273–277
    Dalton's law of, 275
  properties, 178
  real and ideal, 280–281
  volume-pressure relationship, 258–261
    calculations, 263–266
  volume-quantity relationship, 261–262
  volume-temperature relationship, 262–263
    calculations, 266–269
Germanium, 100, 108–109
Graduated cylinder, 33, 34
Graham, Thomas, 279
Ground state, defined, 81
Groups, functional
  defined, 446
  table, 452
  periodic, 101–102
  prediction of properties, 109
  see also Elements

Half-cells, 391–413
Half-life, defined, 433
Half-reactions, defined, 218
   writing, 399–401
Halogen oxyions, 413
Halogens, 235–242, 413–415
   chemical properties, 236–238
   interhalogens, 414
   oxidation numbers, 144
   physical properties, 235–236
   reactions with alkali metals, 230–233
   reactions with halogens, 239–242
   reactions with hydrogen, 237–238
   reactions with silver nitrate solutions, 238–239
   reactions with water, 236–237
   solubility, 235, 236
   uses, 414–415
Hard water, 431
Heat, specific, 169–170
Heat energy, 22, 167
Heat of formation, 232–233
Heat of fusion, 171–172
Heat of vaporization, 176–177
Heterogeneous mixtures, defined, 16
Homogeneous mixtures, defined, 16
Homologous series, defined, 441
Hydrates, 194
Hydration reaction, 194
Hydrocarbons, 441
Hydrogen, 201–202
   oxidation number, 144
   reaction with halogens, 237–238
   in water molecule, 195
   water reaction with, 190–192
Hydrogen bonds, 135
   water, 185–186
Hydronium ions, 191
Hydroscopic, defined, 194

Ideal gas law, 271–273
Ideal gases, defined, 257
   and real gases, 280–281
Immiscible, defined, 306
Inner transition elements, 103, 433–434
Inorganic chemistry, defined, 410

Interhalogens, 414
Intermolecular forces, 133–136
Intramolecular forces, 133–136
Iodine, sublimation of, 10–13
Ion product, 359
Ionic bonds, 117–118, 125–129, 135
   organic chemistry, 438
Ionic compounds, 125–130
   combining ratios, 128–129
   names and formulas, 129–130
Ionization, of acids, 349–350
   of bases, 350–351
   defined, 106
   theory of, 334–335
   of water, 194–195
Ionization constants, 359, 382–383
Ionization energy
   defined, 106
   electronegativity and, 120
   metals, 107
   nonmetals, 108
   as periodic property, 105–106
Ions, complex, 340
   defined, 21
   hydronium, 191
   monatomic, 143–144
   polyatomic, 148–149
   spectator, 336
Isotopes, 56–57
   radioactive, 372
      decay, 433
   defined, 57

Kelvin scale, 167
   conversion, 168
Ketones, 453–454
Kinetic energy, 21–22, 161–162
   formula, 22
   types, 161
Kinetic molecular theory, 160, 161–163
   gas model, 162–163, 255–256
Kinetics, chemical, 211–214, 369, 370–372

Lavoisier, Antoine Laurent, 346
Le Chatlier, Henri, 375
Le Chatlier's principle, 375
Length, 31–33
   bond, 115, 438
Lewis, Gilbert N., 354
Lewis model, 353–355

Light, 22–23, 77–78
Liquids, defined, 160
   properties, 173–178
Lowry, T. M., 349

Macroscopic, defined, 5
Magnetism, 86
Manometer, 165–166
Mass, conservation of, law of, 62
   measurement, 35
Mass number, 56–57
Mass units, atomic, 54, 57
Mathematics review, 460–464
Matter, defined, 2
   electrical nature, 19–21
   states of, 17–19, 160–178
Measurement, 28–43
   defined, 29
   density, 40–42
   length, 31–33
   mass, 35
   metric system, 30
   pressure, 164–165
   temperature, 167
   uncertainty, 31
   volume, 33–34
Melting points, 171
   defined, 173
   see also Freezing points
Mendeleev, Dmitri Ivanovich, 100, 108–109
Meniscus, 33
Metallic bonds, 135–136
Metalloids, 108
Metals, activity series, 189, 243, 394
   alkali, 229–235
      chemical properties, 229–230
      electrolysis, 233–234
      physical properties, 229
      reaction with halogens, 230–233
      reaction with water, 230
      solubility, 233
   defined, 47–48, 107
   electronegativity, 118
   general properties, 47–48, 411
   ionization energy, 107
   reaction with acids, 242–245, 249
   reaction with nonmetals, 246
   reaction with water, 187–188, 248

Methane, 439
Metric system, 30
Meyer, Lothar, 100
Microscopic, defined, 5
Microscopic reversibility, 374
Millikan, Robert Andrew, 54
Miscible, defined, 306
Mixtures, 15–17
  heterogeneous, defined, 16
  homogeneous, defined, 16
Model, defined, 6–7
Molality, 314–315
Molarity, 315–317
Molecular orbital, 121
Molecular theory, kinetic, 160, 161–163
  gas model, 162–163, 255–256
Molecular weight, 60–62
Molecules, defined, 23–24
  diatomic, 53
  polar, 132–133
  polarity of, 440–441
Moles, 136–137
  concept, 291–292
  defined, 59–60
  molecular weight and, 60–62
  ratios, 199–201, 209, 292–294
    calculations, 294–300
  unity terms, 287–291
Monatomic ions, 143–144
Moseley, Henry Gwyn Jeffreys, 100
Multiplication, 462–463

Neutralization, 216–217, 347–348
  Brønsted-Lowry model, 352
  stoichiometry, 361–362
Neutrons, 53–54
Newlands, J. A. R., 99
Nitric acid, 421
Nitrogen family of elements, 419–422
  characteristics, 419–420
  hydrogen compounds, 420
  oxides, 420–421
Noble gases, 106–107, 411–413
  characteristics, 411–412
  chemistry, 412
  uses, 412–413
Nonmetals, defined, 48, 107–108
  general properties, 411
  ionization energy, 108
  reaction with metals, 246
  reaction with water, 188–189
Nonpolar covalent bonds, 116
Normality, acid-base reactions, 357
  oxidation-reduction reactions, 403
Notation, scientific
  mathematical operations, 462–464
  writing, 461–462
Nucleus, 55
Number, atomic, 55, 100–101
  Avogadro's, 58–59, 258
  mass, 56–57
  oxidation, 142–149
    variable, 145–147

Octaves, law of, 99
Octet rule, 122
  combining ratios and, 130
Open system, defined, 7
Operational definitions, defined, 6
Orbital diagrams, 87–89
Orbitals, 84–86
  molecular, 121
  overlapping, 121
Organic acids, 448–451
Organic chemistry, 437–456
  chemical bonds, 438–441
  defined, 437
  functional groups
    defined, 446
    table, 452
Oxidation, defined, 199, 388
Oxidation numbers, 142–149
  variable, 145–147
Oxidation potentials, 395–397
Oxidation-reduction reactions, 388–406
  as energy source, 405–406
  half-cells, 391–393
  stoichiometry, 402–405
Oxides, alkaline earth family, 429–430
  boron family, 426–427
  carbon family, 422–424
  nitrogen family, 420–421
  transition elements, 433
  water reaction with, 189–190, 247–248
Oxidizing agent, defined, 389
Oxycompounds, 417–418

Oxygen, 202–204
  oxidation number, 144
  reactions, 246–247
  in water molecule, 197–198
Oxygen family of elements, 415–418
  characteristics, 415–416
  reactions, 416–417
Oxyions, halogen, 413

Paramagnetic, 86
Partial pressures, 273–277
  Dalton's law of, 275
Particles, 2–4
  attractive forces, 12
  motion, 12
  subatomic, 53–54
  see also electronic structure
Pauling, Linus, 118, 412
Peptides, 455
Percent weight, from formula, 65–66
  from laboratory data, 66–67
Periodic chart, combining ratios and, 130
  defined, 99
  electronic configurations and, 102–103
  groups, 101–102
  historical development, 99–100
  illustrated, inside back cover
  periods, 101
  prediction of properties, 108–110
  structural basis, 101
Periodic law, 104–105
Periodic table, see Periodic chart
Periodicity, 98–110
pH concept, 357–361
pH scale, 360–361
Phase equilibrium, defined, 373
  liquid-gas, 176, 373–374
  solid-gas, 173
  solid-liquid, 172–173
Phases, 18–19, 160–178
Photon, 77
Physical changes and chemicals distinguished, 24
Planck, Max, 77
Plasma, defined, 160
Polar covalent bonds, 116–117, 130–133

Polar molecules, 132–133
Polarity, of bonds, 439–440
  of molecules, 440–441
Polyatomic ions, 148–149
Polymers, 451–452
Potential energy, 22
Potentials, oxidation, 395–397
Precipitate, defined, 215
  dissolving, 338–340
Precipitation reactions,
    215–216, 222–223,
    249–250
Pressure, 163–166
  defined, 163
  gases, 258–261, 263–266
  measurement, 164–165
  partial, 273–277
  and solubility, 312
  units, 164
  vapor, 175–176, 318–319, 377
Products, defined, 9
  determining, 245–250
Properties
  colligative, 308, 329
  defined, 2, 3
  derived, 28
  predicting, 108–110
Proteins, 455–456
Proton transfer reactions,
    220–221
  Brønsted-Lowry model,
    348–353
Protons, 53–54
  defined, 20
Pure sample, defined, 2

Qualitative, defined, 28
Quantitative, defined, 28
Quantization, 77
Quantum theory, 79–80

Radioactive carbon dating,
    424–425
Radioactive isotopes, 372
  decay, 433
  defined, 57
Radioactive tracers, 372
Rare earths, 433
Rate of reaction, 211–214, 369,
    370–372
Ratios, combining, 128–129,
    130
  mole, 199–201, 209, 292–294
    calculations, 294–300

Rayleigh, John William Strutt,
    3d Baron, 411
Reactants, defined, 9
Reaction products, defined, 9
  determining, 245–250
Reactions, acid-base, 345–365
  activation energy, 371
  addition, 445
  classification of, 214, 218–223
  combination, 214–215
  decomposition, 215
  displacement, 215–217
  electron donor, 221–222
    Lewis model, 353–355
  electron transfer, 218–220
  endothermic, 233
  exothermic, 233
  hydration, 194
  neutralization, 216–217
  oxidation-reduction, 388–406
    as energy source, 405–406
    half-cells, 391–393
    stoichiometry, 402–405
    titrations, 404–405
  precipitation, 215–216,
      222–223, 249–250
  proton transfer, 220–221
    Brønsted-Lowry model,
      348–353
  rate of, 211–214, 369,
      370–372
  redox, 388–406
    as energy source, 405–406
    half-cells, 391–393
    stoichiometry, 402–405
    titrations, 404–405
  replacement, 215
Redox equations
  balancing, 397–402
  rules, 401–402
Redox reactions, 388–406
  as energy source, 405–406
  half-cells, 391–393
  stoichiometry, 402–405
Redox titrations, 404–405
Reducing agents, 232, 389
Reduction, auto-oxidation, 237
  defined, 199, 388
  see also Redox reactions
Replacement reactions, 215
Representative elements,
    defined, 103
Reversibility, microscopic, 374
Rigidity, 170

Rule of eight, 122
  combining ratios and, 130
Rutherford, Ernest, 1st Baron
    Rutherford of Nelson,
    54–55

Sample, pure, defined, 2
Saturated solution, 19, 309
Scientific notation
  mathematical operations,
    462–464
  writing, 461–462
Sea water, components, 15
Significant figures, 31, 42–43,
    460–461
Silver nitrate solutions, halogen
    reaction with, 238–239
Solids
  defined, 160
  properties, 170–173
Solubility, 307, 330–333
  alkali metals, 233
  factors affecting, 309–312
  halogens, 235, 236
  tables, 185, 223, 249
Solubility product constant,
    383–384
Solute, defined, 306
Solutions, 16, 305–322
  acids, 242–245, 346
  aluminum, 427–428
  bases, 346
  boiling points, 307–308,
    320–321
  colligative properties, 308,
    329
  concentrations, 313–317
    molality, 314–315
    molarity, 315–317
    weight percent, 313–314
  defined, 16, 305
  electrical conductivity, 307,
    326–328, 333–334
  electrolytes, 307, 326–329
  examples, 306–308
  freezing points, 307–308,
    321–322
  mixed, reactions, 249–250,
    335–338
  properties, 317–322
  rate of dissolution, 312–313
  saturated, 19, 309
  solubility, see Solubility
  supersaturated, 309

Solutions (Continued)
   types, 308–309
Solvent, defined, 306
   electrical conductivity and, 333–334
Specific heat, 169–170
Spectator ions, 336
States of matter, 17–19, 160–178
Stoichiometry, defined, 285
Subatomic particles, 53–54; see also Electronic structure
Sublimation, 11, 171
   of iodine, 10–13
Submicroscopic, defined, 5
Subshells, 82–84
Substance, defined, 2
Sulfur, 417–418
Supersaturated solution, 309
Surface tension, 174
Surroundings, defined, 7
Symbols, for elements, 48–49
   in equations, 11, 49–50
Systems, defined, 7

Temperature, 166–169
   conversions, 168–169
   gases and, 262–263, 266–269
   measurement, 167
   scales, 167
   and solubility, 311–312
Tension, surface, 174
Ternary compounds, 147–148
   naming, 153–156
Theoretical definitions, defined, 7
Thermal energy, 22, 167
   defined, 167
Thomson, Sir Joseph John, 54
Titration, 362–363
   redox, 404–405
Torricelli, Evangelista, 164

Tracers, radioactive, 373
Transition compounds, 92
Transition elements, 431–434
   defined, 103
   inner, 103, 433–434
   oxides, 433
   properties, 431–432
Triads, 99

Uncertainty, 31; see also Significant figures
Units, atomic mass, 54, 57
   formula, 61
   see also Measurement
Unity terms, in calculations, 36–39
   chemical, reviewed, 286–292
   defined, 32–33
   grams per mole, 287, 289
   liters of gas per mole, 287–288, 290–291
   particles per mole, 288, 290
   using, 288–291

Valence electrons, 90–92
Van der Waals forces, 134
Vapor pressure, 175–176, 318–319, 377
Vaporization, heat of, 176–177
Viscosity, 174
Volume, measurement, 33–34
   gases, 258–269

Water, 183–201
   chemical properties, 187–198
   decomposition, 198–199
   halogen reactions with, 236–237
   hard, 431
   hydrate formation, 194
   hydrogen bonds, 185–186
   ionization, 194–195
   phase changes, 183–184
   physical properties, 183–184, 306
   reaction with alkali metals, 230
   reaction with certain salts, 192–193
   reaction of halogens with, 236–237
   reaction with hydrides, 248
   reaction with hydrogen compounds, 190–192
   reaction with metals, 187–188, 248
   reaction with nonmetals, 188–189
   reaction with oxides, 189–190, 247–248
   solvent properties, 184, 185, 186–187
Water molecule, hydrogen in, 195
   oxygen in, 197–198
Weak acids, 245
Weight, atomic, 57–58
   formula, defined, 61
   molecular, 60–62
   percent, 65–67
   of reactants and products, 210
Weight percent (solutions), 313–314
Winkler, Clemens, 100, 108–109
Wöhler, Friedrich, 437

Xenon, 412

Zero, absolute, 167

## Common ions

| | | | |
|---|---|---|---|
| $H^+$ | hydrogen | $F^-$ | fluoride |
| $Na^+$ | sodium | $Cl^-$ | chloride |
| $K^+$ | potassium | $Br^-$ | bromide |
| $Cu^+$ | copper (I) | $I^-$ | iodide |
| $Ag^+$ | silver (I) | $H^-$ | hydride |
| $NH_4^+$ | ammonium | $O^{2-}$ | oxide |
| $Mg^{2+}$ | magnesium | $S^{2-}$ | sulfide |
| $Ca^{2+}$ | calcium | $N^{3-}$ | nitride |
| $Ba^{2+}$ | barium | $OH^-$ | hydroxide |
| $Zn^{2+}$ | zinc | $CN^-$ | cyanide |
| $Cd^{2+}$ | cadmium | $ClO^-$ | hypochlorite |
| $Cr^{2+}$ | chromium (II) | $ClO_2^-$ | chlorite |
| $Mn^{2+}$ | manganese (II) | $ClO_3^-$ | chlorate |
| $Fe^{2+}$ | iron (II) | $ClO_4^-$ | perchlorate |
| $Co^{2+}$ | cobalt (II) | $NO_3^-$ | nitrate |
| $Ni^{2+}$ | nickel (II) | $NO_2^-$ | nitrite |
| $Cu^{2+}$ | copper (II) | $SO_4^{2-}$ | sulfate |
| $Sn^{2+}$ | tin (II) | $SO_3^{2-}$ | sulfite |
| $Pb^{2+}$ | lead (II) | $CO_3^{2-}$ | carbonate |
| $Hg^{2+}$ | mercury (II) | $PO_4^{3-}$ | phosphate |
| $Hg_2^{2+}$ | mercury (I) | $PO_3^{3-}$ | phosphite |
| $Al^{3+}$ | aluminum | $AsO_4^{3-}$ | arsenate |
| $Au^{3+}$ | gold (III) | $AsO_3^{3-}$ | arsenite |
| $As^{3+}$ | arsenic (III) | $CrO_4^{2-}$ | chromate |
| $Sb^{3+}$ | antimony (III) | $Cr_2O_7^{2-}$ | dichromate |
| $Bi^{3+}$ | bismuth (III) | $MnO_4^-$ | permanganate |
| $Cr^{3+}$ | chromium (III) | $O_2^{2-}$ | peroxide |
| $Fe^{3+}$ | iron (III) | $HSO_4^-$ | hydrogen sulfate |
| $Mn^{4+}$ | manganese (IV) | $HCO_3^-$ | hydrogen carbonate |
| $Sn^{4+}$ | tin (IV) | $C_2H_3O_2^-$ | acetate |
| $Pb^{4+}$ | lead (IV) | $HPO_4^{2-}$ | hydrogen phosphate |
| $As^{5+}$ | arsenic (V) | $H_2PO_4^-$ | dihydrogen phosphate |